Methods and Concepts for Designing and Validating Smart Grid Systems

Methods and Concepts for Designing and Validating Smart Grid Systems

Special Issue Editors

Thomas I. Strasser
Sebastian Rohjans
Graeme M. Burt

MDPI • Basel • Beijing • Wuhan • Barcelona • Belgrade

MDPI

Special Issue Editors

Thomas I. Strasser
AIT Austrian Institute of Technology
Austria

Sebastian Rohjans
Hamburg University of Applied Sciences
Germany

Graeme M. Burt
University of Strathclyde
Scotland

Editorial Office
MDPI
St. Alban-Anlage 66
4052 Basel, Switzerland

This is a reprint of articles from the Special Issue published online in the open access journal *International Journal of Neonatal Screening* (ISSN 2409-515X) from 2017 to 2019 (available at: https: //www.mdpi.com/journal/energies/special_issues/smart_grid_systems)

For citation purposes, cite each article independently as indicated on the article page online and as indicated below:

LastName, A.A.; LastName, B.B.; LastName, C.C. Article Title. *Journal Name* **Year**, *Article Number*, Page Range.

ISBN 978-3-03921-648-2 (Pbk)
ISBN 978-3-03921-649-9 (PDF)

Cover image courtesy of AIT Austrian Institute of Technology.

Contents

About the Special Issue Editors

Thomas I. Strasser received Master's and Ph.D. degrees from Vienna University of Technology (TU Wien) and was awarded with the venia docendi (Habilitation) in the field of automation from the same university. For several years, he has been a senior scientist in the Center for Energy of the AIT Austrian Institute of Technology. His main responsibilities involve the strategic development of smart grid automation and validation research projects and mentoring/supervising junior scientist and Ph.D. candidates. Before joining AIT, Dr. Strasser spent several years as a senior researcher investigating reconfigurable automation systems at PROFACTOR and some years at TU Wien as a project assistant working on control-related topics. He is currently active as a senior lecturer (Privatdozent) at TU Wien.

Sebastian Rohjans is a Professor for energy informatics at Hamburg University of Applied Sciences focusing on interoperability aspects for automation systems in distribution grids. Further focal topics of his work are simulation frameworks, data modeling, and ICT-architecture development for future energy systems. The application of ICT standards such as the Common Information Model (CIM), IEC 61850, and the OPC Unified Architecture is a cross-cutting key point of his overall work. In 2012 he received a Ph.D. degree from the University of Oldenburg for his research dealing with "Semantic Service Integration for Smart Grids".

Graeme M. Burt is currently a Professor at Strathclyde, where he is director of the Rolls-Royce University Technology Centre. Prof. Burt leads a significant body of research focused on the grid integration of distributed energy, protection and control for distributed generation, power system modeling and simulation, and active distribution networks. He is a lead academic in the BERR UK Centre for Distributed Generation, and leads EPSRC SUPERGEN "Highly Distributed Energy Futures".

energies

MDPI

Editorial

Methods and Concepts for Designing and Validating Smart Grid Systems

Thomas I. Strasser [1,*]**, Sebastian Rohjans** [2,*] **and Graeme M. Burt** [3,*]

[1]   Electric Energy Systems, Center for Energy, AIT Austrian Institute of Technology, 1210 Vienna, Austria
[2]   Department of Information and Electrical Engineering, Hamburg University of Applied Sciences, 20099 Hamburg, Germany
[3]   Institute for Energy and Environment, Electronic and Electrical Engineering Department, University of Strathclyde, Glasgow G1 1XW, UK
*   Correspondence: thomas.strasser@ait.ac.at (T.I.S.); sebastian.rohjans@haw-hamburg.de (S.R.); graeme.burt@strath.ac.uk (G.M.B.); Tel.: +43-664-2351934 (T.I.S.)

Received: 1 March 2019; Accepted: 13 May 2019; Published: 15 May 2019

Abstract: This Editorial provides an introduction to the Special Issue "Methods and Concepts for Designing and Validating Smart Grid Systems". Furthermore, it also provides an overview of the corresponding papers that where recently published in MDPI's *Energies* journal. The Special Issue took place in 2018 and accepted a total of 19 papers from 19 different countries.

Keywords: design, development and implementation methods for smart grid technologies; modelling and simulation of smart grid systems; co-simulation-based assessment methods; validation techniques for innovative smart grid solutions; real-time simulation and hardware-in-the-loop experiments

1. Introduction

Energy efficiency and low-carbon technologies are key contributors to curtailing the emission of green-house gases that continue to cause global warming. The efforts to reduce green-house gas emissions also strongly affect electrical power systems. Renewable sources, energy storage systems and flexible loads provide new system controls, but power system operators and utilities have to deal with their fluctuating nature, limited storage capabilities and typically higher infrastructure complexity with a growing number of heterogeneous components. In addition to the technological change of new components, the liberalisation of energy markets and new regulatory rules bring contextual change that necessitates the restructuring of the design and operation of future energy systems. Sophisticated component design methods, intelligent information and communication architectures, automation and control concepts, new and advanced markets, as well as proper standards are necessary in order to manage the higher complexity of such intelligent power systems that form the smart grid.

Due to the considerably higher complexity of such cyber-physical energy systems (CPES), constituting power system, automation, protection, information and communication technology (ICT), as well as system services, it is expected that the design and validation of smart grid configurations will play a major role in future technology and system developments. However, an integrated approach for the design and evaluation of smart grid configurations incorporating these diverse constituent parts remains evasive. Validation approaches available today focus mainly on component-oriented methods. In order to guarantee a sustainable, affordable and secure supply of electricity through the transition to a future smart grid with considerably higher complexity and innovation, new design, validation and testing methods appropriate for CPES is required. Papers that present results related to the design and validation of smart grid systems have been targeted by this Special Issue.

2. Content of the Special Issue

The accepted and published papers address a wide range of challenging and interesting methods and concepts in the domain of designing and validating smart grid systems, which can be mainly grouped into six clusters related to the following:

(1)   System design methods;
(2)   Simulation concepts;
(3)   Co-simulation approaches;
(4)   Hardware-in-the-loop (HIL) experiments;
(5)   Laboratory tests; and
(6)   Optimisation techniques.

Table 1 provides a brief overview of the assignment of the papers to the aforementioned clusters.

Table 1. Topics and included papers of the Special Issue.

Topic	References
System design methods	[1–3]
Simulation concepts	[4–7]
Co-simulation approaches	[8,9]
Hardware-in-the-Loop experiments	[10–14]
Laboratory tests	[15,16]
Optimisation techniques	[17–19]

2.1. System Design Methods

Often, the development of new approaches and methods begins with the design of the target system. This forms the basis for the implementation, which in turn can be evaluated in various ways.

For example, in [1], a market design was developed that addresses system balancing products within a web-of-cells (WoC) architecture. The focus is on a solution that is as economically efficient as possible. Based on a literature search, as well as analysis methods, different possibilities were compared. In contrast, the authors in [2] explored ways to transfer approaches from other domains, such as healthcare, to the domain of smart grids. In particular, within the framework of the project "Integration of the Energy System" (IES), Austria, interoperable communication will be the focus of attention. An essential element of this is a standard from the International Electrotechnical Commission (IEC), namely IEC 61850. A third approach to architectural modelling is the smart grid architecture model (SGAM) [3]. The SGAM, which provides a structured basis for the design, development and validation of new solutions and technologies, will be analysed in terms of its past use in Europe and its adaptations to other domains.

2.2. Simulation Concepts

A first and early approach to evaluate new concepts is simulation. With simple models, quick first insights can be gained, which can lead to the successive improvement of the solution. Simulations can be used in many different ways.

In [4], simulation was used to analyse a four-cell reference power system to evaluate a new controller. This controller follows a new strategy for dynamically adjusting the power frequency characteristic based on the imbalance state. The authors in [5] were able to demonstrate through simulations that their proposed effective filtering approach (EFA), to improve network traffic performance for the high-availability seamless redundancy (HSR) protocol, reduces network unicast traffic by up to 80%. Simulation of an unbalanced 12.47-kV feeder with 12,780 households and 1000 electric vehicles (EV) under peak and auxiliary load conditions was conducted in [6] to analyse a three-phase loss allocation procedure for distribution networks. The authors in [7] in turn simulated a well-known IEEE 14 bus test system for analysing a developed data mining algorithm. The data came from phasor measurement units (PMU), and the algorithm pursued the goal of better integration of wind turbines.

2.3. Co-Simulation Approaches

In some cases, it is not possible to determine in advance how the environment will look for a new solution. This also implies that it is not clear which factors influence a new solution. This is where the co-simulation approach has become established. Different simulation environments and models are coupled in order to map complex overall system scenarios.

Since the term co-simulation can be a broad field that takes very different depths of detail into account, the authors in [8] devoted themselves to a systematic structuring of different approaches. The authors in [9] used a specific co-simulation platform to demonstrate that the public mobile telecommunication system 4G long-term evolution (LTE) is applicable for fault location, isolation and system restoration (FLISR) applications in radially- and weakly-meshed medium voltage distribution networks.

2.4. Hardware-in-the-Loop Experiments

However, in comparison to co-simulation, if the deployment environment of a new solution, such as that of a controller, can be clearly defined, HIL-based tests are an obvious choice. Here, the real environment of the controller is simulated, and its behaviour is evaluated.

In the field of energy supply, a specific form of HIL, power hardware-in-the-loop (PHIL), has become established. Since the interface between the virtual part (simulated environment) and the real part (for example, the controller) is of the highest importance here, so far, many interface algorithms (IA) have been developed. The authors in [10] examined which IA were suitable for which PHIL experiments. In [14], two concrete approaches—a conventional PHIL design and a simplified structure based on a quasi-dynamic PHIL approach—were compared. Another important aspect of PHIL experiments is that the simulated system or its behaviour may be heavily dependent on whether the hardware being tested is connected. In particular, the initialisation of a real-time simulator (RTS) can lead to problems here. This problem was addressed by the authors in [12] by introducing a procedure for initialising PHIL simulations for synchronous power systems. Another field of application for PHIL is microgrids or their control; in particular, distributed concepts, e.g., through the implementation of multi-agent systems, and experiments are offered for PHIL in [13]. The simulation of PMU has already been mentioned, and distribution phasor measurement units (D-PMU) and micro-synchrophasors (micro-PMUs) are also addressed in [11]. Specifically, the authors presented a vendor-agnostic PMU real-time simulation and HIL (PMU-RTS-HIL) approach.

2.5. Laboratory Tests

The methods described so far for the validation of new approaches are characterised by a predominant use of software. Solutions that have a higher degree of maturity are then validated in lab-based tests before being field-integrated into real systems.

The authors in [15] dealt again with microgrids. They investigated whether distributed voltage control systems can be a serious alternative to centralised approaches. For this purpose, the development, the structure and the operation of a corresponding hardware-based lab test stand was described. In a further laboratory setup, the behaviour of active distribution networks was analysed, which were strongly penetrated by renewable energies [16]. This was examined by means of various parameters (tap changers of the transformers in the primary substation, reactive power injections of the renewable energy sources and active and reactive power exchanged between adjacent feeders being interconnected through a direct current link), to show what an optimal control can look like.

2.6. Optimisation Techniques

In addition to the development of new solutions, a significant part of research and development in smart grid systems is also concerned with the optimisation of existing approaches and methods. However, these optimisations can only be performed if the system was fully understood in advance. This in turn implies extensive prior validations.

The authors in [17] dealt with an optimisation of the coordination of reserve allocations in multiple-cell-based power systems. The starting point was the implementation of linear decision rule (LDR)-based guidelines. An energy market-based optimisation was pursued by the authors in [18]. The intended goal was to offer flexibility bids for a real-time balancing market. Finally, in [19], the concept of microgrids was focused on again. The authors investigated how microgrids prevent weather-related network outages by creating more robust structures. The result was an optimised power supply.

3. Conclusions and Outlook

All 19 papers (17 articles and two reviews) described exciting new design, validation and testing methods for CPES. Several of them were done in the context of funded research and development projects, especially with support from the European Union framework programmes. They covered approaches from system design to (co-)simulation and HIL, to laboratory tests and optimisation. With those methods in hand, the validation on system level becomes possible. However, future research and technology development are still necessary in order to further develop those approaches, to harmonise them and to provide corresponding frameworks and tools.

Acknowledgments: The guest Editors would like to thank the authors for submitting their excellent contributions to this Special Issue. Furthermore, they would like to express their sincerest gratitude to *Energies'* in-house Editor and reviewers for their wonderful work and effort. Without their support, the efficient handling of all received manuscripts (article average processing time was 40.48 days), it would not have been possible to publish this Special Issue. Moreover, this Editorial has been supported through the European Community's Seventh Framework Program (FP7/2007-2013) under Project "ELECTRA IRP" (Grant Agreement No. 609687), as well as by the European Community's Horizon 2020 Program (H2020/2014-2020) under Project "ERIGrid"(Grant Agreement No. 654113).

Conflicts of Interest: The authors declare no conflict of interest.

Abbreviations

CPES	Cyber-physical energy system
D-PMU	Distribution phasor measurement unit
EFA	Effective filtering approach
EV	Electric vehicle
FLISR	Fault location, isolation and system restoration
HIL	Hardware-in-the-loop
HSR	High-availability seamless redundancy
IA	Interface algorithms
ICT	Information and communication technology
IEC	International Electrotechnical Commission
IEEE	Institute of Electrical and Electronics Engineers
IES	Integrating the energy system
LDR	Linear decision rules
LTE	Long-term evolution
Micro-PMU	Micro-synchrophasor
PHIL	Power hardware-in-the-loop
PMU	Phasor measurement unit
RTS	Real-time simulator
SGAM	Smart grid architecture model
WoC	Web-of-cells

References

1. Bobinaite, V.; Obushevs, A.; Oleinikova, I.; Morch, A. Economically Efficient Design of Market for System Services under the Web-of-Cells Architecture. *Energies* **2018**, *11*. [CrossRef]
2. Gottschalk, M.; Franzl, G.; Frohner, M.; Pasteka, R.; Uslar, M. From Integration Profiles to Interoperability Testing for Smart Energy Systems at Connectathon Energy. *Energies* **2018**, *11*. [CrossRef]
3. Uslar, M.; Rohjans, S.; Neureiter, C.; Pröstl Andrén, F.; Velasquez, J.; Steinbrink, C.; Efthymiou, V.; Migliavacca, G.; Horsmanheimo, S.; Brunner, H.; et al. Applying the Smart Grid Architecture Model for Designing and Validating System-of-Systems in the Power and Energy Domain: A European Perspective. *Energies* **2019**, *12*. [CrossRef]
4. Rikos, E.; Caerts, C.; Cabiati, M.; Syed, M.; Burt, G. Adaptive Fuzzy Control for Power-Frequency Characteristic Regulation in High-RES Power Systems. *Energies* **2017**, *10*. [CrossRef]
5. Tien, N.X.; Rhee, J.M.; Park, S.Y. A Combined Approach Effectively Enhancing Traffic Performance for HSR Protocol in Smart Grids. *Energies* **2017**, *10*. [CrossRef]
6. De Oliveira-De Jesus, P.M.; Rios, M.A.; Ramos, G.A. Energy Loss Allocation in Smart Distribution Systems with Electric Vehicle Integration. *Energies* **2018**, *11*. [CrossRef]
7. Klarić, M.; Kuzle, I.; Holjevac, N. Wind Power Monitoring and Control Based on Synchrophasor Measurement Data Mining. *Energies* **2018**, *11*. [CrossRef]
8. Nguyen, V.H.; Besanger, Y.; Tran, Q.T.; Nguyen, T.L. On Conceptual Structuration and Coupling Methods of Co-Simulation Frameworks in Cyber-Physical Energy System Validation. *Energies* **2017**, *10*. [CrossRef]
9. Garau, M.; Ghiani, E.; Celli, G.; Pilo, F.; Corti, S. Co-Simulation of Smart Distribution Network Fault Management and Reconfiguration with LTE Communication. *Energies* **2018**, *11*. [CrossRef]
10. Brandl, R. Operational Range of Several Interface Algorithms for Different Power Hardware-In-The-Loop Setups. *Energies* **2017**, *10*. [CrossRef]
11. Stifter, M.; Cordova, J.; Kazmi, J.; Arghandeh, R. Real-Time Simulation and Hardware-in-the-Loop Testbed for Distribution Synchrophasor Applications. *Energies* **2018**, *11*. [CrossRef]
12. Guillo-Sansano, E.; Syed, M.H.; Roscoe, A.J.; Burt, G.M. Initialization and Synchronization of Power Hardware-In-The-Loop Simulations: A Great Britain Network Case Study. *Energies* **2018**, *11*. [CrossRef]
13. Nguyen, T.L.; Guillo-Sansano, E.; Syed, M.H.; Nguyen, V.H.; Blair, S.M.; Reguera, L.; Tran, Q.T.; Caire, R.; Burt, G.M.; Gavriluta, C.; et al. Multi-Agent System with Plug and Play Feature for Distributed Secondary Control in Microgrid—Controller and Power Hardware-in-the-Loop Implementation. *Energies* **2018**, *11*. [CrossRef]
14. Ebe, F.; Idlbi, B.; Stakic, D.E.; Chen, S.; Kondzialka, C.; Casel, M.; Heilscher, G.; Seitl, C.; Bründlinger, R.; Strasser, T.I. Comparison of Power Hardware-in-the-Loop Approaches for the Testing of Smart Grid Controls. *Energies* **2018**, *11*. [CrossRef]
15. Almasalma, H.; Claeys, S.; Mikhaylov, K.; Haapola, J.; Pouttu, A.; Deconinck, G. Experimental Validation of Peer-to-Peer Distributed Voltage Control System. *Energies* **2018**, *11*. [CrossRef]
16. García-López, F.D.P.; Barragán-Villarejo, M.; Marano-Marcolini, A.; Maza-Ortega, J.M.; Martínez-Ramos, J.L. Experimental Assessment of a Centralised Controller for High-RES Active Distribution Networks. *Energies* **2018**, *11*. [CrossRef]
17. Hu, J.; Lan, T.; Heussen, K.; Marinelli, M.; Prostejovsky, A.; Lei, X. Robust Allocation of Reserve Policies for a Multiple-Cell Based Power System. *Energies* **2018**, *11*. [CrossRef]
18. Camargo, J.; Spiessens, F.; Hermans, C. A Network Flow Model for Price-Responsive Control of Deferrable Load Profiles. *Energies* **2018**, *11*. [CrossRef]
19. Uski, S.; Forssén, K.; Shemeikka, J. Sensitivity Assessment of Microgrid Investment Options to Guarantee Reliability of Power Supply in Rural Networks as an Alternative to Underground Cabling. *Energies* **2018**, *11*. [CrossRef]

Article

Economically Efficient Design of Market for System Services under the Web-of-Cells Architecture

Viktorija Bobinaite [1,*], Artjoms Obushevs [1], Irina Oleinikova [1] and Andrei Morch [2]

[1] Institute of Physical Energetics, Smart Grid Research Centre, Krīvu iela 11, LV-1006 Rīga, Latvia;
a.obusev@gmail.com (A.O.); irina.oleinikova@edi.lv (I.O.)
[2] SINTEF Energy Research, Sem Saelands vei 11, NO-7034 Trondheim, Norway; andrei.morch@sintef.no
* Correspondence: viktorija.bobinaite@lei.lt; Tel.: +371-67-552-011

Received: 13 February 2018; Accepted: 20 March 2018; Published: 23 March 2018

Abstract: Significant power sector developments beyond 2020 will require changing our approach towards electricity balancing paradigms and architectures. Presently, new electricity balancing concepts are being developed. Implementation of these in practice will depend on their timeliness, consistency and adaptability to the market. With the purpose of tailoring the concepts to practice, the development of a balancing market is of crucial importance. This article deals with this issue. It aims at developing of a high-level economically efficient market design for the procurement of system balancing products within the Web-of-Cells architecture. Literature and comparative analysis methods are applied to implement the aim. The analysis results show that a more efficient balancing capacity allocation process should be carried out in a competitive way with closer allocation time to real-time, especially with increased penetration of renewable energy sources. Bid time units, the timing of the market, procurement and remuneration schemes as well pricing mechanisms are the most decisive elements of the market. Their respective advantages and disadvantages are analyzed in the article, as well as their analysis is done against the selected assessment criteria. The results of the analysis show that seeking to improve the operational efficiency of the market, the sequential approach to the market organization should be selected and short-term market time units should be chosen. It is expected that price efficiency could be improved by establishing an organized market where standardized system balancing products should be traded. The balance service providers, who own capital expenditures (CAPEX) sensitive production units, should be remunerated both for the availability of balancing capacities and for their utilization. Uniform pricing rule and cascading procurement principal should be applied to improve the utilization efficiency.

Keywords: market design; market design elements; Web-of-Cells; procurement scheme; remuneration scheme; pricing scheme; cascading procurement

1. Introduction

The overall Pan-European political goals of decarbonising the power sector have already resulted in substantial growth of generation from renewable energy sources (RES) in Europe. In Germany the share of electric power from RES during the first part of 2017 reached 35% of the total generation mix [1], while in Denmark wind power already accounted for 42% of all electricity generated in 2016 and is expected to cover 48% in 2020 [2]. This process has been reinforced by massive subsidies, channelized via various support schemes as feed-in tariffs, feed-in premiums, quota obligation systems and combinations of these approaches, which are the most commonly applied. Other instruments such as quota systems with tradable green certificates and priority dispatch for RES are also applied in several European countries (for detailed descriptions see [3]). Apart from their positive effects, i.e., increasing the share of RES in the European generation mix, these support schemes combined with some inherited properties of variable renewable energy have also caused several interrelated challenges for the European power sector on the market level (for a complete overview see [3]) as well as the system level. Among latter the following may be mentioned:

- Local grids cannot maintain normal $n-1$ security (transmission systems shall be designed to maintain a $n-1$ security criterion, meaning that the system is in a secure state with all transmission facilities in service and in a satisfactory state under credible contingent events. $n-1$ is a common security standard in many countries. The single contingencies to be considered under an $n-1$ criterion are: loss of a single transmission circuit, loss of a single generator, loss of a high-voltage, direct current (HVDC) pole, loss of a single bus section, loss of an interconnecting transformer, loss of a single shunt connected reactive component) if local generation exceeds local demand and if separation of generation and consumption is insufficient. Restoration after fault has become more complicated and more time consuming (the case of Western Denmark with its high share of wind power [4]).
- Security analysis has become less accurate due to missing information on local generation and the unpredictable nature of wind and solar power.
- Traditional under-frequency load shedding schemes will disconnect both load and generation.
- Growing share of RES, following decommissioning of the nuclear generation and reduced operational time for thermal power plants will lead to declining of system inertia and accordingly system's stability [5].

These challenges have raised the necessity of reconsidering the existing power system architecture. Several initiatives have been working in this direction during the recent years. Depending upon the time horizon and the scope, the ambitions may vary from rather evolutionary, which require minor changes in the system's architecture, to very radical approaches:

- The FP7 project EcoGrid EU (2011–2015) worked on development and testing of new market concept allowing to improve the balancing mechanisms by introducing a 5 min real-time price response that will provide additional regulation power from smaller customers directly to Transmission System Operators (TSOs) [6].
- Moving in the same direction, the H2020 project SmartNet (2015–2018) attempts to rethink the system architecture and corresponding market arrangements. This will improve TSO-DSO interaction the acquisition of ancillary services (reserve and balancing, voltage balancing control, congestion management) from subjects located in the distribution segment (flexible load and distributed generation) [7].

More radical approaches are attempting to redesign the overall system architecture, especially aiming for division of the power system into separated grid areas with various level of autonomy:

- The concept developed by the Danish Cell Project led by Energinet.dk looked at dividing the power system into virtual fully autonomous grid areas in terms of control, so-called cells. The cell concept could be realized through the development and implementation of an advanced monitoring and control system capable of monitoring the state of the cell and—in extreme situations—taking control of its individual units such as circuit breakers, transformers, wind turbines and combined heat and power (CHP) plants [8].
- Division of the grid into semi-autonomous units is studied by H2020 project Fractal Grid [9] and recently started C/sells project [10].
- Among the most recent initiatives one can mention FP7 ELECTRA IRP [11], which addresses the issue of deployment of RES connected to the network at all voltage levels as well as establishes and validates proofs of concepts that utilize flexibility from across traditional boundaries in a holistic manner.

In particular, the FP7 ELECTRA IRP [11,12] observes that beyond 2020 the European power sector will undergo significant developments. Electricity production will shift from traditional fossil fuel-based units to units using intermittent RES. Electricity producing units will be connected both to transmission and distribution networks. A wave of electrification of transport and space heating sectors will rise and strengthen towards 2050, resulting in increasing demand for electricity. Electricity storage technologies will become cost-effective solutions for the provision of balancing services. Extensive amounts of flexible loads having fast activation and short ramping times will be available at all voltage levels. Therefore, the demand for activation of balancing capacities to correct the imbalances caused by forecast errors of intermittent RES and flexible loads will increase. The present power system balancing paradigm will need to be transformed from generation following load to load following generation. The location of imbalance-causing problems, which require activation of balancing capacity and resources, will move from transmission to distribution network level.

For that reason, ELECTRA proposes a new cell-based decentralized control framework named Web-of-Cells. In this view the power system is split into control cells. An ELECTRA Cell is a portion of the power grid able to maintain an agreed power exchange at its boundaries by using the internal flexibility of any type available from flexible generators/loads and/or storage systems. The total amount of internal flexibility in each Cell shall be at least enough to compensate the Cell generation and load uncertainties in normal operation.

In short, the solution is starting from the current control area (Transmission System Operator-based (TSO-based)) balancing approach and applying it at any voltage level/power size with enhanced control through concurrent service deployment and greater autonomy and collaboration at local levels.

An ELECTRA Cell has several specific characteristics [12,13]:

- local problems are usually solved within a Cell where local observables are used to decide on local corrective actions to handle local issues;
- communication and computational complexities are minimised;
- local grid conditions are explicitly taken into consideration when deciding what kind of resources are used;
- provision of a distributed bottom-up approach for the restoring of the system balance;
- focus more on balance restoration—and thereby restoring frequency as well—rather than the current traditional sequence of frequency containment followed by frequency restoration.
- the total amount of internal flexibility in each Cell is at least enough to compensate the Cell generation and load uncertainties in normal operation.

For a detailed description of rules, which define the Cell, see [12]. Moreover, some key roles—Cell operator (CO), balance service providers (BSPs), balance responsible parties (BRPs) and load and generation forecaster (LGF)—are identified for the future electricity market design within the ELECTRA.

In addition, an advanced Web-of-Cells control scheme is developed within the ELECTRA [12,13]. It is characterized by six high-level functionalities—balance restoration control (BRC), adaptive frequency containment control (aFCC), inertia response power control (IRPC), balance steering control (BSC), primary voltage control (PVC) and post-primary voltage control (PPVC)—and the following fundamental characteristics [12]:

- solve local problems locally;
- responsibilisation with local neighbour-to-neighbour collaboration;
- ensuring that only local reserves providing resources will be used whose activation does not cause local grid problems.

The above shortly presented Web-of-Cells architecture shall require reconsidering presently existing electricity market design for the procurement of system balancing products. The aim of this article is to demonstrate a high-level market design for the economically efficient procurement of system balancing products within a new architecture called Web-of Cells.

To implement the aim, the following tasks are set:

- to develop the concept of the market design for the procurement of system balancing products by identifying the most crucial structural elements of market design;
- to determine the assessment criteria of economically efficient market design;
- to assess the alternative solutions of market design elements by considering their advantages and disadvantages and against the assessment criteria;
- to prepare a market design for the procurement of system balancing products within the Web-of-Cells architecture.

There is a large volume of literature [14–27] that discusses the design of balancing markets in the presence of intermittent RES. The novelty of this article beyond the prevailing literature is that it proposes harmonized and unified rules for trading standardized system balancing products intra- and inter-Cell between the identified types of the market participants in an organized market to implement the merit order collection (MOC) and merit order decision (MOD) functions of Web-of-Cells architecture. The market for system balancing products is designed in accordance with the general principles of a competitive market, the principles of a wholesale electricity market and the principles of the market for ancillary services. In addition, the market for system balancing products is developed in accordance to the assessment criteria of "economic efficiency" and in line with the requirements tackling the issue of climate change.

A critical review of scientific publications, research projects, energy market monitoring reports and European Union-wide Directives and following Regulations is done. This allowed identifying the alternatives of market design elements, perceiving the structure of existing European electricity market, its strong sides, which should be respected within the Web-of-Cells architecture, and weak aspects of market design, which should be transformed in a way to implement the MOC and MOD functions of the Web-of-Cells architecture. Comparative analysis method is the secondary method applied to understand the advantages and disadvantages of collected market design elements. Assessment of market design elements as a tertiary method is used to rank the market design elements based on the economic efficiency criteria, to identify the key market design elements (the best alternative from the entire set of possible options) and finally take a decision regarding the particular option of market design element, which well fits to the Web-of-Cells market design.

The remainder of the article is organized as follows: in Section 2, the concept of market design for the procurement of system balancing products is developed based on results of literature analysis. In Section 3, market participants and their roles within the Web-of-Cells architecture are identified. Sections 4 and 5 discuss peculiarities of marketplaces and auctions possibly available for a new power grid structure, respectively. Section 7 analyses options of market design elements to improve operational efficiency. Options of market design elements to enhance price efficiency and utilization efficiency are discussed in Sections 8 and 9, respectively. The concluding remarks are drawn in the final section.

2. Concept of Market Design

2.1. Theoretical Background

There is no single and universal definition of the market design [28–32]. Definitions significantly depend on, and they are related to the practical policy goals for achievement of which market design is developed, peculiarities of economic sector or country under consideration, aspects of market design discussed, analysis method applied, assessment criteria used or scientist's insights and penetration glance into the issue.

Considering to a great variety of factors, which disclose an array of research issues the market design covers, the market design for the procurement of system balancing products under the Web-of-Cell architecture is developed considering three concepts, which are market and its roles, designing and market design. The interrelation between them is given in Figure 1 and discussion is provided below.

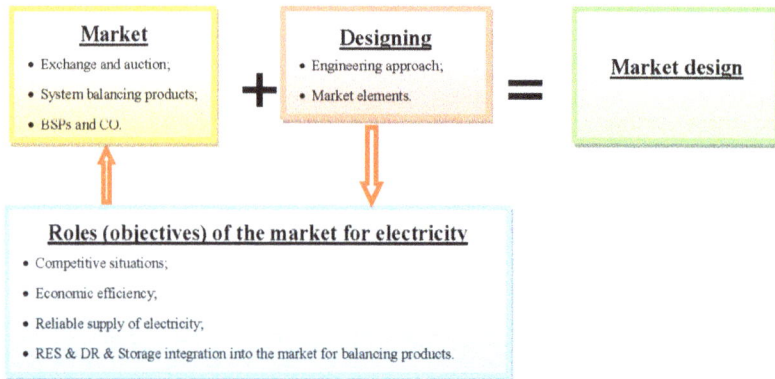

Figure 1. Structural representation of market design concept for the Web-of-Cells architecture.

The primary role (objective) of the electricity market is to establish competitive situations, particularly, preventing the creation or the strengthening of market power or prohibiting the abuse of a position of substantial market power (monopolization) [33].

Competitive situations are established by implementing the competitive electricity market. In [34], five core attributes of a competitive market are segregated. These are ease of market entry and exit, the absence of significant monopoly power, transparency, the absence of market externalities and achievement of public policy goals, such as ensure a reliable supply of electricity, increased production of electricity from RES. The list of attributes of a competitive market could be expanded by including trade in standardized products, which are perfect substitutes for each other [35], great variety of market participants [36], etc. Violation of any attribute of a competitive market hinder the market to achieve both its primary and secondary roles, which are as follows [14,15,37]:

- to promote economic efficiency and lower delivered product costs;
- to foster liquidity. Liquidity seems to be very important, especially for bilateral markets. However, the article does not consider liquidity for the time being, since the concept is very new, but this will be studied further in the subsequent activities;
- to avoid undue barriers to entry for new BSPs;
- to increase the choices to market participants;
- to assure level playing field to all the BSPs;
- to be transparent;
- to be flexible;
- to fit for future;
- to be consumer-oriented, etc.

Within the Web-of-Cells architecture, the market means an auction-based exchange where system balancing products are traded between the market participants, who are the BSPs and the COs (their description and responsibilities in the market are defined in Section 3). This keeps the primary and secondary roles for a well-functioning market for system balancing products. However, this article concentrates on the role of the market to be economically efficient.

Within the Web-of-Cells architecture, the concept of the market presented above is expanded by adding supplementary structural components, called market elements. It is known that at the procurement side of the market for system balancing products attention should paid at balancing service classes, reserve requirements, control system, timing of the markets, activation strategy, bid requirements, program time unit, scope of balance responsibility, gate closure times (GCT), types of balances, closed/opened portfolio positions [16], types of market participants, contracting approach, contract duration, scoring rule, dispatch criteria and cost allocation [17], transaction mechanisms and payments to the BSPs, response time and duration period [18], structure for bids and payments [19], types of procurement and remuneration methods, price caps [19,20], types of auction [21], number of markets, pricing rule, number of bid submissions, scarcity pricing and cascading procurement [22], approach towards markets organization [23] and other. However, with the purpose to establish an economically efficient market design, the following market design elements are considered [16,24–26,38,39]: system balancing product resolution in time, bid time unit, frequency of bidding, frequency of clearance, establishment of Merit Order list, distance to real time of the auction, procurement scheme, remuneration scheme and pricing mechanism.

2.2. Concept of Market Design

By combining the concepts of market and design into a single new concept, the market design is defined as a solution of the process of designing market elements requested within a market, which meets the objectives held for the market [40]. The concept of proposed market design encompasses the institution established for the system balancing products trading, interacting market participants and a set of market design elements.

2.3. Market Design Assessment Criteria

Market design (more precisely separate its elements) is assessed against the following economic efficiency criteria [16]:

- *Operational efficiency* is economic efficiency of handling the transactions related to the administrative process in the market for system balancing products, including of energy schedule and system balancing product bid submission and balance settlement.
- *Price efficiency* involves the cost-reflectivity of system balancing product prices paid to the BSPs. The premise that prices are efficient, to the extent that they already factor in or discount all available information.
- *Utilization efficiency* is defined as the economic efficiency of the utilization of available balancing resources, i.e., the degree to which the least and cheapest balancing resources are used to maintain the system balance.

Utilization, price and operational efficiencies are also criteria against which market elements are assessed. The market elements in relation to the criterion of economic efficiency are summarized in Figure 2.

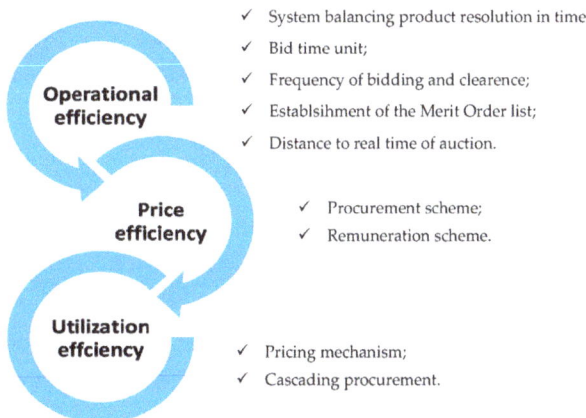

Figure 2. Market elements in relation to criterion of economic efficiency.

3. Roles and Responsibilities in the Market for System Balancing Products

Several key roles are foreseen at the procurement side of the market for system balancing products developed within the Web-of-Cells architecture.

The balance service provider (BSP) is a market participant who is responsible for the provision of system balancing products to the CO if it is in a contract agreement with the CO. In accordance to existing practice and the "Winter package", the list of system balancing products provided by the BSPs is limited to balancing capacity and balancing energy. Within the Web-of-Cells architecture, the BSP has a responsibility to trade in inertia capacity, inertia, balancing energy and balancing capacity for upward or downward regulation.

Within the Web-of-Cells architecture various types of BSPs take part in the market for system balancing products. A detail list of the BSPs in the framework of the Web-of-Cells architecture is given in Figure 3.

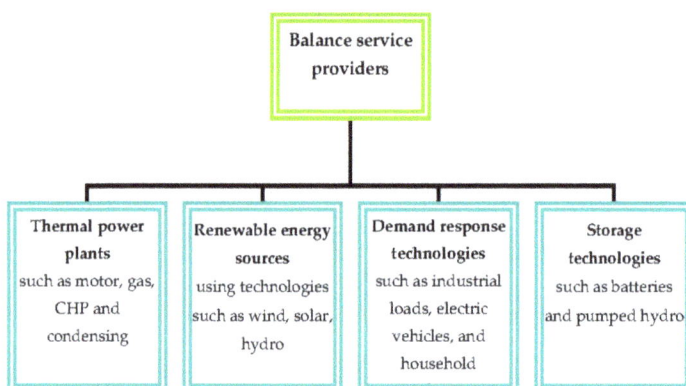

```
                    ┌─────────────────────┐
                    │  Balance service    │
                    │     providers       │
                    └─────────────────────┘
```

Thermal power plants	Renewable energy sources	Demand response technologies	Storage technologies
such as motor, gas, CHP and condensing	using technologies such as wind, solar, hydro	such as industrial loads, electric vehicles, and household	such as batteries and pumped hydro

Figure 3. List of the BSPs.

As the share of intermittent RES significantly increases in future, additional flexibility will become a valuable source to balance real-time generation and consumption. The CO will have to rely on flexible BSPs who could tailor (adapt) their generation and consumption by either producing and consuming above or below their set schedule with the aim to solve the imbalance in the Cell. Thus, in addition to presently acting BSPs, who are centralized gas-fired or hydropower plants, new BSPs will be requested and, indeed, available. They will be found at the distribution level. RES, demand response and storage technologies will be available new BSPs in future. Moreover, seeking to increase the size of the BSPs, an aggregation is believed to be of critical importance. Thus, the aggregators as a separate type of the BSPs will be requested under the Web-of-Cells architecture.

The responsibilities of the BSP in the market are in line with the requirements determined in [26], however, responsibilities are expanded to new types of system balancing products developed within the Web-of-Cells architecture. Therefore:

- the BSP qualifies for providing bids for inertia capacity, inertia, balancing energy or balancing capacity which are procured and activated by the CO;
- each BSP participating in the procurement process for inertia capacity submits and has the right to update its inertia capacity bids before the GCT of the bidding process;
- each BSP participating in the procurement process for balancing capacity submits and has the right to update its balancing capacity bids before the GCT of the bidding process;
- each BSP with a contract for inertia capacity submits to its CO the inertia bids corresponding to the volume, products, and other requirements set out in the inertia capacity contract;

- each BSP with a contract for balancing capacity submits to its CO the balancing energy bids corresponding to the volume, products, and other requirements set out in the balancing capacity contract.

The Cell operator (CO) (TSO/DSO) within the Web-of-Cells architecture is a market participant procuring system balancing products from the BSPs to balance electricity consumption and production in real time. The CO decides on the demanded volume of system balancing products considering to load and generation forecasts provided by Load and Generation Forecaster (LGF) and energy schedules provided by the Balance Responsible Parties (BRPs).

The CO is responsible for the performance of the Cell operation, maintenance and development to ensure electricity supply in a safe, efficient and reliable manner too. More precisely, it is responsible for the secure operation of the Cell, measurement of electricity demand on high voltage (HV), medium voltage (MW) or low voltage (LV) grid, maintenance of HV, MV or LV grid and procurement of the system balancing products for its Cell (based on [41]), which demand arises due to the occurrence of the following events [20]:

- unexpected RES generation variations;
- unexpected consumption variations;
- unplanned outages of generation and consumption capacity and grid elements;
- if discrepancies between the duration of day-ahead/intraday markets periods and real-time settlement periods exists;
- discretization of continuous time in discrete market periods.

Presently, the TSOs are responsible for the preparation of market regulations to the BSPs on the procedure for the procurement of system balancing products. In the Web-of-Cell architecture, this task is under the responsibility of COs, who prepare uniform and harmonized rules for the procurement of system balancing products. The rules are applied to all market participants. For a detailed description of other non-market related responsibilities of the CO, see in [12].

The load and generation forecaster (LGF). Nowadays the TSOs calculate day-ahead, week-ahead, month-ahead and year-ahead forecasts of total load, estimate of the total scheduled generation (MW) and forecast wind and solar power generation (MW) per bidding zone, per each market time unit of the following day [39,42]. Generation units and DSOs located within a TSO's control area provide the TSO with all important information necessary to calculate the load and generation forecasts. This information is used by the TSO to determine volume of system balancing products required to balance electricity consumption and production in real time. In the Web-of-Cells architecture, generation and load forecasts are made by two types of entities. These are the large-scale BRPs, receiving all necessary information from their large-scale generating and load units, and the Aggregator, collecting all important information for this task from the small-scale BRPs who themselves are supplied with data by small-scale generating and load units. The LGF is responsible for provision of load and generation forecasts to the CO.

4. Description of System Balancing Products

A variety of system balancing products, which are procured by the CO and supplied by the BSPs to assure the balance between electricity production and consumption in real time, is suggested within the Web-of-Cells architecture. They differ in functions, technical characteristics, procurement schemes and other requirements.

In Figure 4, the list of system balancing products traded within the Web-of-Cells architecture is presented by categorizing them into classes, directions and types.

In the framework of the Web-of-Cells, four classes of system balancing products in separate sub-markets are traded with the purpose to keep the system frequency target value within the certain limits [43]:

- *Inertia response power control* (IRPC) service where each unit, involved in inertia control, automatically changes its level of inertia power response (synthetic inertia) depending on certain predefined characteristics. Reacts to frequency changes over time.
- *Adaptive frequency containment control* (aFCC) service will not be fundamentally changed compared to today's schemes, except that the resources providing containment reserves will be different: generating units (in the broadest sense) as well as loads and storage distributed across the power grid (within each Cell). Reacts to deviations of the absolute frequency value so as to contain any change and stabilize frequency to a steady-state value.
- *Balance restoration control* (BRC) service initiates the restoration of the Cell balance and load flows based on local information. It is assumed that (almost) all prosumers, that are connected through public communication infrastructure, will be able to offer fast BRC capacity, e.g., through their flexible loads, and possibly local storage. Reacts to absolute frequency deviations in conjunction with the tie line deviations from the scheduled interchanges so as to restore both quantities to their initial values.
- *Balance steering control* (BSC) service will replace the BRC in a more economic manner if this can be done safely or adjust the balance set points. It can as well have pro-active activation based on prognoses. This control deploys resources not only within the Cell but also from neighboring Cells. Regulates power balance within a Cell in order to replace the BRC reserves or mitigate potential imbalances in a cost-effective manner.

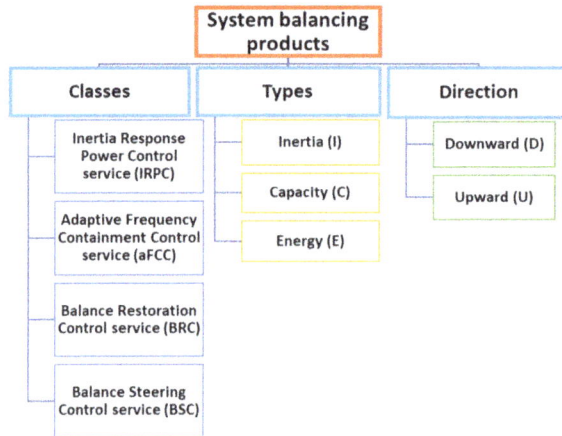

Figure 4. The categorization of system balancing products.

The system balancing products are standard. The minimum requirements are set for them (Table 1).

Table 1. Requirements for the standard system balancing products [44].

Characteristic	IRPC	aFCC	BRC	BSC
Ramping period	<1 MW·s/s	<1 MW/s	<10 MW/min	Same with BRC
Full activation time	<1 s	2–5 s	10–30 s	10–30 s
Minimum and maximum quantity	<1 MW·s	<1 MW	1–5 MW	1–5 MW
Preparation period	<1 s	<5 s	<1 min	<1 min
Deactivation period	<20 s	10–30 s	10–30 s	1–30 s
Minimum and maximum duration of delivery period	15–60 min			
Validity period	15 min	15 min	15 min	15 min
Mode of activation	Merit order	Merit order	Merit order	Merit order

Based on today's operation a clear direct link between the quality of system balancing products and their price is established, showing that the quality of system balancing products is hierarchical in nature. Therefore, the primary control service is a higher quality-balancing product than the secondary control service, which in turn is a higher quality-balancing product than the tertiary control service. Thus, it is reasonable that higher quality balancing products are priced at higher prices. In the Web-of-Cells architecture, the relationship between the quality of system balancing products and their price could be kept too (Figure 5), however, there are some differences.

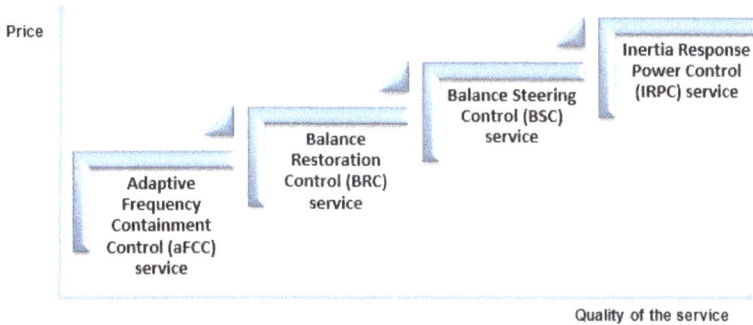

Figure 5. The relation between the inertia, classes of the balancing products in respect to their quality and price.

The IRPC service is the most difficult service, thus, it is expected that it will be the most expensive. The cost (price)—quality of the aFCC and BRC services should not vary, since the type and mechanisms of resources providing these two services are very similar and, if anything, the BRC service could be even more expensive because it is more crucial as a service. The BSC service is not anymore tertiary, hence cheaper, reserves. The BSC involves the change of tie-lines set points only and that is not based on reserves. This means that the commodity here is rather different that the other three services. In any case, the cost are expected to be higher than the BRC's because otherwise, the COs would always tend to modify their set-points and that should be done only in exceptional cases.

The direction of the system balancing products corresponds to upward and downward regulation. Upward regulation means an increase in generation (or decrease in consumption) and down regulation means a decrease in generation (or increase in consumption). Upward regulation is provided by units that are more expensive than the marginal unit of the day-ahead or intraday markets, meaning that the system balancing product of upward regulation is traded at higher price than the day-ahead price. For downward regulation those units that already received payments from the day-ahead and/or intraday markets can save the fuel costs by decreasing generation. Splitting the market for procurement of system balancing products into the sub-markets of "upward" and "downward" and do not setting/keeping the requirement for bid symmetricity creates preconditions and lead to increasing number of the BSPs in the corresponding sub-market. For example, loads are more familiar and capable of reducing load than increasing it. Therefore, if symmetric bids are not required, the available balancing capacity could be better used.

Three types of system balancing products are traded. They are inertia, balancing energy and capacity necessary for balancing and inertia. Balancing energy means energy used by the CO to perform balancing and provided by the BSPs. Balancing capacity means a volume of reserve capacity that the BSP has agreed to hold and in respect to which the BSP has agreed to submit bids for a corresponding volume of balancing energy to the CO for the duration of the contract. In line with the determination of balancing capacity and balancing energy, the capacity necessary for provision of inertia and inertia are determined, respectively.

Separate sub-markets are established to trade a particular class, direction and type of system balancing product. Coding of system balancing products is established to increase the transparency and operational efficiency of the market and with the purpose of quickly directing the BSP and the COs to the appropriate sub-market. The following structure of coding is applied: C-D-T: YMD_H_Q, where C—class, D—direction abnd T—type of the system balancing product and Y, M, D, H, Q—year, month, day, hour and quarter hour of system balancing product delivery. Thus, under the coding aFCC-U-C: 20171026_15_00 the upward regulation (U) balancing capacity (C) for the provision of adaptive frequency containment control (aFCC) service is traded, when the system balancing product has to be delivered on 26 October 2017 at 3 pm.

5. A Marketplace

The Web-of-Cells architecture keeps an idea that all system balancing products, which are needed to operate the Cells are procured in a marketplace. Three types of marketplaces are considered. They are exchange, organized over-the-counter (oOTC) and bilateral over-the-counter (bOTC) [45].

Exchange as a type of organized market is needed. The Web-of-Cells architecture prefers the exchange because of the reasons discussed in [46], the most important of which are the pursuit of transparency, confidentiality, anonymity and publicity. In the exchange, only standard balancing products are traded meaning that contracts are uniform regarding their structure and form. This enables the CO to compare identical system balancing products and activate the most cost-efficient solution. Standardized system balancing products support the integration of RES&DR&Storage technologies into the market. Due date, place of delivery, the time in which the deliveries take place and the conditions for clearing and settlement are standardized too. The set of rules such as the conditions to be admitted to trade on the exchange are made public and are the same for every BSP. Prices and revenues are made public too and this allow the BSPs evaluating the position of their bids on an initial Merit Order list relative to bids from other BSPs. The participation is voluntary and non-discriminatory. Trading partners do not have to be found and the counterparty risk is minimized. Since the trading process is anonymous, the BSPs can keep their strategy in secret. Also, the process to offer flexibility in the market is easy with a low entry barrier.

All the COs and the BSPs use a common platform for trading system balancing products, however, they are not pooling the bids and offers into a single integrated market for system balancing products, except in case of trading the BSC service, when every bid and offer available to one of the Cell is available to other Cells. This comes from the feature of the Web-of-Cells architecture that, basically, local problems should be solved locally and only in case there is a shortage in inertia, balancing capacity and balancing energy, they are procured from the neighboring Cell by procuring the BSC service.

6. An auction

6.1. General Issues

Auction as an instrument promoting competition in procurement of system balancing products is applied within the Web-of-Cells architecture. In addition, within the Web-of-Cells architecture it is used as a mechanism to coordinate efficient production. Moreover, it is chosen as an institution determining the price and conducting the trade in the exchange. It is worth mentioning that within the Web-of-Cells architecture an auction serves as the economic (market) mechanism used to allocate the system balancing products in economically efficient way. The main arguments for the choice of the auction are derived from the following advantages of the auction to be fair, open, transparent, objective, non-discriminatory, and timely process, which well corresponds the determined roles of the market for the procurement of system balancing products to be market facilitating. The trade in system balancing products in the auction is organized in Cell- and inter-Cell levels. The IRPC, aFCC and BRC services are traded within a particular Cell, but the BSC service is traded between the Cells, i.e., the COs organize a common auction to trade in the BSC service.

6.2. Types of Auctions

6.2.1. Closed-Type vs. Opened-Type Auction

In relation to the extent to which information about orders submitted to the auction are provided, closed-type auctions instead of opened- ones are established. In a closed (sealed-bid)-type auction, the BSPs, who are the auction participants during the bidding process, privately submit their bids and offers to the CO, who is the auctioneer, and the CO keeps this information private, such that there is no sharing of bidding information amongst the BSPs. The BSPs are informed whether they won or lost. It is expected that limiting the provision of information works as a constraint on exercising market power and thereby increasing prices of system balancing products.

6.2.2. Uniform Auction vs. Discriminatory Auction

Two types of closed-type auctions differing in bidding formats and pricing mechanisms are considered within the Web-of-Cells architecture. They are uniform and discriminatory auctions. The results of scientific investigations regarding the type of the auction, which is superior another, are controversial and comparison of the auctions is complex, since it depends on the objective of the auction (Table 2) [47]. Thus, there is no unanimous opinion, which type of the auction should be selected.

Table 2. Comparison of uniform and discriminatory auctions [47].

Objective of the Auction	Uniform Auction	Discriminatory Auction
Revenue maximization	√	
Assurance of the continuity of financing and reduction in price volatility		√
Learning about honest valuations		√
Increase in number of bidders and strengthening of the role of the market	√	
Increase in efficiency		√
Prevention of collusion		√
Selling near the market price	√	

However, based on the observations provided in [47] that aim at maximizing revenue, increasing the number of bidders, strengthening the role of the market and selling at the market price, the priority should be given to the uniform auction. In [48], it is supposed that uniform auction is relevant because of its simplicity and effectiveness when responding to the main questions, such as who should receive system balancing products, who should supply them and at what price system balancing products should be traded. Moreover, when the market is competitive the uniform auction has two features, such as short-run and long-run efficiency [48]. In addition, consumers and BSPs are much better off with uniform auction than with other alternatives and the main argument against uniform auction is political but not economical [48]. Research findings of [49] suggest that discriminatory auction may lead to inefficient production. The [50] concludes that discriminatory auctions could reduce volatility, but at the expense of higher average prices. Findings of [51] suggest that switching to a discriminatory auction will not necessarily result in greater competition or lower prices. Considering to advantages of uniform auction and disadvantages of discriminatory auction, the authors of the article have good motives to select the uniform auction design to trade in system balancing products within the Web-of-Cells architecture.

The peculiarities of the uniform auction are as follows: the uniform auction design is selected when at least the one CO and many competing BSPs participate in the market for system balancing product, which has to allocate multiple units of system balancing product. Subject to the uniform auction, the CO collects all the bids and offers from the bidders (BSPs), creates an initial Merit Order list for the system balancing product, and match it with the requested volume of system balancing product. The CO establishes the market-clearing price (MCP) by matching supply and demand of system balancing product. Win the bidders (BSPs) whose bids, or sections of their bids, offered lower price than the MCP. All winners (BSPs) receive the same price, independently on their financial offers.

The advantage of a uniform auction is that it is fair and attracts the participation of small bidders. Since this is a closed-type auction, no information is published before the clearing takes place. It is expected that under the Web-of-Cells architecture the level of competition is achieved high in the market and many small-scale BSPs participate here. Therefore, this type of auction is selected.

6.2.3. One-Sided vs. Two-Sided Auction

In relation to types of market participants involved in the auction both one-sided and two-sided auction designs are relevant for the Web-of-Cells architecture.

The need for one-sided auction comes from the "local problems should be solved locally" feature of the Web-of-Cells architecture. This feature leads to the establishment of a monopsonistic market structure (one-sided), where the CO is a single buyer of the system balancing products and there are many local small-scale BSPs supplying the system balancing products. The peculiarity of the one-sided auction is that only the BSPs bid (peer of price and volume) in the auction and form the price-responsive supply curve of the particular system balancing product. The nature of demand, for establishment of which the CO is responsible, is price-non-responsive. The CO determines the demanded volume of system balancing product but no price is established at which the CO would prefer to procure the system balancing product. The CO is not allowed to bid price because it is a single buyer and can dampen the price, which then could be too low for the BSPs to participate in the market. One-sided auction is established to trade in the IRPC, aFCC and BRC services.

Principally, the two-sided auction design features a number of the COs and the BSPs from different Cells and enables the active participation of supply and demand sides to compete on a level-playing-field basis, where both the BSPs and the COs are allowed to bid. It enables forming the aggregated price-responsive demand curve too, when the COs bid both quantity and price in the auction and in such a way show their preferences. The two-sided auction is established for trading in the BSC service, since cross-Cell trade in the BSC service is allowed and more than one buyer exists. The advantage of the two-sided auction over the one-sided auction is that the former ensures lower transaction cost for the BSPs and the COs. Thus, operational efficiency improves. It helps to control market power, increase price and utilization efficiencies and enhance the social welfare of the market through the establishment of the competitive situations in the market too.

6.3. Consistency of Auctioning

Depending on the consistency of the auctions held during the time, two categories of auctioning (sequential and simultaneous auctioning) are considered for the Web-of-Cells architecture. Based on their peculiarities, the Web-of-Cells architecture develops an idea of the hybrid auctioning, which combines the features of both sequential, which is applied in European countries, and simultaneous, widely used in USA, auctioning.

The hybrid auctioning takes into account the approach of the sequential auction that market for electricity (day-ahead and intraday) products and market for system balancing products are cleared sequentially meaning that the market for balancing energy is cleared after the clearance of day-ahead and intraday electricity markets. Winners in the sequential auctioning are chosen easy by selecting the lowest price-offering participants for each product separately. This type of auctioning is selected as a core auctioning mechanism because of its practical use in European countries. The transformation of electricity markets auctioning to simultaneous auctions shall mean significantly increased cost. Subject to the simultaneous auctioning, the products to bidders are allocated along with the minimization of joint bid cost of providing energy and balancing services. In this approach, it is hard to justify the schedule and pricing of the product.

The Web-of-Cells architecture improves the sequential market mechanism by adding the cascading procurement principle applied in the simultaneous auctions. In result, the system balancing products are procured by the CO in a way that total cost of providing system balancing services products is reduced. That is, the reduction of total cost of system balancing products achieved through the application of the cascading procurement principle, which is described in a more detail in Section 9.2.

7. Analysis of Market Design Variables for Improvement of Operational Efficiency

7.1. Timing of the Market for System Balancing Products

The Web-of-Cells architecture proposes a novel approach towards an organization of timing of the market for system balancing products. The approach is based on the finding of the solutions to reduce the costs of system balancing products provision (see in detail Section 7.2), the rolling schedule of bidding, clearance and the Merit Order list establishment (see in detail Section 7.4), the transparency of rules how the market for system balancing products is organized and the non-discrimination of system balancing products provision against the time criteria. The composition and alignment in point of timing of the market for system balancing products is provided in Figure 6.

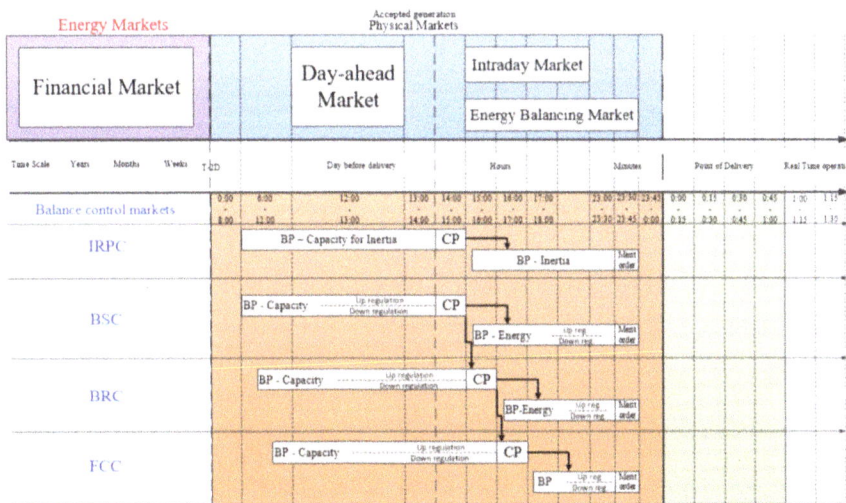

Figure 6. The timing of the market for system balancing products. Here: BP—bidding process; CP—clearance process; Merit order—establishment of Merit Order list.

The market for system balancing products is organized in a way that initially, the BSPs decide on in which sub-market—inertia capacity or balancing capacity—they will take part in. Those BSPs who decide to participate in the sub-market for inertia capacity and whose bids are accepted, are not allowed participating in other sub-markets, except for inertia. The same is valid for the BSPs who bid balancing capacity. Those BSPs who decide to participate in the sub-market for balancing capacity for a particular service (either this is the BSC, BRC or aFCC service), and whose bids are accepted are not allowed participating in other sub-markets, except in the sub-market for balancing energy for this particular service.

The Web-of-Cells architecture assumes that the sub-markets for balancing capacity for the BSC, BRC and aFCC services are organized in a sequential manner. After the sub-market for balancing capacity for the BSC service is cleared, the sub-market for balancing capacity for the BRC service is cleared and later on the sub-market for balancing capacity for the aFCC service is cleared. Such a layout of the system balancing products in time is recommended for several reasons, including to reduce the costs of the system balancing products provision and to limit a "price reversal" effect of lower-quality balancing capacity to be priced at a higher price. Aiming at implementation of aforementioned recommendations, the cascading procurement principle is implemented (see the arrows connecting separate sub-markets for balancing capacity). Therefore, higher-quality balancing capacity is cleared one hour earlier than lower-quality balancing capacity. This hour is required both to move the rejected by the market higher-quality balancing capacity bids (the bids submitted but not accepted by the sub-market) to the sub-market for lower-quality balancing capacity, if they correspond to technical requirements held for the system balancing product of this sub-market and clearing the sub-market. The sub-market for upward and downward regulations are organized in parallel, i.e., they have the same timing.

The sub-markets for inertia capacity and balancing capacity are organized earlier than the sub-markets for inertia and balancing energy, since inertia and balancing energy bids are submitted to the market by the BSPs who won capacity auction and thus have an obligation to keep the capacity for the particular hours in real time. Thus, clear interrelationships of the timing of the sub-markets for inertia capacity and inertia, balancing capacity and balancing energy are established.

Considering to the criteria of transparency and non-discrimination, the length of bidding and clearance processes as well as the length of the Merit Order list establishment is based on the following time-related principles:

- the BSP submits bids for inertia capacity and balancing capacity during 8 h;
- the clearance process takes 1 h;
- if inertia capacity or balancing capacity bids are accepted, the inertia and balancing energy bids are submitted during 6.5–8.5 h depending on the service for the provision of which the inertia capacity and balancing capacity is reserved, respectively;
- the Merit Order list for inertia and balancing energy provision is established during 15 min and 15 min before real time.

7.2. System Balancing Product Resolution in Time

In Europe, both short- and long-term markets for the provision of system balancing products are established. The products (balancing capacity and balancing energy) for primary control are acquired by entering into hourly, daily, weekly, monthly or annually contracts. The system balancing products for secondary and tertiary controls are purchased in the long-term markets by entering into contracts, the duration of which varies from week to year or more [30]. Certainly, in some countries both short- and long-term markets for the provision of secondary and tertiary services are established. The "Winter Package" sets that the contracting period should have a maximum of one day.

Considering to the experience of European TSOs and suggestions provided in the "Winter Package", short and long contracting (procurement cycles) is considered within the Web-of-Cells architecture. Shorter procurement cycles allow new entrants and BSPs with a small portfolio of either generation units and/or schedulable load to participate in the market for system balancing products. The BSPs owning RES units prefer short procurement cycle because RES capacity can be used only for a limited time, for example, 3500 h a year whereas traditional fossil fuel-based power plants can run for much longer period. Therefore, if a system balancing product's resolution (in time) is set long, the BSPs owning RES units will not be technically capable to provide the required capacity. Thus, utilization efficiency of available resources reduces. Since an entry barrier for intermittent low marginal cost technologies is created, the price efficiency reduces too, because these RES technologies do not take part in the market and do not form prices. Thus, prices are tended to be high reflecting marginal cost of traditional production technologies. The BSPs operating peak/marginal plants as well some large consumers are interested in longer procurement cycles to guarantee fixed revenues [27].

The system balancing product resolution (in time) as a market element is essential for the BSPs under the Web-of-Cells architecture, especially for the demand response, RES and storage technologies. The market element's relevance for the BSPs who aggregate consumption units or electric vehicles is highly dependent on the habits of consumers. The amount of reserve they are able to provide is variable and it will be precluded from a market where the time is too long. Demand response providers are smaller than traditional generators and may be limited in the contiguous period in which they deliver the system balancing product. With the aim to increase the number of demand response related participants and competition between the BSPs, shorter balancing product resolution (in time) should be set. However, short system balancing product resolution (establishment of short-term markets) alone could be too volatile and risky to support procurement of the products. Thus, long system balancing product resolution (long-term contracts) could be considered as an alternative too. It is justified for the purpose to protect investments. Moreover, long-term markets could give the BSPs the opportunity to better control risks by fixing the contracted volume and price for long-term. However, from the point of view of Web-of-Cells architecture and the European Commission's approach to long-term contracts [52], longer system balancing product resolution (in time) potentially comes with some drawbacks, for example:

- it reduces competition in future bidding processes since the COs who are beneficiaries holding long-term contracts will effectively be out of the market for the duration of the contract;
- it transfers risks to the BSPs (both the risk that prices of balancing products will rise in future and capacity prices fall, and—as more contracts are signed—the risk that contracted capacity will not be required in future).
- it increases the costs of any future market design transition (operational efficiency reduces), since long-term contracts would in principle need to be honored if in future a new market design will be adopted.

Considering advantages and disadvantages of different length of the system balancing product resolution, the results of EcoGrid EU project (in the EcoGrid EU project a 5 min resolution was initially used, but the overall conclusion was that 15 min would be enough and serve the purpose), EU practice, and requirements set in "Winter Package" for future power systems, the Web-of-Cells architecture establishes the short-term market for system balancing products.

7.3. Bid Time Unit

Within the Web-of-Cells architecture, the bid time unit (BTU) is closely linked to the energy schedule time unit (STU), which divides responsibility between the CO and the BRPs, and imbalance settlement period (ISP), the period for which imbalance of the BRP is calculated. The recommendation to link BTU to STU and ISP comes from the need to harmonize time units to increase operational and price efficiencies. The explanation of how different types of time units are interrelated is derived from the analysis of the following chain of events—information (in terms of prices of the system balancing products) received from a particular BTU is used to price an imbalance of a particular ISP, which is established taking into account energy schedules from a particular STU. Thus, if time units of BTU, STU and ISP are not equalized and harmonized the operational efficiency and price efficiency reduces. Operational efficiency reduces because additional tasks should be performed by the market participants to normalize the results of transactions exercised in one type of time unit (for example, in the BTU) in a way that they could be available for use in other types of time unit (for example, in the ISP). The comparability of information received from time units of different length is complicated. Moreover, more time, qualified personnel and physical infrastructure is required, which increases cost and reduces price efficiency.

The Web-of Cell architecture considers short and long BTU&STU&ISP. Short BTU&STU&ISP is selected of 15 min and long BTU&STU&ISP—of 60 min. It is expected that a short BTU&STU&ISP provides the BRP with a stronger incentive to balance the available energy portfolio than a long BTU&STU&ISP because more accurate information is available on short terms, and deviations from the scheduled energy will be smaller. Thus, balance planning accuracy shall increase. This shall lead to smaller energy imbalances at the Cell level. Thereby, the CO will activate less balancing energy bids, as demand for balancing energy will be lower. Indeed, reduced demand for the balancing energy will lower imbalance price, which in turn will diminish incentives for the BRPs to balance their energy portfolios, as imbalance cost could be smaller than cost if the balancing efforts were placed. However, short BTU&STU&ISP shall raise the transaction cost because energy schedules and balancing energy bids shall be submitted and imbalance shall be calculated frequently. Within a short BTU&STU&ISP, the reduced demand for balancing energy will influence the improvement of criteria of the availability of balancing resources in a way the BSPs will have more opportunities to provide balancing energy, even if they have abundant commitments in the intraday electricity market. Subject to the abundant commitments in the intraday market and long BTU&STU&ISP the BSP may be technically incapable to provide balancing energy for BTU&STU&ISP of 60 min. If only few BSPs are capable to provide balancing energy, they can start using power in the market and offer balancing energy at price not reflecting actual cost. Thus, price efficiency shall reduce. Within the framework of the Web-of-Cells architecture, the BTU&STU&ISP is set of 15 min.

7.4. Time Horizon (Frequency) of Bidding, Clearance and Establishment of Merit Order List

The time horizon of the system balancing products' bidding, clearance and the Merit Order list establishment is short in the Web-of-Cells architecture. This means that bidding and clearance processes as well as the establishment of the Merit Order list for the particular system balancing product is performed hourly instead of daily, quarter-annually, semi-annually or annually and on the rolling schedule.

Short time horizon means that the BSP has a right to take part in 24 auctions per day for the particular system balancing product selling if the CO organizes the auctions. The auction can be held 24 times a day for each coming hour of the day. The BSPs and the CO submit bids for each hour, which inter alia is divided into quarters. These quarters are known as the market time units (MTU). Thus, totally 96 MTU are foreseen per day. Bids that are submitted for a particular MTU can be changed until the clearance process of inertia capacity or balancing capacity starts or the Merit Order list for inertia or balancing energy is established. The rolling schedule means that the proposed timing of the sub-market for system balancing product (see Figure 6) moves forward in a way that the Merit Order lists for the MTUs of the particular hour are established 15 min before the real time.

Short time horizon has sense in terms of improved utilization and price efficiency. For instance, if the sub-market for system balancing capacity is cleared yearly, then only those BSPs who have balancing resources, which are available across the whole year, could offer balancing resources. Indeed, this is an entry barrier for many RES&DER BSPs since they are intermittent and cannot offer capacity for the whole year. Thus, efficiency of RES&DER utilization reduces. Because of limited supply, the price efficiency reduces too [16].

7.5. The Distance to Real Time of the Auction

The general existing practice in Europe is that the TSOs organize auctions or enter into the agreement for the procurement of the balancing capacity for the delivery of the primary control services day- or week-ahead from real time while agreements for the provision of services for the secondary and tertiary control are signed from day- to year-ahead from real time. Long distance to real time is favorable to the TSOs since it reduces the uncertainty of balancing capacity availability. However, the uncertainty increases the risk premium added on the top of balancing capacity procurement price. Therefore, balancing capacity procurement price can be high. During the long time many events (including new lower cost balancing capacity BSPs could appear) could happen. Their impact will not be reflected into the price of balancing capacity agreed in advance. Thus, the real value of the balancing capacity will be rarely reflected in the procurement price. Seeking to avoid such situations, the distance to real time of the auction should be reduced.

The general practice is that the distance to real time of the auction/agreement is linked to the system balancing product resolution (in time). It is well known that long-term contracts are agreed in advance from real time and they are effective far ahead from real time. Indeed, this proclaims that market is "locked" for new BSPs even for several years forward because old annual contracts are valid and additional contracts for the following years have been already agreed. Thus, new BSPs have not even possibility to enter the market until the old contracts expire. And this is a serious barrier, which hinder and inhibits competition in the market. Therefore, the price of balancing capacity will not reflect its true value and the price efficiency will reduce.

What could happen if the existing practice was hold for future? Let's consider the case of the aggregator. Long distance to real time of auction will induce uncertainty on decision making. If the procurement of balancing capacity was made long before the real time, the aggregator would made particular assumptions on the amount of reserve it could provide. For instance, if the procurement of balancing capacity was made one year in advance, the aggregator would submit the bid on the number of units at the time of the bid, and it would not be able to take into account all the potential new aggregated units. Thus, in real time, the utilization efficiency would reduce, since additional balancing capacities would not be requested and accepted by the market.

Thus, for the future market design the distance to real time is reviewed and shortened, as it is proposed in the "Winter Package". The Web-of-Cells architecture proposes to solve the issue of risk premium, reflection of the real value in the price of the system balancing products, reduction in price efficiency and utilization efficiency by organizing auctions day ahead from real time and on rolling schedule.

8. Analysis of Market Design Elements for Improvement of Price Efficiency

8.1. Procurement Scheme

Within the Web-of-Cells architecture, several types of schemes for the procurement of inertia capacity, inertia, balancing capacity and balancing energy are considered (Figures 7–9, respectively).

Organized and bilateral markets are recognized as two main market arrangements for the procurement of the system balancing products within the Web-of-Cells architecture, although alternative procurement schemes (mandatory offers, etc.) are considered too. However, alternative procurement schemes are not discussed in this article as they are usually country-specific.

The peculiarity of the procurement scheme in the organized market is that there is no contract or obligation for the BSPs to offer capacity for inertia, inertia, balancing capacity or balancing energy in the market; the BSPs voluntarily participate here and bid a price, at which they wish to sell, and volume. This is a centralized market, where the CO utilizes a merit order to dispatch the generators and loads in a least cost way. Besides, standardized products with a short duration are exchanged here. The organized market model enhances transparency and fosters competition. The drawback of the market model is high data management costs and availability to facilitate the exercise of market power by some BSPs.

If the BSPs and the CO are allowed negotiating a contract regarding the offered system balancing product (its quantity and quality) and its price, bilateral market is established. Negotiations (through the customization) provide a flexible way to determine prices, quality of the system balancing product, financing terms and other, but they are costly (high transaction cost) and time consuming.

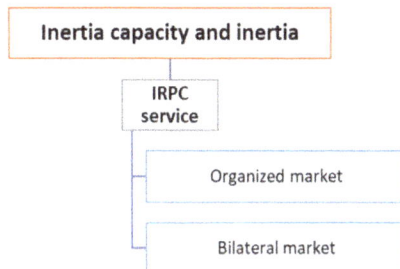

Figure 7. The methods for procurement of inertia capacity and inertia.

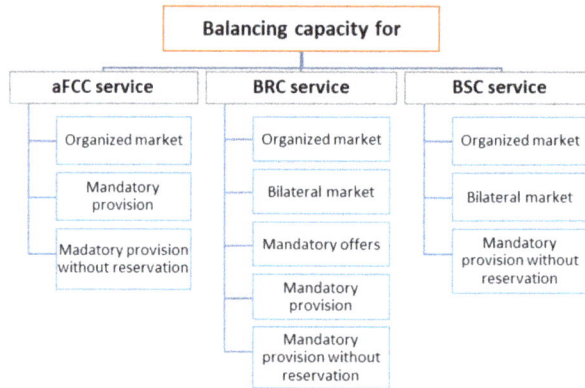

Figure 8. The methods for procurement of the balancing capacity.

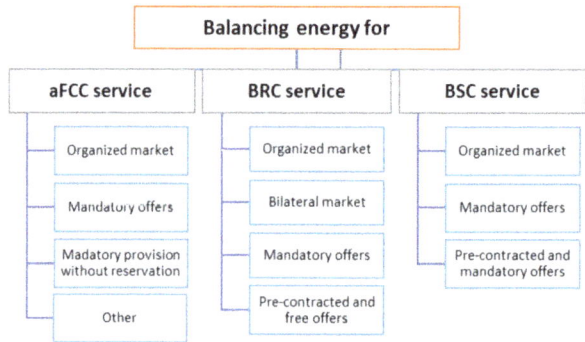

Figure 9. The methods for procurement of the balancing energy.

The CO and the BSPs may want customized contracts that offer flexibility because there are many uncertain factors that have an influence on predictions of electricity consumption, production and price in the future. The flexibility enables them to make adjustments more easily and at lower cost when the factors become better recognized. The economists argue that this type of market arrangements is decisive to the functioning of electricity markets, because they provide parties with price stability and certainty necessary to perform long-term planning and to make rational and socially optimal investments. Bilateral markets are valuable since they protect the BSPs and the COs against price uncertainty and make revenue and payment streams more predictable. As a result, investment decisions are facilitated. Within the Web-of Cells architecture they could be used to provide location-based services where there are potentially insufficient volumes of competition. The bilateral market is a decentralized market, where the CO is constrained in scheduling by negotiated contract price and volume. The advantage of the bilateral market model against the compulsory provision is the fact that the CO procurers only the amount it needs and deals only with the cheapest BSPs. However, the bilateral market model has drawbacks, which are discussed in [19].

Taking into account the advantages and disadvantages of the markets for procurement of system balancing products, as well as the requirements established in the "Winter Package" (particularly, that "the procurement processes ... are transparent while at the same time confidential ... ", " ... the procurement ... is organized in such a way as to be non-discriminatory between the market participants ... " and " ... procurement ... is market-based ... " [26]) the Web-of-Cells architecture accepts organized market for the system balancing products' procurement processes.

8.2. Remuneration Scheme

The Web-of-Cells architecture considers that the system balancing products are procured on commercial basis and the BSPs are remunerated for the provision of them. The COs can apply different remuneration mechanisms (Figure 10).

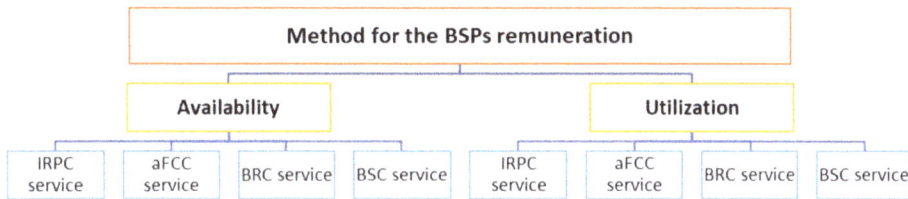

Figure 10. Methods of the BSPs remuneration for system balancing products.

In case of aFCC, BRC and BSC services, the COs pay the BSPs for balancing capacity availability (i.e., for holding balancing capacity) and for its utilization (i.e., for the delivery of balancing energy) (Figure 10):

- the availability payment is made to the BSP in return for the balancing capacity being made available to the CO during the MTU. The availability payment is equal to the price (EUR/MW) at which all the BSPs are paid for holding balancing capacity for the MTU;
- later, when balancing capacity is called upon, in addition to the availability payment the CO pays a utilization price (EUR/MWh) for balancing electricity delivery. The utilization price may be noticeably different from the price the BSP asked for the activation of balancing capacity. The utilization price therefore reflects the prevailing market price at the time of use.

In case of IRPC service the BSPs are remunerated for availability of inertia providing capacity (EUR/(MW × s)) and for its utilization (delivery of inertia) (EUR/kg × m^2).

9. Analysis of Market Design Elements for Improvement of Utilization Efficiency

9.1. Pricing Mechanisms for the System Balancing Products

The Web-of-Cells architecture considers both market-based pricing mechanisms and pricing exercised by the regulating authority or the CO. By definition, the regulated price is set by the regulating authority or the CO and is the same for all the BSPs. The use of regulated price is related to the mandatory provision of the services by a few BSPs (often large producers) since there is no information to choose the BSPs based on their cost [30]. As it is argued in [30], even if the rules allow for new entrants, such as aggregators to propose system balancing products, the selection of the capacity will be made by an administrative rule, which would not allow new participants competing effectively with incumbent market participants. Thus, this form of pricing should be or is used, when market power is an issue in the market. However, a regulated price is not a desirable choice since it reflects very imperfectly the actual cost of providing the balancing service, especially, if cost changes in time and circumstances [19]. This means that regulated price does not take into account the market value of electricity generation [30]. With a fixed guaranteed and unchanging price, a generator can receive cross-subsidies [30].

The Web-of-Cells architecture considers two alternative market-based pricing rules to the regulated price. In particular, it considers uniform and pay-as-bid pricing rules.

Under the uniform pricing rule all market participants with accepted bids are paid a uniform (single) price, which is the market-clearing price (MCP), regardless of their bids. The MCP is determined as the offer price of the highest accepted offer in the market, as it is shown in Figure 11.

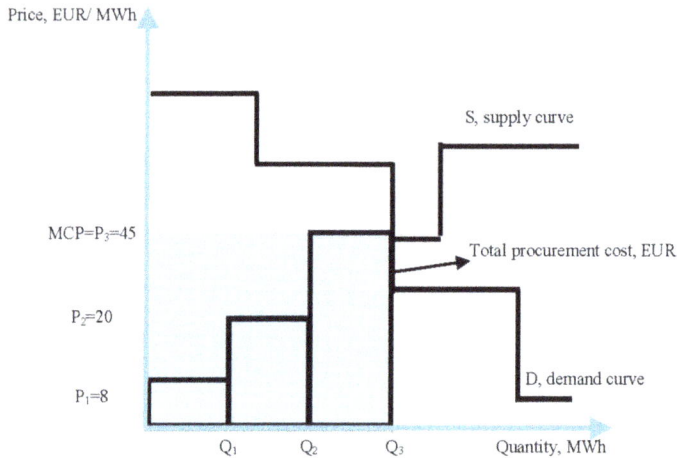

Figure 11. Price determination and total procurement cost subject to uniform pricing rule.

The uniform pricing rule is beneficial to apply because the MCP aligns the actions of the BSPs and COs with maximizing the gains from the trade. The reason for the application of the uniform pricing rule to determine the price in the market for system balancing products is twofold [48]:

- *the incentive to provide for efficient dispatch*. The uniform pricing rule means that the BSPs offer the system balancing products at prices closely reflecting their marginal costs. This results in operationally efficient dispatch meaning that demand is supplied by the lowest cost resources and technologies, which are owned by the BSPs;
- *the optimal investment* means that generation technologies are built in necessary places at required time with the investment risks borne by the investors but not the tariff payers.

In [48], four principal criticisms against the application of the uniform pricing rule to set the MCP (Figure 12) are discussed.

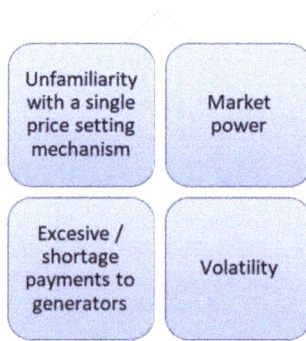

Figure 12. Principal criticism of the uniform pricing rule [48].

In [48], it is argued that the critique of employing the unified pricing rule in electricity market includes a reference to the thoughtful unfamiliarity with the price setting process and application possibilities. However, critics do not notice that the pricing mechanism is successfully implemented in other commodity markets. Nowadays, considering to its wide application in all timeframes of electricity market, the argument is not significantly relevant, but it should be taken into account when introducing uniform pricing in newly developed markets, including market for system balancing products. The potential exercise of market power is an issue too, but if market power is an issue under the uniform pricing rule, it could be also an issue under an alternative pricing rule, i.e., pay-as bid. In [48], it is argued that the shift from a uniform pricing rule does not curb the exercise of market power. A number of measures available to mitigate market power that is consistent with the uniform pricing rule could be used. Since the MCP could set by the expensive offers (for instance of gas-based generators), there is a criticism about the excessive level of payment to baseload resources such as coal and nuclear generators—particularly to owners of old assets those construction cost was lower than presently new construction cost. This means that BSPs owning old baseload generation units receives high operational profit. However, the excessive payments to the BSPs is reduced by allowing of varies technologies, including RES having low marginal cost participating in the market. The current practice shows that subject to the uniform pricing rule and high shares of RES-E in the market, electricity prices are low or even negative [53]. The generation side is threatened by closure of conventional power plants that are presently experiencing decreasing profitability due to low electricity prices and limited number of operating hours [54,55]. However, changes in pricing rule aimed at reducing or increasing profitability to the generating assets are unfair [48]. It would signal to the investors in generating assets that they are exposed [48]. Volatility of prices caused by the unified pricing rule is an issue too [48] but forward contracts could be applied to hedge against price variability and reduce exposure to volatility.

Under the pay-as-bid pricing rule (it is commonly applied at the Nordic regulating power market for so-called "special regulation" e.g., resolving congestions, where location of the resource is important), the prices are set based on a first-come, first-served principle, where best prices, which are the lowest selling prices offered by the BSPs, come first. Figure 13 illustrates price setting and total procurement cost of the pay-as-bid pricing rule.

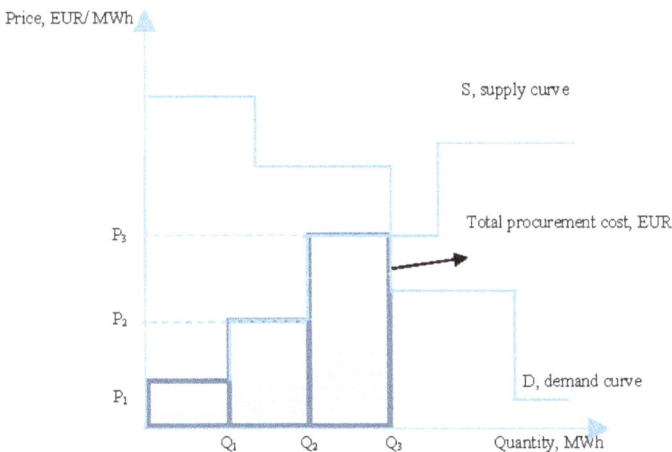

Figure 13. Price determination and total procurement cost subject to pay-as-bid pricing rule.

As it is seen from Figure 13, under the pay-as-bid pricing rule, the BSPs with the accepted bids are paid according to their bids, i.e., prices P_1, P_2 and P_3, respectively. No single MCP is established.

Total procurement cost (TPC) is the sum of individual procurement cost, as it is expressed by Equation (1):

$$TPC = \sum P_n \times Q_n \tag{1}$$

where: n—is the identification BSP; P_n—price of accepted bid of the n-th BSP; Q_n—volume of accepted bid of the n-th BSP.

Thus, total procurement cost, which are defined as the payment by the CO as a representative of demand side, typically involves the demand paying a price that is equal to the BSPs weighted-average price. Using the BSPs-weighted average price ensures that the total payment by the CO equals total paid to the BSPs.

The pay-as-bid pricing rule is typically used as a measure tackling the issues of the unifom pricing. The pay-as-bid pricing rule used for continuous trading meaning that BSPs and COs submit supply and demand bids to a central platform (exchange) and matching bids are continuously cleared on an individual basis. The continuous trading order book is visible to all market participants and contains all submitted bids that have not cleared yet. The BSPs can cancel submitted and not-cleared bids at any time. Continuous trading bids are matched based on price-time priority: orders are matched in sequence of the attractiveness of their price. If offers have identical prices, the time of submission is the decisive factor. It implies that BSPs have to anticipate the clearing price and accordingly mark up their bids. There is no empirical or experimental evidence that pay-as-bid or other alternatives would reduce prices significantly compared to a single market-clearing price design. In fact, some evidence suggests that pay-as-bid would increase prices compared to explicitly setting the single MCP. Moreover, pay-as-bid pricing rule has some significant drawbacks [48,51]:

- When bidding truthfully, non-marginal BSPs may receive smaller remuneration for the system balancing products they supply compared to the BSPs paid at uniform pricing rule. In this case, the CO meets more favorable prices compared to a uniform pricing rule, however, is at risk of losing the opportunity to match bids if it waits too long. Thus, continuous trading with a pay-as-bid pricing rule may incentivize market participants do not bid truthfully. Thus, incorrect demand and supply signals could happen;

- A market based on continuous trading with a pay-as-bid pricing rule includes a certain first-come-first serve characteristic, as matching bids are immediately cleared, which may not lead to a welfare maximization and an optimal allocation of balancing resources;

- An important question regarding the organization of continuous trading with a pay-as-bid pricing rule that is not answered yet includes the optimal number of auctions and their timing, taking into account the impact on liquidity;

- Inefficient dispatch. Under the pay-as-bid pricing rule, a profit maximizing offer involves prediction of the MCP. When uncertainties exist, the forecast is inaccurate and different BSPs will forecast different values of the price. Since the CO uses offers to decide on the dispatch, this means that occasionally a high marginal cost BSPs with a lower offer will be dispatched instead of low marginal cost BSP submitting higher offer, i.e., the efforts to maximize profits will result in inadequate dispatch decision. Inefficient dispatch wastes costly resources and is undesirable;

- Another inefficiency of pay-as-bid pricing rule is the cost of forecasting market prices that it will impose on all BSPs and COs. Under the uniform pricing rule, the BSPs are motivated to bid at marginal cost, which are available to them. If the method is changed to pay-as-bid pricing rule, uncertainty about market price calculation and large cost to forecast it are introduced. Moreover, small companies achieve much higher cost per unit of output, although they put as much efforts to forecast price as large ones.

- Bias against the small BSPs. The basic argument against pay-as-bid pricing rule is that the bsps suffer relatively higher costs in the assessment required to form their offers for pay-as-bid pricing rule than assessment required for offers for uniform pricing rule. Under the pay-as-bid pricing rule, the MCP is forecasted instead of offering a price reflecting the marginal cost. This requires knowledge about the market and more investment in market analysis and data gathering are required. Large BSPs could spread the associated cost across a greater amount of output while this is not possible by small BSPs who as a result may lose their competitive position. From this perspective, pay-as-bid pricing rule is more favorable to large BSPs. A pay-as bid pricing rule make it relatively harder for new entry by small BSPs than entry under the uniform pricing rule.

- Difficulties with market monitoring. The aim of market monitoring is to assess if prices are not competitive. The assessment is done considering the offers. Under a pay-as-bid pricing rule, competitive offers involve markup above the marginal cost. There is no easy to assess if or not such offers are exploiting market power, i.e., market power monitoring becomes impossible under the pay-as-bid pricing rule. Under the uniform pricing rule there is a particular test to monitor market power. This is an assessment whether offers track marginal cost.
- Investment. Under the pay-as-bid pricing rule the prices paid to the bsps would be driven towards recovery of operating cost only. This would fail to provide enough remuneration to cover investment cost. Owners of generating assets will find themselves going bankrupt and no new investment will be available. Thus, pay-as-bid impede capacity expansion, which alongside with a demand-side response is a measure for a better performance of the market. The pay-as-bid impede new entry, which is a measure of mitigating market power.
- Tend to weaken the competition in generation that is the remedy against of monopoly power, which may cause the steady price increase at times of peak demand.

Pricing mechanism for the Web-of-Cells architecture is chosen in a way that selected pricing mechanism performed three important functions [56]:

- *Signaling*, meaning that changes in prices provide information both to the CO and the BSPs about changes in the market conditions, prices reflect scarcities and surpluses,
- *Transmission of preferences* meaning that through the choices the CO send information to the BSPs about the changing nature of needs;
- *Rationing* meaning that when there is a shortage of balancing product, its price will rise and deter some CO from buying the balancing product.

The "Winter Package" foresees that the uniform pricing rule as an advanced method of pricing should be applied for the pricing balancing energy instead of pay-as-bid. Considering to results of analysis of advantages and disadvantages of different pricing mechanism and in accordance to the proposal of "Winter Package", that prices should be formed based on demand and supply and price signals should drive the market to react to shifting energy demands and fluctuating renewable energy generation, the Web-of-Cells architecture suggests that all system balancing products are priced by a uniform pricing rule.

9.2. Cascading Procurement

A sequential auction with uniform pricing rule in each round and without the substitution of a higher-quality system balancing product for a lower-quality product may result in a price reversal. "Price reversal" is the phenomena when prices for lower-quality system balancing products are set higher than prices for higher-quality products. Lower-quality products clear at higher prices than higher-quality products due to lower capacity availability after the initial rounds of procurement. To examine how the price reversal may occur, a behavior of the generator, which keeps small capacity for higher-quality product and as a result maintains large capacity for lower-quality product that is offered at high price, could be discussed. In this case, the CO has to satisfy demand for higher-quality product by purchasing it from other generators. Consequently, the CO has no choice but to pay high price for lower-quality product offered by the first generator since there is no sufficient lower-quality product providers as they sold their capacities for higher-quality product. The lack of competition is increased and it becomes necessary to plan measures preventing the generator to introduce low-quality product for higher prices. In the perfectly competitive markets, the phenomenon shouldn't be found [46]. The Web-of-Cells architecture foresees what is called the cascading procurement to solve the issue of the "price reversal".

To achieve the cascade, the sub-markets of system balancing products are cleared in sequence starting with the highest-quality and applying the rule of substituting the higher-quality lower-cost products for the lower-quality higher-cost products if total procurement cost is reduced. Qualities of reserves are graded by quickness and sureness of their response. The benefit of the variable of market design—the cascading procurement—is both reduced procurement cost and expanded supply [46].

The main question is how the substitution should be conducted? The Web-of-Cells architecture assumes that any surplus of high-quality system balancing product could be automatically transferred by the CO to the market for lower-quality product and so on (see Figure 6). This should increase the efficiency of the markets compared to the absence of a cascade.

It is worth noting that cascading procurement contributes to the reduction of price reversal but it cannot avoid. For example, if there is shortage in the market for low-quality product and totally, but not in markets for high- and medium-quality product, then the price in the market for low-quality product is set higher than in other markets. In such case, high-quality reserve units will try to sell in the market for low-quality product. Indeed, this could result in an inefficient use of reserves [46].

10. Conclusions

The paper investigated the scientific issue of designing the market for power system balancing products within the Web-of-Cells architecture. An integrated approach, combining the concepts of market and its objectives, market elements, designing and market assessment criteria, to the issue of the economically efficient market design for the procurement of system balancing products has been developed.

During the research three-core elements of the market were analyzed. These are exchange, system balancing products and market participants. Thus, the market for system balancing products was determined an auction-based exchange where system balancing products should be traded between BSP and the COs. The research results show that with an increasing volume of intermittent RES integrated into the power markets, new types of BSPs should be requested. Thus, in addition to the centralized thermal power plants, small-scale RES, demand response and storage technologies should be available at the distribution level. With the purpose to increase the size of the BSPs, aggregators should play an important role too. An organized marketplace (exchange), contributing to improvement of operational efficiency, developing preconditions for competition, assuring transparency and level-playing field to all the BSPs and the COs, should be established. Auction as an instrument promoting competition, as an institution determining price and as the economic (market) mechanism used to allocate the balancing products in economically efficient way should be used. With the aim to overcome market failure due to the missing market problem, new classes, types and directions of the system balancing products should be suggested and traded in separate sub-markets. The direct link between the quality of system balancing products and their price should be established through the implementation of the principle of cascading procurement and by considering the distance to real time of auction.

The analysis of market elements improving operational efficiency of the market for system balancing products showed that bid time unit and timing of the market are of pivotal importance. Bid time unit should be selected in relation to the energy schedule and imbalance settlement time units. Moreover, priority should be given to short-term bid time unit instead of long-term since it should increase balance planning accuracy, availability of balancing resources and price efficiency. The sequential approach to the market for procurement of system balancing products organization should be applied meaning that market for lower-quality system balancing product should be closed and cleared after the clearance of the market for higher-quality system balancing product. The length of bidding process should be selected in a manner that sufficient time is left to the BSPs for bidding system balancing products. With the purpose to increase liquidity, price efficiency and utilization efficiency in the market, bidding and clearance of system balancing products should be performed often and time horizon should be reduced. It is suggested that bidding and clearance of system balancing products should be executed hourly. Auctions for the procurement of system balancing capacity should be organized not earlier than one day ahead from real time because more accurate information is available close to market time unit; therefore cost-reflectivity in price increases, more BSPs will participate in the market. Balancing product procurement cycle should be short (for instance, quarter-hour), since it shall contribute to new entries and providers with a small portfolio of either generation units and/or schedulable load should be capable to participate in the market for system balancing products.

The analysis of market design elements for improvement of price efficiency showed that system balancing products should be supplied by the BSPs on voluntary basis and procured by the COs on commercial basis through the organized markets. The organized market should assure transparency, equity, anonymity, clear trading and pricing rules, as well should fosters competition. Moreover, standardized products should be traded here. The BSPs should be remunerated for balancing capacity availability and its utilization in real time. The need to be remunerated for the balancing capacity should come from the market structure, where CAPEX (instead of operating expenditures (OPEX)) intensive power plants are abundant.

The analysis of market design elements for improvement of utilization efficiency showed that pricing mechanisms are of high importance. Seeking to improve utilization efficiency, priority should be given to the market-based pricing mechanisms instead of regulated pricing. Subject to the market-based pricing mechanisms, market forces should influence on the level of price and its changes, but not the regulator who could establish it based on average cost. Moreover, priority should be given to a uniform pricing rule instead of pay-as-bid pricing rule. An incentive to provide for an efficient dispatch and contribution to optimal investments are relevant reasons for an application of a uniform pricing rule. The features of first-come-first-served, high transaction cost, untruthfully bidding and other of pay-as-bid pricing rule send wrong signals to the market, may not lead to welfare maximization, optimal allocation of balancing resources and cause difficulties in monitoring. These features are drawbacks of the pay-as-bid pricing rule and quite good arguments for the uniform pricing rule, which tackles them. The principle of the cascading procurement should be implemented too. Its implementation should serve as an element limiting the behavior of lower-quality balancing products to be priced at higher prices. By implementing the principle, the bids submitted but not accepted by the market of higher-quality balancing products should be shifted to the market for lower-quality balancing products (if they satisfy technical requirements). In such a way, cost effective balancing resources should be utilized and balancing cost reduced.

Acknowledgments: The research leading to these results has received funding from the European Union Seventh Framework Programme (FP7/2007–2013) under grant agreement No. 609687. Any opinions, findings and conclusions or recommendations expressed in this material are those of the authors and do not necessarily reflect those of the European Commission.

Author Contributions: V.B.–development of the concept of the market design for the system balancing products; analysis of alternatives of the marketplaces for the system balancing products trading, analysis of types of the auctions; analysis of alternatives of the market design elements for improvement of price efficiency, utilization efficiency and operational efficiency. A.O.–development of the new concept of the electricity market design for system balancing products within Web-of-Cells architecture; analysis of market design elements and their options in European countries; definition and categorization of requirements for bids and offers of the control products; timing of the market for system balancing products trading. I.O.–energy policy analysis; new electricity balancing concepts development; power system security assessment in terms of functionality; future electricity market design identification; electricity market architecture and its elements; roles in the market specification and description. A.M.–Web-of-Cells architecture development and description; several types of schemes for the procurement of inertia capacity, inertia, balancing capacity and balancing energy consideration; categorization of system balancing products; related EU project analysis.

Conflicts of Interest: The authors declare no conflict of interest.

References

1. Germany Produced Record 35 Percent of Power from Renewables in First Half. Available online: https://www.reuters.com/article/us-germany-energy-renewables/germany-produced-record-35-percent-of-power-from-renewables-in-first-half-idUSKBN19N0GQ (accessed on 8 November 2017).
2. OECD/IEA. Energy Polices of IEA Countries: Denmark 2017 Review 2017. Available online: https://www.iea.org/publications/freepublications/publication/EnergyPoliciesofIEACountriesDenmark2017Review.pdf (accessed on 8 November 2017).
3. Auer, H.; Burgholzer, B. Opportunities, Challenges and Risks for RES-E Deployment in a Fully Integrated European Electricity Market. Technical Report from Market4RES Project 2015. Available online: http://market4res.eu/wp-content/uploads/D2.1_Market4RES_20150217_Final.pdf (accessed on 8 November 2017).
4. Lund, P. The Danish Cell Project. In Proceedings of the Power Engineering Society General Meeting, Tampa, FL, USA, 24–28 June 2007.
5. ENTSO-E. Future System Inertia Technical Report 2016. Available online: https://www.entsoe.eu/Documents/Publications/SOC/Nordic/Nordic_report_Future_System_Inertia.pdf (accessed on 8 November 2017).
6. EcoGrid EU. Available online: http://www.eu-ecogrid.net/ (accessed on 8 November 2017).
7. SmartNet. Available online: http://smartnet-project.eu/ (accessed on 8 November 2017).
8. Lund, P. Cell Controller Overview and Future Perspectives. Energinet.dk 2012. Available online: http://osp.energinet.dk/SiteCollectionDocuments/Engelske%20dokumenter/El/Cell%20Controller%20overview%20and%20future%20perspectives.pdf (accessed on 8 November 2017).
9. FRACTAL GRID. Available online: http://fractal-grid.euhttp://fractal-grid.eu (accessed on 8 November 2017).
10. C/sells Social Web. Available online: http://www.csells.net/en/ (accessed on 8 November 2017).
11. European Liaison on Electricity Committed Towards Long-term Research Activity Integrated Research Programme (ELECTRA IRP). Available online: http://www.electrairp.eu/ (accessed on 8 November 2017).
12. Martini, L.; Brunner, H.; Rodriguez, E.; Caerts, C.; Strasser, T.I.; Burt, G. Grid of the future and need for a decentralized control architecture: The web-of-cells concept. In Proceedings of the 24th International Conference & Exhibition on Electricity Distribution (CIRED), Session3: Operation, Control and Protection, Glasgow, UK, 12–15 June 2017.
13. D3.1 Specification of Smart Grids High Level Functional Architecture. Available online: http://www.electrairp.eu/index.php?option=com_content&view=category&id=63&Itemid=154 (accessed on 8 November 2017).
14. CEDEC. A European Electricity Market Design Fit for the Energy Transition. Position Paper 2015. Available online: http://www.cedec.com/files/default/a-european-electricity-market-design-fit-for-the-energy-transition-cedec-position-paper.pdf (accessed on 15 March 2017).
15. Energy UK. Ancillary Services Report 2017. Available online: https://www.energy-uk.org.uk/publication.html?task=file.download&id=6138 (accessed on 15 March 2017).
16. van der Veen, R.A.C.; Hakvoort, R.A. The electricity balancing market: Exploring the design challenge. *Util. Policy* **2016**, *43*, 186–194. [CrossRef]
17. Kristiansen, T. The Nordic Approach to market-based Provision of Ancillary Services. *Energy Policy* **2007**, *35*, 3681–3700. [CrossRef]
18. Raineri, R.; Rios, S.; Schiele, D. Technical and Economic Aspects of Ancillary Services Markets in the Electric Power Industry: An International Comparison. *Energy Policy* **2006**, *34*, 1540–1555. [CrossRef]
19. Rebours, Y.G.; Kirschen, D.S.; Trotignon, M.; Rossignol, S. A Survey of Frequency and Voltage Control Ancillary Services—Part II: Economic Features 2007. Available online: http://ieeexplore.ieee.org/document/4077136/ (accessed on 10 February 2017).
20. Brijs, T.; Jonghe, C.; Hobbs, B.F.; Belmans, R. Interactions between the design of short-term electricity markets in the CWE region and power system flexibility. *Appl. Energy* **2017**, *195*, 36–51. [CrossRef]
21. Jamalzadeh, R.; Ardehali, M.M.; Rashidinejad, M. Development of modified rational buyer auction for procurement of ancillary services utilizing participation matrix. *Energy Policy* **2008**, *36*, 900–909. [CrossRef]
22. Isemonger, A.G. The Evolving Design of RTO Ancillary Service Market. *Energy Policy* **2009**, *37*, 150–157. [CrossRef]
23. Banshwar, A.; Sharma, N.K.; Sood, Y.R.; Shrivastava, R. Market based procurement of energy and ancillary services from renewable rnergy rources in deregulated environment. *Renew. Energy* **2017**, *101*, 1390–1400. [CrossRef]

24. ELEXON. The Electricity Trading Arrangements: A Beginner Guide 2015. Available online: https://www.elexon.co.uk/wp-content/uploads/2015/10/beginners_guide_to_trading_arrangements_v5.0.pdf (accessed on 15 June 2017).

25. Abbassy, A.; van der Veen, R.A.C.; Hakvoort, R.A. Timing of Markets—the Key Variable in Design of Ancillary Service Markets for Power Reserves 2010. Available online: https://www.sintef.no/globalassets/project/balance-management/paper/timing-of-as-and-short-term-markets_abassy_2010.pdf (accessed on 10 June 2017).

26. Commission Regulation Establishing a Guideline on Electricity Balancing 2017. Available online: https://ec.europa.eu/energy/sites/ener/files/documents/informal_service_level_ebgl_16-03-2017_final.pdf (accessed on 16 July 2017).

27. MacDonald, M.; House, V. Impact Assessment on European Electricity Balancing Market 2013. Available online: https://ec.europa.eu/energy/sites/ener/files/documents/20130610_eu_balancing_master.pdf (accessed on 20 September 2017).

28. Nordic Energy Regulators (NordREG). Development of a Common Nordic Balance Settlement 2006. Available online: http://www.nordicenergyregulators.org/wp-content/uploads/2013/02/Common_Nordic_balance_settlement.pdf (accessed on 17 April 2017).

29. What is the Energy Union and the Market Design Initiative? 2016. Available online: https://www.clientearth.org/energy-union-market-design-initiative (accessed on 20 May 2017).

30. Borne, O.; Korte, K.; Perez, Y.; Petit, M.; Purkus, A. Barriers to Entry in Electricity Reserves Markets: Review of the Status Quo and Options for Improvements 2017. Available online: https://pet2017paris2.sciencesconf.org/139091/document (accessed on 20 November 2017).

31. Ostrovsky, M.; Pathak, P. Market Design. Available online: http://www.nber.org/workinggroups/md/md.html (accessed on 10 March 2017).

32. Wilson, R. Architecture of Power Markets. Graduate School of Business Stanford University, Research Paper, 2001, No 1708. Available online: http://citeseerx.ist.psu.edu/viewdoc/download?doi=10.1.1.201.8737&rep=rep1&type=pdf (accessed on 10 March 2017).

33. OECD Competition Committee. Market Definition. 2012. Available online: http://www.oecd.org/daf/competition/Marketdefinition2012.pdf (accessed on 15 March 2017).

34. Melody, W.H. Liberalising Telecommunication Markets: A Framework for Assessment. Policy Primer Papers, 2006, No 3. Available online: https://idl-bnc-idrc.dspacedirect.org/bitstream/handle/10625/41874/129521.pdf (accessed on 15 March 2017).

35. Perfectly Competitive Market: Introduction. Available online: http://econtutorials.com/blog/perfect-competitive-market-introduction/ (accessed on 15 March 2017).

36. Perfect Competition and Why it Matters. Available online: www.khanacademy.org/economics-finance-domain/microeconomics/perfect-competition-topic/perfect-competition/a/perfect-competition-and-why-it-matters-cnx (accessed on 15 March 2017).

37. Moffatt Associates Partnership. Review and Analysis of EU Wholesale Energy Markets: Evaluation of Factors Impacting on Current and Future Market Liquidity and Efficiency 2008. Available online: https://ec.europa.eu/energy/sites/ener/files/documents/2008_eu_wholesale_energy_market_evaluation.pdf (accessed on 15 March 2017).

38. stCenturyPower.org. Market Evolution: Wholesale Electricity Market Design for 21st Century Power Systems 2013. Available online: http://www.nrel.gov/docs/fy14osti/57477.pdf (accessed on 16 April 2017).

39. ENTSO-E WGAS. *Survey on Ancillary Services Procurement, Balancing Market Design 2015*; ENTSO-E: Brussels, Belgium, 2016.

40. Mishra, D. An Introduction to Mechanism Design Theory 2006. Available online: http://www.isid.ac.in/~dmishra/doc/survey.pdf (accessed on 16 July 2017).

41. Gubina, A. Ancillary Services in the Distribution Network: Where Are the Opportunities? In Proceedings of the Zagreb Energy Conference, Zagreb, Croatia, 10 December 2015.

42. European Parliament and Council. Regulation No 543/2013 on Submission and Publication of Data in Electricity Markets. 2013. Available online: http://eur-lex.europa.eu/LexUriServ/LexUriServ.do?uri=OJ:L:2013:163:0001:0012:EN:PDF (accessed on 7 March 2018).

43. Caerts, C.; Evangelos, R.; Mazheruddin, S.; Efren, G.S.; Merino-Fernández, J.; Rodriguez Seco, E.; Evenblij, B.; Rezkalla, M.M.N.; Kosmecki, M.; Temiz, A.; et al. *Description of the Detailed Functional Architecture of the Frequency and Voltage Control Solution (Functional and Information Layer)*; ELECTRA Deliverable D4.2: WP4 Fully Interoperable Systems; Technical University of Denmark: Lyngby, Denmark, 2016.

44. Oleinikova, I.; Bobinaite, V.; Obushevs, A.; Grande, O.; Rikos, E.; Esterl, T.; Rodriguez Seco, E.; Merino-Fernández, J.; Moneta, D.; Correntin, E.; et al. *Market Design Supporting the Web-of-Cells Control Architecture*; ELECTRA Deliverable 3.2: WP3 Scenarios and Case Studies for Future Power Systems; Technical University of Denmark: Lyngby, Denmark, 2018.

45. KU Leuven Energy Institute. The Current Electricity Market Design in Europe 2015. Available online: https://set.kuleuven.be/ei/images/EI_factsheet8_eng.pdf/ (accessed on 16 July 2017).

46. Stoft, S. *Power System Economics: Designing Markets for Electricity*; IEEE Press: Piscataway, NJ, USA, 2002; ISBN 0-471-15040-1.

47. Monostori, Z. Discriminatory Versus Uniform-Price Auctions. MNB Occasional Papers, No. 111. 2014. Available online: https://www.econstor.eu/bitstream/10419/141995/1/796715769.pdf (accessed on 7 March 2018).

48. Baldick, R. Single Clearing Price in Electricity Markets 2009. Available online: ftp://www.cramton.umd.edu/papers2005-2009/baldick-single-price-auction.pdf (accessed on 20 September 2017).

49. Wolfram, C. Electricity Markets: Should the Rest of the World Adopt the UK Reforms? *Regulation* **1999**, *22*, 48–53.

50. Rassenti, S.; Smith, V.; Wilson, B. Discriminatory Price Auctions in Electricity Markets: Low Volatility at the Expense of High Price Levels. *J. Regul. Econ.* **2003**, *23*, 109–123. [CrossRef]

51. Kahn, A.; Cramton, P.; Porter, R.; Tabors, R. Uniform Pricing or Pay-As-Bid Pricing: A Dilemma for California and Beyond. *Electr. J.* **2001**, 70–79. [CrossRef]

52. Spector, D. The European Commission's Approach to Long-term Contracts: An Economist's View. *J. Eur. Compet. Law Pract.* **2014**, 1–6. Available online: http://www.mapp-economics.com/wbxfile/download/wbxCoreBundle%3AFile/28?filename=Journal+of+European+Competition+Law+%26+Practice-2014-Spector-jeclap_lpu060.pdf (accessed on 7 March 2018). [CrossRef]

53. KU Leuven Energy Institute. Negative Electricity Market Prices 2014. Available online: https://set.kuleuven.be/ei/images/negative-electricity-market-prices (accessed on 7 March 2018).

54. Göß, S. Closure of German Power Plants 2016. Available online: https://blog.energybrainpool.com/en/closure-of-german-power-plants/ (accessed on 7 March 2018).

55. Marcacci, S. Cheap Renewables Keep Pushing Fossil Fuels Further Away from Profitability 2018. Available online: https://www.forbes.com/sites/energyinnovation/2018/01/23/cheap-renewables-keep-pushing-fossil-fuels-further-away-from-profitability-despite-trumps-efforts/#6486e2826ce9 (accessed 7 March 2018).

56. Functions of the Price Mechanism Explained. Available online: https://www.tutor2u.net/economics/reference/functions-of-the-price-mechanism (accessed on 20 September 2017).

energies

MDPI

Article

From Integration Profiles to Interoperability Testing for Smart Energy Systems at Connectathon Energy

Marion Gottschalk [1,*], **Gerald Franzl** [2,*], **Matthias Frohner** [3,*] and **Richard Pasteka** [3,*] and **Mathias Uslar** [1,*]

1   OFFIS e.V., 26121 Oldenburg, Germany
2   AICO EDV-Beratung GmbH, 1030 Vienna, Austria
3   Department of Biomedical, Health and Sports Engineering, University of Applied Sciences Technikum Wien, 1200 Vienna, Austria
*   Correspondence: gottschalk@offis.de (M.G.); franzl@aico-software.at (G.F.); frohner@technikum-wien.at (M.F.); pasteka@technikum-wien.at (R.P.); uslar@offis.de (M.U.); Tel.: +49-441-9722-226 (M.G.); +43-2245-824-48 (G.F.); +43-1333-4077-354 (M.F.); +43-1333-4077-893 (R.P.); +49-441-9722-128 (M.U.)

Received: 31 October 2018; Accepted: 28 November 2018; Published: 2 December 2018

Abstract: The project Integrating the Energy System (IES) Austria recognises interoperability as key enabler for the deployment of smart energy systems. Interoperability is covered in the Strategic Energy Technology Plan (SET-Plan) activity A4-IA0-5 and provides an added value because it enables new business options for most stakeholders. The communication of smart energy components and systems shall be interoperable to enable smooth data exchange, and thereby, the on demand integration of heterogeneous systems, components and services. The approach developed and proposed by IES, adopts the holistic methodology from the consortium Integrating the Healthcare Enterprise (IHE), established by information technology (IT) vendors in the health sector and standardised in the draft technical report ISO DTR 28380-1, to foster interoperable smart energy systems. The paper outlines the adopted IES workflow in detail and reports on lesson learnt when trial Integration Profiles based on IEC 61850 were tested at the first Connectathon Energy instalment, organised in conjunction with the IHE Connectathon Europe 2018. The IES methodology is found perfectly applicable for smart energy systems and successfully enables peer-to-peer interoperability testing among vendors. The public specification of required Integration Profiles, to be tested at subsequent Connectathon Energy events, is encouraged.

Keywords: interoperability; smart energy systems; use cases; IEC 62559; SGAM; TOGAF; integration profiles; IHE; testing; gazelle; connectathon

1. Introduction

The energy transition refers to the migration toward solely renewable energy sources (RES) feeding the power grid. Their diverse size, their fluctuating production potential, their total number and their intrinsic heterogeneity makes it somewhat difficult to meet legacy grid requirements and challenges the reliable grid control. Solar energy drives all renewable sources. The sun enables the growth of plants, propels the weather system and thereby the wind and the water cycle, which in addition to photo-voltaic panels and mirror based systems are utilised to generate green secondary energy by wind turbines, hydroelectric power plants, and bio-gases. However, these primary sources are fluctuating, they are not available in the same amount at all times [1]. In Figure 1, an overview of the central participants of a Smart Grid are shown.

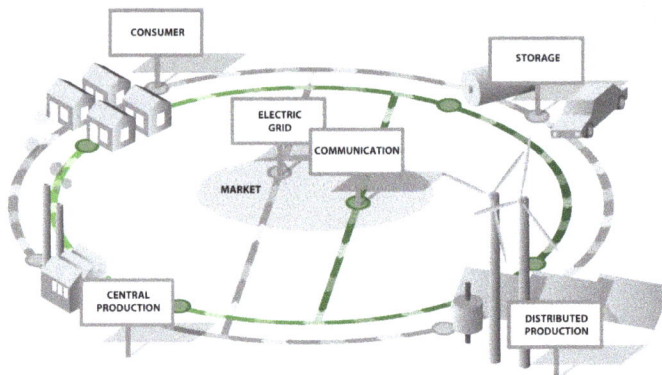

Figure 1. The Smart Grid: Services based on Information and Communication Technology (ICT) Define if appropriate. are added to the established energy management paradigm to support efficient grid control in the presence of distributed energy resources (DER).

Figure 1 also sketches the communication demand among the participants required to safely operate the grid. For safe operation, it is necessary to constantly balance the power grid by assuring that at any time the energy production meets the consumption. To better integrate small and less control-able distributed energy resources (DER), these assets can be yoked into Virtual Power Plants (VPP). A VPP commonly combines many different DER to constitute a virtual plant that can communicate as one entity with energy markets and grid operators, the latter being local Distribution System Operators (DSOs) and possibly transnational Transfer System Operators (TSOs) [2–4]. The aggregation reduces the communication interfaces to external partners and enables more flexible energy production and consumption management in defined areas.

Internal as well as external communication interfaces, data models and operation protocols need to be sensibly specified to allow the integration of DER from different vendors. Thus, interoperability [5,6] is a key factor to successfully integrate distributed RES and smart DER in the energy distribution and grid control, which constantly improved over many decades. Integrating novel energy sources and services is, therefore, a big challenge. The individual components of any smart energy system rely on seamless cooperation, i.e., on exchanged information used to reliably support the operation of physically connected system of systems (SoS) being the power grid.

The deployment of a Smart Grid is possible when vendors and customers work together to establish joint solutions, where smart energy systems' interaction is feasible. Organisations like the IEC and VHPready recognised that and develop *basic application profiles* utilising joint consortia consisting of different vendors (cf. Sections 2.6 and 2.7). The Austrian Project *Integrating the Energy Systems (IES)* [7] adapted the IHE methodology from the healthcare sector to match the energy domain by specifying exemplary use case centric interfaces for energy systems and by establishing an interoperability testing framework, the annual *Connectathon Energy*. The main goal of IES is to provide a workflow for the energy domain that fosters interoperable energy system components required for the communication in Smart Grids. The core contributions of the work performed are:

- Transferring the IHE methodology into the energy domain,
- Providing a test environment to specify and document test cases and test sequences,
- Validation of the IES methodology via exemplary profiles for VPPs based on IEC 61850.

The IES workflow and the results from the Connectathon Energy presented in the following are an extension of the conference poster abstract [8]. The proposed workflow is explained in more detail, from profiling to testing. In the following Section 2, the considered methodologies and some other works in the field of interoperability are outlined. The IES workflow is introduced in Section 3 and Sections 4 and 5 describe the major steps. The validation of the IES workflow is presented and discussed in Section 6. Section 7 concludes the paper providing an outlook on further envisaged and progressing steps toward a general application of the methodology.

2. Related Work

The concept and methods applied by the project *Integrating the Energy System* are based on recent research results and recommendations on achieving interoperable IT solutions and the *Single Digital Market* envisaged by the European Commission [9]. These are briefly outlined in the following subsections.

2.1. The IEC 62559 Use Case Methodology

The IEC has standardised the Use Case Methodology from the European mandate M/490 in the series IEC 62559 [10]. The methodology describes complex systems functionalities, actors and processes in a structured way. The methodology is used to analyse requirements and to detect standardisation gaps; e.g., in data models and protocols. Use Cases are first building blocks in software engineering to specify systems. The system functionality is approached from a static and a dynamic point of view: involved actors and system boundaries are considered static and relations between the actors contributing to a Use Case show dynamic aspects [1,11].

With the IEC 62559 Use Case Methodology, use cases are textually described in a template by application and implementation experts. The template is specified in IEC 62559-2. The other three parts of the standard series classify the operational use of the methodology, possible tool-support, and best practices. For the template, an XSD schema is defined to enable content exchange between different tools [10]. The template contains the eight parts shown in Table 1. The first two parts can be defined by domain experts: it provides the management perspective. The technical requirements are specified by application and implementation experts in parts three to six. The last two parts are jointly completed where necessary. Based on the structured Use Case description, technical specifications can be derived; i.e., architecture models (cf. Section 2.2) and interface specifications (cf. Section 4.2).

Table 1. The eight parts of the Use Case template—content and viewpoint.

	Use Case Template	Content	Viewpoint
1	Description	General info: identifier, name, non-technical description, boundaries, objectives, etc.	management perspective
2	Diagram	Visual representation of the Use Case: e.g., UML Use Case Diagrams, Sequence Diagrams.	
3	Technical details	List of actors (humans/systems) involved in the Use Case with brief role explanation.	ICT experts
4	Step-by-step analysis	Technical description of implementation: data exchanged, information objects, requirements.	
5	Information exchanged	List of information objects exchanged between actors.	
6	Requirements	List of requirements that have to be adhered to actors and information objects.	
7	Common terms and definitions	Glossary of the Use Case: define the terms used.	stakeholders (if necessary)
8	Custom information	Further information/explanation that cannot be addressed elsewhere.	

2.2. The Smart Grid Architecture Model (SGAM)

For visualising connections within Use Cases, the Smart Grid Architecture Model (SGAM) can be applied. The SGAM resulted from the European Mandate M/490, where the GridWise Architecture Council (GWAC) stack became adopted for the Smart Grid domain [12,13]. The SGAM is a three-dimensional visualisation, presented in Figure 2, that helps to identify components and their relations. An example application is later shown in Section 4.1.

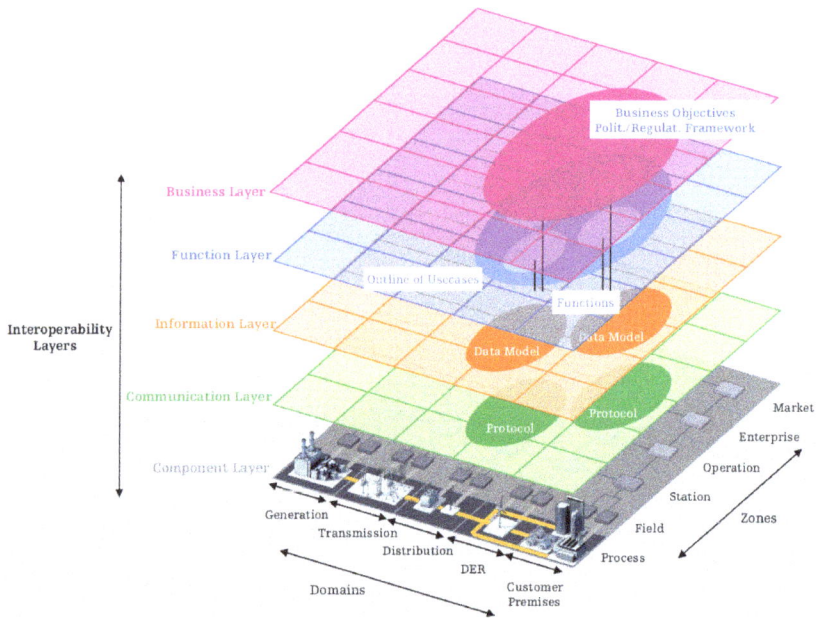

Figure 2. The Smart Grid Architecture Model [12]: a three-dimensional positioning scheme to identify and confine features and systems of smart grids, © CEN/CENELEC, reproduced with permission.

From the top to the bottom, five layers are used to consider different interoperability aspects (viewpoints). It starts with the *Business Layer*, where regulatory and economic structures, business models and processes are positioned. On the next layer, the *functional view*, the functionalities are located. The required functionalities result from analysing different Use Cases. On the *Information Layer*, the used and exchanged information is considered; i.e., the Use Case specific semantic for functions and services to enable an interoperable information exchange. Therefore, information objects and canonical data models are positioned here. On the *Communication Layer*, protocols and mechanisms for the information exchange between actors are positioned; i.e., the data channel specification. The *Component Layer* is the lowest level and covers the individual entities that contribute to the Smart Grid. It includes autonomous systems, connected components, atomic applications, and any kind of smart devices. Each layer spans all domains along the energy conversion chain, from generation to consumption, as well as the control and management zones, from process until market control and management, respectively [14].

The energy conversion chain starts with the *generation* of electric energy that is typically connected to the transmission system. The *transmission* is the infrastructure and organisation that distributes the electricity over long distances to major industry and cities. The *distribution* represents the infrastructure and organisation that distributes the electricity to customers within a specific region. *Distributed Energy Resources* (DER) are comparably small power plants feeding electricity into the distribution grid. At the end of the chain, *customer premises* refers to the industrial, commercial, and residential facilities of energy consumers. The aggregation, separationand utilisation of the information used to manage the power system is governed by organisational and legal rules on participating in the energy sector. In the *market* zone, possible market operations along the energy conversion chain are considered; e.g., energy and capacity trading. The *enterprise* zone covers commercial and organisational processes and services; e.g., customer contracting and billing. The *operational* zone refers to power system control operations for the generation, the transmission and the distribution systems. The *station* zone aggregates data and functions from the *field* zone, which describes the equipment to protect, control, and monitor physical processes and power flows. In the *process* zone, physical, chemical, and spatial transformation of energy, the applied physical/mechanical/electrical equipment, is represented [14].

2.3. The Open Group Architecture Framework (TOGAF)

The concept considering different viewpoints on an architecture is also used by The Open Group in their Architecture Framework (TOGAF) [15]. This logical cyclic process for developing an architecture is depicted in Figure 3. It is a widespread and mature enterprise architecture development framework that covers three high-level viewpoints; i.e., *business, information system* and *technology* architecture [1,15]. The business architecture addresses management and planning; e.g., business strategies, governance, and organisation. The information system architecture can be divided into data and application centric views. The former supports database designers, administrators and systems engineers to structure an organisation's logic, its data assets and resources. The latter addresses system and software engineers describing the logic of the system; i.e., the business processes. The technology architecture describes the hardware and software capabilities to implement business needs; i.e., data and application services [1].

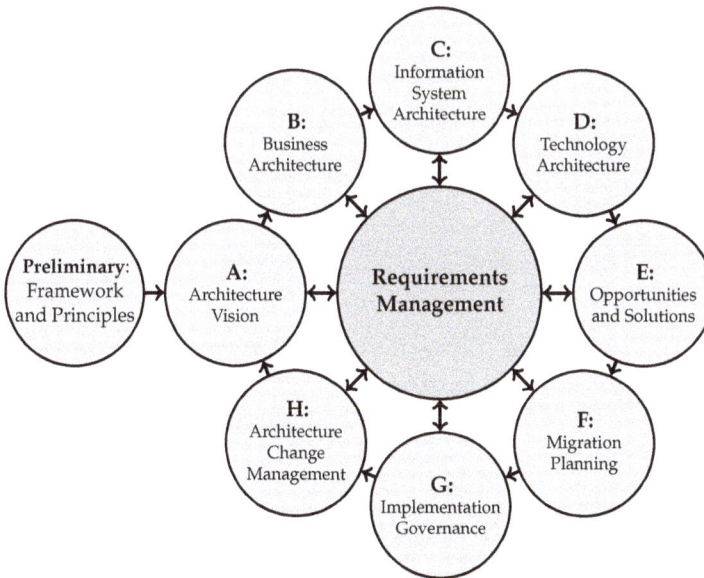

Figure 3. The Open Group Architecture Framework: a logical, cyclic, step-by-step requirements elicitation guideline.

The cyclic TOGAF process consists of eight steps plus a preliminary step and a central constantly used, as shown in Figure 3. A possible application of the steps is described in Sections 4 and 4.2. In brief: A *preliminary* phase takes place to identify enterprise-specific frameworks or principles before the architecture development starts. The first step in the actual architecture development is the *architecture vision*. It is about the definition of the envisioned architecture scope and analysis of stakeholders and their needs. Defined objectives are the basis for architecture viewpoints. Following the first step, the business, the information systems, and the technology architectures are developed. Base thereon the initial implementation is planned in *opportunities and solutions*. During *migration planning*, an implementation and migration plan is developed, which include the definition of working packages and building blocks. *Implementation governance* addresses the supervision of the implementation; e.g., check whether the implementation meets the objectives. The TOGAF process ends with the step *architecture change management*, where based on monitoring and verifying the reactions a subsequent iteration of the process may be triggered. The *requirements management* action is connected to every step because it covers the requirement monitoring necessary in parallel to each step.

2.4. Integrating the Healthcare Enterprise (IHE)

Formed by healthcare professionals and industrial partners the IHE initiative aims to empower secure and coherent transfer of information between involved computer systems. The ability of a system to communicate with other products or systems, whose interfaces are completely understood, without any restrictions, is a basic concept of interoperability. The IHE initiative defines interoperability profiles based on existing standards (e.g., DICOM, HL7), creating a framework for interoperability testing definitions. Each profile contains the definition of involved actors, transactions and their mutual relationship in terms of information transfer. The experts supervised interoperability testing events, called Connectathons, take place annually [16,17]. The IHE process consists of four steps:

1. Critical aspects of information transfers are identified and thereby relevant Use Cases defined.
2. IHE profiles are jointly developed to detail every communication via established standards.
3. Vendors develop systems that implement the communication as specified in IHE profiles.
4. The interoperability of a vendor's system is peer-to-peer tested at Connectathon events.

2.5. Gazelle—A Test Platform for Interoperability Testing

Gazelle is the name of the test bed used at the IHE Connectathons since 2006 [18]. It provides the means for mainly two flavours of software testing: (1) interoperability testing and (2) conformance testing [19] as shown in Figure 4. Meanwhile, the list of features supported by Gazelle has risen. From the test organisation perspective, this tool is used to register the companies, the systems under test (SUT), and their implementer. On the other hand, it provides Gazelle with the Integration Profiles that can be tested (including relevant actors and transactions) and the related test case definitions.

To capture exchanged data, Gazelle provides a proxy that is configured to log all messages exchanged between SUTs participating in a test case. The logged messages (traces) can be validated for syntactic and semantic compliance in a subsequent step using the validation services available in Gazelle or from external sources. The Gazelle Transformation Service facilitates binary to XML transformation by means of DFDL (Data Format Description Language) schemata.

Figure 4. Testing interoperability and validating the conformance with Integration Profiles.

Finally, the Gazelle platform documents the result of every conducted test step and hosts the Connectathon results and a Products Registry for successfully tested products that achieved an Integration Statement.

2.6. IEC Basic Application Profiles

The Smart Grid Mandate M/490 developed a concept to achieve interoperability by using existing standards. The concept is named *Basic Application Profiles* (BAP) [20]. BAPs define elements in modular frameworks for specific Use Cases of energy systems—a standard set for an application. Contrary to common standards, the BAP specification contains only mandatory aspects that have to be implemented to realise the Use Case and enable interoperability [21]. Use Cases address business problems identified by vendors and customers who need a solution for as specific issue in a given environment. To solve it, every feature is divided into elementary parts, which can be recombined to define more complex Use Cases. A profile consists of a profile name, requirements, boundaries and scalability, standards and specifications, communication network topology, and best practices.

Each BAP can be used as building blocks that, in combination with others, realises complex applications (cf. Sections 2.3 and 4.2.1 where the concept of building blocks is also used). An extension of BAPs are Basic Application Interoperability Profiles (BAIOP) [22] based on the V-Model [23]. BAPs provide an implementation strategy and BAIOPs contain an approach for testing on different levels (e.g., unit, integration, system, and acceptance testing) that also considers interoperability. Therefore, further specifications are defined with BAIOPs: device configuration, test configuration, test cases for the BAPs, specific capability descriptions and engineering frameworks for data modelling and communication infrastructures, which together enable the testing of a proposed system.

2.7. VHPready

VHPready e.V. [24] is an alliance, founded by stakeholders of the energy industry targeted to specify an industry-standard enabling interoperable communication in the field of virtual heat and power plants. The communication is based on IEC 60870-5-104 [25] or IEC 61850-7-420 [26]. Based on these standards and identified Use Cases, VHPready defines a set of data points that specify needed data objects either from IEC 61850 or 104. Based on the specified restrictions, VHPready compliant systems can be modelled and implemented. Moreover, the list of data points enables a mapping between the two different standards.

VHPready focuses on the interoperable communication between DER and a VPP operator (aggregator) and does not consider issues with other assets and entities of the energy system that may be needed to fulfil related Use Cases that reach beyond the boundaries of managing a VPP.

3. IES Workflow Overview

The project IES migrates and adopts the established and matured IHE methodology on achieving interoperable IT based healthcare systems. All started with incompatible digital X-ray images (radiology) in 1999 and by today covers many areas and achieved wide acceptance in the US, Europe, and East Asia. For example, the Austrian digital health record database [27] is entirely based on public IHE Integration Profiles. The generality and success of the methodology motivates the migration into the energy sector with similar security demands, market structure and regulatory governance.

The IES methodology realises the four basic steps shown in Figure 5, which evidently split up into many more intermediate steps [28]. A moderate expansion yields:

1. Identify Use Cases where interoperability is an issue and specify these by identifying system borders and requirements [11]:

    - Write a Business Overview (define actors, the environment and the general issue),
    - Describe Business Functions (apply the Use Case Methodology and draw UML use case diagrams),

- Reuse Integration Profiles where possible (save specification and test effort).

2. Jointly identify how interoperability issues can be prevented and specify the requirements normatively as Integration Profile [19]:

- Evaluate which standards can be used to fulfil the Use Case requirements,
- Specify the process to realise a Business Function (UML sequence diagram),
- Define the actors and transactions (decompose Meta-Actors into modules),
- Describe the role of the individual actors (modules),
- Draw an Actors-Transactions Diagram (decompose the process into steps),
- Draw detailed UML sequence diagrams per transaction (steps sequence),
- Specify additional communication and security requirements.

3. Test independent prototype solutions against each other on annual plugfest and improve the Integration Profiles until it is fixed [29]:

- Specify test cases and test sequences according to Integration Profile specification,
- Integrate test cases, procedures, and documentation in test environment (Gazelle),
- Create or integration conformity validation tools (e.g., Schematron),
- Develop simulators as dummy pre-Connectathon test partner (optional),
- Execute test cases with at least two independent peer vendors,
- Validate recorded messages/traces and log passed test results (neutral monitor),

4. Publish successful test results: who has successfully tested which Integration Profile [29],

- Publish which vendors successfully tested an Integration Profile (Result Browser),
- Get written approval for successful implementation (Integration Statements).

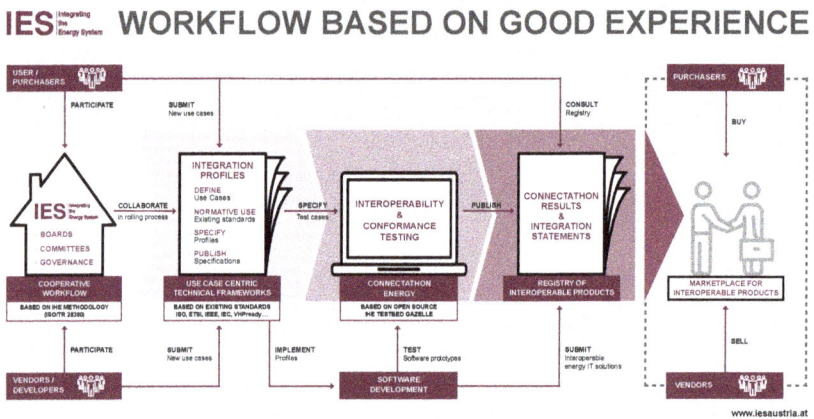

Figure 5. The IES process to interoperability are, in brief, four steps: identify → specify → test → sell.

Integration Profiles shall be living documents that improve and grow until they become stable. The core idea of the methodology is lively cooperation between the users and vendors. All stakeholders shall participate in the process as peers and contribute jointly to the development of demand oriented solutions. Sometimes, interoperability can be achieved most reliably with very simple means that work fine for many. In contrast to certification, where one system is tested against a set of rules or a single reference system, here the implementer from different vendors test their solutions among each other.

All peers participating in a test case have a common goal: they want to ultimately pass the test. A multi-day plugfest provides the environment and time to learn and make corrections prior to the final decisive test. Implementer can talk to each other and jointly identify why something does not work as it should. Such issues are often based on different interpretation of the Integration Profiles, which demands amendment of the text. Comments and errors recorded at the test event are the most valuable and powerful input to improve Integration Profiles. This feedback is practice driven and supports the advancement of the Integration Profiles in becoming a good basis for interoperable systems that can be integrated in the bigger system-of-systems.

3.1. Process Coordination and Control

The committees of IHE are instruments that manage and implement the industry integration process. Their importance may be concealed by results, being the Integration Profiles produced and test events organised. However, it requires coordination and cooperation to achieve interoperability. It is important that a committee is addressable, so tasks can be forwarded to it and its decisions are accepted by all stakeholders. Committees fulfil required roles that demand certain skills, which are presented in this section. Different skills are fundamental to execute the various management tasks that coordinate and control the IES process. Committees, depicted in the centre of Figure 6 as connected circles stating the skills grouping foreseen, manage the topics related to profile preparation and test execution, shown on the left and right sides, respectively.

Figure 6. Functional Committees: coordinating, managing, and executing bodies that drive and manage the IES based process to integrate energy enterprises.

Functional roles are defined to manage the integration process by the committees. Experts can serve in one, two or all roles, depending solely on skills they actually contribute. Participants in committees shall be voluntary delegates from institutions, companies, and enterprises interested in interoperable solutions.

The planning committee and the technical committees are the main working bodies. The planning committee assigns Integration Profiles to Technical Frameworks and is responsible for organising testing events, including the decision whether profiles are ready for testing either as trial or mature Integration Profile. The technical committees are the task forces writing Integration Profiles and specifying test scenarios, i.e., test cases and sequences, and providing tools like the test platform Gazelle (cf. Section 2.5) and its components (proxy, simulators, evaluators, as well as the logging, reporting and documenting means and repository), essential to reliably execute test cases. In addition, the domain coordination board, composed of experts covering all domains, is in between and resides slightly above. It determines which domain a Technical Framework shall be assigned to. Moreover, it supports technical committees by proposing cross-domain reuse of existing Integration Profiles to prevent the development of redundant Integration Profiles. Thereby, the technical solutions landscape is also harmonised across frameworks, simplifying interoperability beyond domain borders.

3.2. Participation Benefits—Implicit and Explicit

The benefits of vendors and customers in contributing to the Integration Profile development is manifold. First of all, the contributing parties can influence the solution design. Thereby, customers can be sure that profiles match their needs. Both customers and vendors, gain an advantage from knowing early what will be specified when new profiles become released—not least, gain vendors and customers trust and respect from working together toward a solution that satisfies both sides.

The publicly available Integration Profiles speed up the development of products because they provide clear answers to questions on possible options. Profile conformity allows start-ups and small companies to offer sub-systems that can be integrated into big systems without the need to develop dedicated interfaces to potentially proprietary systems on the market. Market leaders need only one extra interface to make their solution interoperable.

Testing prototypes at a Connectathon Energy has two main benefits: firstly, vendors profit from testing their prototypes with peers. In the course of the test event, they can identify and possibly even solve problems in an early development stage, with the active help of their peers who also want to solve their issues and pass the tests eventually. Secondly, companies that successfully pass the test cases of an Integration Profile become publicly listed after the event in the public accessible Result Browser. On demand, vendors can get Integration Statements. These list the Integration Profile related test cases that a system (product and version) has successfully passed. Such a statement is essential if a tender requests the successful testing of certain profiles. The latter is a clear advantage for customers because it enables them to precisely specify what they want without going deep into technical details.

Finally, if wished by the vendor, Integration Statements can be added to the publicly accessible Products Registry. Being listed is of particular importance for start-ups and companies that launch new products they might not be known for. The registry is a neutral, still valuable, marketing and advertisement forum. System purchasers can find matching components by comparing the passed tests listed in the Integration Statements.

4. Technical Frameworks

A Technical Framework (TF) is dedicated to a Business Case, e.g., the operation of a Virtual Power Plant (VPP), that belongs to a Business Domain—here the electrical energy production, distribution, regulation, and market area. The structure of a TF shown in Figure 7 is predefined, basically following a top-down approach in accordance with the IEC 62559 Use Case Methodology [10], the SGAM [14], the IHE Global Standards [30], and the V-model [23]. The relation to the architecture development framework TOGAF (The Open Group Architecture Framework) [15] and the European Interoperability Reference Architecture (EIRA) [31] is addressed later on. Figure 7 also shows that a Technical Framework is divided into two parts: a solely informative use cases (Business Functions) identification (volume 1) and a more normative solutions specification (volume 2).

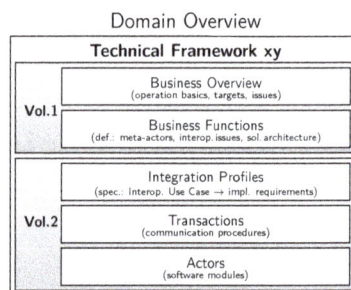

Figure 7. The Document structure of Technical Frameworks: a top-down approach from Use Cases definition to Integration Profiles specification.

4.1. Volume One—Informative

A Technical Framework starts with the *Business Overview* detailing the system environment, possible variants, and functional requirements. For the TF on VPPs, this includes at least the basic architecture, possible archetypes, and operational objectives. These are required to enable the specification of functional requirements. The application centric view corresponds with the Business Layer of the SGAM [14], shown in Figure 2, and step B in TOGAF [15], shown in Figure 3. The three-dimensional SGAM view (cf. Section 2.2) positions a VPP in the *Distribution*, *DER*, and *Customer Premises* realm and shows that the VPP operation covers all layers from *Process* till *Market*.

The next level of detail (one step down in the TF structure) identifies the functionalities required to establish a Business Case, which are called *Business Functions* in accordance with their implementation in the Function Layer of the SGAM. For a remote controlled VPP, functions like *establish the VPP*, *send planned schedules*, *provide measurement values* are specified. The identified Business Functions are documented applying the IEC 62559 Use Case Methodology (cf. Section 2.1). Information exchange sequences between actors comprise of steps from two categories: those that a business actor (meta-actor) performs on its own, and those where more than one actor is involved. The former, called *Operational Use Cases*, do not raise interoperability issues. The latter, called *Interoperability Use Cases*, are essential for interoperability because, for their realisation, more than one actor is needed. To prevent interoperability issues, these latter interactions need to be specified as Integration Profiles (cf. Section 4.2).

An excerpt of the Use Case specification that describes the Business Function *send planned schedule* from the VPP operator to some DEU controller is shown in Figure 8. It consists of the *Use Case identification*, the *Narrative Use Case description* and the *Steps—Scenarios table* as specified in the IEC 62559 Use Case template.

Use case identification		
ID	**Area Domain(s)/ Zone(s)**	**Name of use case**
VPP04 Send planned schedule	IES/DER/Operation,Station,Field,Process	Send planned schedule

Narrative of use case
Short description
Controlling the energy production and load of DEUs is the main purpose of a VPP to participate in the energy market. Based on this agreement, the VPP creates schedules that are transmitted to the DEUOP as a regional schedule or to DEUCs as individual schedules.
Complete description
Based on the agreement achieved on the market (committed schedule sold), the VPPOP splits the schedules into feasible regional schedules, which may be coordinated with the involved DSOs (cf. Use Case VPP-02), and finally transmits individual schedules to the DEUOPs and DEUCs involved. In case a DEUOP is involved, the DEUOP splits the received regional schedule further into individual schedules per managed energy asset and sends these to the DEUCs controlling the different DEUs. Depending on the features of the DEUCs these schedules may be sent as a complete schedule by the VPPOP or as a sequence of adjustment messages by the DEUOP, such that the connected DEUs execute the individual schedules. A DEUOP merges individual local DEUs into one and adds local flexibility (smartness) by enabling the DEUOP to decide locally when which asset shall produce or consume how much energy. Local fluctuations and short-term demands can be compensated/fulfilled locally, without involving the VPPOP, as it is required where the VPPOP communicates directly with the DEUC. Regarding normative operation, no difference is made between direct and indirect control.

Scenario								
Scenario name :	Send schedule directly to the DEUC							
Step No.	**Event**	**Name of process/ activity**	**Description of process/ activity**	**Service**	**Information producer (actor)**	**Information receiver (actor)**	**Information exchanged (IDs)**	**Requirements R-ID**
01	Created schedule	Send schedule	The functional schedule is sent from the VPPOP to the DEUC.	GET	VPPOP	DEUC	225	
02	Got schedule	Control DEU	The DEUC creates control signals based on the functional schedule and uses these to control the DEU.	GET	DEUC	DEU	219	

Figure 8. A Use Case Template excerpt showing three parts: (**top**) Use Case identifier, relations, and name, (**middle**) narrative definition of the Use Case, and (**bottom**) interaction steps and their allocation.

The *Use Case identification* provides a unique label (*ID*) according to a jointly agreed methodology. Next, it points out the placement of the Use Case in the SGAM *area domains and zones* (cf. Section 2.2). Finally, a human readable descriptive *name of the Use Case* is defined. The *narrative Use Case description* is composed as a short still complete textual description, providing a brief explanation of the Use Case. The *complete description* is a comprehensive user viewpoint centric narrative on what happens how, where, when, why, and under which assumptions. The narrative shall be written in terms and a language well understood by non-experts. The *Steps—Scenarios table* details the scenarios of Use Cases. The scenario name in the headline shall be the same as in the *Overview of scenarios* section of the template. The steps of a scenario are listed in consecutive execution order. Every step is identified by a step number and specifies the triggering event that causes the execution of the step. A triggering event is often an activity that also gets a unique name and may contains a short explanation of the procedure taking place, in addition to the name and description of the step it triggers. The *service* column states the nature of the information flow, e.g., get, create, or change, and the columns to the right specify the source, the destination, the information object exchanged, and additional requirements to be met.

The Use Case mapping into the SGAM is shown in Figure 9. The affiliation of the Use Case *Send Planned Schedule* to the VPP Business Case, the used data models and communication standards, and the involved actors are identified at a glance.

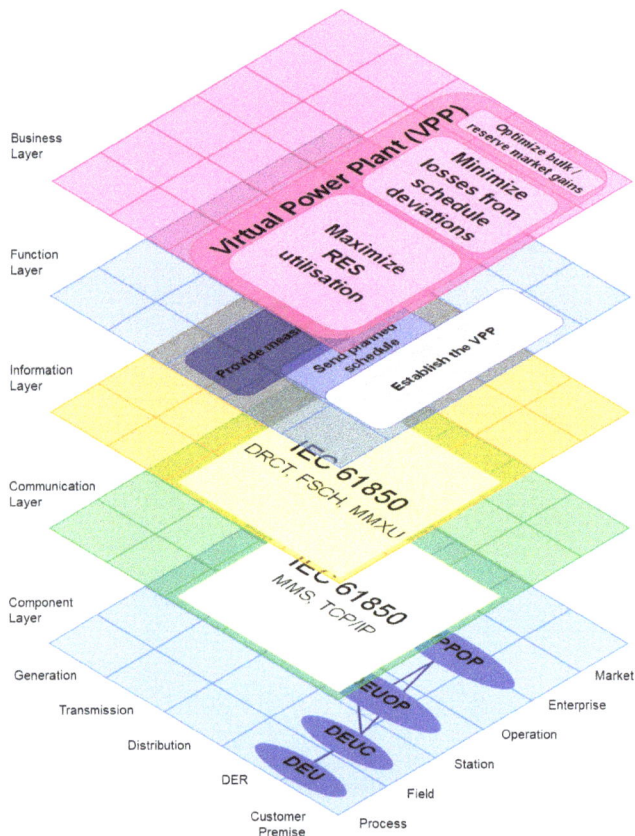

Figure 9. Positioning of features in the three-dimensional SGAM: selected VPP features mapped onto the three reference designation axes.

The result of these first two steps in preparing a TF, writing a Business Overview and identifying the required Business Functions, shall be summarised in a use case diagram. In this visualisation the involved actors (i.e., Meta-Actors that provide the sum of all features required to perform their tasks) get connected by the Business Functions they contribute to. This representation is the basis for the detailed Actors–Transactions Diagram, which shows only the interoperability relevant data exchanges between the actors (software components) the Meta-Actors are composed of. The detailed Actors-Transactions Diagram provides the schematic overview of the Meta-Actors composition at the beginning of Volume Two, summarising the technical functionalities to be specified by the *Transactions* of different Integration Profiles.

The TF–VPP example of a Use Case Diagram on requesting electricity production, i.e., *Send Planned Schedule*, is shown in Figure 10. Managing the information object *functional schedule (FSCH)* specified in IEC 61850 comprises three functions: (a) transferring the requested (intended) schedule from the operator to the production asset; (b) controlling the power generation according to the received (current) schedule; and (c) potential schedule adjustment in between reception and execution, as partially contained also in the template excerpt shown in Figure 8. Only the *Send schedule* function involves more than one actor (cf. Figure 10). Therefore, only this may cause interoperability issues. Interoperability among the involved actors is essential to correctly realise the transfer of the information contained in an FSCH object. The other two functions do not constitute interoperability issues and need not be specified by Integration Profiles and Transactions. Their correct implementation can be left open to the vendor.

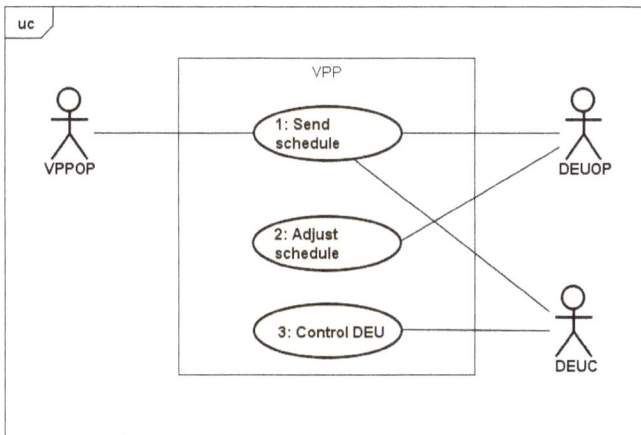

Figure 10. Use Case Diagram: Meta-Actors (here: Virtual Power Plant Operator, Distributed Energy Units Operator, Distributed Energy Unit Controller) connected to Use Case specific functions.

For the Interoperability Use Case *ISend schedule*, an Actors-Transactions Diagram is created, shown in Figure 11. It depicts the schematic view of the interoperability issue to be solved for the Business Function *Send Planned Schedule*. The diagram depicts the interdependence of actors; i.e., the interoperability demand of actors, and thereby the *Interoperability Use Cases* to be normatively specified in Volume Two. In the example, a VPPOP (Virtual Power Plant Operator), a DEUOP (Distributed Energy Unit Operator) and a DEUC (Distributed Energy Unit Controller) are required to realise the Business Function *Send Planned Schedule* that achieves the essential exchange of the information object FSCH, required to manage the power production of remote assets (here DER).

The Actors–Transactions Diagrams (cf. Figure 11) establishes the connection to the technical specifications in Volume Two and concludes the purely informative description of Volume One.

This first part is required to understand why and where interoperability is necessary, being readable to managers and users likewise. Experts confident in the matter may go directly for Volume Two and consult Volume One where questions arise only.

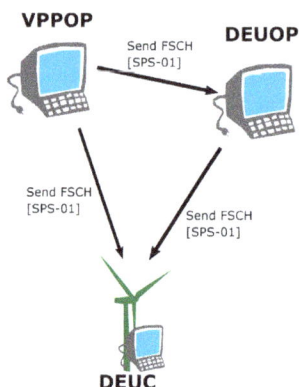

Figure 11. Actors–Transactions Diagram: Meta-Actors connected by the functions to be specified in Volume Two for interoperability.

4.2. Volume Two—Normative

Volume Two of a Technical Framework specifies normatively how the *Interoperability Use Cases*, identified and defined in Volume One, shall be technically solved, optionally including informative implementation examples. The resultant normative parts, each covering a Business Function's interoperability issues, are the *Integration Profiles*, which constitute the core of any TF. They are so called because they specify what is required to achieve the interoperability necessary to pass the system integration test from the V-model [23].

Multiple features, for example operational, technical and security requirements may be specified within one Integration Profile or separately. To enable the latter, Integration Profiles can be grouped (bundled) to realise the complete Business Function, i.e., all its Interoperability Use Cases, via grouped Integration Profiles. Whether or not to split features depends on the reusability of features for other Business Functions, i.e., if the addressed Interoperability Use Case can be solved in the same way for different Business Functions. For example, setting up a secure connection between two actors most likely occurs with many Business Functions and shall therefore be specified in a reusable fashion, as an individual Integration Profile that can be bundled with other Integration Profiles to compose the different actors (software modules) that all use the same method to set up a secure connection if not already established.

Therefore, one of the first normative specifications required to get a complete Integration Profile is a table that specifies which other Integration Profiles need to be bundled with the one specified. Formally, this can be expressed by

$$a/P \mapsto \check{a}/\check{P}, \tag{1}$$

where a identifies an actor (or Meta-Actor) and P an Integration Profile. The generic bundling rule in Equation (1) states that the actor a from Integration Profile P shall be mandatory bundled with the actor $\check{a}$ from Integration Profile $\check{P}$. If all actors of a profile shall include the same actor of some other profile, we can use $*$ as wildcard, i.e., $*/P \mapsto \check{a}/\check{P}$.

This mapping nomenclature may be used with Transactions t and Business Functions f as well, where, for the latter, we need to specify the Technical Framework F instead of an Integration Profile P. For example,

$$f/F \mapsto [*/P, \hat{t}/\hat{P}] \tag{2}$$

states that, for the Business Functions f / F, all actors from Integration Profile P and the transaction $\hat{t}$ from some other Integration Profile $\hat{P}$ shall be grouped. To do so, the actors $\hat{a}_i$ required for the transaction $\hat{t}$ shall be specified in profile $\hat{P}$,

$$\hat{t} / \hat{P} \mapsto (\hat{a}_1, \hat{a}_2, \ldots \hat{a}_n) / \hat{P}, \tag{3}$$

where, in Equation (3), actors from other profiles may as well become included using the nomenclature exemplified in Equation (2). How this formal approach can be extended and formally used to automate the specification of composed Integration Profiles is presented in [32].

In general, the specification of an Integration Profile follows the TOGAF steps C, D, and E, going down the SGAM Model from the Information Layer through the Communication Layer down to the Component Layer, respectively.

1. Specify semantics and syntax for the required Information pieces, i.e., how information shall be converted into data and vice versa, how information shall be handled (protection) and in which format data shall be exchanged among actors.
2. Specify the transactions, i.e., the communication channel and protocol that shall be used to transport data chunks from one actor to the other.
3. Specify the actors, i.e., features and characteristics of sender and receiver components (drivers) to be integrated with actors.

These steps, and the two from Volume One, are reflected in the document structure, where, in Figure 12, the multitude of Technical Frameworks, Building Blocks, Transactions, and Actors is indicated.

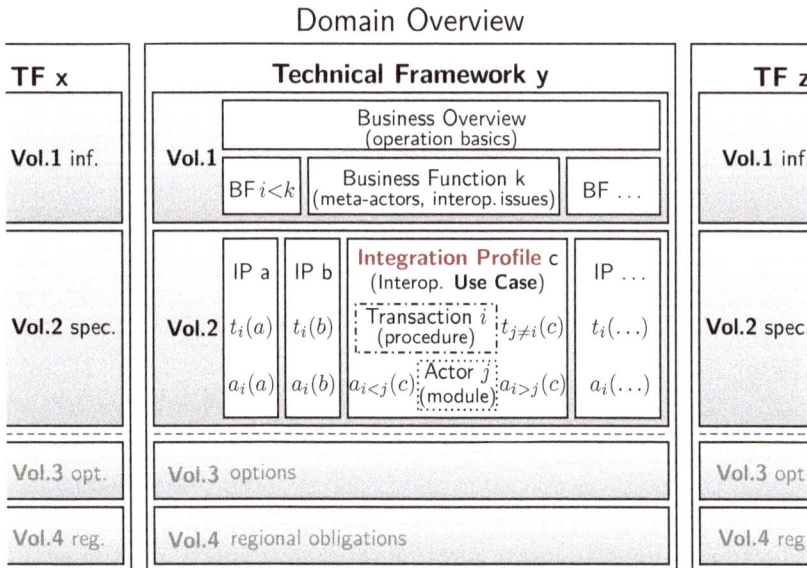

Figure 12. The Extended Document Structure: Technical Frameworks are collections of many parts, each assigned to a layer, that may be individually extended and adjusted following changing demands and the progress of technologies over time.

In each step, on each layer, every interoperability relevant aspect concerning applied standards and available options, is to be normatively specified. This is a balancing act between flexibility and dogmatism because the actual implementation shall be technology-neutral and remains a vendor decision.

Sometimes, for specific customer demands, deviations from the generic approach are required. For example, options provided by a standard that differ from those specified in Volume Two may be implemented on customer or vendor demand. Such options, which do not comply with the generic specifications in Volume Two, shall be specified in Volume Three. Similarly, deviations enforced by regional legislation are to be specified in Volume Four. These extra Volumes are solely provided to allow deviations without spoiling the generic definition. Evidently, they shall be kept to the absolute minimum truly required.

The next step in developing products is combining the individual functional actors specified with the Integration Profiles into the Meta-Actors that realise the Business Functions defined in Volume One (i.e., step F and G in TOGAF and system integration in the V-model). This may be environment and variant dependent, and is consequently not part of the Technical Framework. The same applies for the last step H in TOGAF (and also the first, being A), which is performed by the system planner that decides which features to buy and how to integrate them in the existing system architecture.

4.2.1. Relation to TOGAF and EIRA Architecture Building Blocks

The sequential separation into individual parts is typical for top-down approaches, as are the Use Case Methodology [10] or the TOGAF Architecture Development Method [15,31,33]. What we call Business Functions, or more precisely the individual features a Meta-Actors shall or may implement, are the Architecture Building Blocks (ABBs) in TOGAF/EIRA nomenclature. Their normative specification, the ABB_spec, is alike the IES Integration Profile. The Solution Building Blocks (SBBs) are the chosen standards, the exemplary specified transactions and the provided implementation examples, as shown in Figure 13.

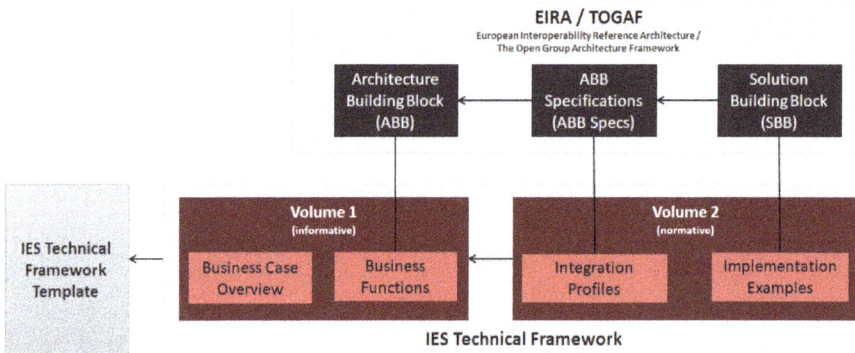

Figure 13. The Open Group Building Blocks: coarse relation of Business Function, Integration Profiles, etc. to Architecture and Solution Building Blocks used by TOGAF/EIRA.

Note that SBBs are always examples only. How a vendor achieves the fulfilment of an ABB_spec, which here implements the specifications stated in an IES Integration Profile, is not relevant for the interoperability. Only what is actually required for interoperability needs normative specification.

Having noticed that modern standards offer a plurality of options to realise certain features, all based on the same technical background but with many options, it appears straightforward to use the features that a standard offers as solution building blocks. These can than be bundled (grouped) with Integration Profiles, as shown in Figure 14. Their usage shall be constrained by the calling Integration Profile, such that interoperability is achieved precisely as required for the related Business Function.

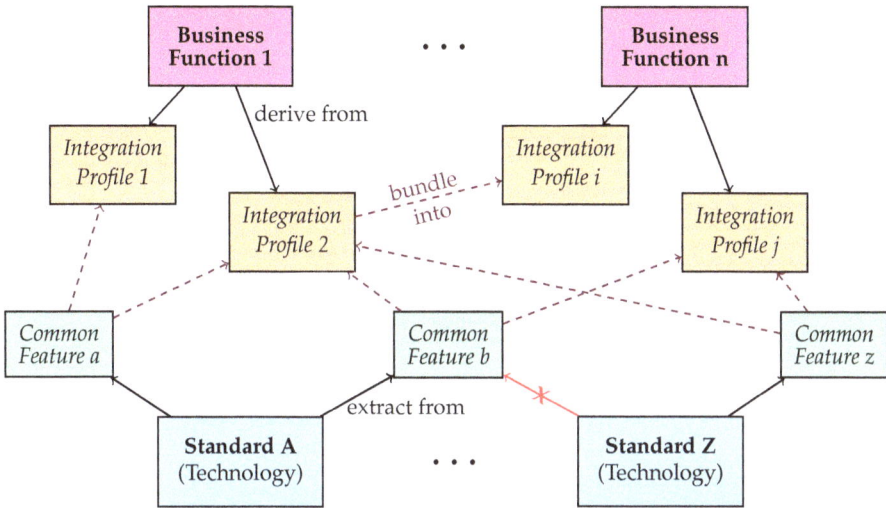

Figure 14. Common Features: bundling highly flexible features as provided by modern standards.

These solution building blocks we call a Common Feature (CF). They represent best practice solutions or excerpts from standards. They may refer to a single standard only, and shall provide the full flexibility available. To specify a feature of a standard as individual CF is only economic if this CF is used by many Integration Profiles. However, if that is the case, CF saves redundant specification of similar usage of the same feature in multiple Integration Profiles.

CF may be used to specify parts of the conformance tests. The more restrictive interoperability tests, and the conformance to the Use Case specific, Business Function related, restrictions, obligations, and constraints, cannot be derived from CF because these are specified in the Integration Profiles only. Therefore, for the remaining discussion, we assume generally that tests are derived from Integration Profiles, presuming that requirements and specifications from bundled in profiles and CF are completely considered for every Integration Profile.

5. Testing the Interoperability and Profile Implementation Conformance

The IES approach on interoperability and Integration Profile conformance testing is the central pillar of the IES methodology, as shown in Figure 15. The Technical Framework, containing the Integration Profile definitions introduced in Section 4, specifies the requirements on how different assets (actors) exchange information to implement the functionality required to realise a Business Function, and thereby enable a business case. The well established IHE methodology underlines the essential importance of peer-to-peer testing to ensure interoperability between systems. Publicly available test results serve as proof of a system's compliance with the specifications evaluated when passing a defined test case. When a software vendor passes all test cases defined for an Integration Profile, an *Integration Statement* can be requested. This document formally proves that the component it is issued for has successfully passed all defined tests to evaluate the implementation's compliance with the listed specifications (Integration Profiles). The Integration Statement enables the vendor to promote its products, makes offers to tenders requesting compliance to listed profiles and convinces customers that the product can be integrated in an existing infrastructure.

Specifications may include uncertainty, such that vendors may implement specifications in divergent ways. To overcome this issue, peer-to-peer interoperability testing between independent developed systems is necessary. This kind of software test is most effective when a system from vendor A is tested directly against a system from an independent vendor B, as shown in Figure 16. To perform this, IHE hosts the Connectathon test events, which are an established and proven means to ensure the interoperability of systems beyond company boundaries [29].

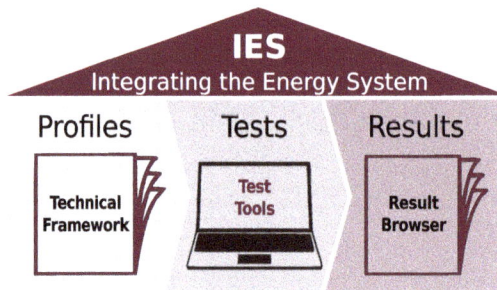

Figure 15. The three pillars of the IES approach: identification and specification, peer-to-peer testing, public available specifications and results (Open Access).

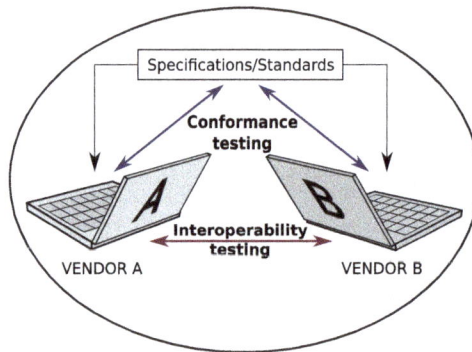

Figure 16. Peer-to-peer interoperability testing: independent implementations get connected and their cooperation is evaluated, including verification of exchanged messages (formats and contents).

The Connectathon test events are scheduled annually in Europe, the US and East Asia. The implementer of different vendors come together for a whole week and test whether information can be exchanged according to IHE Integration Profiles between their prototype implementations. These Connectathon events bring together the implementer for discussions and gives the community a forum for the validation of stated IHE specifications. Technically, IHE provides the test bed *Gazelle* [34]. This web-based tool provides, aside from event organisation functionalities, the features needed for test case management, system management, data traffic logging, message and sequence validation, and test documentation.

Gazelle's validation uses primary XML validation methodologies, i.e., the validation services check the XML payload: if it is well-formed, whether the payload adheres to specifications specified in XML schema files, and if additional formatting rules are fulfilled, using XML-Schematron validation. Since the example VPP profiles specify the use of IEC 61850 compliant data objects exchanged using MMS to generate and transport data, here the payload is binary encoded. To validate the exchanged messages using Gazelle's validation services, the binary encoded data need to be transferred into an XML representation. Therefore, the Data Format Description Language (DFDL) [35] and a Daffodil-Transformation [36] are used for the example. DFDL is an XML schema like a definition language extending classics schema file with additional information on how single bytes, and even bits, shall be interpreted in a form of annotations. Daffodil-Transformation is an open source tool that processes the binary data and generates XML files based on the specifications stated in a DFDL file. Gazelle already includes capabilities for this transformation step and, therefore, only the DFDL files need to be defined to create XML files from the logged messages containing binary encoded information. The resulting XML files can be further validated with the common Gazelle tools.

An essential features of the test-bed Gazelle is its capability to record messages that are exchanged between the systems under test (SUT). To use this feature called Proxy, the SUT needs to be configured not to communicate with test partners indirectly via the Proxy, as shown in Figure 17.

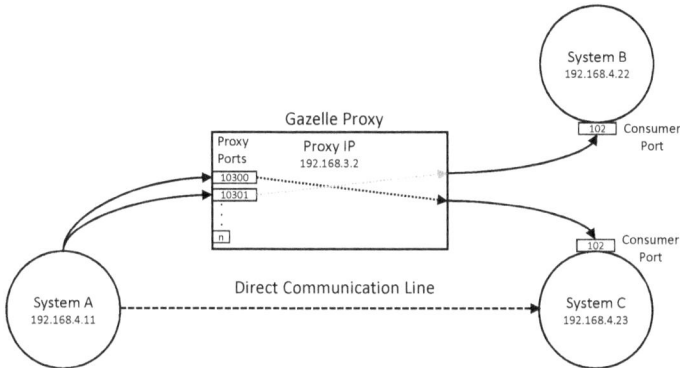

Figure 17. Derailing communication via Gazelle proxy service: port numbers are assigned to the systems prior testing, transferred messages are captured to undergo validation.

The proxy needs to be configured to assign all the corresponding endpoints to test instances and SUTs, such that messages received from a SUT are logged and in parallel forwarded to the correct receiving SUT. Logged messages can be forwarded to Gazelle's validation tools (or external validation options). Only during this last step the Daffodil-Transformation stated above is applied to gain an XML representation of the message content for the actual validation.

6. Realisation of the IES Workflow

First trial peer-to-peer testing sessions have shown that interoperability between vendors is possible by using Integration Profiles demonstrated at the first IES Connectathon in conjunction with the IHE Connectathon 2018 in the Hague, Netherlands. Four different vendors have successfully implemented the following *Transactions* from trial Integration Profiles, and passed the according interoperability tests:

- *Send MMXU*: specifies how measured values are sent from a DEUC to DEUOP or VPPOP using the IEC 61850 logical node MMXU.
- *Send FSCH*: specifies how a functional schedule is sent from a DEUOP or VPPOP to a DEUC using the IEC 61850 logical node FSCH.
- *Send DRCT*: specifies how DER controller characteristics are transferred from a DEUC to a DEUOP or VPPOP according to IEC 61850.

6.1. Test Setup for Interoperability Testing at the Connectathon 2018

Exchanged messages are recorded using the proxy of a dedicated Gazelle instance. The Gazelle instance is installed at the University of Applied Sciences Technikum Wien in Vienna, Austria. The local area network (LAN) used at the Connectathon floor for testing the energy components is managed by the IES team, independent from the IHE floor-LAN. The Gazelle instance running in Vienna is connected using a virtual private network (VPN) connection, as shown in Figure 18. This enables the recording and validation of exchanged messages via the Gazelle instance running on servers in Vienna. Gazelle management and configuration are performed via the Gazelle web interface.

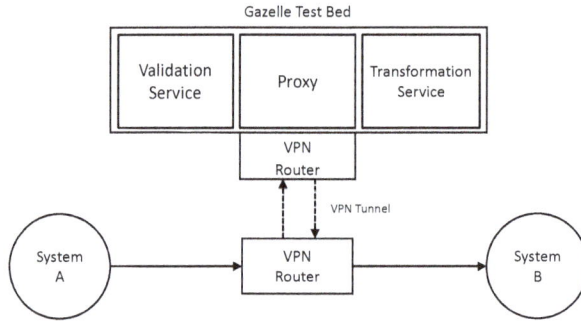

Figure 18. Schematic set-up for interoperability testing: directional communication is routed through the VPN connection to capture, transform and validate messages using the according Gazelle tools.

6.2. Interoperability testing Result—Send Functional Schedule

Test steps are defined for selective interoperability test scenarios. In general, the sequence diagram shows the required information exchange steps that need to be performed during a given testing instance. The transaction *Send Functional Schedule (Send FSCH)* is shown in Figure 19. The according test scenario evaluates the communication between VPPOP or DEUOP (System A) and DEUC (System B):

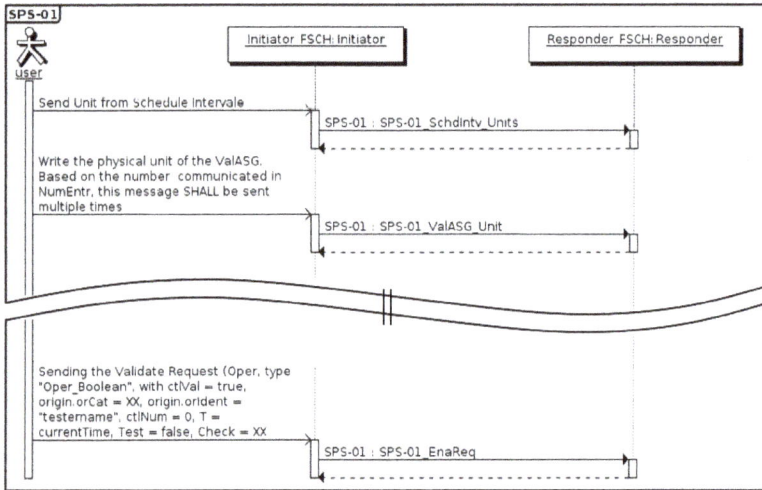

Figure 19. Sequence diagram sketch for the in total eleven test steps to be evaluated for the transaction sends FSCH in accordance with IEC 61850.

The data packages transferred between system A and B are captured via the Gazelle Proxy and undergo transformation into an XML representation, according to a predefined DFDL schema that converts the ASN.1 encoded binary data blocks into ASCII text. Based on this, conformity assessment as shown in Figure 20 (a screenshot from the external evaluation tool) is performed in three steps, each resulting in either *pass* or *fail*. First, the XML document adherence to the XML 1.0 syntax rules (well-formed) is inspected. Secondly, the document is checked against the associated XSD schema. The last step includes content validation applying rules defined in a Schematron.

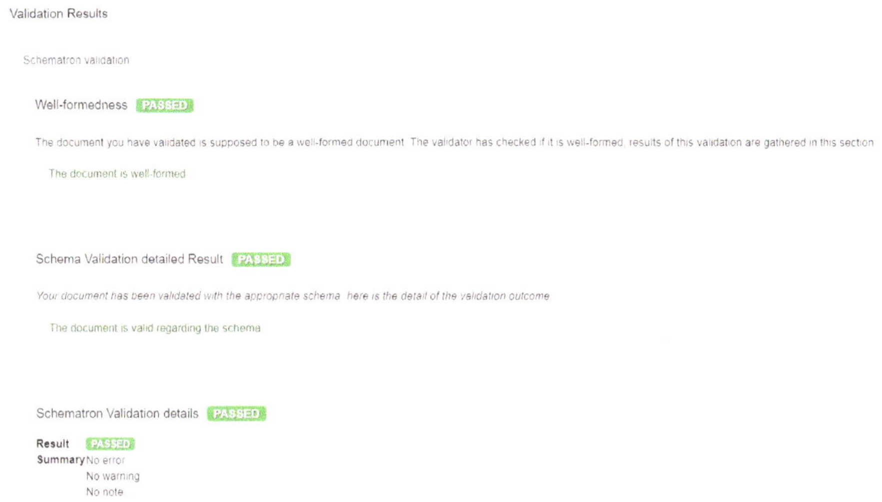

Validation Results

Schematron validation

Well-formedness `PASSED`

The document you have validated is supposed to be a well-formed document. The validator has checked if it is well-formed, results of this validation are gathered in this section

The document is well-formed

Schema Validation detailed Result `PASSED`

Your document has been validated with the appropriate schema, here is the detail of the validation outcome

The document is valid regarding the schema

Schematron Validation details `PASSED`

Result `PASSED`
Summary No error
No warning
No note

Figure 20. External validation service result showing conformity assessment for captured messages.

An overview of the interoperability test-results achieved at the Connectathon 2018 is shown in Figure 21 (screenshot from Gazelle). The systems under test supporting given Integration Profiles are listed: Send Measured Values (SMV), Send Planned Schedule (SPS), and Send Asset Configurations (SAC), Their role is either the Initiator or Responder actor, depending on which entity starts the communication. Information about the required number of independent test-runs for receiving an integration statement and the list of different test instances conducted is shown. Individual steps can be opened in the user interface and manually inspected. Altogether, the interoperability testing in the energy domain performed at the Connectathon 2018 comprised of twenty-one tests instances for the three Integration Profiles. Four DEUC and three VPPOP prototypes/systems from four different manufacturers were tested.

Sys ⇕	Profile ⇕	Actor ⇕	Profile Option ⇕	Type	R/O ⇕	Partners	V ▾
	SMV	Initiator	MMXU	T	1/0		100%
	Test		Meta test				
	SMV-01_V00-01			+	R / 3	3/3	14807 ⚠ 14808 14809
	One test						
	SMV	Responder	MMXU	T	1/0		67%
	Test		Meta test				
	SMV-01_V00-01			+	R / 3	2/3	14807 ⚠ 14811
	One test						
	SPS	Responder	FSCH	T	1/0		67%
	Test		Meta test				
	SPS-01_V00-02			+	R / 3	2/2	14806 ⚠ 14810

Figure 21. Connectathon 2018 test-results: the VPPOP, DEUOP and DEUC communication interfaces from four different manufacturers were tested on interoperability and compliance with the SMV, SPS and SAC Integration Profile (Gazelle screenshot).

7. Conclusions and Outlook

The research presented is based on the design research principle by Hevner et al. [37]. The main principle of the approach taken is to point out the utility of an artefact for a given scope and problem. The scope was to find a meaningful canonical, standardised process to create profiles for interfaces between components of Smart Grids. As of now, no such process existed in the domain of electrical engineering in the context of Smart Grids. There is no standardised way for system-of-systems based interoperability testing. However, a well established design artefact in from of the IHE (Integrating the Healthcare Enterprise) concept existed. The healthcare domain has similar problems motivated form the view of system-of-systems integration. However, it uses different processes, protocols, and data formats and ontology. A particular challenge was to establish these for the Smart Grid domain and to do a first trial of the well-established healthcare originated methodology in the energy domain. Since this particular aspect is not in the scope of the EU and its European Interoperability Framework (EIF), going vertical with established methods, we conducted the first Connectahon Energy using the IHE Gazelle successfully. This has proven to be a first proof of concept and was considered successful form the healthcare experts. Time will show how many interfaces and use cases will emerge with IES compliant profiles to be tested in the coming years.

Based on the experience from the first Connectathon Energy in the Hague 2018, including the preparation time for this test event, one of the lessons learned is strengthening and aligning the test specification within Gazelle with the Integration Profile development. Companies participating in the Connectathon Energy in the Hague reported that some assertions need to be made more restrictive. Besides the refinement of specifications, Gazelle's feature for *pre-Connectathon testing* should be considered; i.e., before a face-to-face test event, vendors want to run conformance tests using the IES Gazelle instance by uploading captured message contents/traces for validation. Feedback gathered at Connectathon Energy events will be considered for upcoming events. Besides updated and new test specifications, web-calls shall be scheduled to support interested and registered companies with detailed technical information on how to prepare best and how Gazelle tools can be accessed and used prior to the event. Concerning the upcoming Connectathon Energy end of January 2019 in Vienna, more trial profiles for interoperability and conformance testing have been prepared and are ready to be tested. The new profiles specify how to communicate metered values and schedules using the IEC 60870-5-104 standard in accordance with specification efforts from VHPready.

The IES activity has resulted in becoming part of work done within the ISGAN initiative. The main objective of ISGAN Annex 6 is to establish a long-term vision for the development of future sustainable power systems. The Annex 6 on Power Transmission & Distribution Systems focuses on system-related challenges, with an emphasis on technologies, market solutions, and policies that contribute to the development of system solutions. The work is carried out by a global network of experts and is managed in four Focus Areas: *Expansion Planning and Market Analysis*, *Technology Trends and Deployment*, *System Operation and Security*, and *Transmission and Distribution System Interactions*. The active dissemination of the IES methodology is an integral part.

IES Europe is listed as activity A4-IA0-5 in the European Strategic Energy Technology-Plan (SET-Plan) as an Implementation Plan *Increase the resilience and security of the energy system* [38]. This listing supports the opportunity to specify IES Technical Frameworks and Integration Profiles in the course of international R&D projects with great visibility.

Author Contributions: All of the authors have read and approved the final manuscript. The individual contributions of the authors are: 1. Introduction, G.F.; 2. Related Work, M.G., G.F., M.F., R.P. and M.U.; 3. IES workflow overview, M.G. and G.F.; 4. Technical Frameworks, M.G. and G.F.; 5. Testing the interoperability and profile implementation conformance, M.F. and R.P.; 6. Realization of the IES workflow, M.F. and R.P.; 7. Outlook, M.F. and M.U.

Funding: This research was funded by the Austrian Climate and Energy Fund (KLIEN) managed by the Austrian Research Promotion Agency (FFG), under contract number 853693 within the e!MISSION program, 2nd call, of the Austrian Federal Ministry for Transport, Innovation and Technology.

Acknowledgments: The presented work is part of the project Integrating the Energy Systems (IES) Austria supported by the Austrian Climate and Energy Fund (KLIEN) managed by the Austrian Research Promotion Agency (FFG). We thank all team members of the IES project for the support to write this contribution.

Conflicts of Interest: The authors declare no conflict of interest.

Abbreviations

The following abbreviations are used in this manuscript:

ABB	Architecture Building Block
ABB_spec	ABB Specifications
ACM	Association for Computing Machinery
BAP	Basic Application Profile
BAIOP	Basic Application Interoperability Profiles
CEN	European Committee for Standardization
CENELEC	European Committee for Electrotechnical Standardization
CF	Common Feature
CG	Coordination Group
CO2	Carbon Dioxide
DER	Distributes Energy Resource
DEU	Distributed Energy Unit
DEUC	Distributed Energy Unit Controller
DEUOP	Distributed Energy Unit Operator
DFDL	Data Format Description Language
DICOM	Digital Imaging and Communications in Medicine
DRCT	DER Controller Characteristics
DSO	Distribution System Operator
DTR	Draft Technical Report
EIRA	European Interoperability Reference Architecture
ETSI	European Telecommunications Standards Institute
FFG	Austrian Research Promotion Agency
FIP	Functional Integration Profile
FSCH	Functional Schedule
GWAC	GridWise Architecture Council
HL7	Health Level 7
ICT	Information and Communication Technology
ID	Identification
IEC	International Electrotechnical Commission
IEEE	Institute of Electrical and Electronics Engineering
IES	Integrating the Energy Systems
IHE	Integrating the Healthcare Enterprise
IP	Internet Protocol
ISGAN	International Smart Grids Action Network
ISO	International Standardization Organization
IT	Information Technology
KLIEN	Austrian Climate and Energy Fund
LAN	Local Area Network
M/490	European Mandate 490
MMXU	Measurements
OASIS	Organization for the Advancement of Structured Information Standards
PHR	Personal Health Record
RA	Reference Architecture
R&D	Research and Development
RES	Renewable Energy Source
SAC	Send Asset Configuration
SBB	Solution Building Blocks

SET	Strategic Energy Technology
SETIS	Strategic Energy Technologies Information System
SG	Smart Grid
SGAM	Smart Grid Architecture Model
SMV	Send Measured Values
SoS	System of Systems
SPS	Send Planned Schedule
SUT	Systems Under Test
TF	Technical Framework
TOGAF	The Open Group Architecture Framework
TSO	Transmission System Operator
UML	Unified Modelling Language
VPN	Virtual Private Network
VPP	Virtual Power Plant
VPPOP	Virtual Power Plant Operator
XML	Extensible Markup Language
XSD	XML Schema Definition

References

1. Uslar, M.; Specht, M.; Daenekas, C.; Trefke, J.; Rohjans, S.; Gonzalez, J.M.; Rosinger, C.; Bleiker, R. *Standardization in Smart Grids*; Springer: Berlin/Heidelberg, Germany, 2013.
2. Ramchurn, S.D.; Vytelingum, P.; Rogers, A.; Jennings, N.R. Putting the 'Smarts' into the Smart Grid: A Grand Challenge for Artificial Intelligence. *ACM* **2012**, *55*, 86–97. [CrossRef]
3. Kroposki, B.; Johnson, B.; Zhang, Y.; Gevorgian, V.; Denholm, P.; Hodge, B.M.; Hannegan, B. Achieving a 100% Renewable Grid: Operating Electric Power Systems with Extremely High Levels of Variable Renewable Energy. *IEEE Power Energy Mag.* **2017**, *15*, 61–73. [CrossRef]
4. Steg, L.; Shwom, R.; Dietz, T. What Drives Energy Consumers? Engaging People in a Sustainable Energy Transition. *IEEE Power Energy Mag.* **2018**, *16*, 20–28. [CrossRef]
5. IEEE Standards Board. *IEEE Standard Glossary of Software Engineering Terminology*; Technical Report; Standards Coordinating Committee of the Computer Society of the IEEE: Los Alamitos, CA, USA, 1990.
6. Merriam-Webster Incorporated. Definition of Interoperability. Available online: https://www.merriam-webster.com/dictionary/interoperability (accessed on 25 May 2018).
7. IES Team. IES—Integrating the Energy System. Available online: www.iesaustria.at (accessed on 30 October 2018).
8. Gottschalk, M.; Franzl, G.; Frohner, M.; Pasteka, R.; Uslar, M. Structured workflow achieving interoperable smart energy systems. *Energy Inform.* **2018**. [CrossRef]
9. European Commission. Digital Single Market. Available online: https://ec.europa.eu/commission/priorities/digital-single-market_en (accessed on 30 May 2018).
10. IEC 62559-2:2015. *Use Case Methodology—Part 2: Definition of the Templates for Use Cases, Actor List and Requirements List*; IEC TC8 Standard; IEC: Geneva, Switzerland, 2015.
11. Gottschalk, M.; Uslar, M.; Delfs, C. *The Use Case and Smart Grid Architecture Model Approach: The IEC 62559-2 Use Case Template and the SGAM Applied in Various Domains*; SpringerBriefs in Energy; Springer: Berlin/Heidelberg, Germany, 2017.
12. CEN-CENELEC-ETSI Smart Grid Coordination Group. *First Set of Standards*; Technical Report; CEN-CENELEC-ETSI Smart Grid Coordination Group: Brussels, Belgium, 2012.
13. Englert, H.; Uslar, M. Europäisches Architekturmodell für Smart Grids-Methodik und Anwendung der Ergebnisse der Arbeitsgruppe Referenzarchitektur des EU Normungsmandats M/490. In Proceedings of the Tagungsband VDE-Kongress, Stuttgart, Deutschland, 5–6 November 2012; Technical Report.
14. SG-CG/RA. *Smart Grid Reference Architecture*; Technical Report; CEN-CENELEC-ETSI Smart Grid Coordination Group: Brussels, Belgium, 2012.
15. The Open Group. *TOGAF Version 9—The Open Group Architecture Framework (TOGAF)*, 9th ed.; The Open Group: San Francisco, CA, USA, 2009.

16. Urbauer, P.; Sauermann, S.; Frohner, M.; Forjan, M.; Pohn, B.; Mense, A. Applicability of IHE/Continua components for PHR systems: Learning from experiences. *Comput. Biol. Med.* **2015**, *59*, 186–193. [CrossRef] [PubMed]

17. Noumeir, R. Integrating the healthcare enterprise process. *Int. J. Healthc. Technol. Manag.* **2008**, *9*, 167. [CrossRef]

18. IHE International. Gazelle—eHealth Test Framework for Interoperability. Available online: https://gazelle. ihe.net/ (accessed on 30 October 2018).

19. Frohner, M.; Gottschalk, M.; Franzl, G.; Pasteka, R.; Uslar, M.; Sauermann, S. Smart Grid Interoperability Profiles Development. In Proceedings of the 2017 IEEE International Conference on Smart Grid Communications, Dresden, Germany, 23–26 October 2017. [CrossRef]

20. Smart Grid Mandate. Methodologies to Facilitate Smart Grid System Interoperability through Standardization, System Design and Testing. Available online: https://www.google.com. tw/url?sa=t&rct=j&q=&esrc=s&source=web&cd=1&cad=rja&uact=8&ved=2ahUKEwie3vD7_ _3eAhXCEnAKHRGNCyoQFjAAegQIAxAB&url=https%3A%2F%2Fwww.researchgate.net% 2Fpublication%2F322752190_Smart_Grid_Interoperability_Testing_Methodology_A_Unified_Approach_ Towards_a_European_Framework_for_Developing_Interoperability_Testing_Specifications&usg= AOvVaw1MICQd3PD5VnSUZxHksKpy (accessed on 30 October 2018).

21. Wolfgang, B.; Heiko, E. Basic Application Profiles for IEC 61850. 2013. Available online: https://docstore. entsoe.eu/Documents/RDCdocuments/BAP_concept_TC57WG10.pdf (accessed on 30 October 2018).

22. Masera, Marcelo. The Role of Interoperability in the Digital Energy Vision. 2017. Available online: https://www.etip-snet.eu/wp-content/uploads/2017/06/2.-The-role-of-Interoperability-Marcelo-Masera.pdf (accessed on 30 October 2018).

23. Forsberg, K.; Mooz, H.; Cotterman, H. *Visualizing Project Management—Models and Frameworks for Mastering Complex Systems*, 3rd ed.; John Wiley & Sons, Inc.: Hoboken, NJ, USA, 2005; pp. 108–116, 242–248, 341–360.

24. Industry Alliance VHPready e.V. VHPready. Available online: https://www.vhpready.com/ (accessed on 30 November 2018).

25. IEC 60870-5-104—Telecontrol Equipment and Systems—Part5-104: Transmission Protocols—Network Access for IEC 60870-5-101 Using Standard Transport Profiles. Available online: https: //www.google.com.tw/url?sa=t&rct=j&q=&esrc=s&source=web&cd=2&cad=rja&uact=8&ved= 2ahUKEwj81rL6_v3eAhUJfnAKHSfIBgYQFjABegQIBhAB&url=https%3A%2F%2Fwebstore.iec.ch% 2Fpublication%2F3746&usg=AOvVaw2LLUacS4XK_It0q8rsKXnt (accessed on 30 October 2018).

26. IEC 61850-7-420:2009—Communication Networks and Systems for Power Utility Automation—Part 7-420: Basic Communication Structure—Distributed eNergy Resources Logical Nodes. Available online: https://www.google.com.tw/url?sa=t&rct=j&q=&esrc=s&source=web&cd=1&cad=rja&uact=8&ved= 2ahUKEwiZo4jE_v3eAhXYQd4KHX35Ab4QFjAAegQIBRAB&url=https%3A%2F%2Fwebstore.iec.ch% 2Fpublication%2F6019&usg=AOvVaw2BLqlPG9vYoFh7SuXlbmGj (accessed on 30 October 2018).

27. Eisl, Hubert. ELGA Erfolgreich Gestartet. 2016. Available online: http://ihe-austria.at/elga-erfolgreich-gestartet/ (accessed on 30 October 2018).

28. Franzl, G.; Frohner, M.; Gottschalk, M.; Reif, V.; Koch, G.; Berger, A. Interoperabilität im Datenaustausch in der Energiewirtschaft—Vom Use Case zum Test der Integrationsprofile. In Proceedings of the Symposium Energieinnovation 2018, Graz, Austria, 14–16 February 2018. [CrossRef]

29. IHE Europe. Whitepaper on Connectathon. The IHE Connectathon. What Is It? How Is It Done? Version 004. Available online: https://www.google.com.tw/url?sa=t&rct=j&q=&esrc=s&source=web&cd= 3&cad=rja&uact=8&ved=2ahUKEwjU-t_q_f3eAhWI62EKHT46BAsQFjACegQICBAC&url=https%3A% 2F%2Fwww.ihe-europe.net%2Fsites%2Fdefault%2Ffiles%2FWhitePaper_Connectathon_2016.pdf&usg= AOvVaw3vNVAnPQmRWdfV5aSS8e5h (accessed on 30 October 2018).

30. ISO DTR 28380-1: Health Informatics—IHE Global Standards Adoption—Part 1: Process. Available online: https://www.google.com.tw/url?sa=t&rct=j&q=&esrc=s&source=web&cd=1&cad=rja&uact=8&ved= 2ahUKEwil7-W3_f3eAhUYIIgKHcYsBFYQFjAAegQIAhAB&url=https%3A%2F%2Fwww.iso.org% 2Fstandard%2F63383.html&usg=AOvVaw1iuYOVXKoGsFrnfKhE6ucA (accessed on 30 October 2018).

31. Europa Analytics. An Introduction to the European Interoperability Reference Architecture (EIRA) v2.1.0. Available online: https://joinup.ec.europa.eu/sites/default/files/distribution/access_url/2018-02/ b1859b84-3e86-4e00-a5c4-d87913cdcc6f/EIRA_v2_1_0_Overview.pdf (accessed on 30 October 2018).

32. Masi, M.; Pavleska, T.; Aranha, H. Automating Smart Grid Solution Architecture Design. In Proceedings of the 2018 IEEE International Conference on Smart Grid Communications, Aalborg, Denmark, 29–31 October 2018.

33. The Open Group. Introduction to Building Blocks. Available online: http://www.opengroup.org/public/arch/p4/bbs/bbs_intro.htm (accessed on 30 May 2018).

34. IHE Europe. Gazelle | IHE Europe. 2018. Available online: https://www.ihe-europe.net/testing-IHE/gazelle (accessed on 30 October 2018).

35. Open Grid Forum. Data Format Description Language (DFDL) v1.0 Specification. Available online: https://www.ogf.org/documents/GFD.207.pdf (accessed 30 October 2018).

36. The Apache Software Foundation. Apache Daffodil (Incubating) | Getting Started. Available online: https://daffodil.apache.org/getting-started/ (accessed on 30 October 2018).

37. Hevner, A.; Chatterjee, S. *Design Research in Information Systems*; Integrated Series in Information Systems 22; Springer: Berlin/Heidelberg, Germany, 2010; pp. 9–21.

38. *Temporary Working Group 4—Increase the Resilience and Security of The Energy System. Strategic Energy Technology Plan—Implementation Plan—Final Version—15.01.2018*; Technical Report; European Commission—SETIS: Brussels, Belgium, 2018.

energies

MDPI

Review

Applying the Smart Grid Architecture Model for Designing and Validating System-of-Systems in the Power and Energy Domain: A European Perspective

Mathias Uslar [1], Sebastian Rohjans [2,*], Christian Neureiter [3], Filip Pröstl Andrén [4], Jorge Velasquez [1], Cornelius Steinbrink [1], Venizelos Efthymiou [5], Gianluigi Migliavacca [6], Seppo Horsmanheimo [7], Helfried Brunner [4] and Thomas I. Strasser [4,8]

[1] OFFIS—Institute for Information Technology, 26121 Oldenburg, Germany; mathias.uslar@offis.de (M.U.); jorge.velasquez@offis.de (J.V.); cornelius.steinbrink@offis.de (C.S.)
[2] Department of Information and Electrical Engineering, Hamburg University of Applied Sciences, 20099 Hamburg, Germany
[3] Center for Secure Energy Informaticss, University of Applied Sciences Salzburg, Urstein Süd 1, 5412 Puch, Salzburg, Austria; christian.neureiter@fh-salzburg.ac.at
[4] AIT Austrian Institute of Technology—Electric Energy Systems, Center for Energy, 1210 Vienna, Austria; filip.proestl-andren@ait.ac.at (F.P.A.); helfried.brunner@ait.ac.at (H.B.); thomas.strasser@ait.ac.at (T.I.S.)
[5] FOSS Research Centre, University of Cyprus, Nicosia 1678, Cyprus; efthymiou.venizelos@ucy.ac.cy
[6] Ricerca sul Sistema Energetico (RSE), 20134 Milan, Italy; gianluigi.migliavacca@rse-web.it
[7] VTT Technical Research Centre of Finland, Knowledge intensive products and services, 02044 Espoo, Finland; seppo.horsmanheimo@vtt.fi
[8] Institute of Mechanics and Mechatronics, Vienna University of Technology, 1060 Vienna, Austria
* Correspondence: sebastian.rohjans@haw-hamburg.de; Tel.: +49-404-2875-8019

Received: 9 December 2018; Accepted: 6 January 2019; Published: 15 January 2019

Abstract: The continuously increasing complexity of modern and sustainable power and energy systems leads to a wide range of solutions developed by industry and academia. To manage such complex system-of-systems, proper engineering and validation approaches, methods, concepts, and corresponding tools are necessary. The Smart Grid Architecture Model (SGAM), an approach that has been developed during the last couple of years, provides a very good and structured basis for the design, development, and validation of new solutions and technologies. This review therefore provides a comprehensive overview of the state-of-the-art and related work for the theory, distribution, and use of the aforementioned architectural concept. The article itself provides an overview of the overall method and introduces the theoretical fundamentals behind this approach. Its usage is demonstrated in several European and national research and development projects. Finally, an outlook about future trends, potential adaptations, and extensions is provided as well.

Keywords: Architecture; Development; Enterprise Architecture Management; Model-Based Software Engineering; Smart Grid; Smart Grid Architecture Model; System-of-Systems; Validation

1. Introduction

The continuously increasing complexity of modern and sustainable power and energy systems leads to a wide range of solutions for operating transmission and distribution grids. Those approaches developed by industry and academia in the context of Smart Grids become increasingly specific to the individual topology of the power grids in which they are to be deployed. At the same time, however, those solutions should be transferable to other topologies, preferably in an easy and cost-efficient way. In addition, competing technical and operational solutions, with their respective costs, needed Technology Readiness Levels (TRL), advantages and disadvantages, are being developed for various problems occurring due to the new operation paradigm in the context of Smart Grids [1]. While it can be argued that energy transition to Smart Grids may also be a sociological problem [2], we take the technological perspective from systems engineering view here.

To evaluate different operational approaches for power grids, a growing number of methods and techniques are introduced and applied today. One problem arising is that Smart Grids must be still considered an emerging topic and transferring solutions from one utility to another with parts and technologies being replaced is usually not an easy task. Techniques usually come in so-called technology experience packages and someone learns about one specific instance in a package [3]. For example, replacing a single technology (e.g., wired communication by power line carrier or wireless communication) in a solution could lead to unexpected results. To learn from previous experience in Research and Development (R&D) projects, demo, or field trials, those solutions have to be thoroughly documented in a meaningful way.

Unfortunately, those solutions could have some characteristics that make knowledge preservation difficult. Typically, the projects use various combinations of runtime environments, software, and algorithms, from different vendors and Original Equipment Manufacturers (OEM). The systems under scope must be considered a System-of-Systems (SoS) with all the implications to the complexity arising in those projects [4]. Finding a definite "best" solution is hard as a lot of contextual knowledge and degrees of freedom has to be known to the team implementing the solution. The knowledge and agreement on requirements, both functional as well as non-functional, become apparently more of a socio-technical than purely technical problem [4]. Various projects have addressed this issue when trying to document knowledge gained from field trials in order to disseminate the results [5]. One of the most important things to get to know to transfer a technical solution is to know its scope and applicability. To assess for this information, so-called tacit knowledge is often needed. Tacit knowledge (as opposed to formal, codified or explicit knowledge) is the kind of knowledge that is difficult to transfer to another person by means of writing it down or verbalizing it [6]. For re-use, certain important aspect of a technical solution given in a procedural context must be made formal.

In modern systems and software engineering, specifications are created and based on some kind of a requirements engineering process. Mostly, this process is used for elicitation of the information needed for creating a solution architecture and implementing and operating it. The architecture of a system is one key element to work towards the common project goal of deploying a product according to the specification. However, this documentation used to be done by mid-sized teams who could communicate a lot on the needs, mostly in-house. With the SoS-based needs, this process should be carried out in a formal and knowledge-intensive manner. Engineering teams are responsible for different components and parts, the knowledge and work is far more fragmented in the process. For the dissemination of a solution and achieving a higher TRL (possibly levels 7–9), the process must be formalized and governed. Typically, (formal) standards are needed at a certain point. Therefore, the whole development and validation process of Smart Grid projects need to be more professional in terms of products, processes, and governance of operations. One part of the solution is to use a method which has proven to be useful over the last couple of years, the so-called reference designation system Smart Grid Architecture Model (SGAM) [7].

The main aim of this review article is to provide a comprehensive overview of the state-of-the-art and related work for the theory, distribution, and usage of the SGAM. The contribution itself focuses on an overview of the overall method, the theoretical fundamentals and foundation as well as current applications of the method in various projects. Finally, an outlook about future trends, potential adaptations, and extensions is provided as well.

The remainder of this review is structured as follows: First, the concept and the history of SGAM are introduced in Section 2. Afterwards, corresponding tools and their usage in different projects and initiatives are discussed in Section 3. Section 4 shows how this architectural model can be applied also to other domains, while Section 5 discusses potential adoptions. Finally, the article is concluded with Section 6 providing the lessons learned and the main findings.

2. Overview of the Smart Grid Architecture Model (SGAM)

The original scope of the SGAM was created in the M/490 mandate of the European Commission (EC) to the European standardization bodies CEN (Comité Européen de Normalisation), CENELEC (European Committee for Electrotechnical Standardization), and ETSI (European Telecommunications Standards Institute) with the focus on finding existing technical standards applicable to Smart Grids as well as identifying gaps in state-of-the-art and standardization. Given the distribution and sheer number of the experts, it apparently became somewhat of a problem to agree on terms, technology, scope, and subjects for discussion [8]. Therefore, the SGAM was used as a tool for reference designation to solve this wicked problem (i.e., the fact that formulating the problem itself is already a problem [9]). Originally, the References Architecture Working Group (RAWG) and the Sustainable Processes (SP) group worked in parallel. One had the aim to create a methodology to elicit applications for future and emerging Smart Grids, while the other to come up with a solutions and blueprint for technical architectures of future technology portfolios. Both groups worked in parallel due to time constraints but shared experts. Therefore, the groups could both apply the IntelliGrid Use Case Template and IEC PAS 62559 for standardizing user stories and use cases to provide a basic documentation from the functional point of view for future Smart Grids as well as a reference designation system to document three main viewpoints of a technical Smart Grid solution. Both methods can be used on their own, but they work seamlessly and elicit data based on a common meta-model shared by the ISO 42010 architecture standards. Therefore, the dynamics of a static architecture with processes and exceptions can be documented in a IEC 62559 template while the individual solutions can be put into the reference designation system [4]. Filling out a use cases provides enough information to get to know the basic information needed to create SGAM models [10]. The following paragraphs elaborate more on the design decisions taken and the rationals behind the current SGAM method as well as taking into account a Systems Engineering perspective.

Considering the Smart Grid from a *Systems Engineering perspective* and following the classification given by Haberfellner et al. [11], it can be categorized as a *complex system*. More precise, it can be argued that the electric power grid evolves from a *massively interconnected, complicated system* into a *complex system*. As depicted in Figure 1, such a system is characterized by its constituent subsystems reflecting a certain level of diversity/variety/scale on the one hand and its structure being subject to a certain dynamic/alterability on the other hand [4].

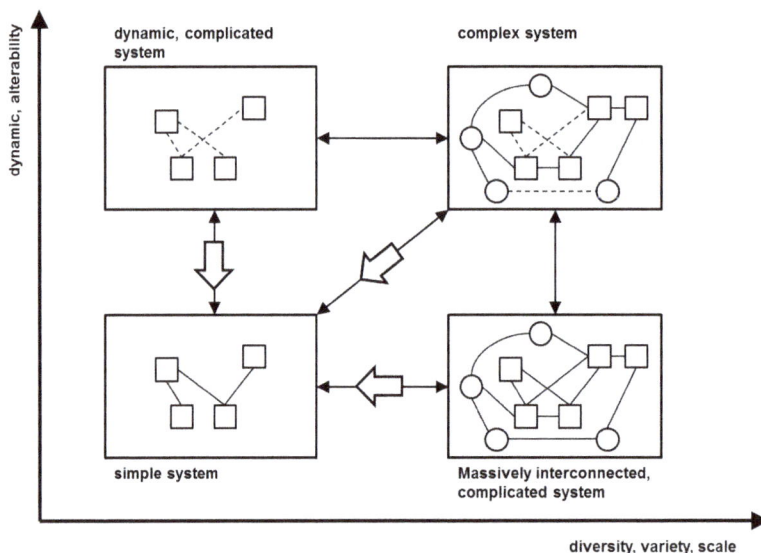

Figure 1. Classification of system types [11].

In the recent past, however, a subset of complex systems have been identified as System-of-Systems (SoS), characterized by at least eight criteria postulated by Maier [12] and DeLaurentis [13]:

1. Operational Independence of Elements
2. Managerial Independence of Elements
3. Evolutionary Development
4. Emergent Behavior
5. Geographical Distribution of Elements
6. Interdisciplinary Study
7. Heterogeneity of Systems
8. Networks of Systems

The SoS perspective is of importance for taking into account interoperability between the constituting systems. This challenge has been identified as mostly a problem of standardization. It became apparent that the integration cost drivers were mostly from unharmonized technical models and semantics [6]. To deal with these issues, standardization bodies issued work on reference architectures and corresponding road-maps [8]. Therefore, the EC issued the M/490 mandate. Within the scope of this mandate, gaps in standardization, needed use cases, security requirements and reference architectures had to be defined [5].

The work acted as initial focal point for basic method engineering research on how to model and document Smart Grid architectures using standardized canonical methods. In addition to the IEC 62559 use case template and methodology for documenting meaningful blueprint solutions for Smart Grid systems of systems to be implemented [10], the SGAM has been created for the purpose of identifying gaps in existing and future standardization. The SGAM acts as a reference designation system [14], providing three main axis for the dimensions of: (i) value creations chain ("Domains"); (ii) automation pyramid ("Zones"); and (iii) interoperability ("Interoperability Layer"). Within this visual representation (cf. Figure 2), systems and their interfaces can be allocated to some point in the reference model, thus providing a categorization and classification of individual parts, data exchanged and interfaces of the system landscape.

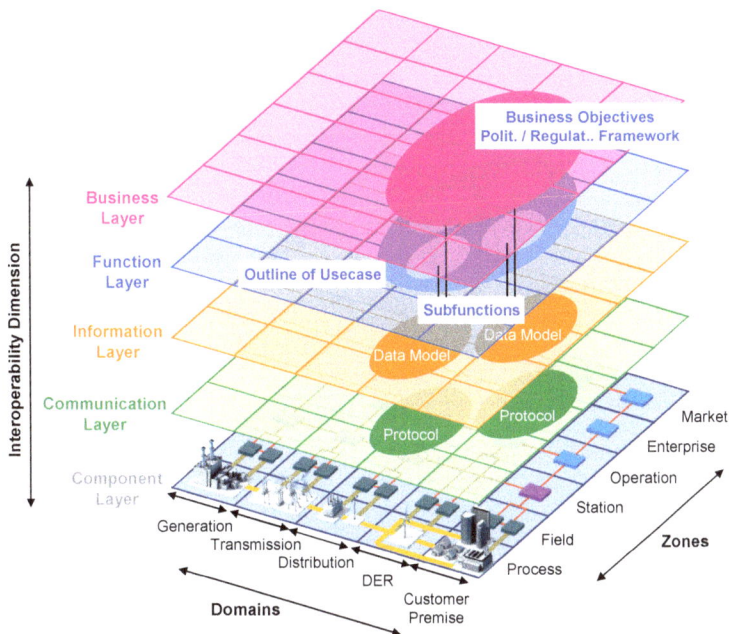

Figure 2. Overview of the Smart Grid Architecture Model (SGAM) [7].

The *Domains* basically represent the energy conversion chain as described in the fundamental and well-known *NIST Conceptual Model* [15]. The individual domains are described as follows [7]:

- *Bulk Generation:* Represents generation of electricity in bulk quantities, such as by fossil, nuclear and hydro power plants, off-shore wind farms, large scale solar power plant (i.e., Photovoltaic (PV) and Concentrated Solar Power (CSP)), which are typically connected to the transmission system.
- *Transmission:* Represents the infrastructure that transports electricity over long distances.
- *Distribution:* Represents the infrastructure that distributes electricity to customers.
- *Distributed Energy Resource (DER):* Represents distributed electrical resources directly connected to the public distribution grid, applying small-scale power generation technologies (typically in the range of 3–10 MW). These distributed electrical resources may be directly controlled by a Distribution System Operator (DSO).
- *Customer Premises:* Host both end users of electricity and producers of electricity. The premises include industrial, commercial and home facilities (e.g., chemical plants, airports, harbors, shopping centers, and homes). In addition, generation in the form of, e.g., PV generation, Electric Vehicles (EV), storage, batteries, micro turbines, etc., are hosted.

The *Zones* are orthogonal to the domains and basically represent the Information and Communication Technology (ICT) based control systems, controlling the energy conversion chain. Based on the automation pyramid, the individual Zones are described as follows [7]:

- *Market:* Reflects the market operations possible along the energy conversion chain, e.g., energy trading, mass market, retail market, etc.
- *Enterprise:* Includes commercial and organizational processes, services and infrastructures for enterprises (utilities, service providers, energy traders, etc.), e.g., asset management, logistics, work force management, staff training, customer relation management, billing, etc.

- *Operation:* Hosts power system control operation in the respective domain, e.g., Distribution Management Systems (DMS), Energy Management Systems (EMS) in generation and transmission systems, microgrid management systems, virtual power plant management systems (aggregating several DER), and EV fleet charging management systems.
- *Station:* Represents the areal aggregation level for field level, e.g., for data concentration, functional aggregation, substation automation, local Supervisory Control and Data Acquisition (SCADA) systems, plant supervision, etc.
- *Field:* Includes equipment to protect, control and monitor the process of the power system, e.g., protection relays, bay controller, and any kind of Intelligent Electronic Devices (IED) that acquire and use process data from the power system.
- *Process:* Includes the physical, chemical or spatial transformations of energy (electricity, solar, heat, water, wind, etc.) and the physical equipment directly involved (e.g., generators, transformers, circuit breakers, overhead lines, cables, electrical loads, any kind of sensors and actuators that are part of or directly connected to the process, etc.).

To maintain interoperability between any two components in the Smart Grid, interoperability needs to be considered on five different *Interoperability Layers*. The first two layers are related to functionality, whereas the lower three layers can be associated with the intended technical implementation. The interoperability layers being used are basically derived by the GridWise Architecture Council (GWAC) interoperability stack [16] and described as follows [7]:

- *Business Layer:* Provides a business view on the information exchange related to Smart Grids. Regulatory and economic structures can be mapped on this layer.
- *Function Layer:* Describes services including their relationships from an architectural viewpoint.
- *Information Layer:* Describes information objects being exchanged and the underlying canonical data models.
- *Communication Layer:* Describes protocols and mechanisms for the exchange of information between components.
- *Component Layer:* Physical distribution of all participating components including power system and ICT equipment.

3. Application of the Smart Grid Architecture Model

SGAM models soon proved to be a useful solution in both standardization and research and development projects in order to document system architectures in a canonical and standardized manner and gained attention in the community over the years to come. In addition, many tools were developed to cope with the graphical representation as well as the procedural application of the method and toolchain. Various funding schemata, such as the German SINTEG (Schaufenster Intelligente Energien), the Austrian "Energieforschung" or the European H2020 LCE calls, have adopted the need to document the research conducted in a standardized way.

However, it became apparent that the model has already outgrown its original purpose of allocating standards to various Smart Grid systems and interfaces as it was envisioned in the mandate M/490 [8]. Cost–benefit analysis, security analysis, technical debt analysis and maturity levels of organizations can be visualized using the SGAM. One missing link as of now has been the coupling of the "higher" interoperability levels such as the ones described in the LCIM [17] in order to cope with conceptual dimension which is relevant for, e.g., simulation purposes. Therefore, in the following sections, an overview of the usage of the SGAM approach is provided, which is divided into the categories: (i) software tools; (ii) European and (iii) national-funded projects; and (iv) further activities.

3.1. Software Support and Tools

3.1.1. SGAM Toolbox

Due to the inherent complexity of Smart Grids, the realization of particular solutions has proven to be a challenging task. Possible concepts for dealing with this challenge can be found in the field of *Systems Engineering* [18,19] with a special focus put on *Model Based Systems Engineering* (MBSE) [20]. This approach targets a consistent understanding of systems as possible approach to manage complexity. To establish such a consistent understanding, the MBSE concept fosters the utilization of different *models* in order to establish well-defined *views* over *well-defined* abstraction levels. This concept has proven to be of value especially in terms of interdisciplinary development.

One of the key concepts in MBSE is the definition of particular *views* on basis of *viewpoints*. According to ISO 42010, it can be said that one viewpoint *governs* one particular view [21]. Furthermore, a viewpoint is intended to *frame* one or more *concerns* associated with one or more different *stakeholder*. An overview on these relations can be seen in Figure 3.

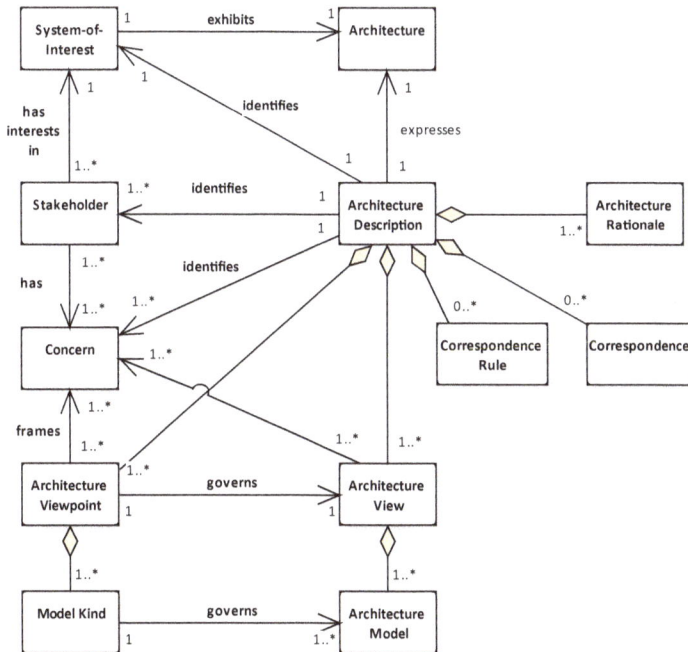

Figure 3. Conceptual model of an architecture description according to ISO 42010 [21].

Considering the structure of the SGAM cube as described in Section 2, one could argue that the individual layers ("*Interoperability Layers*") represent a set of basic viewpoints for Smart Grid architectures. The particular *concerns* addressed in this structure are *business, functional, informational, communication,* and *physical* aspects. Besides the consideration of views, the definition of *abstraction layers* with explicit transformations in between is of very importance. During the research conducted in this area, the concept of *Model Driven Architecture* (MDA) [22] has been chosen as main paradigm. The MDA concept basically aims at a separation of *functionality* and *technology*. To individually address these different aspects, the following abstractions were defined:

- *Computation Independent Model (CIM)*: The CIM can be interpreted as "System Level" describing a system from its outside perspective, which means focus is put on the delivered *functionality* rather than on the *technology*. Please note that the IEC 61970/61968 series [23,24], also known as CIM (Common Information Model), which is an ontology for defining objects and relations to model power system, is indeed a PIM in the sense of MDA.
- *Platform Independent Model (PIM)*: This layer can be seen as "Architecture Level". It aims at focusing on the decomposition of the system without considering detailed technical aspects of individual components.
- *Platform Specific Model (PSM)*: The PSM describes the technical aspects for realizing the individual components. Thus, it can be seen as *Detailed Design Layer*.
- *Platform Specific Implementation (PSI)*: This last layer represents the realized implementation. In case of one artifact being realized as software, this can be seen as the source code created.

By aligning the MDA concept with the SGAM, it could be argued that the CIM is represented by the *Business* and *Function Layer* viewpoints as they analyze and define the systems functionality. Furthermore, the PIM can be associated with the lower three layers (*Information, Communication, and Component Layer*). According to this mapping the SGAM can be used to describe the overall architecture to a level that considers the constituent components as black-boxes. Keeping the described system-focus in mind, a domain-specific description (as the SGAM delivers) appears suitable.

For the detailed design of the individual components (PSM) and their implementation (PSI), the modeling language of choice should rather be associated with the "type" of the component (software could be designed by means of UML, embedded systems with SysML, control functionality with linear algebra, etc.) than domain specific concepts. Considering this changing perspective, the transformation between the architectural level and the design level typically also represents the handover between, for example, a DSO and an OEM from the supply chain. An overview on the alignment between the SGAM and MDA as described can be seen in Figure 4.

When it comes into practical application of these concepts, the need for an appropriate modeling language and corresponding tools arises. In the fields of Software and Systems Engineering, the utilization of so-called *Object Modeling Languages* is widely accepted. Especially the *General Purpose Languages* (GPL) UML [25] and SysML [26] are state of the art. However, as these languages anticipate object oriented patterns such as instantiation, inheritance and others, they are rather hard to understand by non-software educated stakeholders and, thus, their acceptance outside the software/systems community is rather low. To enable the utilization of consistent models on the one hand and to provide interdisciplinary understanding on the other hand, the application of so-called *Domain Specific Languages* (DSL) can be considered. In terms of object modeling, for example, some domain specific aspects can be put on top of GPLs. Thus, well-established concepts such as "traceability" or existing tools with a high maturity can be made accessible to domain stakeholders.

In the field of Smart Grid Engineering, from 2012 to 2017, such a DSL has been developed by utilization of *UML Profiles*, a UML specific concept for lightweight extensions. Developing a DSL on basis of standardized UML profiles brings the benefit that the DSL is tool-independent. However, a drawback of UML profiles is the limitation that does not provide capabilities for, e.g., automation mechanisms such as model transformations. To overcome this shortcoming, a dedicated Add-In for the widely spread modeling tool *Enterprise Architect* has been implemented and made publicly available as *SGAM Toolbox* (www.sgam-toolbox.org) [27].

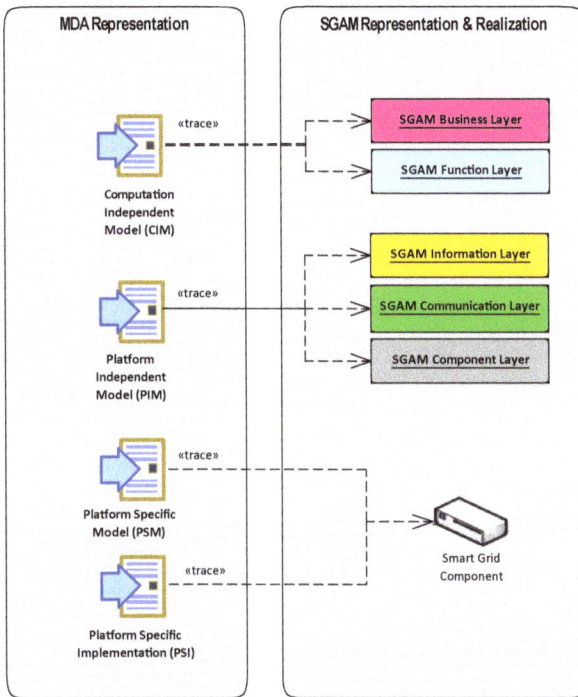

Figure 4. Mapping between MDA and SGAM.

Besides the technical implementation of this toolbox, an appropriate process model has been specified to give users guidance on the application [27]. It reflects the concepts from ISO 15288 [28] and has been tailored by application of the guidance delivered in the IEC TR 24748 guidelines [29].

Since the SGAM Toolbox has been released it has proven its value in several research and real-life projects [30]. For example, by using the SGAM Toolbox, the American NIST Logical Reference Model (NIST LRM) [31,32] could have been successfully modeled in the context of the SGAM framework and, thus, compatibility between the American and the European concepts could have been demonstrated [27,33].

However, despite the already demonstrated value, there are still several aspects to be considered. Besides, some necessary alignments and improvements as discussed in [27], especially the integration into a holistic tool-chain, is the focus of present research. The capability of such a tool-chain as envisioned in [34] comprises sophisticated interoperability between various repositories, tools and standards on the one hand and some additional functionality for model validation (e.g., Co-Simulation) on the other hand. Both topics are the subjects of ongoing research.

3.1.2. 3D Visualisation

Based on the work in the DISCERN project [5], it became apparent that to implement the concept needed to exchange knowledge, the typical way to create SGAM models in PowerPoint was not enough. The overall cube picture lacks visibility for certain layers, therefore the initial models used to be created with five individual 2D planes and tabs that were combined using a Visual Basic Macro (VBM); however, it soon became obvious that a tool using Microsoft Visio would be even more beneficial as stencils, semantics and exporting into XML could be used. In addition, the browser-based 3D SGAM viewer was created in order to manipulate the view for a given standardized SGAM file and model. Figure 5 shows an example of this tool for viewing SGAM models. The files can be exchanged with both the Use Case Management Repository (UCMR) and the SGAM Toolbox.

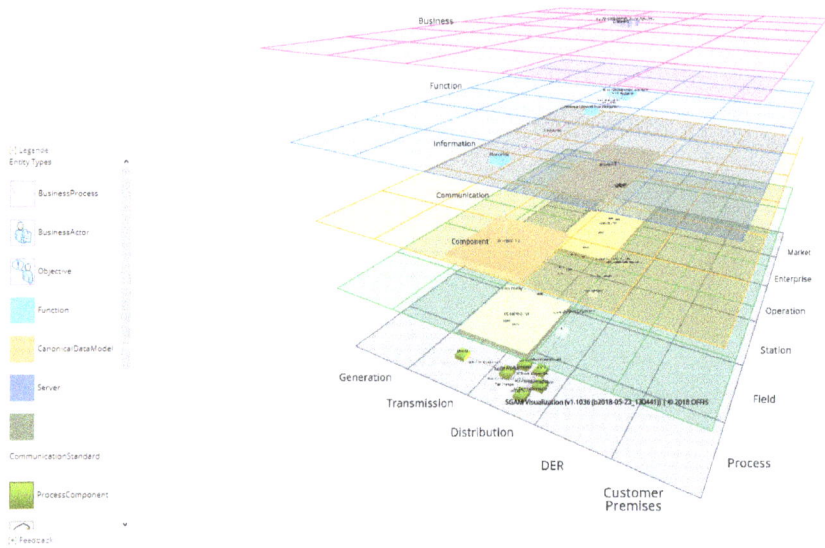

Figure 5. Visual import of a SGAM model for Substation Automation in the 3D SGAM viewer.

3.1.3. Power System Automation Language (PSAL)

The main intention with the Power System Automation Language (PSAL) is to provide a formal domain specific language for SGAM compatible use case design [35]. At the same time, another focus of PSAL is to allow rapid development of automation, control, and ICT functions for power system applications [36]. Therefore, although possible, PSAL is not directly intended to be used for development of high-level use case descriptions. Instead, it offers specific tools for detailed use case design that can be used in further steps for generation of code and configuration.

Although the core of PSAL is based on SGAM, it also introduces an extra abstraction layer, containing a System and an application. The System consists of definitions for the component and the communication layers of SGAM and the application contains definitions for the business, function, and information layers. One benefit of this is that it allows the user to define an application independently from the System. Consequently, solutions developed as an application for one System can be easily ported to another. One main difference, compared to the SGAM Toolbox, is that PSAL is a textual language. Figure 6 shows a UML representation of the PSAL meta-model as well as example implementations of an application and a system.

As mentioned above, one of the main ideas with PSAL is that it should allow rapid generation of code and configurations, such as executable IEC 61499 code and IEC 61850 configurations [36].

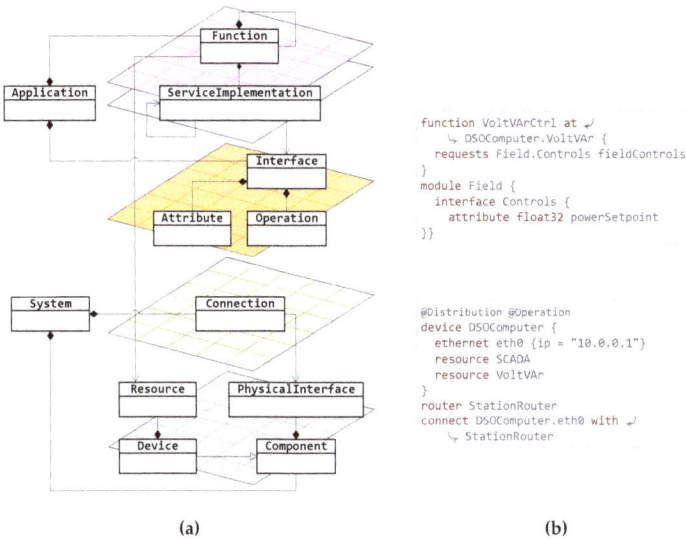

```
function VoltVArCtrl at ↵
    ↳ DSOComputer.VoltVAr {
    requests Field.Controls fieldControls
}
module Field {
    interface Controls {
        attribute float32 powerSetpoint
}}


@Distribution @Operation
device DSOComputer {
    ethernet eth0 {ip = "10.0.0.1"}
    resource SCADA
    resource VoltVAr
}
router StationRouter
connect DSOComputer.eth0 with ↵
    ↳ StationRouter
```

(a)   (b)

Figure 6. The PSAL engineering approach: (**a**) UML (Unified Modeling Language) representation; and (**b**) example code [36].

3.2. European-Funded Projects and Activities

3.2.1. FP7 DISCERN (Distributed Intelligence for Cost-Effective and Reliable Solutions)

During the DISCERN project, the SGAM and IEC 62559 use was strongly enforced to deal with a methodological approach of the so-called 3L (Leader, Learner, and Listener [37]) concept agreed upon by the partners, as shown in Figure 7. The overall target of the DISCERN project was to somehow assess the optimal level of intelligence in the distribution networks, and, in addition, to determine a set of so-called replicable technological options (e.g., basic Smart Grid solutions for operation of distribution grids) that would allow for a both cost-effective and reliable enhancement of both observability and controllability of distribution grids.

After starting the project, it became apparent that, to exchange knowledge between the diverse stakeholders, more formal aspects of this knowledge documentation and exchange had to be dealt with. One particular aspect to cope with the leaders being organizations who have already implemented a solution, testing them out in operations and planning large scale roll-out is that they must find a way of documenting their knowledge, fallacies which occurred and important context of the operation of a Smart Grid solution. Learners have already decided to implement the solution, but listeners still struggle to find the business benefit. Therefore, different information on context and CBA (Cost-Based Analysis) is of importance. This challenge led to the DISCERN approach for documenting using the IEC 62559 template as well as blueprints for architectural documentation using SGAM.

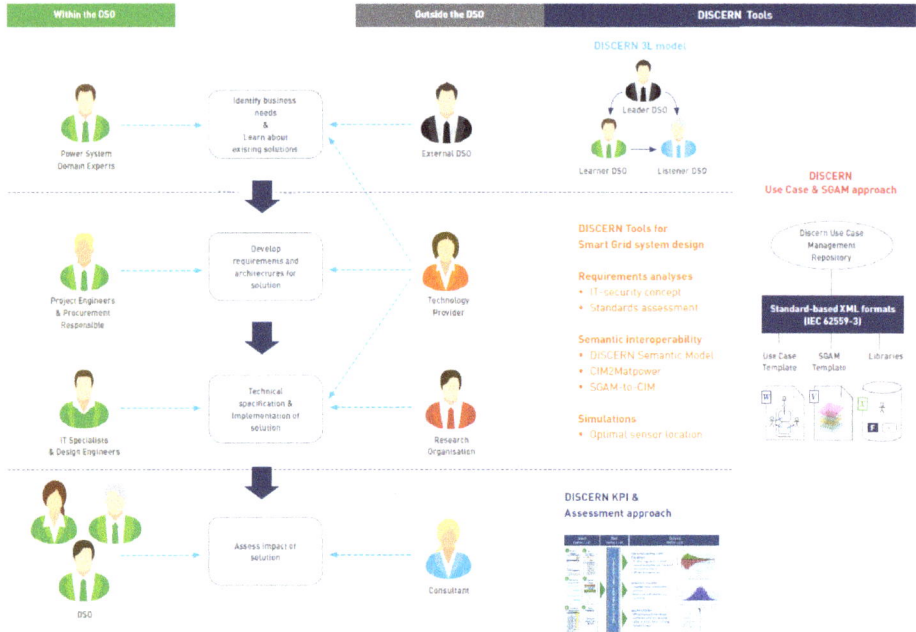

Figure 7. Overview on DISCERN concept for modeling solutions for 3L [5].

In DISCERN, the SGAM visualization template in PowerPoint, Visio and browser plug-in, in addition to providing an intuitive high-level representation of the solutions, enables DSOs to [5]:

- Identify existing interoperability issues in the used systems that implement a particular sub-functionality or functionality. Available standards and standardization gaps for each individual solution can be represented in the SGAM visualization template.
- Describe the real-life physical distribution of the components used in the field (e.g., software based applications, devices and communication elements deployed). In the use case template, it is possible to define which existing and future actors are involved in a functionality, but it is currently not possible to represent how these actors are actually implemented in the physical layer of the system.
- Establish clear relationships between the business use cases and business objectives that explain the benefits derived by the (leader) company with the functionality, the technical functions that are required to realize such functionality, the information exchanges between the individual functions, the standards used for communication and data models that enable the information exchange, and the physical components that implement the technical functions.
- Carry out an impact analysis, analysis for security compliance, find risk elements, compatibility/comparison at DSO level and the future specification of new features.

The contribution in [37] presents the implemented tool support developed within DISCERN in order to manage use cases and SGAM models. Use cases define the requirements for Smart Grid solutions, whereas SGAM models describe so-called high-level Smart Grid architectures for the solutions and portfolios. Both methodologies have been used during DISCERN project with the aim of facilitating knowledge sharing among DISCERN partners and also outside the project. The objective of the tool support is to promote the adoption of these methodologies in the context of large Smart Grid projects by improving re-usability of use case and SGAM descriptions. The existing tool support consists of [38]:

- Some standard-based formats to exchange items such as use cases, SGAM models, and libraries of terms used in the corresponding descriptions (Actors, Functions, and Requirements);
- Enhanced templates with standards-based XML export functionality to export use cases, SGAM models, and libraries in those standard-based 62559 compliant formats; and
- Web-based repository to store and manage elicited use cases, SGAM models, and libraries, managing access rights and, thus, enabling multi-editing of the defined descriptions.

The tools developed in DISCERN were used during the project to store and edit the elicited DISCERN use cases. This was particularly useful for learner's use cases, since they might evolve during the project after receiving inputs from the simulations carried out. In addition, given that the tools rely on international standards from IEC and architectural frameworks, they can be leveraged in other projects with the aim of facilitating sharing of Smart Grid requirements and architectures. This is particularly useful in the context of larger Smart Grid projects for the future, in which partners from different areas of expertise and different countries need to exchange information on Smart Grid solutions and portfolios with each other.

The tools from DISCERN are complete versions, including required functionalities for exchanging and managing Smart Grid requirements and architectures in a collaborative environment. It was planned to enhance the tools during the project to include further features based on the feedback received. The basis of the overall concept was to utilize the experience of major European DSOs with innovative future-proof technological solutions for a more efficient monitoring and control. The complementary nature of the demonstration sites with regard to the specific challenges as well as technological and operational solutions served as knowledge and trial resources. Overall, SGAM proved to be a meaningful solution for implementing the 3L model envisioned.

3.2.2. FP7 ELECTRA IRP (European Liaison on Electricity Committed towards Long-Term Research Activity Integrated Research Programme)

The main aim of the ELECTRA project was the development of a new real-time control concept that can handle the massive integration of renewable generators and flexible loads in a more effective way compared to today's approaches [39]. As a result of the developments, the so-called "Web-of-Cells" (WoC) control architecture has been introduced where the whole power system over all voltage levels is divided into smaller control areas. This division allows solving local problems locally using the flexibility of the local resources (i.e., distributed generators and loads).

For the development of the WoC approach, the SGAM and corresponding tools have been used in different ways. First, SGAM and the use case modeling approach was applied for the development of the WoC control schemes and corresponding functions, which led to a proposal for changing the SGAM itself in order to address the needs of the ELECTRA IRP project [40]. In addition, the security analysis of the WoC concept was performed with the NISTViz! tool and method developed by OFFIS [40–42]. For the time being, the methodological toolchain was extended for the WoC concept. In a final step, SGAM was also applied for proof-of-concept evaluation of the WoC approach. In the following, some details about the usage of SGAM in ELECTRA are provided.

In a first step, the locations in the SGAM plane of the actors from the identified six main ELECTRA use cases related to voltage, frequency/power balance and inertia control have to be defined. Based on mapping the identified WoC control schemes and functions onto the plane, a categorization of the interfaces was done, defining if the interface is either operation.

Additionally, the Logical Interface Category (LIC) for each communication interface was determined with a standardized procedural approach. This mapping work has been done using a simplified mapping tool. It has been setup to simplify the editing of an individual SGAM plane, thus focusing on the functions for ELECTRA actors in the deliverables of the project dealing with the use cases. Instead of creating a fully complete SGAM model, the tool can focus on the very interfaces (and their classes) among the actors, functions and systems. One main aspect of communication security and meaningful mitigation on the interface can be easily analyzed using this approach.

Based on a shared common overview provided by the NIST LRM model from the SGAM Toolbox Sparx add-in, system classes are assigned to the functions of the use cases, thus mapping them onto the NIST classification of high-level systems classes and their corresponding generic existing interfaces. This process makes for a rather easy assignment of the logical interface classes based on NISTIR 7628. This way of modeling combines a low-learning curve with the simplified assumptions done towards architectural modeling of the so-called Web-of-Cells approach invented in the project. The SGAM Toolbox with its (meta)data model would have relied on some information that was not available for modeling through the green field approach taken in ELECTRA as certain systems (respective functions) were new and could not be mapped onto the generic existing ones.

In Heussen et al. [41], changes for the axes of SGAM in order to cope with the web-of-cells approach are also discussed. The contribution relfects on needed change to to DSO level cells interacting and, thus, making the TSO and generation zones somehow obsolete from the modeling point of view as the top-down distribution is replaced by an heterarchical system approach of (generation) cells. A mapping is provided on how the new concepts could be mapped onto the archetype SGAM model.

Finally, for the proof-of-concept, the SGAM was used in order to identify suitable laboratories for testing selected control schemes and functions [43]. Therefore, lab capabilities have been mapped to the different SGAM layers in a first step. Afterwards, mapping of the control schemes with the lab capabilities have been carried out in order to identify and select the most appropriate lab for the proof-of-concept evaluation. For the definition of the corresponding test cases and performance measures, the holistic testing approach motivated by SGAM, which has been developed in the ERIGrid project (see Section 3.2.5), has been used.

3.2.3. H2020 SmartNet: Smart TSO-DSO Interaction Schemes, Market Architectures and ICT Solutions for the Integration of Ancillary Services from Demand Side Management and Distributed Generation

In the SmartNet project, new TSO-DSO coordination schemes for Ancillary Services (AS) were developed and tested in multiple field test pilots. These new coordination schemes consider different market scenarios with different coordination patterns, roles, and market design. Summarized, the coordination schemes show different approaches for how an AS market can be designed and coordinated between TSO and DSO. Centralized options were considered, where the TSO operates a market for resources connected at both transmission and distribution levels, without extensive involvement of the DSO. This is similar to a traditional market. New and distributed approaches were also considered, where the balancing responsibility is shared between TSO and the DSO in different manners and both common and local markets are investigated [44].

ICT is playing an important role in the future TSO-DSO coordination schemes to enable information exchange between different market players and to ensure quality and operability of the grid. To ensure this, the first step in the SmartNet project was to capture ICT requirements for the interactions between market players and to design a common ICT architecture model supporting new ancillary services. To capture relevant ICT requirements and to compare different coordination schemes, an iterative and incremental design and analysis process was developed. Outcomes of this were the reference ICT architecture model for ancillary services with ICT requirements, and recommendations for pilots and practical system realizations. The process, depicted in Figure 8, is divided into three stages (white blue-framed boxes) involving four internal iteration cycles (blue circles). The three analysis stages are simply referred to as first stage, second stage, and third stage.

The base of the process is the SGAM approach. To identify ancillary service interactions and associated ICT requirements in different SGAM layers, the use case design methodology introduced in SGAM and the IEC 62559 use case template were used. Furthermore, the SGAM Toolbox mentioned in Section 3.1.1 was used as the main tool for the design and management of the ICT model. The model specifications in the electronic form significantly eased the design work during iteration cycles and, later, the delivery of specifications to the system implementation work packages.

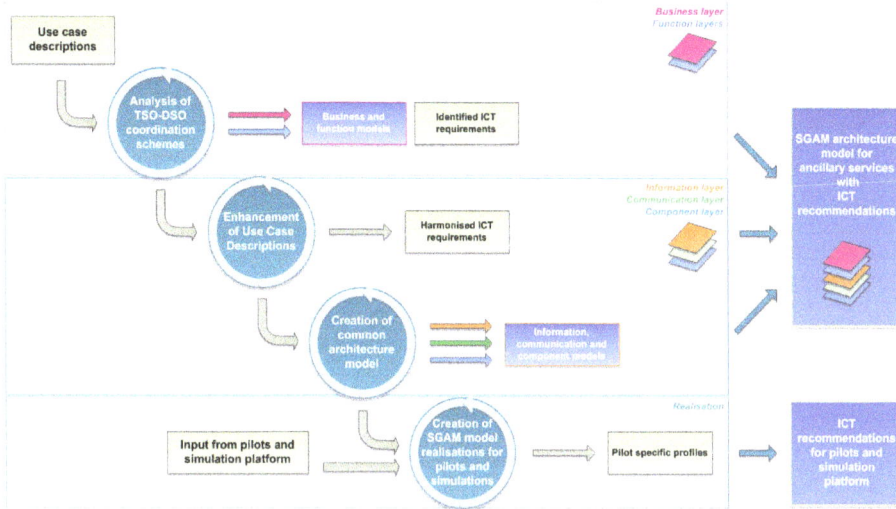

Figure 8. The design and analysis process for capturing ICT requirements for ancillary services in different TSO-DSO coordination schemes in the SmartNet project [45].

The first stage in Figure 8 focuses on defining classification metrics for ICT requirements and implementing business and function layers for the architecture model. ICT requirements are captured and prioritized in top-down and bottom-up manners. During the top-down step, each coordination scheme and use case is closely analyzed to understand better the interactions between stakeholders and to create the business and function layers in the SGAM model. In the bottom-up step, the first analysis of information exchanges between physical system components is performed.

The second stage concentrates on harmonizing ICT requirements across the coordination schemes and extending the SGAM model to include information, network, and component layers with ICT requirements. During this stage, the coordination scheme specific requirements are compiled into a common ICT requirement table and uploaded to the SGAM model. In our case, harmonized information exchange events, information objects, ICT requirement classes, protocols, and system actors were collected and harmonized. Once integrated into the SGAM model, the goal was not to create an entirely new ICT architecture design, but to implement a model with adjustable parameters to fit to the selected TSO-DSO coordination schemes and beyond. Within the model, the data flows and changing ICT requirements in different coordination schemes were analyzed. For each connection, the properties of the link (i.e., the exchanged information objects) were investigated to create different types of graphical presentations. Using this approach, it was possible to create a visual understanding of the characteristics and differences in the TSO-DSO coordination schemes. The diagrams in Figures 9 and 10 are examples from the analysis.

Figure 9 shows an example, where wired connections are presented in black (a stringent requirement for latency or security) and wireless connections with less strict requirements in green. The analysis of links was done by examining each conveyed object and associated latency, security, and cost requirements. Figure 10 shows how much stakeholders would be willing to invest on the sufficient Quality of Service (QoS). For this, the average investment cost of each connection was calculated.

The last stage of the analysis process is to apply it to a real implementation. In the SmartNet project, the process was tested with three planned pilots using available design specifications. The purpose of including them in the design and analysis process was to benefit from the early alignment of design and implementation work and to support planning and implementation of the pilot realizations.

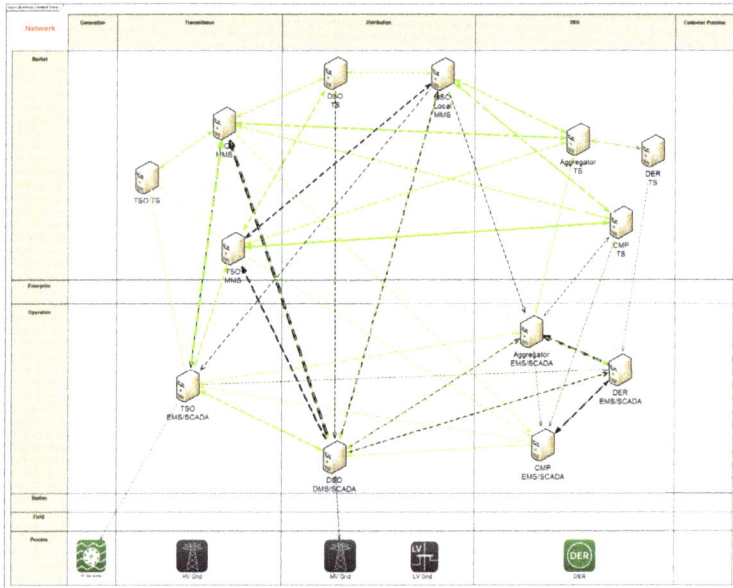

Figure 9. ICT analysis in the SmartNet project: Communication layer with different network types: wireless connections are shown in green, and wired connections are shown in black in the SGAM toolbox.

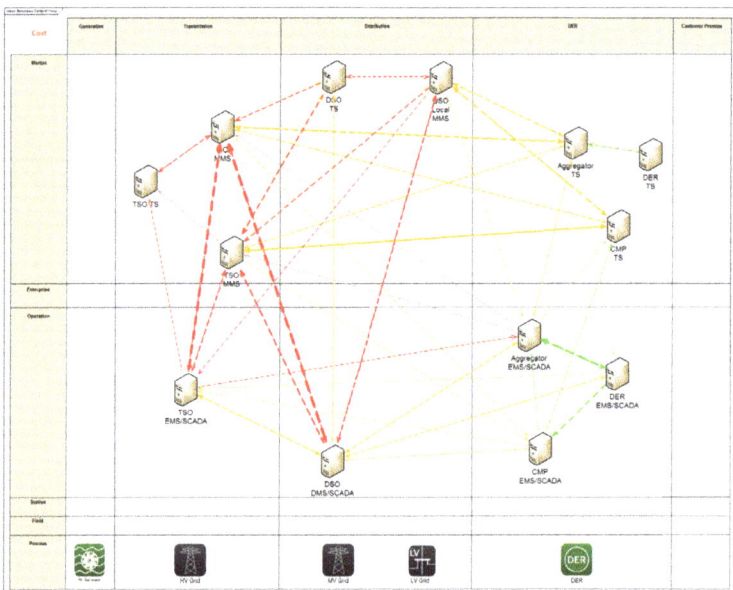

Figure 10. Communication links with different costs: high cost (red), medium cost (orange), and low cost (green).

A conclusion from the SmartNet project is that the outcome of the analysis depends on how precisely ICT requirements can be defined for each connection link. ICT requirements and their thresholds (e.g., for latency, reliability, and security) keep changing as energy systems and markets evolve.

Therefore, the model needs to be re-configurable and in a machine-readable form. In the SmartNet SGAM model, the values for ICT requirements can be altered for each information exchange event as well as the parameters for analysis (e.g., the cost function).

3.2.4. H2020 TDX-Assist: Coordination of Transmission and Distribution Data Exchanges for Renewables Integration in the European Marketplace through Advanced, Scalable and Secure ICT Systems and Tools

The TDX-Assist project has the overall goal to design and develop new so-called ICT tools and technologies that facilitate especially scalable and secure information systems and data exchange between both TSO and DSO. Three novel aspects of ICT tools and techniques shall be focused on in the TDX-Assist project: (i) scalability, the ability to deal with new (end) users and increasingly larger volumes of both information and data; (ii) security, protection against external threats and attacks, thus providing better mitigation upfront; and (iii) interoperability, information exchange and communication based on existing and emerging international Smart Grid ICT standards. One of the main pillars to achieve those goals is to use both SGAM and the IEC 62559 use case methodology.

Figure 11 depicts the used toolchain in the project and its relation to future international standardization activities driven by the partners. One particular focus of the project is to test methods engineering in the context of Smart Grids systems engineering. Therefore, different work packages in TDX-Assist try out different parts of a holistic toolchain, one work package takes into account the full taxonomy of use cases from IEC 62559 to distinguish among business use cases, high-level use cases and system use cases. The modeling is done directly with a plugin in Sparx Enterprise Architect, focusing on the aspect of model-driven development. Later, the use cases are exported according to the IEC 62559-3 XML format or into Microsoft Word documents for the deliverables of the project. On the other hand, the second work package start with a light-weight approach, starting with the word IEC 62559 basic template, refining a subset of use cases for the full template and then going into the repository and creating SGAM models there.

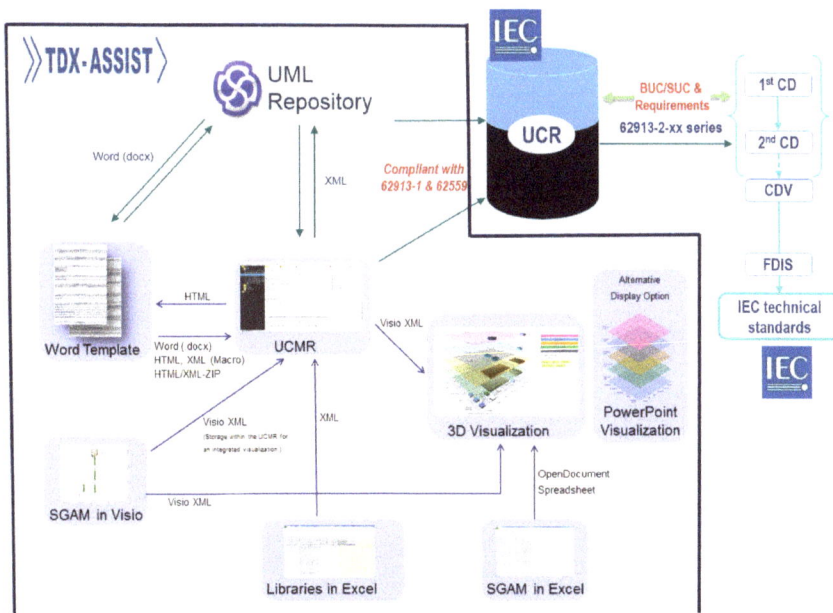

Figure 11. Toolchain used in H2020 TDX-Assist (source: EDF).

The results will focus on the applicability of the use case method for eliciting architectural requirements and bringing them into development and documentation. As the projects strictly cooperates with IEC TC 57, many results and use cases will be fed back into the community.

3.2.5. H2020 ERIGrid: European Research Infrastructure Supporting Smart Grid Systems Technology Development, Validation and Roll Out

The ERIGrid project is focused on improvement of tools and procedures for the testing and validation of novel Smart Grid technologies on system-level [46]. Parts of the project serve the research and practical demonstration of cutting-edge hardware and software validation platforms. This work is accompanied by the specification of a formal process that guides the development, documentation, and implementation of complex Smart Grid system-level test cases as outlined in Figure 12 [47,48]. This process is called the *Holistic Test Description* (HTD) process and spans across different stages of test planning. Users start out by outlining a holistic Test Case (TC) in a given template, providing an abstract and high-level view on the validation problem. In this stage, the purpose of the investigation and the system functions under test, among others, are specified. Following the TC, it is to be split up into several sub-tests to tackle the partial investigations contributing to the complex TC. Each sub-test is documented in a Test Specification (TS) template. After that, an experiment implementation is to be derived from each TS, specifying tools for the realization of the test endeavors. This implementation is documented in an Experiment Specification (ES) template.

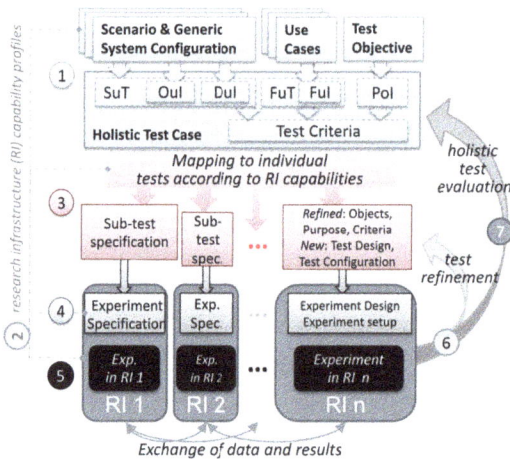

Figure 12. Overview of the ERIGrid holistic testing approach [47].

All of the specifications may be refined iteratively based on further insight gained along the HTD process. The benefits provided by the HTD include a common understanding of different validation concepts among all its practitioners. Furthermore, its separation of methods and implementation allow for tool-independent communication between research partners. This communication advantage is necessary when complex, interdisciplinary TCs need to be handled that require several sub-tests and involved domain experts.

The development of the HTD has been inspired by the SGAM and the IEC 62259 use case methodology. Nevertheless, a mapping of the TC system onto SGAM is currently not considered in the HTD process but might be subject for further investigations.

One major reason for this is the difference of scope between SGAM and the HTD. While the former seeks to limit the view on the Smart Grid to its most fundamental domain, the latter tries to especially involve new developments and technologies that may need validation. Therefore, the HTD requires a potentially broader view on the tested system in order to avoid limiting the documentation capabilities for its users. Instead, use cases representing the SGAM view on the system are considered as input information for the specification of TCs. Thus, a new workflow can be envisioned that starts with a SGAM model and use case based representation of a desired Smart Grid (technical) setup and has users derive TCs from it, following the HTD until the experiment implementation, resulting in the validation of all crucial parts of the system. Obviously, guideline material is needed to help users transition from the use cases to the HTD. Establishing such guidelines is a matter of future work.

3.2.6. H2020 TwinPV: Stimulating Scientific Excellence through Twinning in the Quest for Sustainable Energy (TwinPV)

The aim of the TwinPV project is to generate strong working R&D relations, build collaborative work sharing knowledge/resources and address gaps and shortcomings between different research institutions in Europe aiming to develop ways of softening their negative results and strengthening their collective impact as twinned institutions [49].

The core activities of this project are related to improving research and development in the domain of solar PV as well as grid integration and developing Smart Grid infrastructure and systems in Cyprus. In different training activities and summer schools, the integration of PV systems in distribution grids has been taught where one of the courses was related to teaching the SGAM approach together with the holistic testing approach of the "ERIGrid" project.

3.3. National-Funded Projects and Activities

3.3.1. Austrian ICT of the Future OpenNES: Open and Interoperable ICT Solution for Integration of Renewables

The main aim of the OpenNES project was the development of an interoperable ICT infrastructure for the integration of inverter-based DER devices [50]. For the collection of scenarios, uses cases, and test cases, a SGAM-based development approach has been applied. In a first step, various use cases have been collected and analyzed using the UCMR (see Section 3.1).

Based on the the the outcome of this phase, a flexible controller architecture using a virtual functional bus based approach—motivated from the AUTOSAR automotive controller environment—has been specified. For achieving interoperability on functional and communication layer according to SGAM, a strict decomposition of application and communication-related content has been carried out. For the development of control applications, the above-mentioned PSAL approach has been developed.

3.3.2. Austrian ICT of the Future MESSE: Model-Based Engineering and Validation Support for Cyber-Physical Energy Systems

When use case methodologies such as SGAM are used properly, the results are structured use case descriptions and diagrams. Furthermore, when used as intended, SGAM descriptions often contain a lot of information with different level of details. With the currently available tools (see Section 3.1), there is now also software support available that can help Smart Grid engineers during design and development. Nevertheless, even with these new tools current engineering approaches require a significant amount of avoidable manual work during the different engineering phases, such as implementation, validation, and deployment.

In the MESSE project, these issues are being addressed with the main focus to develop a concept for an automated, model-based engineering, and validation framework. Compared to the current tool support, MESSE especially provides support for validation and deployment and operation of Smart Grid applications. The main methodology of MESSE consists of three main parts [51]:

- *Specification and use case design*: For this phase, a formal specification and use case analysis method is defined. It is based on SGAM, IEC 62559, and PSAL. Various levels of detail can be addressed during the design. High-level use case descriptions as well as more detailed specifications of functionality, communication, and information models are possible. The information defined in this phase act as the main input for the automatic engineering and validation.

- *Automated engineering*: Based on the specifications and use case design, different types of configurations are being generated. In MESSE, approaches for three different domains are developed: executable code for field devices, ICT configurations and Human–Machine Interface (HMI) configurations. HMI configurations are used to define the layout of visualizations as well as to configure how user actions should be interpreted and executed.
- *Automated validation and deployment*: Automated testing for software development has been common practice for several years. However, similar approaches for Smart Grid systems are currently missing. In MESSE, a methodology for the automatic testing of Smart Grid systems is being developed. Based on the scenarios and specifications from the engineer, appropriate tests are generated. Apart from pure software testing, tests can be a combination of software, hardware, and simulations. For manual hardware setups, guidelines for the user are generated.

The concept developed in the MESSE project is applicable to architecture and system development for many different Smart Grid applications. Starting with a design and specification phase, where the SGAM modeling approach is the main foundation, the model-based engineering concept fosters a formalized and systematic comparison of different development options. Furthermore, based on a set of scenarios and test specifications, the test and validation framework generates test cases for each development option under investigation. Using this approach, many steps can be automated that are traditionally carried out manually [51].

3.3.3. German SINTEG Project Enera: The Next Big Step in the Energy Transition

The project enera demonstrates how the infrastructure of the energy system can be innovated in such a way that, despite the new requirements and the variety of technologies used at the same time, it shows a high degree of resilience. In addition, enera demonstrates how markets and digitization can significantly reduce network expansion costs and create opportunities for innovative business models. This opens up new avenues and opportunities for political control of the progress of the energy transition in Germany.

As one of the so-called SINTEG regions in Germany, the project is also recommended to use IEC 62559 as well as SGAM for documentation purposes of their work. In enera, one focus was set on the application of the use case method to create blue-prints for Smart Grid solutions in large scale. The first results from nearly two years of elicitation of use cases, information objects and creating SGAM models can be found in [52].

3.4. Further Projects, Activities, and Applications

Many other examples of SGAM application can be found in the literature. In most cases, the SGAM is used to describe newly designed architectures or to analyze existing architectures. One area in which SGAM is frequently used is risk management or risk identification and analysis. There the SGAM is used to develop models representing both current and near-future European Smart Grid architectures [53], to establish a national Smart Grid ICT reference model that provides the starting point for a Smart Grid cyber security risk assessment [54], to identify the target of evaluation in a Smart Grid for a risk analysis [55], and to define use cases for identifying data privacy issues [56].

In addition to assessing risks, the SGAM is also used to evaluate and analyze architectures and solutions for other criteria. In [57], it is argued that the SGAM does not provide a way to perform economic analysis. The authors provided a SGAM-based method to assess the economic feasibility of new commercial services. These services include, e.g., demand side management and trade of electricity. The developed method does so by computing standard decision investment techniques such as Net Present Value (NPV) and Internal Rate of Return (IRR). Beyond cost assessment, in [58], an approach is presented that aims to determine the strategic value of Smart Grid projects in terms of their importance and effectiveness. Based on a method in the field of design science, a framework is developed that consists of the following three components: the SGAM as the reference model, the adapted Bedell's method as the assessment method and a DSS (Decision Support System) to perform assessments. The framework, helping companies to tackle challenges other than economic issues such as energy efficiency and CO_2 emissions, is evaluated within a blockchain-inspired project [59].

A scoring scheme for interoperability assessment within Advanced Metering Infrastructure (AMI) for Demand Side Management (DSM) is presented in [60]. It aims to quantify the capacity of components of interest to interoperate with each other. Here, the SGAM is used to map components, communication protocols and information models. Furthermore, a TVT (Test Verdict Tracing) scheme is used for determining the layer where lack of interoperability takes place.

Further examples can be found where the SGAM has been used to describe use cases. The following list gives a first overview of the manifold application possibilities:

- Requirements analysis for *Virtual Power Plants* (VPP) and their mapping onto standards as IEC 61850 and IEC 61970/61968 [61];
- Identification of involved actors, equipment, communications and processes for *Electric Vehicles (EV)* charging control [62];
- *ICT planning* approach that can be used in combination with distribution network planning processes and tools [63];
- Development of a *railway energy management system* by using the SGAM model and methods [64];
- Design of an architecture of a distribution grid automation system focusing on PMU-based monitoring functions accommodating for key dynamic *information exchange between TSOs and DSOs* [65]; and
- SGAM-based explanation of Smart Grids in order to present *Big Data* analytics [66].

Since the development of SGAM, it has been increasingly used by various research and development projects. Some of them have already been described in detail here. In the following, further important projects are presented briefly. The SmarterEMC2 project aims at ICT integration with power systems for enhancing various Smart Grids services. The main objective of the project is to propose business models and to develop the necessary ICT tools to support Customer Side Participation (CSP), increase Renewable Energy Sources (RES) penetration, and foster the participation in the electricity market [67]. Virtual Power Plant and Demand Response cases are presented, proposing a general architecture, as well as analyzing core functionalities, information and communication requirements, along with relative standards and technologies. The SGAM was used for the methodological perspective and it turned out to be extremely helpful by providing a more systemic view of the applications and exposing a number of interoperability and operational issues that would otherwise endanger the robustness of the ICT tools to be implemented [67].

In [68], some results from the FINSENY (Future Internet for Smart Energy) project are presented. FINSENY is one of eight usage area projects within the FI-WARE project and aims to define Smart Energy Systems using the generic enablers developed by FI-WARE. The focus is on combining adaptive intelligence with reliability and cost-efficiency to sustainably meet the demands of a highly dynamic energy landscape. Therefore, a methodology has been developed that allows investigating use cases, ICT requirements and a functional Future Internet architecture for distribution systems. To derive both the functional and ICT requirements on the Information and Communication layer and to identify data models and interfaces, the use cases are analyzed along an adaptation of the SGAM.

The integration of novel Smart Grid solutions and services to enable energy flexibility markets, with enhanced demand response schemes and active prosumer participation is the overall goal of the Nobel Grid project [69]. For developing an according architecture, the SGAM Toolbox has been used as introduced in Section 3.1.1. The authors stated that the traceability across the different interoperability layers, enabling the choice of system elements to be rationalized in a top-down and bottom-up manner, is a major benefit of SGAM.

In [70], results of creating an Austrian Reference Architecture for Smart Grids as part of the RASSA project (Reference Architecture for Secure Smart Grids in Austria) are presented. The main contribution of the project is the creation of the reference architecture on an available modeling framework plugin while keeping all stakeholders on board of the process to form a common understanding of its growing importance in the future. Again, the SGAM Toolbox has been used to model the projects architectures, whereas the focus throughout the project is security.

EMPOWER (Local Energy Retail Markets for Prosumer Smart Grid Power Services) is a project with a special focus on the local energy markets development [71]. The goal of the project is to create a trading platform for local energy exchange in local markets. Therefore, a SGAM-based system architecture is developed describing an ICT platform to manage the flexibility of generation, load and storage units at distribution level. To take advantage of these flexibilities, innovative business models are being proposed setting the operational rules of the local markets operation. The architecture has then been analyzed in terms of resilience by combining SGAM with a data structured diagram, the Entity Relationship Model (ERM) [72]. The architecture of smart distribution grids is analyzed through SGAM. Then, their technical characteristics and functionalities are defined and represented in a ERM diagram. Finally, the attributes or properties of the system components are used to formulate resilience indicators against different types of disturbances.

Finally, it can be observed that the SGAM is now also applied beyond Europe. Examples include application development for device management and control as well as system state monitoring for residential demand response ancillary services based on graph database modeling and high-availability web services in the USA [73], the development of an effective but efficient approach to risk assessments for Smart Grid projects in Australia [55] and modeling an off-grid rural village microgrid as a multi-agent nodal system and therefore formulating distributed market-based transactive control as a discrete-time system in South Africa/the USA [74].

4. Transfer to Other Domains

In addition to being used in the power and energy domain, the model has been taken over by other disciplines as one way to document reference architectures as well. In the following, the adoption of SGAM to other domains is discussed.

4.1. Industrial Automation

The Reference Architecture Model for Industry 4.0 RAMI (RAMI 4.0) is probably the most sophisticated derivative of the SGAM as of today, originally developed by ZVEI in Germany and taken up by standardization as IEC 63088—a status the original SGAM has not yet been achieved. Based on the German Industrie 4.0 concept, the main aspect for the RAMI is the re-use of the existing GWAC interoperability stack. In addition to business, function, information, communication and asset representing component, a completely new layer called integration is introduced [75,76]. The domain and zone axis are not custom taxonomies but are based on the known IEC 62890 value stream chain or the IEC 62264/61512 hierarchical levels for automation, respectively, as outlined in Figure 13.

The main purpose of the RAMI model is defined by IEC respective ZVEI and its stakeholders as follows: "The model shall make for the harmonization of different user perspectives on the scope and provide a common understanding of the relations and attributes between individual components for Industrie 4.0 solutions". Different (industrial) branches such as automation, engineering and process engineering have a common view on the overall systems engineering and life-cycle landscape. The SGAM principle of having the main scope of locating and assessing standards is re-used in the RAMI paradigm, it is also using a reference designation system. The next steps for proceeding with a holistic modeling paradigm is to come up with basic examples for Industrie 4.0 solutions in the RAMI (similar to system use cases), providing proper means for the devices and components to be identified and allow for a discovery service modeling those devices, harmonizing both syntax and semantics of the data and focus on the main aspect of the new integration layer which was introduced in order to model the communication requirements in factory automation for the administration shell concept.

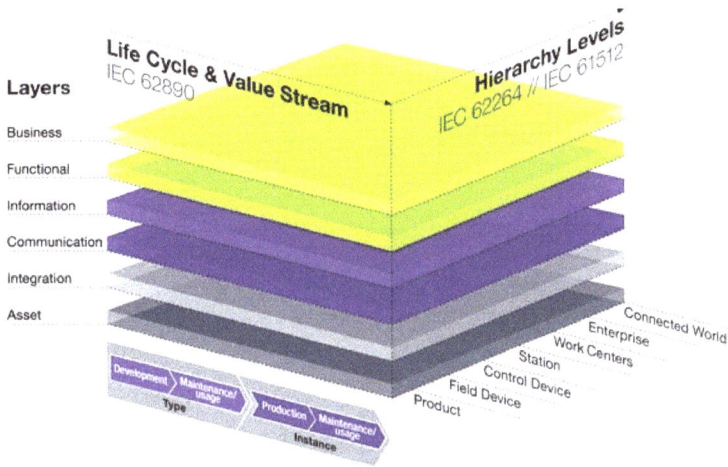

Figure 13. RAMI 4.0 (Reference Architecture Model I4.0) by ZVEI [75].

As in the original SGAM, a 3D visualization has been created on the very same technical basis as the SGAM browser manipulator [77]. The MBSE approach can be taken as developed in the SGAM [78]; the most crucial part is the modeling of the so-called administration shell on the integration layer, which mixes (from the formal perspective) some parts originally separated in the SGAM model viewpoints. Figure 14 provides an overview on how the different methods and tools can be applied in either the Smart Grid or Industrial Automation domain.

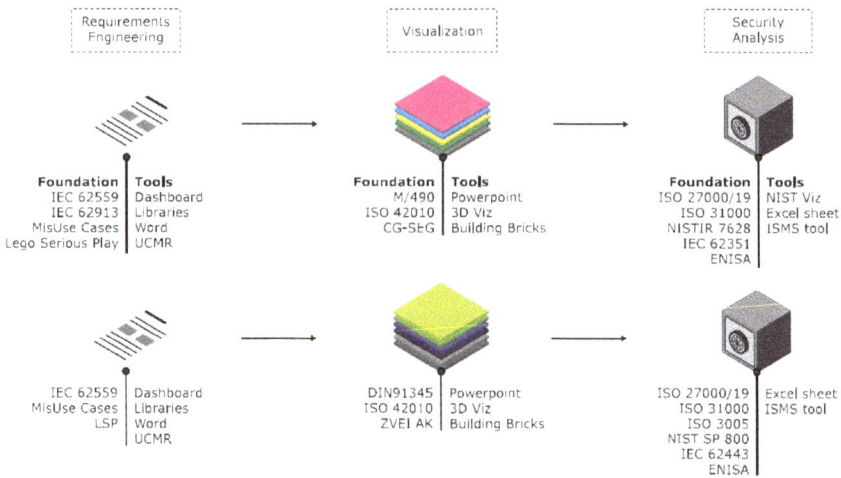

Figure 14. RAMI 4.0 vs. SGAM tooling [79].

4.2. Legislation

As a co-evolutionary method and tool to the aforementioned RAMI 4.0 model from the previous section, the so-called legal reference architecture model 4.0 (ju-RAMI 4.0) has been developed in the very context of the national German AUTONOMIK Industrie 4.0 funding scheme [10]. One of the main aspects of having more and more complex systems interacting with each other is the separation of the individual organizations participating (as actors and stakeholders) and operators of system-of-system parts in operating mode. Legal aspects come into play in terms of aspects such as liability for parts including components and products and copyright for processes or solutions deployed.

The ju-RAMI 4.0 model aims at providing a simple access to both terms and wording used in the legal domain to lower an entry barrier for technical aspects and implementations to take into account legal risks and challenges at the development time for a given new solution. One particular aspect is the visual representation of various dimensions of legal requirements in order to structure those aspects during the lifetime and cycle of the product. Different, so-called legal domains (privacy, intellectual property, liability, etc.) are addressed. While the model itself cannot address all legal issues from the jurisdictional point of view, it provides a useful visualization of key terms for starting the discussion on legal aspects of the product development phase and viewpoint as well as the inherent attributes of the so-called intelligent product in Industrie 4.0.

As the domain of law is not a natural science but more or less interpreted, provided solutions in the ju-RAMI can only hint to needed aspects to be discussed with legal departments later on. The authors aimed to provide compliance barriers by defining risks and liability involved. This can lead to a better understanding of legal aspects of industrial solutions covered by RAMI 4.0 models [10].

The axes of the ju-RAMI 4.0 are briefly defined as follows. The vertical axis covers the defined and needed legal domains (e.g., intellectual property rights, data protection law, workers protection law, civil rights, etc.); the horizontal axis entails the actors participating in the development based on the original RAMI 4.0 model and clusters them in four areas of actors groupings; and the third axis covers the risks involved when a certain legal requirement is not met or taken into account [10].

4.3. Automotive Domain

The Reference Architecture Model Automotive (RAMA) is a graphical representation to model most vehicular information and communication technology of future (connected) vehicles, as outlined in Figure 15. It models aspects of the integration of vehicles in their (contextual) environment as well as the complete vehicular life-cycle from the beginning of the development of the vehicle until the phase of scrapping. The RAMA model is an adaptation of the existing RAMI 4.0 for the automotive domain. It was designed to model the aspects of use cases, functional behavior of systems, data created, data flows and physical components in a common graphical notation scope. Thus, RAMA is primary constructed to model technical behavior, but automotive business issues can be depicted as well.

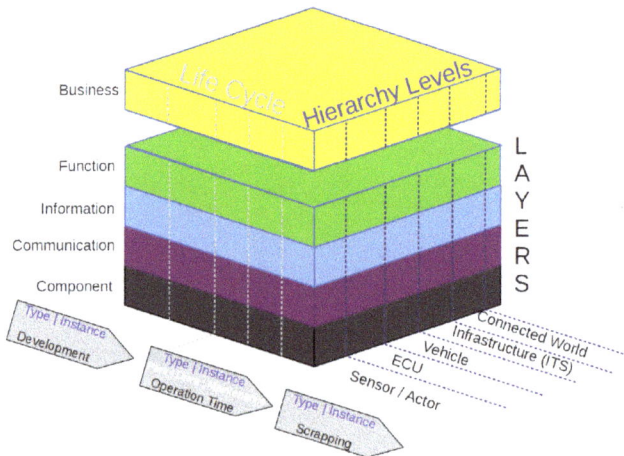

Figure 15. RAMA (Reference Architecture Model Automotive) [80].

The model was designed within the Security Sub-Working Group "Connected and Automatic Driving" of the German Governmental Department of Transport and Infrastructure (BMVI) in 2016. it is intended to be applied for a security threat analysis concerning automated and connected vehicles in particular [80].

4.4. Maritime Domain

The Maritime Architecture Framework (MAF) describes another new derivation from the original SGAM model. It is created for the maritime sector [81] in particular. The MAF has been discussed and developed during the project *EfficienSea2* as an early architecture framework to depict information exchanges between various maritime actors and services in a so-called Maritime Cloud environment.

The Maritime Cloud concept is a framework that provides both standardized protocols and functions for the aspects of identity and role management, authentication, encryption, service discovery, and bandwidth efficient messaging in a spatial context for the maritime domain. Therefore, the information exchanges between the Maritime Cloud environment and additional actors from the domain has to be created interoperable. Actors typically are various software systems on-board a ship, offshore as well as onshore, and personal devices for personnel, such as smart phones and tablets.

The construction of the interoperability layer has been done similar to the SGAM—apart from the Regulation and Governance Layer, which replaces the Business Layer. For the domains and zones, new dimensions are developed that match the maritime sector. The defined domains are based on the International Maritime Organization (IMO) e-navigation strategy and divide the architecture into so-called ship-side and shore-side view [82,83]:

- Ships and other maritime traffic objects are actors that are at sea; they can be vessels, or cargo or passenger ships.
- The link describes the existing connection between actors from the ship-side to the shore side with telecommunication methods and protocols. This additionally includes actors such as radio towers and transmission masts.
- Actors on the shore are sea ports, docks, halls, and third-parties where ships land or which organize the shiploads.

Similar to the other architecture models based on SGAM, the zones describe both the hierarchy and aggregation of management and control systems [82,83]. The defined zones are as of now (see Figure 16):

- All components and systems which can execute a physical action are depicted in the Transport Objects zone, e.g., ship, crane, port, and transmission masts.
- The Sensors and Actuators zone includes all the components that are needed for receiving or sending data, such as antenna, transceiver, ISO 11898, etc.
- Single services are shown in the Technical Services zone, e.g., IEC 61162 and NMEA (National Marine Electronics Association) 2000.
- Actors, information objects and protocols for operating and control services are displayed in the Systems zone, e.g., the Vessel Traffic Service (VTS).
- In the zone Operations, the operating and control units from global, regional, national or local perspective are depicted, e.g., the VTS center.
- In the Fields of Activity zone, systems are described which support markets and eco-systems along the maritime domain, e.g., the traffic message broadcast.

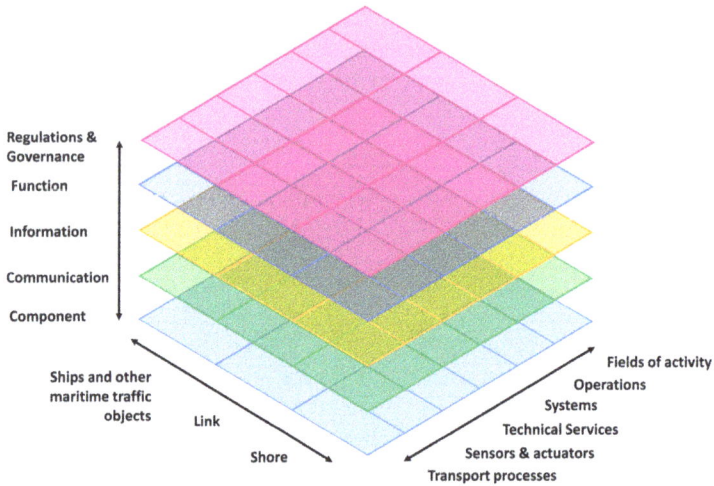

Figure 16. MAF (Maritime Architecture Framework) [83].

4.5. Smart Cities

Motivated by the good experiences made with MDE in the context of Smart Grid systems and SGAM, Neureiter et al. [84] contributed a concept on how to adopt the SGAM and the Smart Grid Model Driven Engineering approach for the development of Smart City systems. Therefore, the underlying new DSL is extended on basis of a proposal for a so-called Generic Smart City Architecture Model (GSCAM), as shown in Figure 17.

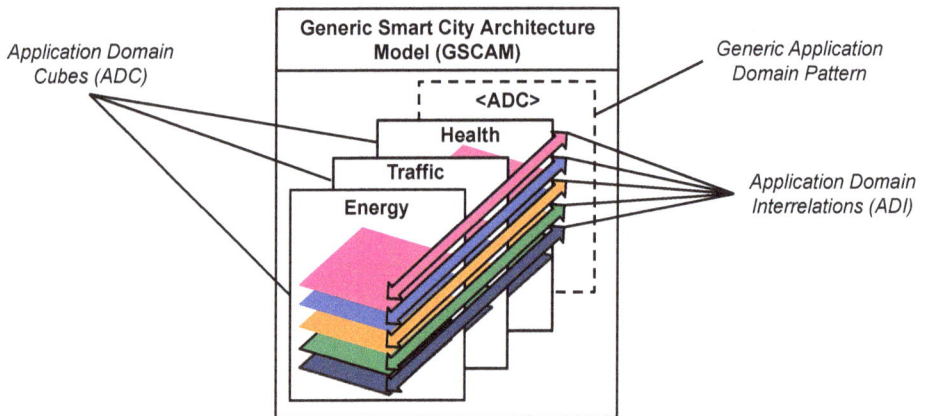

Figure 17. Overview of the Generic Smart Cities Architecture Model (GSCAM).

The contribution of Neureiter et al. [84] describes the existing approaches for the utilization of MDE in the Smart Grid as basis for a DSL. The proposed concept for extending the SGAM to the domain of Smart Cities is discussed in the context of its applicability while extending the original SGAM. The approach in [84] utilizes the SGAM concepts for the development of a GSCAM as holistic basis for Smart Cities.

The GSCAM also addresses one particular aspect of the SGAM: having more than one utility or domain interacting. Typically, the focus of SGAM is to model a DSO and/or TSO interacting with its environment using the market dimension for communication with third parties. It is difficult to model more than one DSO in one SGAM cube because of the complex graphical notation. Hence, this concept could also contribute to this modeling challenge.

4.6. Further Adoptions

In addition to the domains already mentioned, the SGAM also forms the basis for new reference architectures in other areas. The following approaches can be found in the literature. In [85,86], the coupling of Building Energy Management Systems (BEMS) with the Smart Grid (SG) is addressed. It is argued that this requires a framework that takes into account the critical control blocks of both sides. To meet this requirement, the SG-BEMS was developed as a reference architecture, which is characterized in particular by the extension of a building consumer domain and a reduction of the other domains to the distribution grid. An agent-based solution is also presented to enable optimal integrated operation of the distribution grid and the BEMS. Finally, a computational intelligence technique, i.e., Particle Swarm Optimization (PSO), is used to maximize both comfort and energy efficiency for the consumer.

Based on the SG-BEMS, Shafiullah et al. [87] introduced the Smart Grid Neighborhood Energy Management System (SG-NEMS) Framework. The focus shifts from the building to the neighborhood. It is stated that SG-NEMS aims to improve the following points: SG-BEMS only expands the control standard for component layer and function layers and it does not specify the information and communication layers; SG-BEMS only considers the prosumers as a member of a single building; and SG-BEMS does not specifically describe the interoperability among the buildings within the neighbourhood. Because of this and the fact that the related approaches were developed independently of each other, it is argued that it is necessary to have a modular framework for integrating the existing approaches. Compared to the SG-BEMS, the SG-NEMS redefines the prosumer domain as neighborhood domain.

Another adaptation of the SGAM can be found in the field of electric mobility. In [88], the E-Mobility Information System Architecture (EM-ISA) is presented. The authors gave two main reasons the SGAM has to be adjusted. On the one hand, SGAM is not designed for describing informational interactions between devices and human and, on the other hand, the level of detail is too high, e.g., in the zones, and leads to unnecessary complexity. In addition, the following requirements are set for the development of the EM-ISA: domain specificity, supporting inter-organizational information flow, applicable in many countries, configurable for organizations, and support user-interaction. As a result, the EM-ISA has a similar layer structure to the SGAM and simplified zones (here called "scope") by omitting field and process. The main difference lies in the domains, of which there are only two in the EM-ISA: immobile (infrastructure) and mobile (user of infrastructure).

In [89], a cloud-based IoT platform based on REST-APIs is presented. Basically, the concept and functional blocks behind the idea of a virtualization platform with RESTful interfacing is introduced. The link between the SGAM and the virtualization platform is given by the layer-oriented structure.

A four-layer architecture model of Peer-to-Peer (P2P) energy trading (people can generate their own energy from Renewable Energy Sources and share it with each other locally) is designed in [90]. Following the three-dimensional SGAM approach, here the first dimension is the time-scale of P2P energy trading (bidding, exchanging, and settlement). The second dimension shows the size of the P2P energy trading users: single premises, microgrids, cells and regions. In the third dimension, the hierarchical process of P2P energy trading is categorized into four interoperability layers for management which comprise business layer, control layer, ICT layer and power grid layer.

In addition to the GSCAM already discussed in Section 4.5, there are two other approaches to Smart Cities. On the one hand, the Smart City Infrastructure Architecture Model (SCIAM) is introduced in [91]. In comparison to the SGAM, the interoperability layers are taken over, whereby it is discussed whether the business layer can be replaced by an action layer. Furthermore, the zones from the SGAM are adapted without modification. However, the domains of the SGAM have been completely redefined, so that the following domains can be found on the new axis: Supply/Waste Management, Water/Waste Water, Mobility Transport, Healthcare/AAL, Civil Security, Energy, Building and Industry.

Encouraged among other things by the GSCAM and the SCIAM, another model for Smart Cities is presented in [92], which also consists of three axes. It is stated that a principle gap in the existing approaches is the consensus on a common language/taxonomy and a set of Smart City architectural principles. The focus of this model is on the newly defined interoperability layers: Functional (key concepts, component, functionalities), Collaboration (configuration of interoperable communications), Semantic (semantic of the common language), Information (syntax of the common language) and Communication (data exchange interfaces). In the case of the zones, the ICT is addressed so that the areas Sources, Local Solution and Smart City Platform result. The domains are application-oriented and include, e.g., Building, Street, Home, Lighting, Community and Mobility.

The last derivative to be named was developed as part of a methodology that enables DSOs to implement advanced technical solutions which improve their market responsiveness, process adaptability and meet their business needs [93]. This methodology extends the SGAM in the area of transactive energy and Multi-Agent Systems (MAS). The interoperability layers remain unchanged. However, the zones and domains are fundamentally changed. Since the approach only applies to DSOs, the domains are replaced by the objectives of the agents. Specifically, these are demand elasticity, demand flexibility schedule and demand flexibility dispatch. To achieve these objectives, the following strategies—which replace the zones—can be chosen: improved responsiveness, process adaptability, and customer enabling.

5. Future Perspectives

As previously motivated, the architectural framework SGAM allows studying the system across multiple interoperability layers spanning across several domains (e.g., electrical power systems and communication infrastructures). However, the use of this reference architecture is not sufficient when it is important to determine the dynamic behavior of the system under study and the impact of the interaction of the actors involved in a particular use case. With this in mind, further work is required regarding integration of multi-domain tools and software as well as the improvement and development of the current framework. In the following, there is a overview of the most prominent challenges to overcome to strengthen the current architectural model.

5.1. Supporting Tools and Software

There are several efforts to improve and extend the functionalities of the SGAM framework. For this reason, different supporting tools and software have been developed. They are divided into two groups: the first set consist of approaches dedicated to improving the usability and capabilities of SGAM, and the second set seeks to increase the interoperability of the framework in order to increase its functionality and to extend the analysis of Smart Grid use cases from a static overview of the system to the study of the dynamic behavior of the electrical power system components and its associated ICT infrastructure.

As mentioned above, there are at least two developments on the usability of SGAM. The first one is the so-called SGAM Toolbox. In here, this artifact offers tool support, which not only allows the modeling of architectures but also its analysis. Another important aspect is the possibility to include several analysis tools that are commonly employed in the study of use cases from the perspective of Systems Engineering and Model Driven Engineering. Moreover, the authors of [94] motivated an approach to handle the analysis of complex distributed systems as a part of a Model-Driven-Architecture.

Another method used for improving Smart Grid solutions is co-simulation. The aim here is to integrate a heterogeneous setup of simulators with each other in order to use their respective strengths [95]. Thus, it is possible to analyze complex impact relationships. One possible implementation of this concept is the mosaik framework [96]. An essential part of the co-simulation is the design of the scenarios. These specify how the architecture is defined for the simulation, i.e., it is determined which grid component is represented by which simulator and how the links of the simulators look to each other. Thus, an environment is created to validate and verify modeled architectures.

Important to realize is that, although the aforementioned integration of SGAM and simulation tools is useful for an architectural and systematic assessment of complex systems, there are still several challenges to overcome in order to have a more grounded approach that enables to address pressing research questions around the interaction of components within electrical power systems. In the following, three challenges are identified and discussed in order to close the integration gap previously mentioned.

The first point is related to functional modeling. This means here that is important to have a clear and formal functional description (which includes the actors and its interaction) as well as the definition of scenarios and network events that are helpful to have a complete road-map (or preliminary analysis) in order to perform the simulations in a systematic manner and with the appropriate context.

Secondly, a well-defined tool interfacing is required in order to fully utilize the architectural structure that the SGAM framework can provide when analyzing the interaction of components across different domains, zones and interoperability layers. Correspondingly, the main concern is to shift from the requirement engineering analysis (e.g., use case management repository) to the architectural model in SGAM and finally to the simulated environment in a seamless and meaningful manner.

The final challenge is the automatic configuration of the simulation setup. Once the requirement engineering and the architecture development analysis are completed, it is possible to prepare a list of required simulators as well as the definition of a number of components to be instantiated considering a number of operational scenarios and possible network events according to the predefined use case under study. This would result in combining the above-mentioned Systems Engineering and Model Driven Engineering approach and co-simulation framework that are able to expand the already defined SGAM capabilities.

5.2. Design and Engineering

One particular aspect that is currently in the focus is the elicitation of individual information needed to create SGAM models (and models similar to SGAM). Originally, the meta-model was designed to work alongside an IEC 62559 compliant use case template, thus making sure that all relevant information for a static architecture model is gathered. As of today, for example the MAF has used a different model (RAMS) and certain partners create SGAM models from scratch based on their TOGAF or Archimate Models. Processes start at different gates in order to create the models, therefore, it is not easy to assess which information can later be used for engineering or code generation purposes. In particular, it is obvious that SysML, TOGAF ADM, reqIF, Archimate, FMI, STIX 2.0, and TAXII as well as other formats and tools shall be harmonized and evaluated for meta-model overlaps and mappings with the SGAM in order to foster the future use of the useful architecture models in the context of those methods. This has already been done for the part of security engineering and risk assessment where the original toolchain was changed and extended with uses like misuse cases [97].

The current project Integrating the Energy System (IES) Austria recognizes interoperability as key enabler for the deployment of so-called smart energy systems. Interoperability is covered by the Strategic Energy Technology Plan (SET-Plan) activity A4-IA0-5 and provides an added-value because it enables new and future business options for most stakeholders in the domain. The communication of smart energy components and systems shall be interoperable to enable smooth data exchange, and thereby, the on-demand integration of heterogeneous systems, components and services. The approach developed and proposed by IES, adopts the holistic methodology from the consortium Integrating the Healthcare Enterprise (IHE), established by information technology (IT) vendors in the health sector and standardized in the draft technical report ISO DTR 28380-1, to foster interoperable smart energy systems. The paper [98] outlines the adopted IES workflow in detail and reports on lesson-learned when trial Integration Profiles based on IEC 61850 were tested at the first Connectathon Energy instalment, organized in conjunction with the IHE Connectathon Europe 2018 in The Hague, Netherlands. The IES methodology is found perfectly applicable for smart energy systems and successfully enables peer-to-peer interoperability testing among vendors [99] based on use cases, profiles and SGAM models.

5.3. Validation and Testing

A common challenge in simulation-based testing of Smart Grid systems is the selection of the most appropriate simulation tool for a given task. Due to the interdisciplinary character of the domain, researchers often seek to reuse existing and established tools instead of developing new ones from scratch. This will at times leave experts with the choice between several models. The usual approach in such a situations is to go with the solution that is as simple as possible yet as complex as necessary (*Occam's razor*). However, this concept becomes insufficient when the simulated system is so complex that co-simulation needs to be employed. In this context, a number of simulation tools have to be selected that on the one hand serve their purpose in the overall setup and on the other hand are able to interact with one another on a syntactical a semantical level. In other words, researchers need to be able to assess whether a simulation tool can be employed in a co-simulation and, if yes, in which context. While standards such as FMI [100,101] provide a first approach at facing this challenge, many problems in simulator interoperability are yet to be solved, e.g., model interaction in terms of temporal resolution, accuracy or synchronization needs.

An important future task will thus be to extend or harmonize the SGAM with a sense of interoperability in the context of simulation models. Such an approach must involve classifications or metrics for model validity and accuracy as well as concepts for the semantical interaction of component models. Furthermore, synchronization capabilities of co-simulation frameworks (e.g., mosaik) have to be comparable with synchronization needs of simulators as stated in [102].

Another important challenge is the modeling and testing of cyber-physical energy systems. The ultimate goal is the validation of the use case and architectural analysis via the use of laboratory experiments that involves the use of among others intelligent and coordinated devices as well as real components commonly used in emerging Smart Grid scenarios. The authors of [103] presented a holistic testing methodology that includes research infrastructures (e.g., experiment design and set-up) as a comprehensive analysis of modern electrical power systems and its associated ICT components. Furthermore, the authors of [65] presented a detailed description of the interaction of several components across the different interoperability layers in order to accommodate for key dynamic information exchange between DSO and TSO. However, although the current architecture is able to take into account the information about the components involved in the experiment and its communication infrastructure, a meaningful description that is able to automatically configure the hardware under study and its interconnections is still not fully developed. Further work is required in the transition from the SGAM-based analysis into the hardware-in-the-loop or pure hardware testing applications.

5.4. Wide Usage in R&D Projects

The European Technology and Innovation Platform (ETIP) European Technology and Innovation Platform Smart Networks for Energy Transition (SNET) in supporting R&D work and evolution of Smart Grid infrastructures and systems [104]. It is within the core activities of the corresponding Working Group (WG) 5 "Innovation implementation in the business environment" to find ways of reporting, analyzing, and supporting R&D projects. To this effect, work is ongoing to build the following platform as a multi-functional collaborative platform to fulfil the targeted objectives as outlined in Figure 18.

As can be seen in Figure 18, the build-up architecture is SGAM-based and is working to adapt best practices developed in European and national-funded projects as introduced and discussed above.

Typical requirements towards such a platform are:

- Standard-based formats to design and exchange use cases, SGAM models, and libraries of terms and data;
- Enhanced templates with standards-based XML export functionality to export use cases, SGAM models and libraries following standard-based IEC 62559 compliant formats;

- Web-based repository to store and manage elicited use cases, SGAM models, and libraries, managing access rights and enabling multi-editing of the defined descriptions;
- Automated engineering to the highest degree possible that will be continuously enhanced; and
- Automated validation and deployment through collaborative simulation work possibilities.

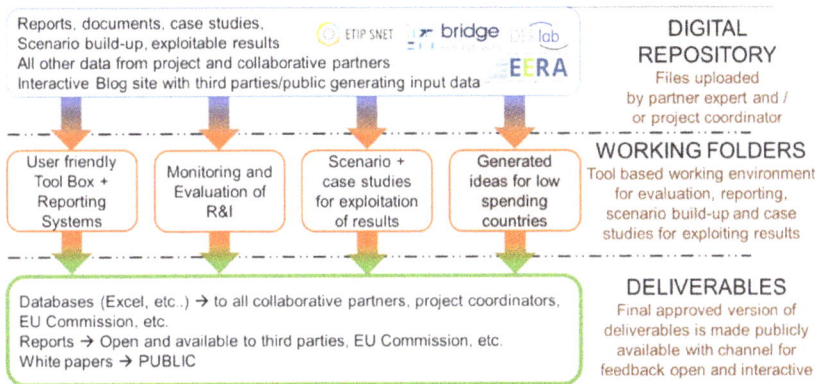

Figure 18. Overview of the ETIP SNET multi-functional collaborative platform [104].

Developed solutions (e.g., [104]) will be fundamental in the day-to-day operation of the platform. The above will be open to all and it will be particularly useful for Learner's Use Cases, less experienced researchers that are keen to build on current best practices using real data and means for simulating and testing. In addition to this, given that the tools rely on international standards from IEC and architectural frameworks, they can be leveraged in other projects with the aim of facilitating sharing of Smart Grid requirements and architectures. This is particularly useful in the context of larger Smart Network projects for the future, in which partners from different areas of expertise and different countries need to exchange information on Smart Network solutions and portfolios with each other.

As was already postulated in the DISCERN project, the tools to be provided are complete versions, including required functionalities for exchanging and managing Smart Grid/Networks requirements and architectures in a collaborative environment. The intention is to build this platform for the possibility of being a one-stop facility that links active researchers with all existing knowledge in the field, collaborative test facilities, online tools, etc. that can be supporting services to their targeted work. All these are to be SGAM-based to offer the required interoperability and functionality. Progress in the field will be updated and, hence, offer long-term sustainability. Attempts will be made to link the platform to solutions/facilities such as the HTD of the ERIGrid project aiming to offer testing possibilities very early in the development of project concepts and ideas to support the R&D process of all Europeans including the low-spending countries in R&D.

6. Discussion, Lessons Learned, and Conclusions

This review shows that the SGAM currently has already a wide spread use and has outgrown its original purpose for documenting gaps in standardization. Various projects provided experiences to using SGAM in the context of architecture management, requirements engineering and security and risk analysis. Even though work and projects are progressing, constantly new uses and application areas can be found for the systems engineering part of using the SGAM and derivatives.

The concept of SGAM has already been disseminated widely as a well-known concept, but there is still no official definition of a standardized document by IEC SyC Smart Energy. Even though different domains have taken up the concept, such as industrial automation, only SGAM derivatives are official standards as of now. As of recently, interactions between Smart Grid systems and heat/gas systems are becoming increasingly necessary. Therefore, an official definition of SGAM and its expansion to potentially include heat/gas systems is required. An official definition of SGAM should be provided by SyC Smart Energy as a SRD (System Reference Document), associated with a formal ontology provided in a textual format as well as with code components. The work will be conducted so that generic elements that could be used by different domains are separated from the specific application of the generic elements to the smart energy grid domain.In addition, methods other than elicitation from use cases have to be tried, e.g., based on Lego Serious Play [105]. One particular feedback is that using SGAM is easier for stakeholders who are already into systems engineering thinking. Most stakeholders then only need to transfer the new glossary and vocabulary into their known everyday routines and can benefit from the overall method. However, tooling better than Powerpoint figures was needed for collaboration.

Within this review, we have outlined the current use as well as the basics of the so-called Smart Grid Architecture Model SGAM and its corresponding methodology. Despite having been developed with the narrow purpose of using it for reference designation of technical standards in the scope of the EC M/490 mandate, the SGAM has outgrown its original purpose and has gained more and more attention. Different tools have been developed in order to elicit requirements needed to create SGAM models from a canonical and standardized use case process as well as plug-ins to state-of-the-art UML tooling. The SGAM has evolved into a kind of systems engineering approach for the scope of Smart Grid infrastructures. Projects have successfully used the approach and gained knowledge on refining and extending the method for additional benefits and gains. The successful application did not go unnoticed, and different domains have taken over and refined the approach for architectural reference designation of their individual domains. Different approaches have been even more successful in terms of standardization than the original SGAM model. Taking into account standards from different domains that have proven to be useful because they are based on system engineering principles has also proven very useful, e.g., from logistics to avionics [106,107]. Established practices are tailored and profiled, e.g., according to the IES Austria Process.

This review briefly presents the most important ones. In addition, the broader use of the SGAM is covered by a literature research on state-of-the-art projects that are presented in research publications. Finally, future perspectives are discussed and presented.

Author Contributions: M.U., S.R. and T.I.S. developed the main structure of this review article. M.U. contributed to the Abstract, Introduction, the SGAM overview and its usage in DISCERN, ELECTRA IRP, TDX-Assist, IES Austria and enera. He also contributed to all sections on the transfer to other domains with RAMI, etc. He created revision two and the camera-ready version of this paper. S.R. provided inputs and analysis to the Further Projects and State-of-the-Art, The Future Perspectives and the Conclusions. C.N. mainly elaborated the design, implementation and application of the SGAM Toolbox and he contributed to the description of the structure of SGAM and its relation to Systems Engineering. F.P.A provided the description of PSAL and contributed to the applications in OpenNES, MESSE, and SmartNet. J.V. contributed to the Future Perspectives Section, specifically to the Supporting Tools and Software and Validation and Testing Subsections. C.S. contributed to the description of related work in the context of ERIGrid as well as to parts of the validation and testing outlook for future perspectives. V.E. contributed to the TwinPV description as well as the wide usage in R&D projects. G.M. and S.H. contributed to the SmartNet description. H.B. provided inputs to ELECTRA IRP and TwinPV. T.I.S. contributed to the Abstract and provided inputs to the projects ELECTRA IRP, ERIGrid, OpenNES and MESSE. He also contributed to future perspectives. Finally, all authors contributed by proofreading this paper.

Funding: Parts of this work received funding in the European Community's Horizon 2020 Program (H2020/2014-2020) under projects "TDX-Assist" (Grant Agreement No. 774500), "ERIGrid" (Grant Agreement No. 654113), "SmartNet" (Grant Agreement No. 691405), "TwinPV" (Grant Agreement No. 692031), the ERA-Net Smart Grids Plus "uGrip" (Grant Agreement No. 77731), and the Austrian ICT of the Future "MESSE" (Grant Agreement No. 861265).

Conflicts of Interest: The authors declare no conflict of interest.

Abbreviations

The following abbreviations are used in this manuscript:

3L	Leader, Learner, Listener
ju-RAMI	juristisches Referenzarchitekturmodell Industrie 4.0

reqIF	Requirements Interchange Format
AAL	Ambient Assisted Living
ADM	Architecture Development Method
AMI	Advanced Metering Infrastructure
API	Application Programming Interface
AS	Ancillary Services
AUTOSAR	AUTomotive Open System ARchitecture
BEMS	Building Energy Management System
BMVI	Bundesministerium für Verkehr und digitale Infrastruktur
CBA	Cost based analysis
CEN	Comité Européen de Normalisation
CENELEC	European Committee for Electrotechnical Standardization
CIM	Computational Independent Model
CO_2	Carbon dioxide
CSP	Concentrated Solar Power or Customer Side Participation
DER	Distributed Energy Resource
DISCERN	Distributed Intelligence for Cost-effective and Reliable Solutions
DMS	Distribution Management System
DSL	Domain Specific Language
DSM	Demand Side Management
DSO	Distribution System Operator
EC	European Commission
ELECTRA	European Liaison on Electricity Committed Towards long-term Research Activity
EM-ISA	E-Mobility Information System Architecture
EMS	Energy Management System
EMPOWER	Local Energy Retail Markets for Prosumer Smart Grid Power Services
ERIGrid	European Research Infrastructure supporting Smart Grid Systems Technology Development, Validation and Roll Out
ERM	Entity Relationship Model
ETIP	European Technology and Innovation Platform
ETSI	European Telecommunications Standards Institute
EV	Electric Vehicle
ES	Experiment Specification
FINSENY	Future Internet for Smart Energy
FMI	Functional Mock-up Interface
FP7	Framework Program 7
GPL	Generalized Programming Language
GSCAM	Generic Smart City Architecture Model
GWAC	GridWise Architecture Council
H2020	Horizon 2020
HTD	Holistic Test Description
HMI	Human Machine Interface
ICT	Information and Communication Technology
IEC	International Electrotechnical Commission
IED	Intelligent Electronic Devices
IMO	International Maritime Organization
IRR	Internal Rate of Return
ISO	International Organization for Standardization
LCE	Low-Carbon Energy
LCIM	Levels of Conceptual Interoperability Model
LIC	Logical Interface Class
LRM	Logical Reference Model
MAF	Maritime Architecture Framework
MAS	Multi-Agent System
MBSE	Model-Based Systems Engineering
MDA	Model-Driven Architecture
MESSE	Model-based Engineering and Validation Support for Cyber-Physical Energy Systems
NEMS	Neighborhood Energy Management System
NIST	National Institute of Technology
NMEA	National Marine Electronics Association
Nobel Grid	New Cost Efficient Business Models for Flexible Smart Grids
NPV	Net Present Value
OEM	Original Equipment Manufacturer
OpenNES	Open and Interoperable ICT Solution for Integration of Renewables
PAS	Publicly Available Specification
PIM	Platform Independent Model
PMU	Phasor Measurement Unit

PSAL	Power System Automation Language
PSI	Platform Specific Implementation
PSM	Platform Specific Model
PSO	Particle Swarm Optimization
PV	Photovoltaic
P2P	Peer-to-Peer
QoS	Quality of Service
RAMA	Reference Architecture Model Automotive
RAMI	Reference Architecture Model for Industry 4.0
RAMS	Reliability, Availability, Maintainability, Safety
RASSA	Reference Architecture for Secure Smart Grids in Austria
RAWG	Reference Architecture Working group
R&D	Research and Development
RES	Renewable Energy Source
REST	Representational State Transfer
SCIAM	Smart City Infrastructure Architecture Model
SCADA	Supervisory Control and Data Acquisition
SG	Smart Grid
SGAM	Smart Grid Architecture Model
SINTEG	Schaufenster Intelligente Energie
SmarterEMC2	Smarter Grid:Empowering SG Market ACtors through Information and Communication Technologies
SNET	European Technology and Innovation Platform Smart Networks for Energy Transition
SoS	System of Systems
SP	Sustainable Processes
SRD	System Reference Document
STIX	Structured Threat Information eXpression
SysML	System Markup Language
TAXII	Trusted Automated eXchange of Indicator Information
TC	Test Case or Technical Committee
TDX-Assist	Coordination of Transmission and Distribution data eXchanges for renewables integration in the European marketplace through Advanced, Scalable and Secure ICT Systems and Tools
TOGAF	The Open Group Architecture Framework
TR	Technical Report
TRL	Technology Readiness Level
TS	Test Specification
TSO	Transmission System Operator
TwinPV	Stimulating scientific excellence through twinning in the quest for sustainable energy
UCMR	Use Case Management Repository
UML	Unified Modeling Language
VPP	Virtual Power Plant
VTS	Vessel Traffic Service
WG	Working Group
WoC	Web-of-Cells
XML	Extensible Markup Language
ZVEI	Zentralverband Elektrotechnik- und Elektronikindustrie

References

1. Farhangi, H. The path of the smart grid. *IEEE Power Energy Mag.* **2010**, *8*, 18–28. [CrossRef]
2. Fraune, C. The politics of speeches, votes, and deliberations: Gendered legislating and energy policy-making in Germany and the United States. *Energy Res. Soc. Sci.* **2016**, *19*, 134–141. [CrossRef]
3. Birk, A. *A Knowledge Management Infrastructure for Systematic Improvement in Software Engineering*; Fraunhofer-IRB-Verlag: Stuttgart, Germany, 2001.
4. Uslar, M.; Engel, D. Towards generic domain reference designation: How to learn from smart grid interoperability. *DA-Ch Energieinform.* **2015**, *1*, 1–6.
5. Santodomingo, R.; Uslar, M.; Goring, A.; Gottschalk, M.; Nordstrom, L.; Saleem, A.; Chenine, M. SGAM-based methodology to analyse Smart Grid solutions in DISCERN European research project. In Proceedings of the 2014 IEEE International Energy Conference (ENERGYCON), Cavtat, Croatia, 13–16 May 2014; pp. 751–758.
6. Uslar, M. Semantic interoperability within the power systems domain. In Proceedings of the First International Workshop on Interoperability of Heterogeneous Information Systems IHIS, Bremen, Germany, 31 October–5 November 2005; pp. 39–46.

7. Smart Grid Coordination Group. *Smart Grid Reference Architecture*; Technical Report; CEN-CENELEC-ETSI: Brussels, Belgium, 2012.
8. Englert, H.; Uslar, M. *Europäisches Architekturmodell für Smart Grids-Methodik und Anwendung der Ergebnisse der Arbeitsgruppe Referenzarchitektur des EU Normungsmandats M/490*; Tagungsband VDE-Kongress: Berlin, Germany, 2012.
9. Rittel, H.W.J.; Webber, M.M. Dilemmas in a general theory of planning. *Policy Sci.* **1973**, *4*, 155–169. [CrossRef]
10. Gottschalk, M.; Uslar, M.; Delfs, C. *The Use Case and Smart Grid Architecture Model Approach—The IEC 62559-2 Use Case Template and the SGAM Applied in Various Domains*; Springer: Heidelberg, Germany, 2017.
11. Reinhard, H.; de Weck, O.L.; Fricke, E.; Vössner, S. *Systems Engineering. Grundlagen und Anwendung*; Orell Füssli: Zurich, Switzerland, 2012.
12. Maier, M.W. Architecting principles for systems-of-systems. *Syst. Eng.* **1999**, *1*, 267–284. [CrossRef]
13. DeLaurentis, D. Understanding Transportation as a System-of-Systems Design Problem. In Proceedings of the 43rd AIAA Aerospace Sciences Meeting and Exhibit, Reno, Nevada, 26–27 September 2005.
14. Uslar, M. Energy Informatics: Definition, State-of-the-Art and New Horizons. In Proceedings of the ComForEn 2015—6th Symposium Communications for Energy Systems, OVE-Schriftenreihe Nr. 80 Oesterreichischer Verband fuer Elektrotechnik Austrian Electrotechnical Association, Vienna, Austria, 7 September 2015; Volume 5, pp. 15–26.
15. Office of the National Coordinator for Smart Grid Interoperability. *NIST Framework and Roadmap for Smart Grid Interoperability Standards, Release 3.0*; Technical Report; National Institute of Standards and Technology (NIST): Gaithersburg, MD, USA, 2014.
16. The GridWise Architecture Council. *GridWise Interoperability Context-Setting Framework*; Technical Report; NIST: Gaithersburg, MD, USA, 2008.
17. Uslar, M.; Schulte, J.; Babazadeh, D.; Schlögl, F.; Krüger, C.; Rosinger, M. Simulation: A case for interoperability based on LCIM—The uGrip approach. In Proceedigns of the 2017 IEEE 15th International Conference on Industrial Informatics (INDIN), Emden, Germany, 24–26 July 2017; pp. 492–497. [CrossRef]
18. Lightsey, B. *Systems Engineering Fundamentals*; Technical Report; Department of Defense—Systems Management College: Fort Belvoir, VA, USA, 2001.
19. INCOSE International Council on Systems Engineering. *Systems Engineering Handbook. A Guide for System Life Cycle Processes and Activities, Version 3*; INCOSE International Council on Systems Engineering: San Diego, CA, USA, 2006.
20. INCOSE Technical Operations. *Systems Engineering Vision 2020, Version 2.03*; Technical Report; INCOSE: San Diego, CA, USA, 2007.
21. International Standards Organisation. *ISO/IEC/IEEE Systems and Software Engineering—Architecture Description*; Technical Report; International Standards Organisation: Geneva, Switzerland, 2011.
22. Soley, R. *Model Driven Architecture (MDA)*; Technical Report; Object Management Group: Needham, MA, USA, 2000.
23. Uslar, M.; Specht, M.; Rohjans, S.; Trefke, J.; González, J.M. *The Common Information Model CIM: IEC 61968/61970 and 62325-A Practical Introduction to the CIM*; Springer Science & Business Media: Berlin, Germany, 2012.
24. Uslar, M.; Specht, M.; Rohjans, S.; Trefke, J.; Gonzalez, J. *The Common Information Model CIM—IEC61968/61970/62325 CIM*; China Electric Power Press (CEPP): Beijing, China, 2016.
25. Object Management Group. *OMG Unified Modeling Language (OMG UML), Superstructure*; Technical Report; Object Management Group: Needham, MA, USA, 2009.
26. Object Management Group. *OMG Systems Modeling Language (OMG SysML) Version 1.2*; Technical Report; Object Management Group: Needham, MA, USA, 2010.
27. Neureiter, C. *A Domain-Specific, Model Driven Engineering Approach for Systems Engineering in the Smart Grid*; MBSE4U—Tim Weilkiens: Hamburg, Germany, 2017.
28. International Standards Organisation. *ISO 15288:2015 Systems Engineering—System Life Cycle Processes*; Technical Report; International Standards Organisation: Geneva, Switzerland, 2015.
29. International Standards Organisation. *ISO/IEC TR 24748-2 Systems and Software Engineering—Life Cycle Managemen—Part 2: Guide to the Application of ISO/IECO/IEC 15288 (System Life Cycle Processes)*; Technical Report; International Standards Organisation: Geneva, Switzerland, 2011.

30. Hu, R.; Hu, W.; Chen, Z. Research of smart grid cyber architecture and standards deployment with high adaptability for Security Monitoring. In Proceedings of the 2015 International Conference on Sustainable Mobility Applications, Renewables and Technology (SMART), Kuwait City, Kuwait, 23–25 November 2015; pp. 1–6. [CrossRef]

31. The Smart Grid Interoperability Panel—Cyber Security Working Group. *NISTIR 7628—Guidelines for Smart Grid Cyber Security Volume 1–3, Revision 2*; Technical Report; National Institute of Standards and Technology (NIST): Gaithersburg, MD, USA, 2014.

32. Neureiter, C.; Eibl, G.; Engel, D.; Schlegel, S.; Uslar, M. A concept for engineering smart grid security requirements based on SGAM models. *Comput. Sci.-Res. Dev.* **2016**, *31*, 65–71. [CrossRef]

33. Neureiter, C.; Engel, D.; Uslar, M. Domain specific and model based systems engineering in the smart grid as prerequesite for security by design. *Electronics* **2016**, *5*, 24. [CrossRef]

34. Neureiter, C.; Engel, D.; Trefke, J.; Santodomingo, R.; Rohjans, S.; Uslar, M. Towards Consistent Smart Grid Architecture Tool Support: From Use Cases to Visualization. In Proceedings of the IEEE Innovative Smart Grid Technologies (ISGT) 2014, Istanbul, Turkey, 12–15 October 2014; pp. 1–6.

35. Zanabria, C.; Andrén, F.P.; Strasser, T.I. Comparing Specification and Design Approaches for Power Systems Applications. In Proceedings of the 2018 IEEE PES Transmission Distribution Conference and Exhibition—Latin America (T&D-LA), Lima, Peru, 18–21 September 2018; pp. 1–5. [CrossRef]

36. Pröstl Andrén, F.; Strasser, T.; Kastner, W. Engineering Smart Grids: Applying Model-Driven Development from Use Case Design to Deployment. *Energies* **2017**, *10*, 374. [CrossRef]

37. Santodomingo, R.; Göring, A.; Gottschalk, M.; Valdenmayer, G. *D2-3.2 Tool Support for Managing Use Cases and SGAM Models*; Technical Report; RWE: Essen, Germany, 2014.

38. Santodomingo, R.; Rosinger, M.; Uslar, M. *DISCERN D11.1 Functional Description of the Comprehensive Smart Grid Data Repository*; RWE: Essen, Germany, 2015.

39. Martini, L.; Brunner, H.; Rodriguez, E.; Caerts, C.; Strasser, T.; Burt, G. Grid of the future and the need for a decentralised control architecture: The web-of-cells concept. *Open Access Proc. J.* **2017**, *2017*, 1162–1166. [CrossRef]

40. Heussen, K.; Uslar, M.; Tornelli, C. A use case methodology to handle conflicting controller requirements for future power systems. In Proceedings of the 2015 International Symposium on Smart Electric Distribution Systems and Technologies (EDST), Vienna, Austria, 8–11 September 2015.

41. Uslar, M.; Heussen, K. Towards Modeling Future Energy Infrastructures—The ELECTRA System Engineering Approach. In Proceedings of the 2016 IEEE PES Innovative Smart Grid Technologies Conference Europe (ISGT-Europe), Ljubljana, Slovenia, 9–12 October 2016; pp. 1–6.

42. Uslar, M.; Rosinger, C.; Schlegel, S.; Santodomingo-Berry, R. Aligning IT Architecture Analysis and Security Standards for Smart Grids. In *Advances and New Trends in Environmental and Energy Informatics*; Springer: Cham, Switzerland, 2016; pp. 115–134.

43. Syed, M.H.; Guillo-Sansano, E.; Blair, S.M.; Burt, G.; Strasser, T.; Brunner, H.; Gehrke, O.; Rodríguez-Seco, J.E. Laboratory infrastructure driven key performance indicator development using the smart grid architecture model. *Open Access Proc. J.* **2017**, *2017*, 1866–1870. [CrossRef]

44. Gerard, H.; Rivero, E.; Six, D. *Basic Schemes for TSO-DSO Coordination and Ancillary Services Provision, D1.3*; Technical Report; SmartNet Consortium: San Jose, CA, USA, 2016.

45. Horsmanheimo, K.-T.; Kuusela, T.; Dall, J.; Pröstl, A.; Stephan, K.; Baut, G. *ICT Architecture Design Specification, D3.2*; Technical Report; SmartNet Consortium Homepage: San Jose, CA, USA, 2017.

46. Strasser, T.; Pröstl Andren, F.; Widl, E.; Lauss, G.; Jong, E.D.; Calin, M.; Sosnina, M.; Khavari, A.; Rodriguez, E.; Kotsampopoulos, P.; et al. An Integrated Pan-European Research Infrastructure for Validating Smart Grid Systems. *Elektrotechnik und Informationstechnik* **2018**, *135*, 616–622. [CrossRef]

47. Blank, M.; Lehnhoff, S.; Heussen, K.; Bondy, D.M.; Moyo, C.; Strasser, T. Towards a foundation for holistic power system validation and testing. In Proceedings of the 2016 IEEE 21st International Conference on Emerging Technologies and Factory Automation (ETFA), Berlin, Germany, 6–9 September 2016; pp. 1–4.

48. van der Meer, A.A.; Steinbrink, C.; Heussen, K.; Bondy, D.E.M.; Degefa, M.Z.; Andrén, F.P.; Strasser, T.I.; Lehnhoff, S.; Palensky, P. Design of experiments aided holistic testing of cyber-physical energy systems. In Proceedings of the 2018 Workshop on Modeling and Simulation of Cyber-Physical Energy Systems (MSCPES), Porto, Portugal, 10 April 2018; pp. 1–7.

49. TwinPV Consortium. *Stimulating Scientific Excellence Through Twinning in the Quest For Sustainable Energy*; TwinPV Consortium: Brussels, Belgium, 2018.

50. Pröstl Andren, F.; Strasser, T.; Langthaler, O.; Veichtlbauer, A.; Kasberger, C.; Felbauer, G. Open and Interoperable ICT Solution for Integrating Distributed Energy Resources into Smart Grids. In Proceedings of the 21th IEEE International Conference on Emerging Technologies and Factory Automation (ETFA'2016), Berlin, Germany, 6–9 September 2016.

51. Pröstl Andren, F.; Strasser, T.; Seitl, C.; Resch, J.; Brandauer, C.; Panholzer, G. On Fostering Smart Grid Development and Validation with a Model-based Engineering and Support Framework. In Proceedings of the CIRED Workshop 2018, Ljubljana, Slovenia, 7–8 June 2018.

52. Clausen, M.; Apel, R.; Dorchain, M.; Postina, M.; Uslar, M. Use Case methodology: A progress report. *Energy Inform.* **2018**, *1*, 19. [CrossRef]

53. Kammerstetter, M.; Langer, L.; Skopik, F.; Kastner, W. Architecture-driven smart grid security management. In Proceedings of the 2nd ACM Workshop on Information Hiding and Multimedia Security, Salzburg, Austria, 11–13 June 2014; pp. 153–158.

54. Langer, L.; Skopik, F.; Smith, P.; Kammerstetter, M. From old to new: Assessing cybersecurity risks for an evolving smart grid. *Comput. Secur.* **2016**, *62*, 165–176. [CrossRef]

55. Yesudas, R.; Clarke, R. A framework for risk analysis in smart grid. In *International Workshop on Critical Information Infrastructures Security*; Springer: Berlin, Germany, 2013; pp. 84–95.

56. Holles, S.; De Capitani, J.; Keel, T.; Rechsteiner, S.; Dizdarevic-Hasic, A.; Hettich, P.; Stocker, L.; Mathis, L.; Galbraith, L.; Koller, J. *Datenschutz für Smart Grids: Offene Fragen und mögliche Lösungsansätze (Arbeitspaket 3)*; Study for the BfE: Bern, Switzerland, 2014.

57. Razo-Zapata, I.S. A Method to Assess the Economic Feasibility of New Commercial Services in the Smart Grid. In Proceedings of the IEEE 10th International Conference on Service-Oriented Computing and Applications (SOCA), Kanazawa, Japan, 22–25 November 2017; pp. 90–97.

58. Razo-Zapata, I.S.; Shrestha, A.; Proper, E. An assessment framework to determine the strategic value of IT architectures in smart grids. In Proceedings of the 28th Australasian Conference on Information Systems (ACIS 2017), Hobart, Australia, 4–6 December 2017.

59. Mihaylov, M.; Jurado, S.; Van Moffaert, K.; Avellana, N.; Nowé, A. NRG-X-Change-A Novel Mechanism for Trading of Renewable Energy in Smart Grids. In Proceedings of the 3rd International Conference on Smart Grids and Green IT Systems, Barcelona, Spain, 3–4 April 2014; pp. 101–106.

60. Poursanidis, I.; Andreadou, N.; Kotsakis, E.; Masera, M. Absolute Scoring Scheme for Interoperability Testing of Advanced Metering Infrastructure on Demand Side Management. In Proceedings of the Ninth International Conference on Future Energy Systems, Karlsruhe, Germany, 12–15 June 2018; pp. 391–392.

61. Etherden, N.; Vyatkin, V.; Bollen, M.H. Virtual power plant for grid services using IEC 61850. *IEEE Trans. Ind. Inform.* **2016**, *12*, 437–447. [CrossRef]

62. Lanna, A.; Liberati, F.; Zuccaro, L.; Di Giorgio, A. Electric vehicles charging control based on future internet generic enablers. In Proceedings of the IEEE International Electric Vehicle Conference (IEVC), Florence, Italy, 17–19 December 2014; pp. 1–5.

63. Böcker, S.; Geth, F.; Almeida, P.; Rapoport, S.; Wietfeld, C. Choice of ICT infrastructures and technologies in smart grid planning. In Proceedings of the 23rd International Conference on Electricity Distribution, Lyon, France, 15–18 June 2015.

64. Khayyam, S.; Ponci, F.; Goikoetxea, J.; Recagno, V.; Bagliano, V.; Monti, A. Railway energy management system: Centralized–decentralized automation architecture. *IEEE Trans. Smart Grid* **2016**, *7*, 1164–1175. [CrossRef]

65. Hooshyar, H.; Vanfretti, L. A SGAM-based architecture for synchrophasor applications facilitating TSO/DSO interactions. In Proceedings of the IEEE Power & Energy Society Innovative Smart Grid Technologies Conference (ISGT), Washington, DC, USA, 23–26 April 2017; pp. 1–5.

66. Zhang, Y.; Huang, T.; Bompard, E.F. Big data analytics in smart grids: A review. *Energy Inform.* **2018**, *1*, 8. [CrossRef]

67. Messinis, G.; Dimeas, A.; Hatziargyriou, N.; Kokos, I.; Lamprinos, I. ICT tools for enabling smart grid players' flexibility through VPP and DR services. In Proceedings of the 13th International Conference on the European Energy Market (EEM), Porto, Portugal, 6–9 June 2016; pp. 1–5.

68. Pignolet, Y.A.; Elias, H.; Kyntäjä, T.; de Cerio, I.M.D.; Heiles, J.; Boëda, D.; Caire, R. Future Internet for smart distribution systems. In Proceedings of the 3rd IEEE PES International Conference and Exhibition on Innovative Smart Grid Technologies (ISGT Europe), Berlin, Germany, 14–17 October 2012; pp. 1–8.

69. Piatkowska, E.; Bayarri, L.P.; Garcia, L.A.; Mavrogenou, K.; Tsatsakis, K.; Sanduleac, M.; Smith, P. Enabling novel smart grid energy services with the nobel grid architecture. In Proceedings of the 2017 IEEE Manchester PowerTech, Manchester, UK, 18–22 June 2017; pp. 1–6.

70. Wilker, S.; Mcisel, M.; Piatkowska, E.; Sauter, T.; Jung, O. Smart Grid Reference Architecture, an Approach on a Secure and Model-Driven Implementation. In Proceedings of the 2018 IEEE 27th International Symposium on Industrial Electronics (ISIE), Cairns, QLD, Australia, 13–15 June 2018; pp. 74–79.

71. Bullich-Massagué, E.; Aragüés-Penalba, M.; Olivella-Rosell, P.; Lloret-Gallego, P.; Vidal-Clos, J.A.; Sumper, A. Architecture definition and operation testing of local electricity markets. The EMPOWER project. In Proceedings of the International Conference on Modern Power Systems (MPS), Cluj-Napoca, Romania, 6–9 June 2017; pp. 1–5.

72. Lloret-Gallego, P.; Aragüés-Peñalba, M.; Van Schepdael, L.; Bullich-Massagué, E.; Olivella-Rosell, P.; Sumper, A. Methodology for the Evaluation of Resilience of ICT Systems for Smart Distribution Grids. *Energies* **2017**, *10*, 1287. [CrossRef]

73. Smidt, H.; Thornton, M.; Ghorbani, R. Smart application development for IoT asset management using graph database modeling and high-availability web services. In Proceedings of the 51st Hawaii International Conference on System Sciences, Waikoloa Village, HI, USA, 3–6 January 2018.

74. Prinsloo, G.; Dobson, R.; Mammoli, A. Synthesis of an intelligent rural village microgrid control strategy based on smartgrid multi-agent modelling and transactive energy management principles. *Energy* **2018**, *147*, 263–278. [CrossRef]

75. Heidel, R.; Hankel, M.; Döbrich, U.; Hoffmeister, M. *Basiswissen RAMI 4.0: Referenzarchitekturmodell und Industrie 4.0-Komponente Industrie 4.0*; VDE Verlag: Berlin, Germany, 2017.

76. Binder, C.; Neureiter, C.; Lastro, G.; Uslar, M.; Lieber, P. Towards a Standards-Based Domain Specific Language for Industry 4.0 Architectures. In Proceedings of the International Conference on Complex Systems Design & Management, Paris, France, 18–19 December 2018; Springer: Cham, Switzerland, 2018; pp. 44–55.

77. Uslar, M.; Göring, A.; Heidel, R.; Neureiter, C.; Engel, D.; Schulte, S. *An Open Source 3D Visualization for the RAMI 4.0 Reference Modell*; VDE Kongress 2016, Mannheim; VDE Verlag GMBH: Berlin, Germany, 2016; Volume 1, pp. 1–6.

78. Uslar, M.; Hanna, S. Model-driven Requirements Engineering Using RAMI 4.0 Based Visualizations. In Proceedings of the Modellierung 2018—AQEMO: Adequacy of Modeling Methods, Braunschweig, Germany, 21 February 2018; Volume 2060, pp. 21–30.

79. Clausen, M.; Gottschalk, M.; Hanna, S.; Kronberg, C.; Rosinger, C.; Rosinger, M.; Schulte, J.; Schütz, J.; Uslar, M. Smart grid security method: Consolidating requirements using a systematic approach. In Proceedings of the CIRED Workshops 2018, Ljubljana, Slovenia, 7–8 June 2018; Volume 1.

80. Security Sub-Working Group "Connected and Automatic Driving" of the Governmental Department of Transport and Infrastructure (BMVI). *TFCS-08-05: Reference Architecture Model Automotive (RAMA)*; BMVI: Berlin, Germany, 2017.

81. Weinert, B. *Ein Framework zur Architekturbeschreibung von Sozio-Technischen Maritimen Systemen*; MBSE Press: Hamburg, Germany, 2018.

82. Weinert, B.; Uslar, M.; Hahn, A. System-of-systems: How the maritime domain can leam from the Smart Grid. In Proceedings of the 2017 International Symposium ELMAR, Zadar, Croatia, 18–20 September 2017.

83. Weinert, B.; Uslar, M.; Hahn, A. Domain-Specific Requirements Elicitation for Socio-Technical System of Systems. In Proceedings of the IEEE 13th System of Systems Engineering Conference—SoSE 2018, Paris, France, 19–22 June 2018.

84. Neureiter, C.; Engel, S.; Rohjans, S.; Dänekas, C.; Uslar, M. Addressing the complexity of distributed smart city systems by utilization of model driven engineering concepts. In Proceedings of the VDE-Kongress 2014—Smart Cities, Frankfurt, Germany, 20–21 October 2014; pp. 1–6.

85. Hurtado, L.; Nguyen, P.; Kling, W. Smart grid and smart building inter-operation using agent-based particle swarm optimization. *Sustain. Energy Grids Netw.* **2015**, *2*, 32–40. [CrossRef]

86. Mocanu, E.; Aduda, K.O.; Nguyen, P.H.; Boxem, G.; Zeiler, W.; Gibescu, M.; Kling, W.L. Optimizing the energy exchange between the smart grid and building systems. In Proceedings of the 49th International Universities Power Engineering Conference (UPEC), Cluj-Napoca, Romania, 2–5 September 2014; pp. 1–6.

87. Shafiullah, D.; Vo, T.; Nguyen, P.; Pemen, A. Different smart grid frameworks in context of smart neighborhood: A review. In Proceedings of the 52nd International Universities Power Engineering Conference (UPEC), Heraklion, Greece, 28–31 August 2017; pp. 1–6.

88. Schuh, G.; Fluhr, J.; Birkmeier, M.; Sund, M. Information system architecture for the interaction of electric vehicles with the power grid. In Proceedings of the 10th IEEE International Conference on Networking, Sensing and Control (ICNSC), Evry, France, 10–12 April 2013; pp. 821–825.

89. Meloni, A.; Atzori, L. A cloud-based and restful internet of things platform to foster smart grid technologies integration and re-usability. In Proceedings of the IEEE International Conference on Communications Workshops (ICC), Kuala Lumpur, Malaysia, 23–27 May 2016; pp. 387–392.

90. Zhang, C.; Wu, J.; Cheng, M.; Zhou, Y.; Long, C. A bidding system for peer-to-peer energy trading in a grid-connected microgrid. *Energy Procedia* **2016**, *103*, 147–152. [CrossRef]

91. Gottschalk, M.; Uslar, M.; Delfs, C. *Smart City Infrastructure Architecture Model (SCIAM)*; Springer: Cham, Switzerland, 2017; pp. 75–76.

92. Frascella, A.; Brutti, A.; Gessa, N.; De Sabbata, P.; Novelli, C.; Burns, M.; Bhatt, V.; Ianniello, R.; He, L. A minimum set of common principles for enabling Smart City Interoperability. *J. Technol. Archit. Environ.* **2018**, 56–61. [CrossRef]

93. Babar, M.; Nguyen, P. Analyzing an Agile Solution For Intelligent Distribution Grid Development: A Smart Grid Architecture Method. In Proceedings of the IEEE Innovative Smart Grid Technologies-Asia (ISGT Asia), Singapore, 22–25 May 2018; pp. 605–610.

94. Dänekas, C.; Neureiter, C.; Rohjans, S.; Uslar, M.; Engel, D. Towards a Model-Driven-Architecture Process for Smart Grid Projects. In *Digital Enterprise Design & Management*; Springer: Cham, Switzerland, 2014.

95. Steinbrink, C.; Schlögl, F.; Babazadeh, D.; Lehnhoff, S.; Rohjans, S.; Narayan, A. Future Perspectives of Co-Simulation in the Smart Grid Domain. In Proceedings of the 2018 IEEE International Energy Conference (Energycon), Limassol, Cyprus, 3–7 June 2018.

96. Schütte, S.; Scherfke, S.; Tröschel, M. Mosaik: A framework for modular simulation of active components in Smart Grids. In Proceedings of the 2011 IEEE First International Workshop on Smart Grid Modeling and Simulation (SGMS), Brussels, Belgium, 17 October 2011; pp. 55–60.

97. Uslar, M.; Rosinger, C.; Schlegel, S. Security by Design for the Smart Grid: Combining the SGAM and NISTIR 7628. In Proceedings of the 2014 IEEE 38th International Computer Software and Applications Conference Workshops (COMPSACW), Vasteras, Sweden, 21–25 July 2014; pp. 110–115.

98. Gottschalk, M.; Franzl, G.; Frohner, M.; Pasteka, R.; Uslar, M. Structured workflow achieving interoperable Smart Energy systems. *Energy Inform.* **2018**, *1*, 25. [CrossRef]

99. Gottschalk, M.; Franzl, G.; Frohner, M.; Pasteka, R.; Uslar, M. From Integration Profiles to Interoperability Testing for Smart Energy Systems at Connectathon Energy. *Energies* **2018**, *11*, 3375. [CrossRef]

100. Blochwitz, T.; Otter, M.; Arnold, M.; Bausch, C.; Clauß, C.; Elmqvist, H.; Junghanns, A.; Mauss, J.; Monteiro, M.; Neidhold, T.; et al. The Functional Mockup Interface for Tool independent Exchange of Simulation Models. In Proceedings of the 8th International Modelica Conference 2011, Dresden, Germany, 20–22 March 2011; pp. 173–184. [CrossRef]

101. Chilard, O.; Boes, J.; Perles, A.; Camilleri, G.; Gleizes, M.P.; Tavella, J.P.; Croteau, D. The Modelica language and the FMI standard for modeling and simulation of Smart Grids. In Proceedings of the 11th International Modelica Conference, Versailles, France, 21–23 September 2015.

102. Mirz, M.; Razik, L.; Dinkelbach, J.; Tokel, H.A.; Alirezaei, G.; Mathar, R.; Monti, A. A Cosimulation Architecture for Power System, Communication, and Market in the Smart Grid. *Complexity* **2018**, *2018*, 7154031. [CrossRef]

103. van der Meer, A.A.; Palensky, P.; Heussen, K.; Bondy, D.E.M.; Gehrke, O.; Steinbrink, C.; Blank, M.; Lehnhoff, S.; Widl, E.; Moyo, C.; et al. Cyber-Physical Energy Systems Modeling, Test Specification, and Co-Simulation Based Testing. In Proceedigns of the 2017 Workshop on Modeling and Simulation of Cyber-Physical Energy Systems (MSCPES), Pittsburgh, PA, USA, 21 April 2015.

104. ETIP SNET. European Technology and Innovation Platform for Smart Networks for Energy Transition. Available online: https://www.etip-snet.eu/ (accessed on 9 December 2018).

105. Uslar, M.; Hanna, S. Teaching Domain-Specific Requirements Engineering to Industry: Applying Lego Serious Play to Smart Grids. In Proceedings of the SE 2018—ISEE 2018: 1st Workshop on Innovative Software Engineering Education, Ulm, Germany, 6–8 March 2018; Volume 2066, pp. 36–37.

106. Francesco, E.D.; Francesco, R.D.; Leccese, F.; Paggi, A. The ASD S3000L for the enhancement of "in field" avionic measurements. In Proceedings of the 2014 IEEE Metrology for Aerospace (MetroAeroSpace), Benevento, Italy, 29–30 May 2014; pp. 174–179. [CrossRef]

107. Francesco, E.D.; Francesco, E.D.; Francesco, R.D.; Leccese, F.; Cagnetti, M. A proposal to update LSA databases for an operational availability based on autonomic logistic. In Proceedings of the 2015 IEEE Metrology for Aerospace (MetroAeroSpace), Benevento, Italy, 4–5 June 2015; pp. 38–43. [CrossRef]

energies

MDPI

Article

Adaptive Fuzzy Control for Power-Frequency Characteristic Regulation in High-RES Power Systems

Evangelos Rikos [1,*]**, Chris Caerts** [2]**, Mattia Cabiati** [3]**, Mazheruddin Syed** [4] **and Graeme Burt** [4]

[1] Centre for Renewable Energy Sources and Saving, Pikermi Attiki 19009, Greece
[2] Flemish Institute for Technological Research, Mol 2400, Belgium; chris.caerts@vito.be
[3] Ricerca sul Sistema Energetico-RSE S.p.A., Milano 20134, Italy; mattia.cabiati@rse-web.it
[4] Institute for Energy and Environment, University of Strathclyde, Glasgow G1 1XQ, UK;
    mazheruddin.syed@strath.ac.uk (M.S.); graeme.burt@strath.ac.uk (G.B.)
* Correspondence: vrikos@cres.gr; Tel.: +30-210-660-3368

Received: 17 May 2017; Accepted: 6 July 2017; Published: 12 July 2017

Abstract: Future power systems control will require large-scale activation of reserves at distribution level. Despite their high potential, distributed energy resources (DER) used for frequency control pose challenges due to unpredictability, grid bottlenecks, etc. To deal with these issues, this study presents a novel strategy of power frequency characteristic dynamic adjustment based on the imbalance state. This way, the concerned operators become aware of the imbalance location but also a more accurate redistribution of responsibilities in terms of reserves activations is achieved. The proposed control is based on the concept of "cells" which are power systems with operating capabilities and responsibilities similar to control areas (CAs), but fostering the use of resources at all voltage levels, particularly distribution grids. Control autonomy of cells allows increased RES hosting. In this study, the power frequency characteristic of a cell is adjusted in real time by means of a fuzzy controller, which curtails part of the reserves, in order to avoid unnecessary deployment throughout a synchronous area, leading to a more localised activation and reducing losses, congestions and reserves exhaustion. Simulation tests in a four-cell reference power system prove that the controller significantly reduces the use of reserves without compromising the overall stability.

Keywords: adaptive control; fuzzy logic; cell; frequency containment control (FCC); power frequency characteristic; droop control

---

1. Introduction

Environmental as well as economic considerations constitute principal motivators towards adopting ever-increasing green technologies, namely renewable energy systems (RES) for the electrification of power systems. The higher the RES penetration, the more the system's operation challenges will be expected due to unpredictability, intermittency and the vast dispersion, all intrinsic characteristics of this type of energy resources. The operation challenges are to be further intensified due to the high penetration targets that energy policies have set. For example, the target for green-house gasses (GHG) reduction at the European level is set to at least 40% by 2030, and between 80 and 95% by 2050 compared to the 1990 figures [1]. In addition, from the same report the minimum requirement of energy covered by RES by the year 2030 is 27%. More ambitious studies such as [2,3] show that even higher RES levels can be achieved. For example, e-Highway 2050 [2] predicts that one of the possible pathways for RES development involves a RES energy penetration as high as 100% by 2050. Furthermore, analyses like [3] predict possible high-RES penetration scenarios, namely up to 60% of energy covered by RES by 2030.

Regardless of the approach or the levels that will eventually be reached, RES penetration is expected to substantially increase in the next decades, in a fashion that will have some implications with regard to the best exploitation of the generated energy as well as the security of supply, the latter being a prerequisite for maximizing the former exploitation. To this end, an operation paradigm shift from today's to future power systems is required in order to host the planned RES as effectively as possible. For instance, due to the high degree of dispersion that is expected in RES, not only should the energy exploitation be as local as possible but also operating responsibilities and challenges are issues better addressed locally. In this respect, awareness of the local grid status through increased observability combined with activation of the involved resources is of vital importance due to the high amounts of small-scale distributed energy resources (DER) units that can provide the required control services. In addition, scheduling and operation approaches based on a generation that largely follows load are to be revised due to the unpredictability of RES as well as the increased load flexibility which fosters a paradigm in which load mainly follows generation. Because of this paradigm shift, scheduling of resources may entail high peaks of power flows that occur rather locally, leading parts of the grid to operate close to their limits by contrast to current operating schemes in which load levelling is normally pursued.

However, despite the fact that large amounts of energy exchanges will happen locally, namely at Low Voltage (LV) and Medium Voltage (MV) distribution level, a substantial amount of energy will still be produced and transferred by means of High Voltage (HV) transmission grids, either due to centralised RES power plants, e.g., offshore wind farms or other bulk generation e.g., hydroelectric power plants. The above-described assumptions impose the need for new operating scheme approaches. All in all, research approaches regarding operation of high-RES penetration systems can be distinguished into three main pathways, namely the reconfiguration of roles and responsibilities in operating power systems, invention of new optimal automatic control strategies and reconsideration of operating requirements, especially in terms of frequency stability, namely less stringent frequency limits.

In this paper, one exemplary approach of power systems structured and operated as "cells" which constitute a web-of-cells (WoC) [4] is briefly presented in Section 2. Each cell incorporates the operation responsibilities and capabilities of modern systems' Control areas (CAs) but with enhanced control capabilities at lower, i.e., distribution level, thus unlocking the great potential of distributed generators (DGs) as well as flexible loads and storage elements in the provision and utilisation of ancillary services. Cells are equipped with novel control strategies that optimally exploit flexibility of generation/consumption so much so that maximisation of RES in the grid can be achieved. To this end, the second and main part of this study is concerned with a novel strategy for adaptive frequency containment control (Adaptive FCC) which, as part of the cell control, aims to reduce the contribution of primary frequency reserves within the cell and all over a synchronous area, without jeopardising the overall stability. This way a number of benefits that are analysed in Section 3 are obtained. Section 4 is concerned with the simulation results of the implementation of such a control strategy in a power system consisting of four cells.

2. The "Cell" and "Web-of-Cells" Concepts

For the purpose of control, the main objective of a cell is to ensure stability in the whole synchronous area in a distributed but, at the same time, coordinated fashion. The concept of cell presents a generic applicability and, hence, it may as well be used in transmission and distribution networks, also with a diverse geographical scalability. In fact, the basic idea behind this concept is to engage distribution system operators (DSOs) as active participants of the balance/frequency control process, control actions which are hitherto the sole responsibility of transmission system operators (TSOs) nominated as CA operators in present-day power systems. The use of this conventional approach in power systems with a high amount of DER has some intrinsic disadvantages because, in order for a TSO to effectively use largely distributed reserves, the operator must be aware of characteristics and operating conditions of distribution grids and have control access to all involved resources. On the other hand, in this paradigm DSOs are concerned with issues like voltage control which is highly influenced by the use of balance reserves. In this process, a number of conflicts can emerge due to different technical, regulatory and economic objectives of the involved actors.

As an alternative paradigm, the concept of cells resolves such issues in the sense that an operator like a DSO becomes also an operator of balance/frequency control, thereby reducing the complexity and conflicts caused by the conventional approach. The cell concept, however, does not incapacitate TSOs in their key role as balance/frequency operators as well. It is rather an extrapolation of the balance/frequency control concept to lower levels of the power system, relieving TSOs from the burden of managing a large amount of reserves. Many connected cells are combined to shape a WoC as illustrated in Figure 1. In this approach, one cell cannot encapsulate other cells, thus making cells independent in terms of energy balancing and control objectives since, apart from the overall synchronous area balance, each cell pursues its own goals.

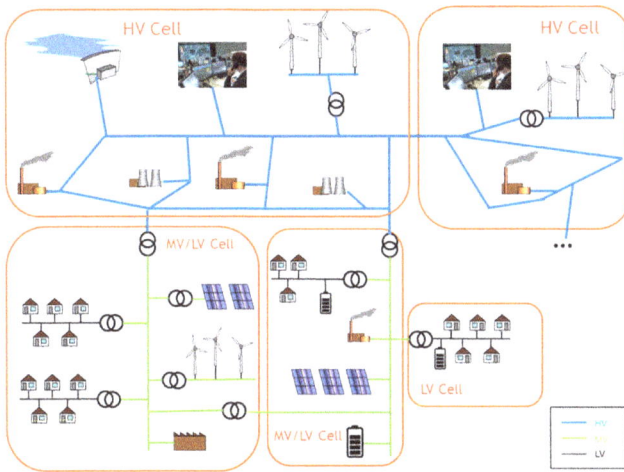

Figure 1. Schematic example of the proposed "web-of-cells" (WoC) architecture.

In terms of control architecture, there is a novel control approach defined for use within cells. In particular, a cell is equipped with a hierarchical control scheme, in which functions regarding planning, scheduling and optimising reserves for balance, frequency and voltage control are located at cell operation control centres together with activation functions. While the cell operates mainly, but not exclusively, at secondary and tertiary timescales, primary time scales, such as provision of virtual inertia, frequency containment and control of nodes voltage are mostly dealt with at single-unit or aggregated (power plant) levels. The main operation functionalities of a cell in this respect are virtual inertia provision, containment of frequency deviations, restoration of balance and frequency, and steering of balance schedules at optimal values. In terms of voltage control, a cell is concerned with the fast-responding primary control locally at nodes and the coordinated optimal voltage control at system level that aims at improving the voltage stability of the cell as a whole by selecting the optimal voltage combinations across the grid, taking into account its specific characteristics [4].

Compared with other architectural approaches, such as microgrids [5–8], which is perhaps the most widely accepted, mature and well-established concept for hosting large amounts of distributed RES, in terms of DER integration, the cell concept presents some advantages such as more generic voltage level applicability (whilst microgrids are confined mainly to LV distribution grids), more than one connection (tie-line) to other cells at any voltage level, or no self-sufficiency in terms of energy balance which means that a cell can largely rely on imports/exports, etc. On top of that, cells ensure their provision of reserves that serve primarily their own stability whilst microgrids, when in grid-connected mode, act rather as aggregators providing ancillary services to DSOs. However, by no means should cells be considered as a substitute of microgrids, since both architectures can coexist in a power system.

3. Proposed Adaptive Frequency Containment Control

3.1. Analysis of the Control Requirements

Within the present-day system operation, primary frequency control is responsible for containment of frequency after imbalances that lead to acceleration/deceleration of synchronous generators (SGs). The response of a system to a frequency event is partly determined by the network power frequency characteristic (NPFC), a measure of the total power deviation in a synchronous area for a specific amount of steady-state frequency change [9]. The classic containment of frequency is a strategy based on the droop slope of the governors of SGs as well as the self-regulation of load during frequency changes. This way, it is easy to specify the response of a synchronous area by setting the appropriate droop in the governors or selecting the appropriate number/capacity of generators to provide such a service. Good though this control strategy may be for present-day power systems, it is not so advantageous in future power systems hosting high amounts of distributed RES. The reasons are the following:

- The conventional approach of the NPFC specification presumes that all CAs, even remotely located, contribute proportionally to the incident. Even though this approach is effective for power systems with centralised power production at transmission level, in power systems with high degree of dispersion, where reserves can be located in LV grids far away from the area of imbalance, the LV distribution grids may experience sizeable changes in power flows. This can potentially lead to congestion, over/under voltages and increased power losses in the grid.
- If DER units such as photo voltaics (PVs) and wind generators (WGs) participate in the provision of FCC, they may be required to curtail their production based on remotely located incidents. Thus, this curtailment would lead to reduction of useful RES generation from areas not responsible for the imbalance incident.
- Operating frequency limits will most probably be reconsidered and relaxed so that operation would require less demanding frequency control and reduction in the corresponding containment reserves.
- Apart from their principal operating goal, DER can be used in multiple-purpose ancillary services provision, particularly the provision of virtual inertia as a primary requirement of future frequency control [10,11]. The fixed droop provision by these DER units in conjunction with the extra control strategies they provide may, under specific circumstances, lead to output saturation and incapacitation of the reserves.

To this end and on the grounds of the "web-of-cells" concept, we propose an adaptive control strategy for FCC capable of detecting the cells' imbalance state, which is the result of any deviation from the scheduled generation and consumption within the cell. This detection is done in real time at cell level and, by doing so, the controller is capable of modifying (curtailing) the droop contribution of the cell so as to reduce its frequency containment reserves (FCR) activation. It should be pointed out that the proposed adaptive control is strongly connected to the WoC concept since it concerns the cell power-frequency characteristic (CPFC) value of a cell and the implementation is done at a cell-level. By contrast, the frequency response control (i.e., droop characteristic of a DER) is a purely decentralised approach and, hence, not specifically connected to the WoC only, but it can be applied to any other architectural approach as well. The scope of the proposed adaptation is to reduce the FCR when the imbalance takes place not inside the cell but somewhere else in the power system, whilst for the case where imbalance happens inside the cell, the slope of power frequency characteristic of it should essentially remain unchanged. Also, in an alternative approach, the slope of the characteristic of a cell could be increased during the incident in order to obtain a narrower frequency deviation and a more constant NPFC across the synchronous area; however, this approach is out of the scope of the present study. The basic benefits of this approach are:

- Mainly, local activation of primary reserves.
- Reduction in power losses, congestion, over/under voltage issues caused in other cells.
- Less conflicts in usage of resources, especially RES and DER used by other ancillary services controllers.
- Potential scheduling at cell level which leads to reduced computational burden and increased accuracy. This is due to the fact that the power frequency characteristic is determined at cell level based on the cell's rating. The calculation of the NPFC at Synchronous Area level is then based on the energy yield of the cells and their contribution to the NPFC instead of considering all units individually.

It is worth noting that adaptive droop control is a method already considered in several studies such as [12–18], which, however, focus on local operation levels such as the operation of single devices, e.g., inverters or implementation at microgrid level at most. Besides that, various research studies such as [19,20] address the issue of the optimal restoration rather than the frequency containment itself in an optimal way. For the above-mentioned reasons, present study is important not only because of its novelty but also for the simplicity of its implementation.

3.2. Fuzzy Logic Controller Selection and Design

The proposed control strategy is briefly described in Figure 2, in which the equivalent of a power system area is assumed as a transfer function. This model was selected not as the most representative example of a cell, although a cell can also be a HV transmission area with the characteristics of Figure 2's model. The reason for selecting this model is due to its simplicity. Based on this power model, it is more convenient to explain the input and output configuration of the proposed controller. In this diagram, cell i's frequency response to imbalances is represented by the transfer function $G_{pi}(s)$, and it involves the parameters of inertia constant H_i as well as load self-regulation D_i in the following relationships:

$$K_{pi} = \frac{1}{D_i}, \ T_{pi} = \frac{2H_i}{f^0 D_i} \tag{1}$$

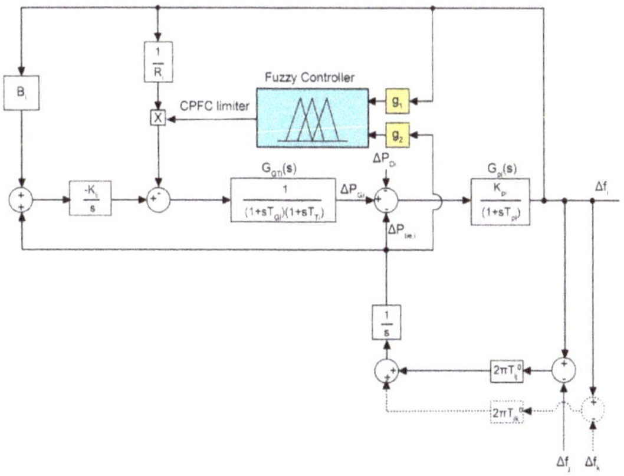

Figure 2. Block diagram of the adaptive control implementation for cell i.

The transfer function $G_{GT}(s)$ is used to represent the time delays of the SG's governor-turbine system assumed for this cell. The specific system is interconnected with other cells (j, k, etc.). The interconnection is reflected on the tie-line error calculation, which depends on the instantaneous frequency deviations of the adjacent cells. Table 1 provides an overview of this model's main parameters and input signals. The adaptive control in our case is obtained by means of a fuzzy logic controller that receives frequency (Δf_i) and tie-line power deviation ($\Delta P_{tie,i}$), generally given by:

$$\Delta P_{Tie,i} = \sum_{j=1}^{m} \Delta P_{Tie,ij} \qquad (2)$$

as inputs and derives a curtailment ratio for the full-scale droop control of the cell. Fuzzy controllers provide numerous capabilities in terms of control logic based on multiple input signals and they are used in various power systems control studies [21]. In our case, the controller uses the combination of both Δf_i and $\Delta P_{tie,i}$ in order to identify the cell's state.

Table 1. Explanation of the assumed model parameters.

Symbol	Description
H_i	Inertia constant in [s]
D_i	Load self-regulation in [pu/Hz]
K_{pi}	Area's constant in [Hz/pu]
T_{pi}	Area's time constant [s]
$G_{pi}(s)$	Transfer function of cell i
T_{ij}	Tie-line static limit [pu/rad]
R_i	Droop slope in [Hz/pu]
B_i	Frequency bias [pu/Hz]
T_{Gi}	Governor's time constant in [s]
T_{Ti}	Turbine's time constant in [s]
$G_{GTi}(s)$	Transfer function of the combined governor-turbine system in cell i
K_{Ii}	Integrator's gain
g_1, g_2	Fuzzy controller input gains for frequency and tie-line power errors
$\Delta f_i, \Delta f_j, \Delta f_k$	Frequency error in cells i, j and k respectively
$\Delta P_{tie,i}$	Tie-line power error in cell i
$\Delta P_{Gi}, \Delta P_{Di}$	Deviation of generation and demand in cell i

The sum in Equation (2) represents the individual contribution of each adjacent cell j to the tie-line error in cell i. The premise of this approach is that when an imbalance incident happens in one cell, the frequency initially increases/decreases based on the imbalance sign and, by the same token, and based on the sign convention for the power production/consumption, the tie-line error aggregate follows an opposite-to-the-frequency course, i.e., the tie-line error increases when frequency decreases. Thus, by detecting the combination of signs as well as sizes of the two errors, it is possible to adjust the droop slope of the cell when the incident takes place outside the cell or maintain its maximum value when the incident concerns the specific cell. The membership functions of the selected controller are depicted in Figure 3. In this exemplary approach, the input and output signals consist of triangular functions. The maximum frequency range is from 48 to 52 Hz (± 2 Hz) and the maximum tie-line error varies from -1 to $+1$ pu. However, these limits are easily adjustable by means of the gains g_1 and g_2 in the input signals. In any case, the output CPFC$_r$ consists of four membership functions which produce a gradual CPFC reduction from 100% to 0%.

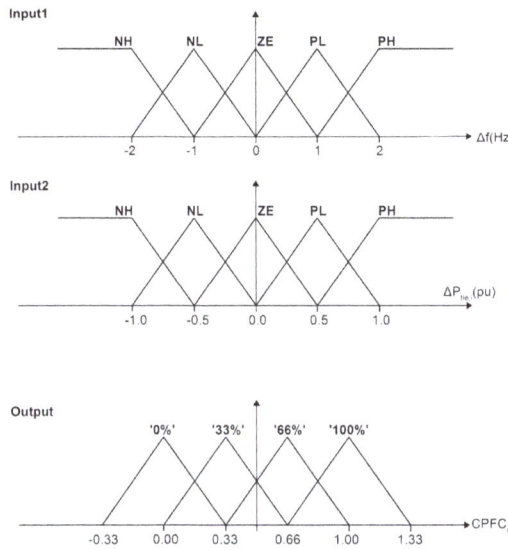

Figure 3. Fuzzy logic controller membership functions selected for the adaptive FCC control strategy.

The rules used for the proposed adaptive control are summarised in the rule table below (Table 2). The symbolic names shown in Figure 3 and Table 2 stand for "Negative High" (NH), "Negative Low" (NL), "Zero" (ZE), "Positive Low" (PL), and "Positive High" (PH). As can be seen in this table, the occurrence of error signals with opposite signs implies incidents inside the cell, thus the CPFC coefficient remains unchanged. By contrast, there is a reduction in the CPFC value whenever the signs of the errors are the same, with a specific reduction selection based on the size of the error. For example, in Table 2 a combination of input signals such as $\Delta f_i = -1$ Hz and $\Delta P_{tie,i} = -0.5$ pu ($g_1 = g_2 = 1$) yields a $CPFC_r = 66\%$.

Table 2. Rule table for the adaptive frequency containment control (FCC) controller.

$\Delta P_{tie,i}$ \ Δf_i	NH	NL	ZE	PL	PH
NH	0%	33%	100%	100%	100%
NL	33%	66%	100%	100%	100%
ZE	66%	100%	100%	100%	66%
PL	100%	100%	100%	66%	33%
PH	100%	100%	100%	33%	0%

It is worth noting that the calculation of the crisp output value is obtained by means of the centre-of-gravity (CoG) method, which takes into account the chopped membership functions of the input signals:

$$CPFC_{coef.}{}^{crisp} = \frac{\sum_i b_i \int \mu(i)}{\sum_i \int \mu(i)} \tag{3}$$

where, b_i is the CoG point of the output membership function $\mu(i)$. In our case the CoG points corresponding to Figure 3 are $b_1 = 0.00$, $b_2 = 0.33$, $b_3 = 0.66$ and $b_4 = 1.00$ respectively.

4. Simulation Results

For the validation of the proposed adaptive control, the power system shown in Figure 4 was selected and implemented in Matlab/Simulink/SimScape Power. This system is based on the CIGRE

Medium Voltage (MV) reference grid presented in [22]. For our analysis, we assumed that the specific power system consists of four cells interconnected in the configuration shown in Figure 4. The number of cells was selected to obtain a sufficient meshed topology in the system with a sufficient number of cells (above two) which leads to a better assessment of the controller. During the tests, two adaptive controllers were used, one in cell 1 and one in cell 2. Cell 3 was not assumed to have DGs and, therefore, FCC whatsoever. However, imbalances happening in cell 3 were implemented to investigate the effect on the controllers of cells 1 and 2. Last, but not least, none of the three MV cells were equipped with control to restore frequency, which for these tests was assumed to be the task of the HV cell only. It should be pointed out here that, in connection with the parameters of Figure 2, the main system parameters for cells 1–3, such as inertia constant H, load damping factor D, time delays T_G and T_T, as well as integrators' gains K_I were all set zero. This assumption does not affect the test results since the presence of these parameters is not a prerequisite for the proposed controller to operate. Also, the nature of the resources in the three cells, all of which are assumed as inverter-based units, makes the absence of inertia and time delays a reasonable assumption. The latter entails that in terms of dynamics, the frequency of all three MV cells is the same and it depends on the parameters of the HV cell. The parameters that were fixed throughout the tests for this cell are shown in Table 3. Other important parameters such as the droop slope of each area were changed based on each scenario. It is noteworthy that the selection of these parameters was not based on a systematic approach but it was done based indicative literature values.

Table 3. Basic parameters values of the High Voltage (HV) cell during the tests.

Parameter	Value
H	5 s
D	0.01 pu/Hz
K_p	100 Hz/pu
T_p	20 s
B	1 pu/Hz
T_G	0.008 s
T_T	0.03 s
K_I	0.01

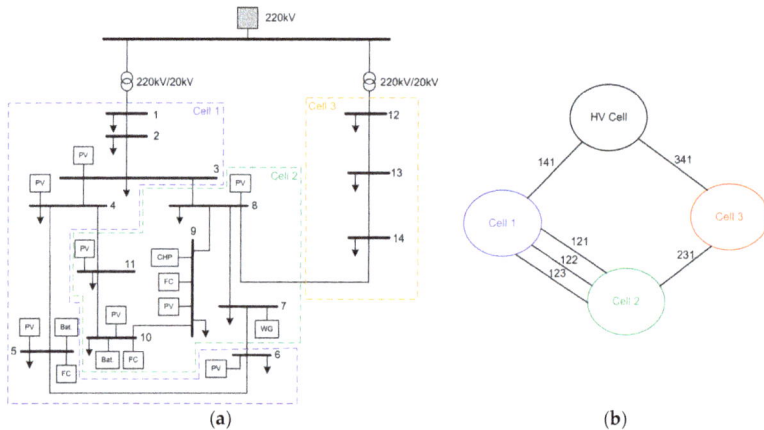

(a)   (b)

Figure 4. Reference power grid and simplified WoC diagram: (**a**) Electrical diagram of the reference power system; (**b**) Selected WoC configuration.

4.1. Scenario A: Short-Term Analysis

This scenario includes the investigation of the adaptive controller's qualitative behaviour for short-term imbalance incidents without the presence of frequency restoration control. The latter was omitted for the sake of clarity of results in order to have a clear view of the fuzzy controller's response to input signals. As a consequence, due to the absence of any frequency restoration, all test results in this scenario present a steady-state frequency and imbalance deviation. In order to assess the ability of the controller to discern the location of the imbalance, three different imbalances were implemented in each of the three MV cells. Each imbalance was a load change of 3 MW located at nodes 2, 9 and 12, respectively.

Also, each imbalance takes place at $t = 100$ s and the controller gains g_1 and g_2 were set to 5 and -50 respectively. The results in Figure 5 show the response of the frequency and tie-line errors in conjunction with the CPFC curtailment for two of the three investigated imbalances. In this case, it is self-evident that the two controllers successfully pinpoint the imbalance location resulting in CPFC reduction only in cell 2 when the imbalance takes place in cell 1, or in both, cells 1 and 2, when the imbalance takes places in cell 3. Furthermore, for the case that the imbalance happens in cell 2 the results are similar to the ones shown in Figure 5a. That means that the CPFC ratio of cell 2 remains unchanged, whereas the CPFC value of cell 1 is curtailed. It is also noteworthy that the variation of CPFC in all these tests does not go below 50%. This is due to the limited imbalance deviation for the specific incident compared to the base power of 50 MVA. With the selection of a gain value equal to -50 for the imbalance error, the latter is amplified enough to see a significant reduction of CPFC to about 50%. By further increasing the gain, or by changing the membership functions and/or the rule table, it is possible to increase the output range of the CPFC curtailment to even lower values. Since it is not easy to predict the exact values of the input errors in a power system model like the one used, the controller has been tested and validated in its full input/output range in a stand-alone set of tests with fully controllable inputs.

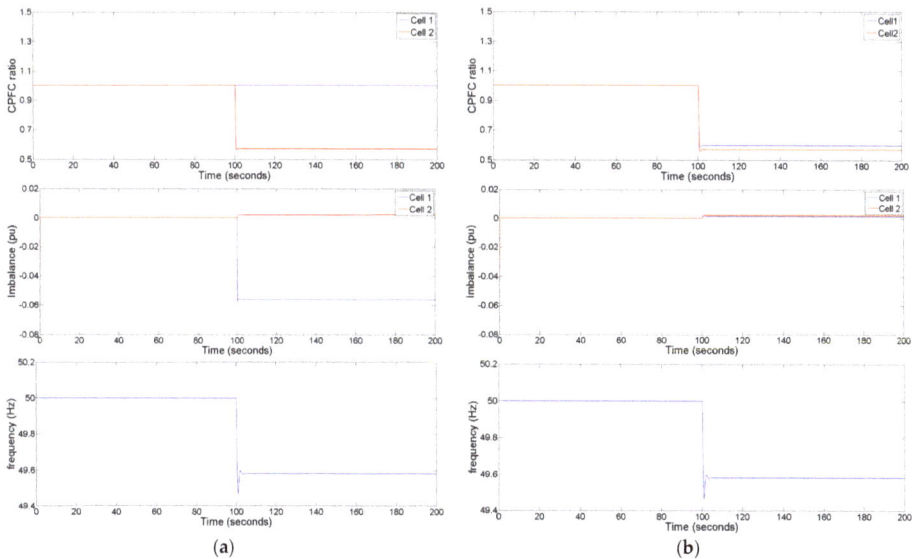

Figure 5. Simulation results for the short-term analysis of the controller. In both columns, the diagrams show the Cell Power-Frequency Characteristic (CPFC) variation (top), the imbalance (middle), and the frequency response (bottom diagrams): (**a**) Imbalance incident in cell 1; (**b**) Imbalance incident in cell 3.

4.2. Scenario B: Long-Term Analysis

This scenario is used to investigate the behaviour of the control scheme in a 24 h operation of the above-described network. To this end, the power profiles shown in Figure 6 were used. These profiles have a sampling rate of 15 min. The input data shown in this diagram were scaled down and used as input signals to variable loads/generators in the model. In order to do this, the original data were divided by the maximum power of each profile in order to calculate the values in pu.

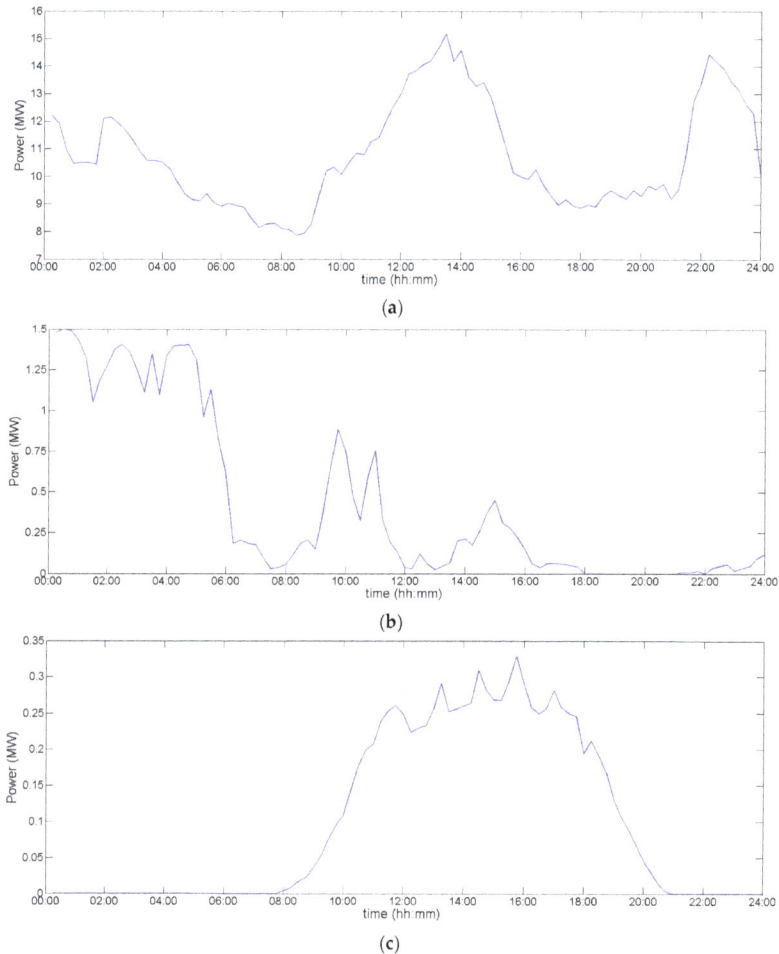

(a)

(b)

(c)

Figure 6. Input data used for the 24 h tests: (**a**) Load profile; (**b**) Wind generation profile; (**c**) photo voltaics (PV) generation profile.

Once the pu values of load were obtained, the minimum load value was used as base-load in the system (fixed load at each bus) and the variable load was introduced as dynamic loads at each bus. In order to have differentiation among the buses, the dynamic load profile of each bus was modified using a random number generator. Finally, to get the actual power at the buses, each pu load profile was scaled-up by a factor based on the nominal load of each bus given by the CIGRE model. The same scaling-down and -up strategy was also used for the PVs and WG profiles, ignoring, however, the differentiation among PVs (all PVs produced the same profile of power). Furthermore, of the four cells making up the specific test grid, only two were equipped with the adaptive control. In order to highlight the impact of these controllers on the overall stability as well as to better assess the impact on the energy usage, the HV cell in this model was assumed to incorporate a relatively small droop slope corresponding to 1 MW/Hz. The contribution of the DER units in the droop slope was selected so that it would reflect their capability of varying their output power within specific margins. Thus, RES such as PVs and WG were assumed to be able only to reduce their output power (curtailment of generation) in case of production surplus. Also, other DER units such as batteries were considered to have capability of both increasing/decreasing their power, responding to any frequency variation. Due to this asymmetric contribution of DER, the scheduled droop slope of each unit should be selected such that a symmetric overall CPFC can be achieved. In our case, however, in order to investigate the impact that an asymmetric CPFC could have on the overall stability, the aggregated maximum droop slope or CPFC of each cell was selected −233 kW/Hz (negative delta *f*) and −300 kW/Hz (positive delta f) for cell 1 and −314 kW/Hz (negative delta f) and −610 kW/Hz (positive delta f) for cell 2.

Apart from the significant asymmetry in the aggregated CPFC, the part of the characteristic concerned with the positive frequency deviations provided by cells 1 and 2 is approximately equal to the HV droop slope which is fixed. Also, since this part of the CPFC is delivered by RES and since one of the main objectives of the test is to show the curtailment reduction caused on such kind of useful energy, the disturbance scenarios included stepwise reductions in the scheduled load profile of Figure 6. Specifically, each timeframe (15 min) one-step change in the power consumption of each bus is implemented. Each change is selected to −10% of the actual power at the moment of reduction. This way, the resulting frequency disturbances lead to increase of frequency and activation of the controller part mainly related to the PVs and WG of the power system. For the quantification of the resulting reserves reduction, two types of mathematical formulation were used. The first one corresponds to the absolute energy usage expressed by the formula:

$$ABS = \int_0^\infty \left[\left| \Delta P_{droop,1} \right| + \ldots + \left| \Delta P_{droop,n} \right| \right] dt \tag{4}$$

The second formulation is an expression of the cost of the usage of these reserves expressed as:

$$Cost = \int_0^\infty \left[\left(\Delta P_{droop,1} \right)^2 + \ldots + \left(\Delta P_{droop,n} \right)^2 \right] dt \tag{5}$$

Last, but not least, the other HV area parameters used in this scenario were K_p = 100 Hz/pu and H = 5 s. Based on the afore-mentioned assumptions, the 24 h simulation test with and without the use of adaptive FCC control shows that a significant reduction in the use of FCC reserves can be achieved in total, but also for the individual types or RES reserves as well. The results obtained for this scenario are summarized in Tables 4 and 5, respectively. Table 4 illustrates the energy/cost saving for all DER when the proposed control is used (left column) in contrast with the classic fixed droop (right column). The results in this table show a significant overall reduction in both the energy (19.7%) and the cost (26.7%) by means of the adaptive modification of CPFC in the system. Similarly, a more specific table (Table 5) shows the impact on the adaptive control on the RES of the system. It is evident that with the use of the proposed controller a significant reduction in RES energy loss is achieved. It is worth noting that the significant reduction in the use of reserves does not compromise the overall stability since the maximum frequency deviation for a fixed-droop FCC is 52.50 Hz, whereas the implementation of the adaptive control increases only slightly the deviation to 52.54 Hz. Likewise, the minimum frequency deviations with and without the use of adaptive control are 49.42 Hz and 49.51 Hz, respectively. The relatively high frequency deviation in both versions of the controller is due to the low overall droop slope of the system. Last, but not least, the system remains stable despite the asymmetric droop due to the type of reserves. Figure 7 shows the frequency response of the two cells over time for the case of adaptive FCC control. It is worth noting that by implementing a slightly modified controller version with an increase of the CPFC ratio above 1 when the incident takes places inside the corresponding cell, it would be possible to maintain the frequency deviation equal to the fixed-droop version of the controller. In such a case, the local character of the reserves activation remains the same, with all the related benefits of this approach.

Table 4. Comparison of aggregated usage of FCC reserves with and without adaptive control.

Function	With Adaptive FCC	With Fixed Droop
ABS (Reduction)	8.59×10^7 (Ws) (-19.7%)	1.07×10^8 (Ws)
Cost	6.94×10^{12} (W^{2}s)	9.46×10^{12} (W^{2}s)
(Reduction)	(-26.7%)	-

The drawback of such a variation, however, is the requirement of a higher amount of reserved power. Thus, in our study we only investigated the curtailment of CPFC with the concession that a slight increase of frequency deviation is acceptable. An operator, however, could select the alternative method if the availability of extra reserves is not an issue.

Table 5. Comparison of curtailment in the renewable energy systems (RES) production for the two FCC versions.

RES	With Adaptive FCC	With Fixed Droop
Wind Generation (Reduction)	1.24×10^8 (Ws) (-18.0%)	1.51×10^8 (Ws)
Photovoltaic Generation	8.97×10^6 (Ws)	9.56×10^6 (Ws)
(Reduction)	(-6.2%)	-

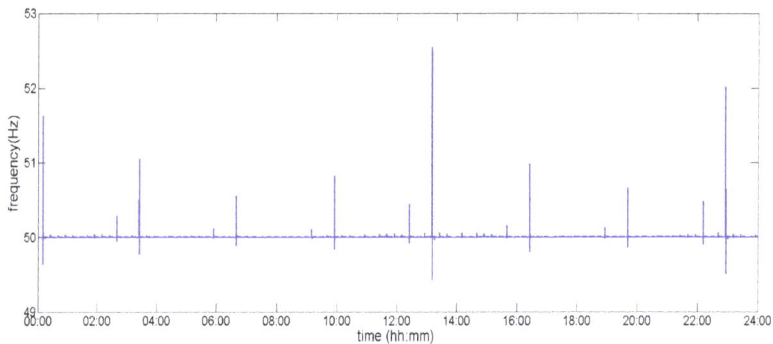

Figure 7. Frequency response during the 24 h simulation test with the use of adaptive FCC control.

5. Conclusions

This study deals with the use of an adaptive control method for FCC in future power systems. The basic idea behind the control method is that the system consists of interconnected cells, entities capable of dealing with imbalance issues even at distribution level, and makes use of the frequency and tie-line error signals in order to obtain a secure imbalance estimation in real time. Based on the imbalance location and size, the controller curtails the droop contribution of the cell so that reduced use of FCR is induced. This results in a number of benefits linked to efficient, improved local operation of reserves. This avoids congestions, over-voltages and conflicts of use with other control schemes and increases the utilisation of the RES energy from areas located far away from the imbalance incident. The method was developed based on a fuzzy controller and it was implemented and tested using a 4-cell interconnected system. The results showed that not only is the proposed controller capable

Energies **2017**, *10*, 982

of reducing the overall usage of FCR, but it can also maintain a quite satisfactory level of stability without deteriorating the overall system behaviour since it only slightly influences the peak deviations of frequency and tie-line power.

Acknowledgments: The research leading to these results has received funding from the European Union Seventh Framework Programme (FP7/2007–2013) under grant agreement No. 609687. Any opinions, findings and conclusions or recommendations expressed in this material are those of the authors and do not necessarily reflect those of the European Commission.

Author Contributions: Chris Caerts conceived the idea of adaptive FCC control. Evangelos Rikos designed and implemented the fuzzy controller. Chris Caerts, Evangelos Rikos and Graeme Burt conceived the WoC concept. Evangelos Rikos and Mattia Cabiati conceived and designed the simulation tests. Evangelos Rikos, Mattia Cabiati and Mazheruddin Syed analysed the test results. Evangelos Rikos, Chris Caerts, Mazheruddin Syed and Mattia Cabiati wrote the paper.

Conflicts of Interest: The authors declare no conflict of interest.

1. European Commission. *Strategic Energy Technology (SET) Plan-Towards an Integrated Roadmap: Research and Innovation Challenges and Needs of the EU Energy System*; European Commission: Rome, Italy, 2014; pp. 1–2.
2. e-Highway 2050. Available online: www.e-highway2050.eu/e-highway2050/ (accessed on 20 October 2016).
3. ENTSO-E. *10 Year Network Development Plan 2014*; ENTSO-E: Brussels, Belgium, 2014; pp. 6–16. Available online: https://www.entsoe.eu/major-projects/ten-year-network-development-plan/tyndp2014/Documents/TYNDP%202014_FINAL.pdf (accessed on 8 May 2017).
4. D'hulst, R.; Fernández, J.M.; Rikos, E.; Kolodziej, D.; Heussen, K.; Geibel, D.; Temiz, A.; Caerts, C. Voltage and Frequency Control for Future Power Systems: The ELECTRA IRP Proposal. In Proceedings of the International Symposium on Smart Electric Distribution Systems and Technologies (EDST 2015), Vienna, Austria, 8–11 September 2015; pp. 1–6.
5. Martin-Martínez, F.; Sánchez-Miralles, A.; Rivier, M. A literature review of Microgrids: A functional layer based classification. *Renew. Sustain. Energy Rev.* **2016**, *62*, 1133–1153. [CrossRef]
6. Su, W.; Wang, J. Energy Management Systems in Microgrid Operations. *Electr. J.* **2012**, *25*, 45–60. [CrossRef]
7. Hatziargyriou, N. *Microgrids: Architectures and Control*; Wiley/IEEE Press: London, UK, 2014.
8. Parhizi, S.; Lotfi, H.; Khodaei, A.; Bahramirad, S. State of the Art in Research on Microgrids: A Review. *IEEE Access* **2015**, *3*, 890–925. [CrossRef]
9. ENTSO-E. *Continental Europe Operation Handbook- Appendix 1, Load Frequency Control and Performance*; ENTSO-E: Brussels, Belgium, 2004; pp. 1–9. Available online: https://www.entsoe.eu/publications/system-operations-reports/operation-handbook/Pages/default.aspx (accessed on 8 May 2017).
10. Vogler-Finck, P.J.C.; Früh, W.-G. Evolution of primary frequency control requirements in Great Britain with increasing wind generation. *Electr. Power Energy Syst.* **2015**, *73*, 377–388. [CrossRef]
11. Teng, F.; Mub, Y.; Jia, H.; Wuc, J.; Zeng, P.; Strbac, G. Challenges on primary frequency control and potential solution from EVs in the future GB electricity system. *Appl. Energy* **2017**, *194*, 353–362. [CrossRef]
12. Oureilidis, K.; Demoulias, Ch. A decentralized impedance-based adaptive droop method for power loss reduction in a converter-dominated islanded microgrid. *Sustain. Energy Grids Netw.* **2016**, *5*, 39–49. [CrossRef]
13. Chaudhuri, N.R.; Chaudhuri, B. Adaptive Droop Control for Effective Power Sharing in Multi-Terminal DC (MTDC) Grids. *IEEE Trans. Power Syst.* **2013**, *28*, 21–29. [CrossRef]
14. Marzbanda, M.; Moghaddamb, M.M.; Akoredec, M.F.; Khomeyranib, G. Adaptive load shedding scheme for frequency stability enhancement in microgrids. *Electr. Power Syst. Res.* **2016**, *140*, 78–86. [CrossRef]
15. Liu, H.; Hu, Z.; Song, Y.; Lin, J. Decentralized Vehicle-to-Grid Control for Primary Frequency Regulation Considering Charging Demands. *IEEE Trans. Power Syst.* **2013**, *28*, 3480–3489. [CrossRef]
16. Amin Ghasemi, M.; Parniani, M. Prevention of distribution network overvoltage by adaptive droop-based active and reactive power control of PV systems. *Electr. Power Syst. Res.* **2016**, *133*, 313–327. [CrossRef]
17. Hoseinzadeh, B.; Faria da Silva, F.M.; Bak, C.L. Adaptive Tuning of Frequency Thresholds Using Voltage Drop Data in Decentralized Load Shedding. *IEEE Trans. Power Syst.* **2015**, *30*, 2055–2062. [CrossRef]
18. Ahmadi, S.; Shokoohi, S.; Bevrani, H. A fuzzy logic-based droop control for simultaneous voltage and frequency regulation in an AC microgrid. *Electr. Power Energy Syst.* **2015**, *64*, 148–155. [CrossRef]

19. Mai Ersdal, A.; Imsland, L.; Uhlen, K.; Fabozzi, D.; Thornhill, N.F. Model predictive load—Frequency control taking into account imbalance uncertainty. *Control. Eng. Pract.* **2016**, *53*, 139–150. [CrossRef]
20. Hota, P.K.; Mohanty, B. Automatic generation control of multi-source power generation under deregulated environment. *Electr. Power Energy Syst.* **2016**, *75*, 205–214. [CrossRef]
21. Papadimitriou, C.; Vovos, N. A Fuzzy Control Scheme for Integration of DGs into a Microgrid. In Proceedings of the 15th IEEE Mediterranean Electrotechnical Conference Melecon, Valletta, Malta, 26–28 April 2010; pp. 872–877.
22. CIGRE. *Benchmark Systems for Network Integration of Renewable and Distributed Energy Resources*; CIGRE Task Force C6.04.02; CIGRE: Paris, France, 2009.

energies

MDPI

Article

A Combined Approach Effectively Enhancing Traffic Performance for HSR Protocol in Smart Grids

Nguyen Xuan Tien [1], Jong Myung Rhee [1] and Sang Yoon Park [2],*

[1]   Department of Information and Communications Engineering, Myongji University, 116 Myongji-ro, Yongin, Gyeonggi 17058, Korea; tiennguyen@mju.ac.kr (N.X.T.); jmr77@mju.ac.kr (J.M.R.)
[2]   Department of Electronic Engineering, Myongji University, 116 Myongji-ro, Yongin, Gyeonggi 17058, Korea
*   Correspondence: sypark@mju.ac.kr; Tel.: +82-31-330-6751

Received: 4 August 2017; Accepted: 5 September 2017; Published: 8 September 2017

Abstract: In this paper, we propose a very effectively filtering approach (EFA) to enhance network traffic performance for high-availability seamless redundancy (HSR) protocol in smart grids. The EFA combines a novel filtering technique for QuadBox rings (FQR) with two existing filtering techniques, including quick removing (QR) and port locking (PL), to effectively reduce redundant unicast traffic within HSR networks. The EFA filters unicast traffic for both unused terminal rings by using the PL technique and unused QuadBox rings based on the newly-proposed FQR technique. In addition, by using the QR technique, the EFA prevents the unicast frames from being duplicated and circulated in rings; the EFA thus significantly reduces redundant unicast traffic in HSR networks compared with the standard HSR protocol and existing traffic filtering techniques. The EFA also reduces control overhead compared with the filtering HSR traffic (FHT) technique. In this study, the performance of EFA was analyzed, evaluated, and compared to that of the standard HSR protocol and existing techniques, and various simulations were conducted to validate the performance analysis. The analytical and simulation results showed that for the sample networks, the proposed EFA reduced network unicast traffic by 80% compared with the standard HSR protocol and by 26–62% compared with existing techniques. The proposed EFA also reduced control overhead by up to 90% compared with the FHT, thus decreasing control overhead, freeing up network bandwidth, and improving network traffic performance.

Keywords: smart grids; substation automation system (SAS); high-availability seamless redundancy (HSR); seamless communications; traffic reduction technique

---

1. Introduction

The smart grid is a new concept of the next-generation electric power system that has emerged to address challenges of the existing power grid. The smart grid is a modern electric power grid infrastructure for enhanced efficiency, reliability and safety, with smooth integration of renewable and alternative energy sources, through automated control and modern communications technologies [1]. Communication architecture of smart grid including requirements, topologies, and protocol stack is required for smart grid environments [2]. Smart grids provide more electricity to meet rising demand, increase reliability and quality of power supplies, increase energy efficiency, be able to integrate low carbon energy sources into power networks [3]. To realize the smart grid technology, legacy substations are retrofitted by substation automation systems (SASs), allowing the robust control and communication tasks [4]. The SAS system is used to control, supervise, and protect substations that are strategic nodes in power networks, which consist of large numbers of switchgears and measuring devices [5]. Functions of an SAS are control and supervision, as well as protection and monitoring of the primary equipment and of the grid. The functions can be assigned to three levels: the station level, the bay level, and the process level. Devices of an SAS may be physically installed on different functional levels [5]. The substation should continue to be operable if any SAS communications component fails [6]. Communication networks in the SAS are defined by the International Electrotechnical Commission (IEC)'s IEC 61850 standard [5,6]. To ensure the operations of substations, seamless communication with fault-tolerance can be provided for SASs by using redundancy protocols. In most redundancy protocols, the communication is recovered from a fault in the network; this recovery takes some time, and even a short time period could be unacceptable for certain time-critical applications, including SAS [7]. To help with this problem, the IEC developed and standardized two protocols that provide seamless communication with fault tolerant capability, including the parallel redundancy protocol (PRP) and high-availability seamless redundancy (HSR) [8]. Both these protocols implement redundancy in the nodes and provide seamless communication based on the duplication of frames sent over redundant paths from a source to a destination. Unlike the PRP that uses two identical local area network (LAN) of similar topology, the HSR is applied to a single LAN to provide seamless redundancy for the network. With respect to PRP, HSR allows to roughly halve the network infrastructure [8]. HSR is one of the redundancy protocols selected to provide seamless communication for SASs. HSR is mainly used ring topologies, including single-ring and connected-ring networks. In HSR, a node has two ports operated in parallel, called a doubly-attached node with HSR protocol (DANH). Singly-attached nodes (SANs), such as maintenance laptops or printers, cannot be inserted directly into HSR rings. SANs communicate with HSR ring devices through a redundancy box (RedBox) that acts as a proxy for the SANs attached to it. Quadruple port devices (QuadBoxes) are used to connect DANH rings in connected-ring networks. A single DANH ring consists of DANHs, each having two ring ports, interconnected by full-duplex links. When a source DANH sends a unicast frame to a destination DANH in a DANH ring, the source DANH prefixes the frame by an HSR tag and sends the tagged frame over each port. In the failure-free case, the destination DANH receives two identical frames from each port, removes the HSR tag of the first frame before passing it to its upper layers, and discards any duplicate. The standard HSR protocol works very effectively in single-ring networks. The HSR protocol, however, generates too much redundant unicast traffic in connected-ring networks. This drawback is caused by the following issues:

1. Issue 1: Duplicating and circulating frames in all the rings, except the destination DANH ring;
2. Issue 2: Forwarding unicast frames into all DANH rings;
3. Issue 3: Forwarding unicast frames into all QuadBox rings.

Several traffic filtering techniques have been proposed to solve the drawback and improve the network performance in HSR networks, including the quick removing (QR) technique [9], the traffic control (TC) technique [10], the port locking (PL) technique [11], the hybrid QR and PL approach (QRPL) [12], the enhanced port locking (EPL) technique [13], and the filtering HSR traffic (FHT) technique [14]. However, most of the techniques (QR, TC, PL, QRPL, and EPL) do not solve all the HSR issues, whereas the FHT solves all the issues but generates additional control overhead in HSR networks. Several dual paths-based techniques have been proposed to reduce redundant unicast traffic in HSR networks based on pre-established paths. These techniques discover and establish dual paths between a source and a destination in an HSR network before forwarding unicast traffic frames from the source to the destination through the dual paths. Dual paths-based techniques include the dual virtual paths (DVP) [15] technique, which was then extended as extended dual virtual paths (EDVP) [16], the ring-based dual paths (RDP) [17] technique, and the dual separate paths (DSP) [18] technique. These dual paths-based techniques significantly reduce redundant unicast traffic in HSR networks. The main drawback of the techniques, however, is to generate additional control overhead in the networks because they exchange control messages to discover and establish dual paths. In addition, there are other techniques for reducing redundant unicast traffic in HSR networks, including the HSR SwitchBox technique [19] and the integration of HSR and OpenFlow (HSE + OF) [20]. The HSR SwitchBox technique defines a new switching node in HSR networks that forwards HSR frames based on looking up of media access control (MAC) tables instead of flooding the frames. The HSE + OF approach aims to manage HSR networks by means of the software defined networking (SDN) paradigm. The approach defines new HSE + OF nodes whose control plane is managed by an OpenFlow controller [20]. In other words, this approach is an implementation of HSR in SDN.

In this paper, we propose a very effectively filtering approach (EFA) to significantly reduce redundant unicast traffic in HSR networks by combining two existing filtering techniques with a newly-proposed filtering technique. We propose a new traffic filtering technique for QuadBox rings (FQR) to filter unicast traffic for unused QuadBox rings. The FQR technique is then combined with two existing techniques, including PL and QR to provide the combined approach EFA that effectively reduces redundant unicast traffic in HSR networks. The proposed EFA filters unicast traffic for both unused DANH rings by using the PL technique and unused QuadBox rings by using the newly-proposed FQR technique. The EFA also prevents the unicast traffic frames from being duplicated and circulated in rings by using the QR technique. The EFA, thus, significantly reduces redundant unicast traffic in HSR networks compared with the standard HSR protocol and other traffic-filtering techniques, such as QR, PL, QRPL, and EPL. In addition, although the EFA, FHT, and dual paths-based techniques exhibit similar network traffic performance, the EFA generates less control overhead than the FHT technique and existing dual paths-based techniques. In other words, the motivation of this paper is to propose a novel approach that reduces more redundant unicast traffic than the existing filtering techniques and generates less control overhead than the FHT and existing dual paths-based techniques. The proposed approach, therefore, saves network bandwidth and improves network performance.

The rest of this paper is organized as follows: Section 2 describes several existing traffic filtering techniques, while Section 3 introduces the proposed EFA for effectively reducing unicast traffic in HSR networks. In Section 4, the performance of the proposed approach is analyzed, evaluated, and compared to that of the standard HSR protocol and existing techniques. Section 5 describes several simulations and their results in order to evaluate and validate the performance analysis. Finally, Section 6 provides several conclusions drawn from this work.

2. Related Work

This section introduces several existing traffic filtering techniques including QR, PL, QRPL, EPL, and FHT that were proposed to reduce redundant unicast traffic in HSR networks. To demonstrate the operations of these existing filtering techniques, we consider an HSR network with eight DANH rings and three QuadBox rings as shown in Figure 1.

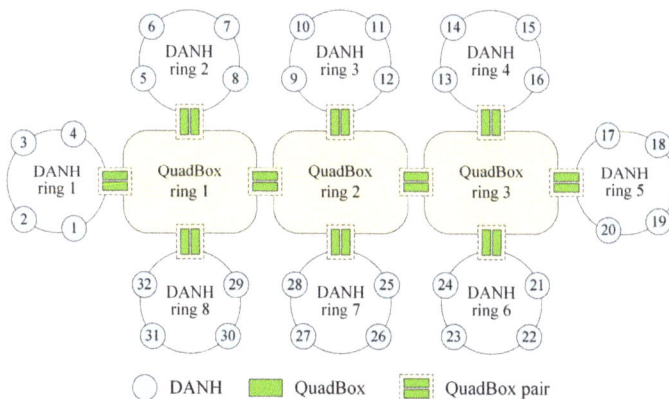

Figure 1. A high-availability seamless redundancy (HSR) network with eight doubly-attached node with HSR protocol (DANH) rings.

Under the standard HSR protocol, when source node 1 sends a unicast frame to destination node 10, the frame is duplicated and circulated in all rings, except the destination DANH ring as shown in Figure 2. This flooding process generates excessively redundant unicast traffic in the HSR network.

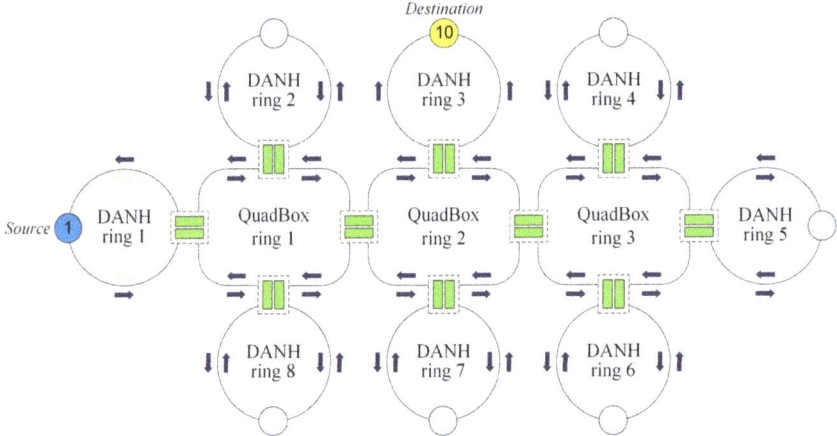

Figure 2. Forwarding a unicast frame using the standard HSR protocol.

2.1. QR

QR reduces redundant traffic in HSR networks by solving the HSR issue 1. In other words, the QR technique prevents traffic frames from being duplicated and circulated in rings, as shown in Figure 3. When a node receives an HSR frame for the first time, the node floods the frame over all its ports except the received port from which the frame has been received. Later, if the node continues to receive other copies of the frame, the node will discard the duplicate copies.

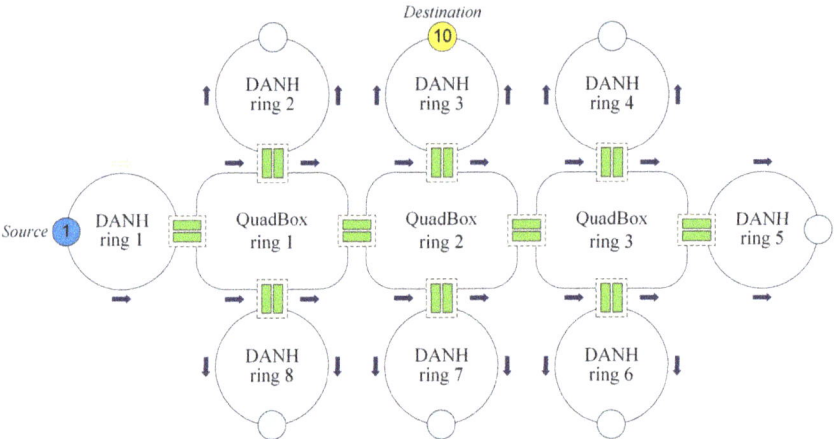

Figure 3. Forwarding a unicast frame using the quick removing (QR) technique.

QR is the simplest traffic filtering technique. The technique can filter any traffic. The main disadvantage of QR, however, is that it does not filter unicast traffic for unused rings.

2.2. PL

PL reduces redundant unicast traffic in HSR networks by filtering unicast traffic for unused DANH rings. PL divides each access QuadBox into two sides: a DANH side that is connected to a DANH ring and a QuadBox side that is connected to a QuadBox ring. The technique does not forward a unicast frame to DANH rings that do not contain the destination DANH by locking DANH sides connecting the DANH rings. When the source DANH sends unicast frames to the destination DANH, the PL technique uses the first sent frame to check if DANH rings contain the destination DANH and then lock DANH rings that do not contain the destination DANH. After the phase, the PL does not forward unicast frames into the locked DANH rings, as shown in Figure 4.

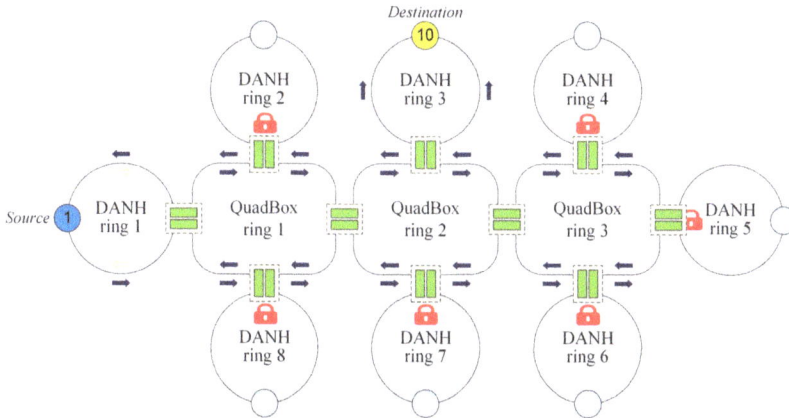

Figure 4. Forwarding a unicast frame using the port locking (PL) technique.

The drawbacks of PL, however, are that it does not solve the HSR issues 1 and 3.

2.3. QRPL

QRPL is a hybrid approach that combines the QR technique with the PL technique. The QRPL approach uses the PL technique to filter unicast traffic frames for DANH rings and the QR technique to remove duplicated and circulated frames from rings, as shown in Figure 5.

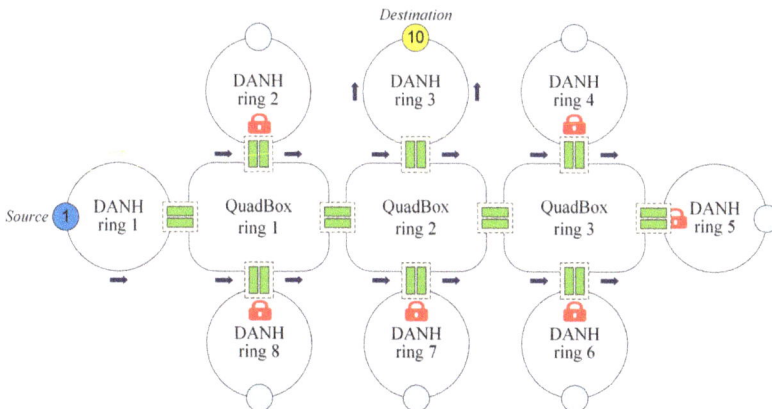

Figure 5. Forwarding a unicast frame using the hybrid QR and PL (QRPL) approach.

The main drawback of the QRPL approach, however, is that it does not filter unicast traffic for QuadBox rings, which results in redundant unicast traffic traveling in these rings.

2.4. EPL

EPL is the improved version of the PL technique that filters unicast traffic for both unused DANH and QuadBox rings, as shown in Figure 6. The EPL technique works with the same locking concept as the PL technique.

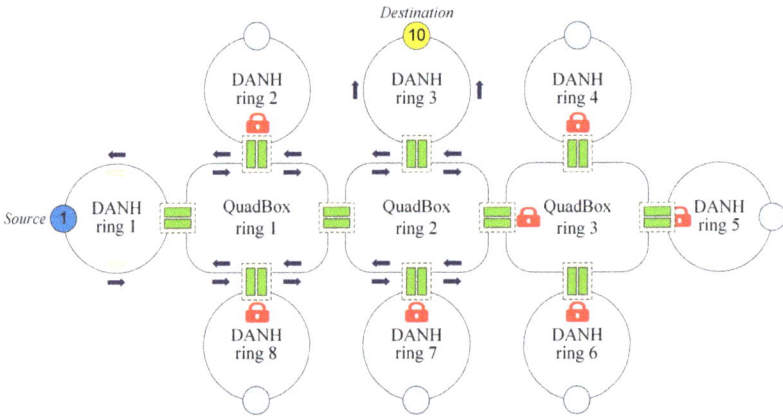

Figure 6. Forwarding a unicast frame using the enhanced port locking (EPL) technique.

The EPL technique, however, does not prevent unicast frames from being duplicated and circulated in active rings.

2.5. FHT

FHT is the only technique that solves all the HSR issues abovementioned in Section 1. The FHT technique filters unicast frames for both unused DANH and QuadBox rings by using MAC tables. The technique learns MAC addresses of DANHs and builds MAC tables by exchanging control messages. In addition, the FHT prevents unicast frames from being duplicated in rings. The process of forwarding unicast frames under the FHT technique is shown in Figure 7.

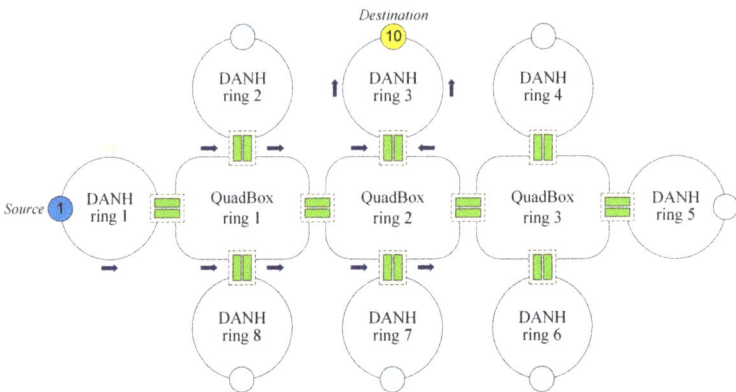

Figure 7. Forwarding a unicast frame using the filtering HSR traffic (FHT) technique.

The main drawback of the FHT technique is that it generates additional control overhead in HSR networks due to the use of control messages.

2.6. Comparisons

As described above, several existing traffic filtering techniques have advantages and disadvantages. While QR prevents unicast frames from being duplicated and circulated in rings, it does not filter unicast traffic for unused rings. The PL and EPL filter unicast traffic for unused rings. They, however, do not remove duplicated unicast traffic from rings. The QRPL is a hybrid approach that combines the QR with the PL in order to filter unicast traffic for unused DANH rings and to prevent the traffic from being duplicated and circulated in rings. The QRPL, however, does not filter the unicast traffic for unused QuadBox rings. The FHT filters unicast traffic for both unused DANH and QuadBox rings, as well as removing duplicated traffic from rings. FHT does generate a certain degree of control overhead for learning and building MAC tables, however. The characteristics of these filtering techniques are summarized in Table 1.

Table 1. Characteristics of traffic filtering techniques.

Features	QR	PL	QRPL	EPL	FHT
Filter traffic for DANH rings	-	√	√	√	√
Filter traffic for QuadBox rings	-	-	-	√	√
Remove duplicated traffic	√	-	√	-	√
Generate control overhead	No	No	No	No	Medium

3. The Proposed EFA

Three types of QuadBoxes are defined in this paper: access QuadBoxes, trunk QuadBoxes, and QuadBox pairs:

- The access QuadBox is a QuadBox that connects to at least one DANH ring;
- The trunk QuadBox is a QuadBox that does not connect to any DANH ring; in other words, trunk QuadBoxes are used to connect QuadBox rings;
- A QuadBox pair refers to two QuadBoxes that are used as a pair to connect two rings to prevent a single point of failure.

The main purpose of the proposed EFA is to filter unicast traffic frames for both unused DANH rings and unused QuadBox rings, as well as preventing the unicast traffic frames from being duplicated and circulated in rings of HSR networks.

3.1. Removing Duplicated Unicast Traffic

To solve issue 1 of HSR, the EFA uses the QR technique to prevent unicast traffic frames from being duplicated and circulated in all rings of HSR networks. By using the QR technique, each HSR node, such as a DANH node or a QuadBox node, forwards a unicast frame over all its ports (except the received port) when the node has received the frame for the first time. Later, if the node receives copies of the frame, the node will discard the duplicated copies.

3.2. Filtering Unicast Traffic for DANH Rings

The EFA filters unicast traffic for any DANH rings that do not contain the destination by applying the PL technique. For a communication session between source node 1 and destination node 10, when the source sends the first unicast frame, each access QuadBox forwards the frame into its DANH ring to check if its DANH ring contains the destination node. If not, then the access QuadBox locks its DANH side to prevent the frame from being forwarded into the DANH ring.

3.3. Filtering Unicast Traffic for QuadBox Rings

To solve issue 3 of HSR (as noted above), we propose in this paper a novel filtering technique called "filtering for QuadBox rings" (FQR) to filter unicast traffic for unused QuadBox rings. FQR divides each trunk QuadBox into two sides, both of which consist of two ports connected to a QuadBox ring. For a communication session between a source and a destination in an HSR network, FQR sets each side of each trunk QuadBox to either the source side or the destination side. The source side is closer to the source than the destination side, as shown in Figure 8. The source and destination sides of each trunk QuadBox are only locally significant in each communication session.

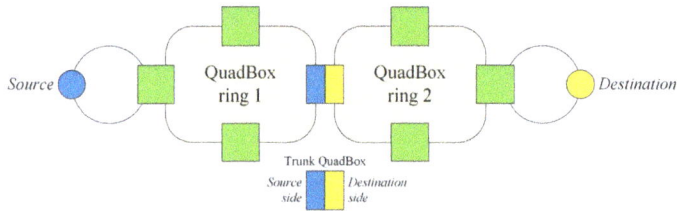

Figure 8. Source side and destination side of a trunk QuadBox.

FQR filters unicast traffic for any QuadBox rings that are not used to deliver unicast frames in a communication session between a source and a destination in an HSR network by locking the destination sides of the QuadBoxes. The FQR operation consists of (1) the setting step and (2) the locking step, as described below.

3.3.1. Setting Step

In this step, each side of each trunk QuadBox is set to either the source side or the destination side. This step is initiated once a communication session between a source and a destination is started. In other words, the step is performed while the first traffic frame is delivered from the source to the destination of the communication session. When the source sends the first HSR frame to the destination, the frame is flooded into all the rings in the network (as occurs in the standard HSR protocol). When a trunk QuadBox receives the first frame of the communication session, the trunk QuadBox sets the side from which the frame is received to the source side and sets the other side to the destination side, as shown in Figure 9.

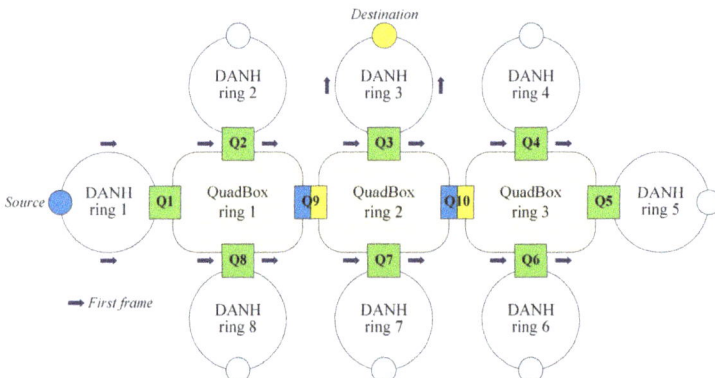

Figure 9. The process of the setting step.

3.3.2. Locking Step

This step locks the destination sides of any QuadBoxes that are not used to deliver unicast traffic frames from the source to the destination. Once the destination receives the first HSR frame sent by the source, the destination initiates the locking step by sending a "locking" message back to the source. Only trunk QuadBoxes process and forward the locking message, whereas access QuadBoxes discard the message. When a trunk QuadBox receives a locking message, it checks if the side from that the locking message has been received is the source side or the destination side:

- If the received side is the destination side, then the QuadBox does not lock its destination side.
- If the received side is the source side, then the QuadBox locks its destination side.

The trunk QuadBox then forwards the received locking message over its ports, except for the received port and the locked ports, which belong to the locked destination side. Figure 10 shows the process of checking and locking destination sides based on receiving the locking message at trunk QuadBoxes.

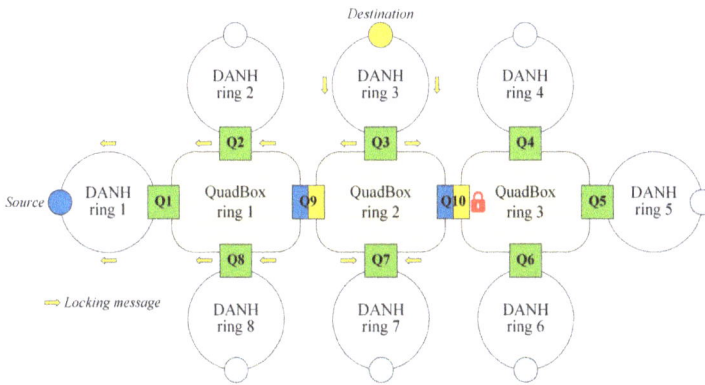

Figure 10. The process of the locking step.

3.4. Operations of the Proposed EFA

The EFA consists of two phases as shown in Figure 11, including locking phase and filtering phase. The locking phase occurs during the forwarding process of the first frame. In this phase, the EFA checks and locks access and trunk QuadBoxes that connect to unused DANH and QuadBox rings, respectively. The locking phase is followed by the filtering phase that starts when the second frame is sent. In the filtering phase, the EFA does not forward unicast frames into unused DANH and QuadBox rings.

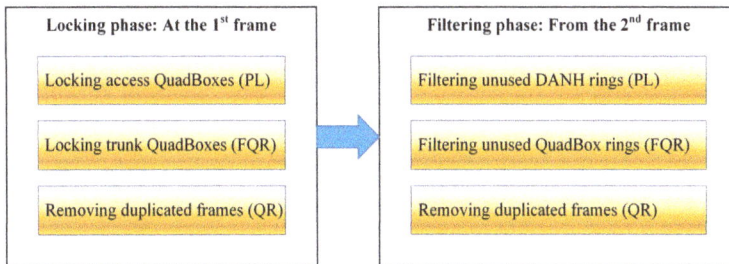

Figure 11. The operational diagram of effectively filtering approach (EFA).

At the first frame, the network will work in the same way as the standard HSR protocol. From the second frame, unicast frames will be forwarded and filtered under the EFA effect. In the locking phase, when the source sends the first frame of a communication session to the destination, the frame is flooded in the whole network. During delivery of the first frame, access QuadBoxes use the PL technique to check their DANH rings and lock their DANH sides if their DANH rings do not contain the destination. Once the destination receives the first unicast frame, the destination sends a locking message back to the source. Upon receiving the locking message, trunk QuadBoxes use the FQR technique to check and lock their destination sides, if needed. Access QuadBoxes discard the locking message they have received. At that time, QuadBoxes have locked sides that are not used to forward unicast frames from the source to the destination. In the filtering phase, the next frames will not be forwarded into any DANH and QuadBox rings that are connected to locked DANH sides and locked destination sides. In addition, the EFA uses the QR technique to remove any duplicated unicast frames from rings. Figure 12 illustrates the process of forwarding unicast frames from source 1 to destination 10 in the sample HSR network.

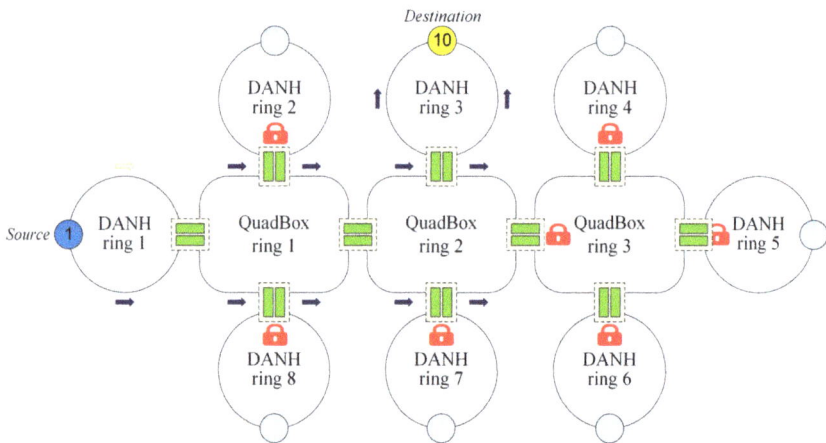

Figure 12. Forwarding a unicast frame using the EFA approach.

4. Performance Analysis

In this section, network traffic performance of EFA is first analyzed and compared with that of the standard HSR protocol and existing techniques. Control overhead performance of EFA is then analyzed and compared with that of the FHT technique.

4.1. Network Traffic Performance

This section describes the network traffic performance analysis of EFA compared to the standard HSR and existing techniques, including QR, PL, QRPL, and FHT.

In this paper, to analyze and evaluate traffic performance, network traffic was chosen as a performance metric. Network traffic is defined as the total number of frame copies that travel on links and that are received by nodes in the network. When a unicast frame is sent from a source to a destination in a network, the network traffic is the total number of the frame's copies that are delivered and received in the network.

4.1.1. Under the Standard HSR Protocol

Let nt^1_{HSR} be network traffic when a source node sends a unicast frame to a destination node under the standard HSR protocol. nt^1_{HSR} is calculated as follows:

$$nt^1_{HSR} = \sum_{i \in NR} 2n_i - n_D \tag{1}$$

where n_D is the number of nodes in the destination DANH ring, n_i is the number of nodes in the ith ring, and NR is a set of all rings in the network.

Generally, network traffic when a source node sends N unicast frames to a destination node under the standard HSR protocol, denoted by nt_{HSR}, is determined by:

$$nt_{HSR} = N \left(\sum_{i \in NR} 2n_i - n_D \right) \tag{2}$$

4.1.2. Under the QR Technique

When a source node sends a unicast frame to a destination node, network traffic under the QR technique, denoted by nt^1_{QR}, is calculated by:

$$nt^1_{QR} = \sum_{i \in NR} n_i \tag{3}$$

Generally, network traffic when a source node sends N unicast frames to a destination node under the QR technique, denoted by nt_{QR}, is determined as follows:

$$nt_{QR} = N \sum_{i \in NR} n_i \tag{4}$$

4.1.3. Under the PL Technique

Network traffic when a source node sends the first unicast frame to a destination node under the PL technique, denoted by nt^1_{PL}, can be calculated using Equation (1):

$$nt^1_{PL} = \sum_{i \in NR} 2n_i - n_D \tag{5}$$

Network traffic when the source node sends the kth unicast frame ($k \geq 2$) to the destination node under the PL technique, denoted by nt^k_{PL}, is determined as follows:

$$nt^k_{PL} = \sum_{i \in QR} 2n_i + 2n_S + n_D \tag{6}$$

where n_S is the number of nodes in the source DANH ring and QR is a set of all QuadBox rings in the network.

Generally, network unicast traffic when a source node sends N unicast frames to a destination node under the PL technique, denoted by nt_{PL}, is determined as follows:

$$nt_{PL} = \sum_{i \in NR} 2n_i + (N-1) \left(2n_S + \sum_{i \in QR} 2n_i \right) + (N-2)n_D \tag{7}$$

4.1.4. Under the QRPL Approach

Network traffic when a source node sends the first unicast frame to a destination node in an HSR network under the QRPL approach, denoted by nt^1_{QRPL}, is similar to that under the QR technique and is calculated using Equation (3):

$$nt^1_{QRPL} = \sum_{i \in NR} n_i \tag{8}$$

As with PL, starting from the second unicast frame sent in the HSR network, QRPL locks and filters all DANH rings except the destination DANH ring.

Network traffic when the source node sends the *k*th unicast frame ($k \geq 2$) to the destination node the QRPL, denoted by nt^k_{QRPL}, is determined as follows:

$$nt^k_{QRPL} = \sum_{i \in QR} n_i + n_S + n_D \tag{9}$$

Generally, network unicast traffic when a source node sends N unicast frames to a destination node under the QRPL, denoted by nt_{QRPL}, is:

$$nt_{QRPL} = \sum_{i \in DR} n_i + N \sum_{i \in QR} n_i + (N-1)(n_S + n_D) \tag{10}$$

4.1.5. Under the FHT Technique

Network traffic when a source node sends a unicast frame to a destination node under the FHT technique, denoted by nt^1_{FHT}, is calculated as follows:

$$nt^1_{FHT} = n_S^{DN} + n_D^{DN} + \sum_{i \in IR} n_i \tag{11}$$

where n_S^{DN} and n_D^{DN} are the number of DANH nodes in the source and destination DANH rings, respectively, and IR is a set of intermediate rings that connects the source and destination DANH rings.

Generally, network unicast traffic when a source node sends N unicast frames to a destination node under the FHT technique, denoted by nt_{FHT}, is calculated as follows:

$$nt_{FHT} = N \left(n_S^{DN} + n_D^{DN} + \sum_{i \in IR} n_i \right) \tag{12}$$

4.1.6. Under the EFA

For the first unicast frame under the EFA, the process of delivering the frame is similar to that of the standard HSR protocol. During the delivering process, access and trunk QuadBoxes check and lock their DANH sides and destination sides, respectively, if the sides are not used to forward unicast frames from the source to the destination. Under the EFA, when a source node sends the first unicast frame to a destination node, the frame is flooded and doubled in all rings except the destination DANH ring. Network traffic, denoted by nt^1_{EFA}, is calculated as follows:

$$nt^1_{EFA} = \sum_{i \in NR} 2n_i - n_D \tag{13}$$

where n_D is the number of nodes in the destination DANH ring, n_i is the number of nodes in the *i*th ring, and NR is a set of all rings in the network.

For the next unicast frames, EFA locks all DANH rings except the destination DANH ring and any unused QuadBox rings. In addition, any duplicated frames are also removed from active rings.

Network traffic when the source node sends the kth unicast frame ($k \geq 2$) to the destination node under the EFA, denoted by nt_{EFA}^k, is determined as follows:

$$nt_{EFA}^k = n_S^{DN} + n_D^{DN} + \sum_{i \in AR} n_i \tag{14}$$

where n_S^{DN} and n_D^{DN} are the number of DANH nodes in the source and destination DANH rings, respectively, and AR is a set of active QuadBox rings that are used to deliver unicast frames between the source and the destination.

Generally, network traffic when a source node sends N unicast frames to a destination node under the EFA, denoted by nt_{EFA}, is:

$$nt_{EFA} = nt_{EFA}^1 + (N-1)nt_{EFA}^k \tag{15}$$

By substituting nt_{EFA}^1 in (13) and nt_{EFA}^k in (14) into the above nt_{EFA} equation, nt_{EFA} can be calculated as:

$$nt_{EFA} = \sum_{i \in NR} 2n_i - n_D + (N-1)\left(n_S^{DN} + n_D^{DN} + \sum_{i \in IR} n_i\right) \tag{16}$$

4.2. Control Overhead Performance

This section analyzes and evaluates the control overhead generated in HSR networks under the FHT and EFA.

4.2.1. Under the FHT Technique

The FHT filters unicast traffic for DANH and QuadBox rings based on MAC tables. To learn MAC addresses and build MAC tables, the FHT uses control messages, including Hello, ACK, and MAC messages.

Hello messages are sent by QuadBoxes and forwarded in DANH rings; DANH nodes forward the Hello messages, whereas QuadBoxes do not. The number of Hello messages sent and forwarded in the HSR network, denoted by co_{FHT}^{Hello}, is determined as follows:

$$co_{FHT}^{Hello} = \sum_{i \in DR} (n_i + 1) + \sum_{i \in QR} 2n_i \tag{17}$$

where n_i is the number of nodes in the ith ring, DR is a set of all DANH rings, and QR is a set of all QuadBox rings in the network.

Upon receiving a Hello message, each DANH node sends an ACK message back to the QuadBox that sent the Hello message. The number of ACK messages sent and forwarded in the HSR network, denoted by co_{FHT}^{ACK}, is determined as follows:

$$co_{FHT}^{ACK} = \sum_{i \in DR} \left(\frac{n_i^{DN}}{2} + 1\right)\frac{n_i^{DN}}{2} \tag{18}$$

where n_i^{DN} is the number of DANH nodes in the ith DANH ring.

Once an access QuadBox has built its MAC1 table, it sends a MAC message that contains all the MAC addresses of the access QuadBox's MAC1 table to its QuadBox ring. The trunk QuadBoxes connected to the QuadBox ring receive the MAC message, update their MAC2 tables, and then forward the MAC message. The number of MAC messages sent and forwarded in the HSR network, denoted by co_{FHT}^{MAC}, is calculated as follows:

$$co_{FHT}^{MAC} = \sum_{i \in AQ} n_i^{QR} \tag{19}$$

where AQ is a set of all access QuadBoxes in the network and n_i^{QR} is the number of QuadBoxes in the QuadBox ring connected to the ith access QuadBox.

The total number of control messages sent and forwarded in the HSR network, denoted by co_{FHT}, is determined as follows:

$$co_{FHT} = \sum_{i \in DR} \left(n_i + \sum_{i \in DR} \left(\frac{n_i^{DN}}{2} + 1 \right) \frac{n_i^{DN}}{2} \right) + \sum_{i \in QR} 2n_i + \sum_{i \in AQ} n_i^{QR} \quad (20)$$

4.2.2. Under the EFA

The EFA uses only one control message (the Locking message) to filter unicast traffic for QuadBox rings. Once the destination node receives the first frame sent by the source, the node sends a locking message back to the source. The locking message is then sent to the source through active rings (usually QuadBox rings) that are used to delivered unicast frames from the source to the destination. The locking message is not forwarded to the DANH rings.

Since the destination node sends the locking message once and the message is only forwarded in the destination DANH ring and active QuadBox rings, the total number of locking messages generated and forwarded in the HSR network, denoted by co_{EFA}, is calculated as follows:

$$co_{EFA} = n_D + \sum_{i \in AR} n_i \quad (21)$$

where n_D is the number of nodes in the destination DANH ring and AR is a set of active QuadBox rings that connect the source and destination DANH rings.

It is clear that the EFA generates less control messages in HSR networks than the FHT.

5. Simulations and Discussion

To validate the analyzed performance and to evaluate the performance of the proposed EFA, various simulations were carried out using the OMNeT++ network simulator [21]. There are several common network simulation tools used to evaluate research results in communication networks, including Network Simulator 2 (ns-2) [22], Network Simulator 3 (ns-3) [23], Riverbed Modeler (formerly referred to as OPNET) [24], and OMNeT++. Each simulation tool has both benefits and drawbacks. Selecting a network simulator as a tool to evaluate results of a new research depends on the target of the research, the supporting ability of each network simulator, and the programming skills of each researcher. The primary application area of these network simulators is communication networks. All of these simulators can be used to analyze and evaluate the network performance of the traffic reduction techniques in HSR network. In this paper, OMNeT++ is selected as the network simulation tool to simulate and evaluate the proposed approach because the tool is one of the best network simulators and we are familiar with it.

In the simulations, the HSR network shown in Figure 1 was considered. The network consists of eight DANH rings and three QuadBox rings. Each DANH ring includes four DANHs.

5.1. Simulation Description

5.1.1. Simulation 1

The objective of the simulation was to validate and compare the network traffic performance of the EFA to that of the standard HSR protocol and existing traffic filtering techniques. In the simulation, source node 1 sends N ($N = 10, 20 \ldots 100$) unicast frames to destination node 10. Network traffic frames were recorded to validate and compare with the analytical results.

5.1.2. Simulation 2

The objective of the second simulation was to validate and compare the overhead performance of the EFA to that of the FHT technique. In this simulation, the sample HSR network shown in Figure 1 was used. Source node 1 sends N ($N = 10, 20 \ldots 100$) unicast frames to destination node 10. The total number of control messages generated in the network was recorded to validate and compare with the analytical results.

5.2. Simulation Results

5.2.1. Simulation 1

Figure 13 shows a comparison of the network traffic performance for the EFA, the standard HSR protocol, and several existing techniques, including QR, PL, QRPL, and FHT.

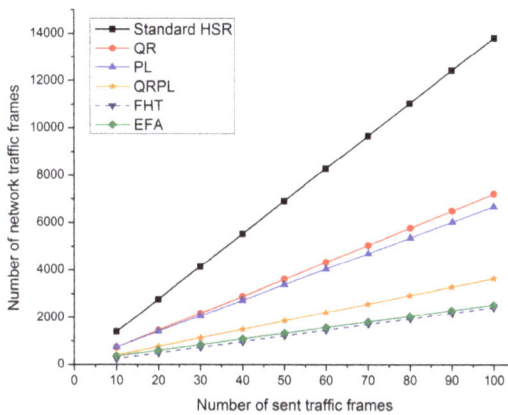

Figure 13. A comparison of network traffic performance.

The line graph in Figure 13 illustrates the number of network traffic frames recorded under traffic reduction techniques. The horizontal axis of the graph shows the number of unicast frames sent from the source to the destination, and the vertical axis shows the number of network traffic frames generated and delivered in the network when the source node sends the unicast frames to the destination node.

5.2.2. Simulation 2

Figure 14a shows a comparison of the control overhead performance between the EFA and the FHT technique; Figure 14b illustrates the control overhead messages to network traffic frames ratio of the EFA and FHT.

The line graph in Figure 14a illustrates the number of control messages recorded under the FHT and the EFA. The horizontal axis of the graph shows the number of unicast frames sent from the source to the destination, whereas the vertical axis shows the number of control messages generated and delivered in the network. The line chart in Figure 14b shows the control messages to network traffic frames ratio under the FHT and the EFA.

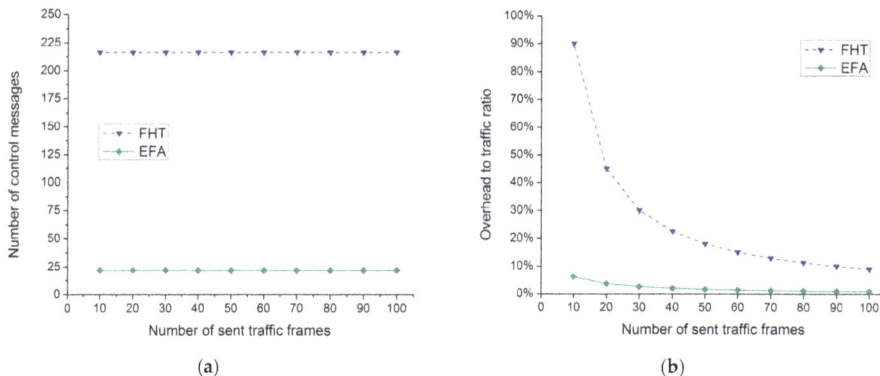

Figure 14. A comparison of overhead performance between the EFA and FHT. (**a**) Number of control messages; and (**b**) control messages to traffic frames ratio.

5.3. Discussion

The line graph in Figure 13 illustrates a comparison of network traffic of the EFA, the standard HSR protocol, and existing traffic-reduction techniques. Unlike other traffic-filtering techniques such as QR, PL, QRPL, and EPL—which either remove duplicate frames, and/or filter unicast traffic for DANH rings, or filter unicast traffic for both DANH and QuadBox rings—the EFA filters the unicast traffic for both DANH and QuadBox rings, as well as removing the duplicate frames from rings, as in the FHT technique. The EFA, thus, significantly reduces redundant unicast traffic compared with the standard HSR protocol and the existing QR, PL, and QRPL techniques. Numerically, for the sample HSR network, EFA reduces unicast network traffic by 80% compared with the standard HSR protocol, by ~62% compared with the QR technique, by ~60% compared with the PL technique, and by ~26% compared with the QRPL approach.

The simulation results also demonstrate that the network traffic performances of the proposed EFA and the FHT are quite similar. The main drawback of EFA is that it still generates a few control messages that are used to check and lock trunk QuadBoxes connecting to unused QuadBox rings, resulting in additional control overhead in HSR networks. The EFA, however, has better control overhead performance than the FHT, as shown in Figure 14a,b. The line chart in Figure 14a shows that for the sample HSR network, the number of control messages generated by the FHT is ten times higher than that generated by the EFA.

Table 2 shows a comparison of filtering features of various traffic filtering-based techniques. It is clear from the table that the proposed EFA is one of the most effective solutions for reducing redundant unicast traffic in HSR networks.

Table 2. A comparison of features of traffic filtering-based techniques.

Features	QR	PL	QRPL	EPL	FHT	EFA
Filter traffic for DANH rings	-	√	√	√	√	√
Filter traffic for QuadBox rings	-	-	-	√	√	√
Remove duplicated traffic	√	-	√	-	√	√
Generate control overhead	No	No	No	No	Medium	Low

6. Conclusions

In this paper we proposed a novel approach called EFA for effectively reducing redundant unicast traffic in HSR networks. The EFA integrates a novel filtering technique for QuadBox rings in HSR networks called the FQR technique with two existing filtering techniques, QR and PL, in order to filter redundant unicast traffic for both DANH and QuadBox rings, as well as to prevent unicast frames from being duplicated and circulated in rings. The EFA thus demonstrates the best network traffic performance among the various available filtering techniques. The analytical and simulation results showed that, for the sample network, EFA reduced network unicast traffic by 80% compared with the standard HSR protocol, by ~62% compared with the QR technique, by ~60% compared with the PL technique, and by ~26% compared with the hybrid approach of QRPL. In addition, among the best traffic performance techniques, the EFA demonstrates the best overhead performance. For the sample network, the number of control messages generated by the EFA is one tenth that of the FHT. In conclusion, the EFA is a very efficient approach that reduces more redundant unicast traffic in HSR networks than existing filtering techniques and generates less control overhead in the networks than the FHT. The EFA, thus, saves network bandwidth and improves network traffic performance in HSR networks.

Our future work will develop and implement the EFA in hardware devices.

Acknowledgments: This work was partly supported by Institute for Information & communications Technology Promotion (IITP) grant funded by the Korea government (MSIT) (No. 2014-0-00501, A Study on the Key Technology of Optical Modulation and Signal Processing for Implementation of 400 Gb/s Optical Transmission) and Basic Science Research Program through the National Research Foundation of Korea (NRF) funded by the Ministry of Science and ICT (grant number: 2017R1A2B4003964).

Author Contributions: Authors Nguyen Xuan Tien, Jong Myung Rhee, and Sang Yoon Park conceived and developed the ideas behind the research. Nguyen Xuan Tien carried out the performance analysis and simulations, and wrote the paper under supervision of Jong Myung Rhee. Jong Myung Rhee, and Sang Yoon Park finalized the paper.

Conflicts of Interest: The authors declare no conflict of interest.

References

1. Güngör, V.C.; Sahin, D.; Kocak, T.; Ergüt, S.; Buccella, C.; Cecati, C.; Hancke, G.P. Smart grid technologies: Communication technologies and standards. *IEEE Trans. Ind. Inform.* **2011**, *7*, 529–539. [CrossRef]
2. Sauter, T.; Lobashov, M. End-to-end communication architecture for smart grids. *IEEE Trans. Ind. Electron.* **2011**, *58*, 1218–1228. [CrossRef]
3. ABB. What Is a Smart Grid? ABB Smart Grids. Available online: http://new.abb.com/smartgrids/what-is-a-smart-grid (accessed on 2 August 2017).
4. Zeynal, H.; Eidiani, M.; Yazdanpanah, D. Intelligent Substation Automation Systems for Robust Operation of Smart Grids. In Proceedings of the 2014 IEEE Innovative Smart Grid Technologies—Asia (ISGT Asia), Kuala Lumpur, Malaysia, 20–23 May 2014.
5. *IEC 61850-1 Standard: "Communication Networks and Systems in Substations—Part 1: Introduction and Overview"*; IEC/TR 61850-1; The International Electrotechnical Commission: Geneva, Switzerland, 2003.
6. *IEC 61850-3 Standard: "Communication Networks and Systems for Power Utility Automation—Part 3: General Requirements"*; IEC 61850-3 Ed.2; The International Electrotechnical Commission: Geneva, Switzerland, 2010.
7. Araujo, J.A.; Lázaro, J.; Astarloa, A.; Zuloaga, A.; García, A. High Availability Automation Networks: PRP and HSR Ring Implementations. In Proceedings of the 2012 IEEE International Symposium on Industrial Electronics (ISIE), Hangzhou, China, 28–31 May 2012.
8. *IEC 62439-3 Standard: "Industrial Communications Networks—High Availability Automation Networks—Part 3: Parallel Redundancy Protocol (PRP) and High-Availability Seamless Redundancy (HSR)"*; IEC 62439-3 Ed.2; The International Electrotechnical Commission: Geneva, Switzerland, 2012.
9. Nsaif, S.A.; Rhee, J.M. Improvement of high-availability seamless redundancy HSR traffic performance for smart grid communications. *J. Commun. Netw.* **2012**, *14*, 653–661. [CrossRef]
10. Shin, M.; Joe, I. Performance Improvement for the HSR Ring Protocol with Traffic Control in Smart Grid. In Proceedings of the 2012 Future Generation Information Technology Conference (FGIT), Gangwon, Korea, 16–19 December 2012.
11. Abdulsalam, I.R.; Rhee, J.M. Improvement of High-Availability Seamless Redundancy (HSR) Unicast Traffic Performance Using Port Locking. In Proceedings of the Fourth World Congress on Software Engineering, Hong Kong, China, 3–4 December 2013.

12. Altaha, I.R.; Rhee, J.M. Improvement to high-availability seamless redundancy (HSR) unicast traffic performance using a hybrid approach, QRPL. *J. Satell. Inf. Commun.* **2016**, *11*, 29–35.
13. Altaha, I.R.; Rhee, J.M.; Pham, H.A. Improvement of high-availability seamless redundancy (HSR) unicast traffic performance using enhanced port locking (EPL) approach. *IEICE Trans. Commun.* **2015**, *E98-D*, 1646–1656. [CrossRef]
14. Tien, N.X.; Rhee, J.M. FHT: A novel approach for filtering high-availability seamless redundancy (HSR) traffic. *Energies* **2015**, *8*, 6249–6274. [CrossRef]
15. Nsaif, S.A.; Rhee, J.M. DVP: A novel high-availability seamless redundancy (HSR) protocol traffic-reduction algorithm for a substation automation system network. *Energies* **2014**, *7*, 1792–1810. [CrossRef]
16. Hong, S.; Lim, D.; Joe, I. Extended dual virtual paths algorithm considering the timing requirements of IEC61850 substation message type. *IEICE Trans. Inf. Syst.* **2016**, *E99.D*, 1563–1575. [CrossRef]
17. Tien, N.X.; Kim, S.; Rhee, J.M. A novel ring-based dual paths approach for reducing redundant traffic in HSR networks. *Comput. Netw.* **2016**, *110*, 338–350. [CrossRef]
18. Tien, N.X.; Kim, S.; Rhee, J.M.; Park, S.Y. A novel dual separate paths (DSP) algorithm providing fault-tolerant communication for wireless sensor networks. *Sensors* **2017**, *17*, 1699. [CrossRef] [PubMed]
19. Tien, N.X.; Rhee, J.M. Developing a new HSR switching node (SwitchBox) for improving traffic performance in HSR networks. *Energies* **2016**, *9*, 36. [CrossRef]
20. Molina, E.; Jacob, E.; Toledo, N.; Astarloa, A. Performance enhancement of high-availability seamless redundancy (HSR) networks using OpenFlow. *IEEE Commun. Lett.* **2015**, *20*, 364–367. [CrossRef]
21. OMNeT++ v4.6 Discrete Event Simulator. Available online: http://www.omnetpp.org/ (accessed on 30 May 2017).
22. Network Simulator ns-2. Available online: https://www.isi.edu/nsnam/ns/ (accessed on 18 August 2017).
23. Network Simulator ns-3. Available online: https://www.nsnam.org/ (accessed on 18 August 2017).
24. River Modeler. Available online: https://www.riverbed.com/products/steelcentral/steelcentral-riverbed-modeler.html (accessed on 18 August 2017).

energies

MDPI

Article

Energy Loss Allocation in Smart Distribution Systems with Electric Vehicle Integration

Paulo M. De Oliveira-De Jesus *, Mario A. Rios and Gustavo A. Ramos

Department of Electrical and Electronic Engineering, School of Engineering, Los Andes University,
Bogotá 111711, Colombia; mrios@uniandes.edu.co (M.A.R.); gramos@uniandes.edu.co (G.A.R.)
* Correspondence: pm.deoliveiradejes@uniandes.edu.co; Tel.: +57-318-256-6629

Received: 24 June 2018; Accepted: 17 July 2018; Published: 28 July 2018

Abstract: This paper presents a three-phase loss allocation procedure for distribution networks. The key contribution of the paper is the computation of specific marginal loss coefficients (MLCs) per bus and per phase expressly considering non-linear load models for Electric Vehicles (EV). The method was applied in a unbalanced 12.47 kV feeder with 12,780 households and 1000 EVs under peak and off-peak load conditions. Results obtained were also compared with the traditional roll-in embedded allocation procedure (pro rata) using non-linear and standard constant power models. Results show the influence of the non-linear load model in the energy losses allocated. This result highlights the importance of considering an appropriate EV load model to appraise the overall losses encouraging the use and further development of the methodology

Keywords: power loss allocation; plug-in electric vehicle; smart grid; locational marginal prices

---

1. Introduction

Electrical distribution systems are immersed in a deep process of transformation becoming very different from what they used to be. The increasing penetration of distributed generators, the expected connection of a large amount of plug-in electric vehicles (EV) and the adoption of advanced metering and communication infrastructure (AMI) are creating new challenges for regulators. The widespread integration of EVs into existing distribution networks will increase feeder demands and therefore will produce rising energy losses [1]. Moreover, due to the different nature of EV loads (slow and fast battery charging stations), one-phase and two-phase connections may increase system unbalance producing additional losses. A recent study about the impact of the placement of fast charging stations in distribution systems showed a power loss increase of 85% [2] with respect to a base layer with no EV integration. Therefore, some conceptual and regulatory questions can be raised about the EV impacts on the increase of energy losses in distribution networks:

1. How much should an EV load pay for the incremental losses in the grid [3]?
2. Should incremental losses produced by EVs connected to fast and slow charging stations be allocated in a proportional manner among all distribution loads [3]?
3. Can a price signal for losses (sent in real-time via AMI and smart metering) force the EV loads to provide volt/var support in order to improve voltage profile and reduce system losses [4]?

These regulatory aspects can be addressed by means of a cost-reflective energy loss allocation procedure in order to send economical signals to consumers and producers with the aim of improving overall performance of the system. The energy loss allocation is not new issue in electricity markets. It has been widely treated in the literature mainly at transmission systems [5] and more recently in distribution systems considering increasing integration of distributed generators [6]. In general, the majority of the loss allocation procedures discussed in the literature are based upon positive-sequence power flow models with balanced power injections where all loads are modeled using standard constant real and reactive power (PQ) models [7–9]. In this case, load demands are not affected by voltage fluctuations (constant power load models).

Power loss allocation constitutes an important strategy to determine locational prices at distribution level in order to send efficiency economical signals to demands [10,11] and distributed generators [12] at distribution level. Recent contributions are devoted to extending positive-sequence power loss allocation procedures into unbalanced three phase domain [13,14]. Based upon the calculation of sensitivity loss factors in the context of the optimal power flow problem, some work is carried out to assess locational marginal pricing and EV charging management [15].

However, previous loss allocation methods discussed in literature only consider constant power loads and do not take into account non-linear nature of EV loads [16]. System unbalance produced by fast charging stations with single- and two-phase connections as well as the above-mentioned voltage dependence justify the development of detailed three-phase loss allocation procedure to assess the impact of EV loads on incremental losses.

To fill the research gap, this paper presents three-phase loss allocation procedure for distribution networks that expressly incorporates a non-linear load model. This model can be adjusted as exponential, constant power, current and impedance depending on EV load parametrization. The proposed procedure is based on the computation of specific marginal loss coefficients (MLCs) per bus and phase.

The method is illustrated in a unbalanced 12.47 kV feeder with 12,780 residential customers. Daily energy losses were allocated considering five levels of EV penetration: 200, 400, 800 and 1000 units corresponding to 5%, 10%, 15%, 20% and 25% of consumption without EV presence. Two operational scenarios with two different type of charging stations are studied. A 3.75 kW slow battery charger from 0:00 to 8:00 and a 7.5 kW fast charger from 18:00 to 22:00. Results obtained were also compared with traditional roll-in embedded allocation method (pro rata) [17]. Finally, a sensitivity analysis was performed to compare the results with ones obtained using a standard constant power model.

This paper is organized as follows. Section 2 is devoted to present the proposed method. Section 3 defines the case study. Section 4 discusses the results. Section 5 draws the conclusions.

2. The Energy Loss Allocation Model

The system model is based upon a typical n buses three-phase unbalanced distribution network with two types of loads connected at each bus i: residential loads and non linear EV loads, as shown in Figure 1.

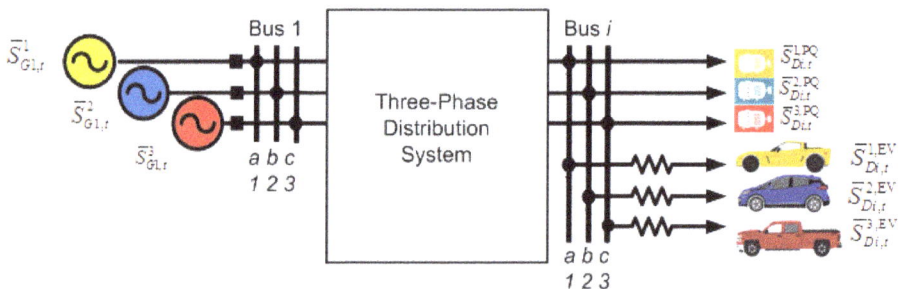

Figure 1. Three-phase distribution system with Electric Vehicles (EV) loads.

In this paper, we assess the energy loss allocation problem among all loads considering a passive network with EV integration. The model can be extended to active networks with distributed generation and bidirectional EV injections. However, our purpose here is to analyze technical and economical impacts of EV's connected to slow and fast charging stations under peak and off-peak conditions.

2.1. PQ and EV Load Modeling

Distribution loads characteristics depend on the share of different demand types (industrial, commercial and residential) and can be modeled in a more complex way as a mix of different voltage–current models as constant impedance, constant current and constant power. For the sake of simplicity, in this paper, all residential demands are regarded as constant loads. This means that constant loads such as induction motors are predominant at demand side. In the reminder of the paper, these constant power loads are denoted as constant real and reactive power (PQ) loads. Constant power loads do not depend on voltage fluctuations.

Then, at given time t of a period T, the apparent power of a PQ load at bus i and phase p is denoted as:

$$\overline{S}_{Di,t}^{p,PQ} = P_{Di,t}^{p,PQ} + jQ_{Di,t}^{p,PQ} \quad i = 2,\ldots,n \quad p = 1,2,3 \quad t = 1,\ldots,T \tag{1}$$

where $P_{Di,t}^{p,PQ}$ and $Q_{Di,t}^{p,PQ}$ are the real and the reactive power of a PQ load. No loads are connected at source bus $i = 1$.

The second type of load considered in the formulation is the aggregation of a number of EVs connected to a given bus i. There are several load models for EVs. Many parametric models are based on real power injections [18,19] considering a power factor equal to 1. Without loss of generality, due to regulatory and operational reasons [20], EVs can be requested to provide voltage and reactive power support. In this case, apparent, real and reactive power are modeled using a non-linear function [21] and a fixed power factor angle $\phi_{Di,t}^p$, respectively:

$$\overline{S}_{Di,t}^{p,EV} = P_{Di,t}^{p,EV} + jQ_{Di,t}^{p,EV} \quad i = 2,\ldots,n \quad p = 1,2,3 \quad t = 1,\ldots,T \tag{2}$$

where the real power demanded by aggregate EVs at time t, bus i and phase p is given by

$$P_{Di,t}^{p,EV} = P_0[a + b(\frac{V_i^p}{V_0})]^\alpha \quad i = 2,\ldots,n \quad p = 1,2,3 \quad t = 1,\ldots,T \tag{3}$$

where $a + b = 1$.

The reactive power demanded by EVs at time t, bus i and phase p is given by

$$Q_{Di,t}^{p,EV} = P_{Di,t}^{p,EV} \tan\phi_{Di,t}^p = \sqrt{(S_{Di,t}^{p,EV})^2 - (P_{Di,t}^{p,EV})^2} \quad i = 2,\ldots,n \quad p = 1,2,3 \quad t = 1,\ldots,T \tag{4}$$

Parameters a, b and α depend on EV charger characteristics and the equivalent resistance R between each connection outlet at low voltage and the system bus i. P_0 and V_0 are the nominal power and nominal voltage. Some values of the EV parameters can be found in [21]. Note that, if $\alpha = 0$, the model reflects a PQ load; if $\alpha = 1$, the model reflects a constant current load; and, if $\alpha = 2$, the model reflects a constant impedance load. In general, EV load parametrization leads to negative values of α as indicated by [21].

At given time t, the system power balance is given by:

$$\sum_{p=1}^{3} \overline{S}_{G1,t}^p = \Delta\overline{S}_t + \sum_{i=2}^{n}\sum_{p=1}^{3} \overline{S}_{Di,t}^{p,PQ} + \sum_{i=2}^{n}\sum_{p=1}^{3} \overline{S}_{Di,t}^{p,EV} \quad t = 1,\ldots,T \tag{5}$$

where $\overline{S}_{Gi,t}^p$ is the power injected at reference bus 1 at Phases 1, 2 and 3, and $\Delta\overline{S}_t$ is the total apparent losses. Total apparent losses can be split into real losses $\Delta P_t = Re(\Delta\overline{S}_t)$ and reactive losses $\Delta Q_t = Im(\Delta\overline{S}_t)$.

Total real energy consumed by PQ and EV loads at each bus i during a period T, i.e., 24 h are given by:

$$W_{Di}^{PQ} = \sum_{t=1}^{T}\sum_{p=1}^{3} P_{Di,t}^{p,PQ}; \quad W_{Di}^{EV} = \sum_{t=1}^{T}\sum_{p=1}^{3} P_{Di,t}^{p,EV} \quad i = 2,\dots,n \tag{6}$$

Total real energy delivered by the source bus 1 is given by:

$$W_1 = \sum_{t=1}^{T}\sum_{p=1}^{3} P_{G1,t}^{p} \tag{7}$$

The system real energy balance is given by:

$$W_1 = \sum_{i=2}^{n} W_{Di}^{PQ} + \sum_{i=2}^{n} W_{Di}^{EV} + \Delta W \tag{8}$$

and, then total real energy losses are

$$\Delta W = \sum_{t=1}^{T}\sum_{p=1}^{3} P_{G1,t}^{p} - \sum_{t=1}^{T}\sum_{i=2}^{n}\sum_{p=1}^{3} P_{Di,t}^{p,PQ} - \sum_{t=1}^{T}\sum_{i=2}^{n}\sum_{p=1}^{3} P_{Di,t}^{p,EV} \tag{9}$$

where ΔW is the system real energy losses to be allocated between all network users (PQ and EV loads) during a defined time interval T.

2.2. Evaluation of Power Losses to Be Allocated among the Network Users

The power and losses to be allocated can be evaluated from the solution of the standard three-phase power flow problem. The solution comprises all system voltages (magnitude and angle) except the voltages fixed at reference. There are several methods to solve this issue. In general, when all loads are regarded as PQ constant, Newton–Raphson method can be applied either by its complete formulation [22] or decoupled formulation [23]. Other Gauss-based methods suitable to be applied at distribution level can be used instead [24,25].

For non-linear loads (such as EVs), if the three-phase admittance matrix (Y_{BUS}) is known, the power flow solution can be obtained at given operating point (time t) from a set of $3(n-1)$ equations and $3(n-1)$ unknowns:

$$P_{i,t}^{p} = -P_{Di,t}^{p,PQ} - P_{Di,t}^{p,EV}(V_{i,t}^{p}) = V_{i,t}^{p}\sum_{k=1}^{n}\sum_{p=1}^{3}\sum_{m=1}^{3} V_{k,t}^{m}[G_{ik}^{pm}\cos\theta_{ik,t}^{pm} + B_{ik}^{pm}\sin\theta_{ik,t}^{pm}] \quad \forall i \neq 1, p \tag{10}$$

where $\theta_{ik,t}^{pm} = \theta_{i,t}^{p} - \theta_{k,t}^{m}$.

$$Q_{i,t}^{p} = -Q_{Di,t}^{p,PQ} - Q_{Di,t}^{p,EV}(V_{i,t}^{p}) = V_{i,t}^{p}\sum_{k=1}^{n}\sum_{p=1}^{3}\sum_{m=1}^{3} V_{k,t}^{m}[G_{ik}^{pm}\sin\theta_{ik,t}^{pm} - B_{ik}^{pm}\cos\theta_{ik,t}^{pm}] \quad \forall i \neq 1, p \tag{11}$$

where $P_{Di,t}^{p,PQ}$ and $Q_{Di,t}^{p,PQ}$ are constant parameters, and $P_{Di,t}^{p,EV}$ and $Q_{Di,t}^{p,EV}$ are non-linear functions depending on its own voltage magnitude as given in Equations (3) and (4). At reference bus 1, voltage magnitude and angle are known for all phases: $\overline{V}_1^{1} = V_1 e^{j0}$, $\overline{V}_1^{2} = V_1 e^{j\frac{-2\pi}{3}}$ and $\overline{V}_1^{3} = V_1 e^{j\frac{2\pi}{3}}$ where V_1 is the voltage magnitude at reference bus 1. Then, once the power flow algorithm is applied to solve the set of Equations (10) and (11), the solution $\mathbf{x_t} = [V_{1,t}^{1},\dots,V_{i,t}^{p},\dots,V_{n,t}^{3};\theta_{1,t}^{1},\dots,\theta_{i,t}^{p},\dots,\theta_{n,t}^{3}]$ is evaluated in the following expression in order to get the real system power losses:

$$\Delta P_t = \sum_{i=1}^{n}\sum_{k=1}^{n}\sum_{p=1}^{3}\sum_{m=1}^{3} V_{i,t}^{p}V_{k,t}^{m}[G_{ik}^{pm}\cos\theta_{ik,t}^{pm} + B_{ik}^{pm}\sin\theta_{ik,t}^{pm}] \quad t = 1,\dots,T \tag{12}$$

The G_{ik}^{pm} and B_{ik}^{pm} entries correspond to the conductance and susceptance terms of the admittance matrix (Y_{BUS}) between phase p at bus i and phase m at bus k.

The total real energy to allocate among network users is given by:

$$\Delta W = \sum_{t=1}^{T} \Delta P_t \tag{13}$$

2.3. Energy Loss Allocation Procedures

We consider two procedures to allocate energy losses among network PQ loads and non-linear EV loads:

1. The proposed marginal allocation procedure per bus and per phase
2. The standard pro rata or proportional allocation for comparison purposes [17]

2.3.1. Marginal Loss Allocation

Distribution losses can be allocated among network users is means of the sensitivity factors also known as marginal loss coefficients (MLCs) [26]. This allocation process yields on different charges depending on the effect of each user on overall losses. Thus, the power losses allocated or assigned to PQ loads located at bus i, phase p, at time t are:

$$L_{Di,t}^{M,p,PQ} = k_{r,t} ||MLC_{Di,t}^{p}|| P_{Di,t}^{p,PQ} \quad i = 2,\ldots,n \quad p = 1,2,3 \quad t = 1,\ldots,T \tag{14}$$

and power losses allocated to EV loads located at bus i, phase p, at time t are:

$$L_{Di,t}^{M,p,EV} = k_{r,t} ||MLC_{Di,t}^{p}|| P_{Di,t}^{p,EV}(V_{i,t}^{p}) \quad i = 2,\ldots,n \quad p = 1,2,3 \quad t = 1,\ldots,T \tag{15}$$

It must be highlighted that the application of MLCs produce an over-recovery of losses [27]. This is due to the nonlinear nature (quadratic) of losses. To reconcile the total power losses, i.e., recover the exact amount of grid losses, it is necessary to multiply the allocated power losses by a reconciliation factor $k_{r,t}$. This factor avoids a over recovery of power losses at each time t:

$$k_{r,t} = \frac{\Delta P_t}{\sum_{i=2}^{n} \sum_{p=1}^{3} ||MLC_{Di,t}^{p}|| P_{Di,t}^{p,EV}(V_{i,t}^{p})} \quad t = 1,\ldots,T \tag{16}$$

The total real energy losses allocated to loads at bus i and phase p are:

$$A_{Di}^{M,p,PQ} = \sum_{t=1}^{T} L_{Di,t}^{M,p,Q}; \quad A_{Di}^{M,p,EV} = \sum_{t=1}^{T} L_{Di,t}^{M,p,EV} \quad i = 2,\ldots,n \quad p = 1,2,3 \tag{17}$$

The total real energy losses allocated to PQ and EV loads under proposed marginal approach are:

$$A_D^{M,PQ} = \sum_{t=1}^{T} \sum_{i=2}^{n} \sum_{p=1}^{3} A_{Di,t}^{M,p,PQ}; \quad A_D^{M,EV} = \sum_{t=1}^{T} \sum_{i=2}^{n} \sum_{p=1}^{3} A_{Di,t}^{M,p,EV} \tag{18}$$

Considering that losses are recovered using a 24-h day-ahead spot price ρ_t in USD/MWh, the payments per losses of loads at bus i and phase p are:

$$\Omega_{Di}^{M,p,PQ} = \sum_{t=1}^{T} \rho_t L_{Di,t}^{M,p,PQ}; \quad \Omega_{Di}^{M,p,EV} = \sum_{t=1}^{T} \rho_t L_{Di,t}^{M,p,EV} \quad i = 2,\ldots,n \quad p = 1,2,3 \tag{19}$$

Global energy loss payments under the marginal approach are:

$$\Omega_D^{M,PQ} = \sum_{t=1}^{T}\sum_{i=2}^{n}\sum_{p=1}^{3} \rho_t L_{Di,t}^{M,p,PQ}; \quad \Omega_D^{M,EV} = \sum_{t=1}^{T}\sum_{i=2}^{n}\sum_{p=1}^{3} \rho_t L_{Di,t}^{M,p,EV} \tag{20}$$

Determining the three-phase MLCs: To get the marginal loss coefficients, we can solve the network stating an optimization problem as follows:

$$\min \Delta P_t = \sum_{i=1}^{n}\sum_{k=1}^{n}\sum_{p=1}^{3}\sum_{m=1}^{3} V_{i,t}^{p} V_{k,t}^{m} [G_{ik}^{pm}\cos\theta_{ik,t}^{pm} + B_{ik}^{pm}\sin\theta_{ik,t}^{pm}] \quad t=1,\dots,T \tag{21}$$

subject to:

$$P_{i,t}^{p} = -P_{Di,t}^{p,PQ} - P_{Di,t}^{p,EV}(V_{i,t}^{p}) = V_{i,t}^{p}\sum_{k=1}^{n}\sum_{p=1}^{3}\sum_{m=1}^{3} V_{k,t}^{m}[G_{ik}^{pm}\cos\theta_{ik,t}^{pm} + B_{ik}^{pm}\sin\theta_{ik,t}^{pm}] \quad \forall i \neq 1 \tag{22}$$

$$Q_{i,t}^{p} = -Q_{Di,t}^{p,PQ} - Q_{Di,t}^{p,EV}(V_{i,t}^{p}) = V_{i,t}^{p}\sum_{k=1}^{n}\sum_{p=1}^{3}\sum_{m=1}^{3} V_{k,t}^{m}[G_{ik}^{pm}\sin\theta_{ik,t}^{pm} - B_{ik}^{pm}\cos\theta_{ik,t}^{pm}] \quad \forall i \neq 1 \tag{23}$$

As the formulation has the same number of equations and unknowns, the optimization problem is determined. The results coincide with the power flow solution. However, it should be highlighted that the Lagrange multiplier associated with Equation (22) for bus i and phase p is just the marginal loss coefficient $MLC_{Di,t}^{p}$:

$$MLC_{Di,t}^{p} = \frac{\partial \Delta P_t}{\partial P_{i,t}^{p}} \quad i = 2,\dots,n \quad p=1,2,3 \quad t=1,\dots,T \tag{24}$$

Lagrange multipliers are usually provided by any optimization package. In the test case we used the *fmincon* optimization solver of Matlab (version R2017, v.9.2) to get the MLCs and to illustrate the application of the method.

2.3.2. Pro Rata or Proportional Allocation

Pro rata method describes a proportionate allocation of losses among all loads according the amount of power demand at each bus and phase. It consists of assigning an amount to a fraction according to its share of the whole [17]. Thus, the power losses allocated to PQ loads at bus i, phase p and time t are:

$$L_{Di,t}^{P,p,PQ} = \Delta P_t \frac{P_{Di,t}^{p,PQ}}{\sum_{i=2}^{n}\sum_{p=1}^{3} P_{Di,t}^{p,PQ} + \sum_{i=2}^{n}\sum_{p=1}^{3} P_{Di,t}^{p,EV}(V_{i,t}^{p})} \quad i=2,\dots,n \quad p=1,2,3 \quad t=1,\dots,T \tag{25}$$

and power losses to be allocated to EV loads located at bus p, phase p, at time t are:

$$L_{Di,t}^{P,p,EV} = \Delta P_t \frac{P_{Di,t}^{p,EV}(V_{i,t}^{p})}{\sum_{i=2}^{n}\sum_{p=1}^{3} P_{Di,t}^{p,PQ} + \sum_{i=2}^{n}\sum_{p=1}^{3} P_{Di,t}^{p,EV}(V_{i,t}^{p})} \quad i=2,\dots,n \quad p=1,2,3 \quad t=1,\dots,T \tag{26}$$

The total real energy losses to be allocated to loads at bus i and phase p are:

$$A_{Di}^{P,p,PQ} = \sum_{t=1}^{T} L_{Di,t}^{P,p,Q}; \quad A_{Di}^{P,p,EV} = \sum_{t=1}^{T} L_{Di,t}^{P,p,EV} \quad i=2,\dots,n \quad p=1,2,3 \tag{27}$$

The total real energy losses allocated to PQ and EV loads under pro rata approach are:

$$A_D^{P,PQ} = \sum_{t=1}^{T}\sum_{i=2}^{n}\sum_{p=1}^{3} A_{Di,t}^{P,p,PQ}; \quad A_D^{P,EV} = \sum_{t=1}^{T}\sum_{i=2}^{n}\sum_{p=1}^{3} A_{Di,t}^{P,p,EV} \tag{28}$$

Considering that losses are recovered using a uniform price ρ in USD/MWh, the payments per losses of loads at bus i and phase p are:

$$\Omega_{Di}^{P,p,PQ} = \rho A_{Di}^{P,p,PQ}; \quad \Omega_{Di}^{P,p,EV} = \rho A_{Di}^{P,p,EV} \quad i = 2,\ldots,n \quad p = 1,2,3 \tag{29}$$

Global energy loss payments under the pro rata approach are:

$$\Omega_D^{P,PQ} = \sum_{i=2}^{n}\sum_{p=1}^{3} \rho A_{Di}^{P,p,PQ}; \quad \Omega_D^{PEV} = \sum_{i=2}^{n}\sum_{p=1}^{3} \rho A_{Di}^{P,p,EV} \tag{30}$$

3. Case Study

The proposed energy loss allocation procedure was applied in the well-known 21-bus Kersting NEV test system [28]. This system has a three-phase main feeder connected to an ideal 12.47 kV (line-to-line) source. The feeder has 1828.8 m (6000 ft) long and an average pole span of 91.44 m (300 ft). The original test case has a unique load concentrated at the ending node. We modified the loading scheme by introducing a uniformly increasing load in each phase from bus 2 to bus 21 according to Table 1. The loading scheme considers a substation with four main feeders in a high density area. In this case, according to [29], the load increase is linear with respect to the distance. Then, source bus 1 has no load and the last bus 21 has the highest load value. For the sake of simplicity, only the main feeder is considered for the proposed analysis. Single phase derivations and laterals are neglected.

Table 1. Base load: No EV connected, only PQ loads.

Bus	Total		Phase 1		Phase 2		Phase 3	
	W_{Di}^{PQ}	P_{Di}^{PQ}	$W_{Di}^{1,PQ}$	$P_{Di}^{1,PQ}$	$W_{Di}^{2,PQ}$	$P_{Di}^{2,PQ}$	$W_{Di}^{3,PQ}$	$P_{Di}^{3,PQ}$
	MW·h/day	kW	MW·h/day	kW	MW·h/day	kW	MWh/day	kW
2	0.6	0.04	0.2	0.01	0.2	0.02	0.2	0.01
3	1.2	0.08	0.4	0.03	0.5	0.03	0.4	0.02
4	1.8	0.12	0.6	0.04	0.7	0.05	0.5	0.04
5	2.4	0.16	0.8	0.05	0.9	0.06	0.7	0.05
6	3.0	0.20	1.0	0.07	1.1	0.08	0.9	0.06
7	3.7	0.24	1.2	0.08	1.4	0.09	1.1	0.07
8	4.3	0.28	1.4	0.09	1.6	0.11	1.2	0.08
9	4.9	0.32	1.6	0.11	1.8	0.12	1.4	0.10
10	5.5	0.37	1.8	0.12	2.1	0.14	1.6	0.11
11	6.1	0.41	2.0	0.14	2.3	0.15	1.8	0.12
12	6.7	0.45	2.2	0.15	2.5	0.17	2.0	0.13
13	7.3	0.49	2.4	0.16	2.7	0.18	2.1	0.14
14	7.9	0.53	2.6	0.18	3.0	0.20	2.3	0.15
15	8.5	0.57	2.8	0.19	3.2	0.21	2.5	0.17
16	9.1	0.61	3.0	0.20	3.4	0.23	2.7	0.18
17	9.7	0.65	3.2	0.22	3.6	0.24	2.9	0.19
18	10.3	0.69	3.4	0.23	3.9	0.26	3.0	0.20
19	11.0	0.73	3.6	0.24	4.1	0.27	3.2	0.21
20	11.6	0.77	3.8	0.26	4.3	0.29	3.4	0.23
21	12.2	0.81	4.0	0.27	4.6	0.30	3.6	0.24
Total	127.8	8.53	42.5	2.84	47.8	3.19	37.5	2.50

The last row of Table 1 corresponds to the sum of all energy consumptions at substation (bus 1) and the sum of all coincident demands flowing at main feeder (between buses 1 and 2). Total peak power flowing by the main feeder (bus 1) is 8526 kW at 20:00. Total three-phase load consumption is 127.8 MW·h/day, corresponding to 12,780 customers (each household consumes 10 kW·h/day, 300 kW·h/month with load factor 0.62). The 24-h real power load curve in p.u. for all buses and phases is depicted in Figure 2. For simplicity, all loads $P_{Di,t}^{p,PQ}$, $i = 2, \ldots, 21$, $p = 1, 2, 3$, $t = 1, \ldots, 24$ have the same load curve. Then, all maximum demands are coincident at 20:00 but with different real power values per phase and bus (as shown in Table 1) ensuring unbalanced operation.

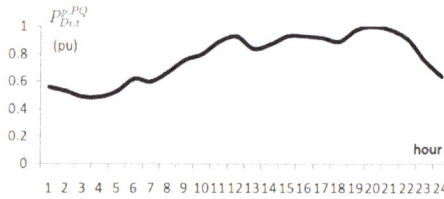

Figure 2. Base load curve: No EV connected, only constant real and reactive power (PQ) loads.

The network structure was scripted in OpenDSS (version 7.6.5.52, Electric Power Research Institute, Inc., Palo Alto, CA, USA) [30] (included in the Appendix A to extract the three-phase network model (admittance matrix). Power flow solution at base layer (with no EV penetration) showed that total peak power losses reach 115 kW at 20:00. Total energy losses are 1.33 MW·h/day (approximately 1.04% of total). The worst voltage drop is 3.69% at node 21 phase *a*.

In this paper, we do not emphasize on voltage profile results since our objective is to illustrate from conceptual viewpoint the proposed three-phase loss allocation procedure under specified operation battery charging schemes. There are other type studies, e.g., hosting capacity [31], where realistic operation schemes is addressed using Monte Carlo simulations with stochastic EV demands [32–36]. Thus, this work does not intend to replicate the probabilistic behavior of EV connection in a given period. Further research can be conducted to assess realistic loss allocation payments in a city and a country with specific patterns of consumption.

The procedure was tested considering five levels of EV load integration ($k = 5$), corresponding to the connection of 200, 400, 600, 800 and 1000 EVs at Phases 1, 2 and 3 according to the scheme presented in Table A1 (included in the Appendix A). Level 0 correspond to the base case with no EV units connected to the grid. Each EV has a battery of 30 kW·h capacity, and then the integration Levels 1–5 correspond to an increase of demand consumption of 5%, 10%, 14%, 19% and 25% with respect the base case, respectively, as shown in Table 2.

Table 2. Base load, EV load, total load, and share at each level.

Level	Base Load	EV Load	Total Load	Share
	MW·h/day			%
Level 0—000EV	127.8	0	127.8	0%
Level 1—200EV	127.8	6	133.8	5%
Level 2—400EV	127.8	12	139.8	10%
Level 3—600EV	127.8	18	145.8	14%
Level 4—800EV	127.8	24	151.8	19%
Level 5—1000EV	127.8	30	157.8	25%

Total load at level k is the sum of the base load (level 0) and the EV loaf at level k.

4. Results

Two operational scenarios for EV's battery charging are considered in the application of the proposed energy loss allocation method:

1. Slow charging at off-peak load conditions: 3.75 kW (16 A) 8 h.
2. Fast charging at peak load conditions: 7.50 kW (32 A) 4 h.

The parameters of the EV load model are $a = 0.9537$, $\alpha = -2.324$, and $b = 0.0463$ and were taken from [21] for a resistance $R = 1.0$ ohm.

The same amount of energy required by aggregated slow and fast EV's battery chargers is integrated under peak and off-peak conditions for comparison purposes. The illustrative example allows us to assess how a progressive integration of EV (with a share from 0% to 25% of total energy) will affect the overall energy losses of the grid and the corresponding allocation results. Results are discussed under peak and off-peak load conditions for the marginal-based approach proposed in Section 2.3.1 and the standard roll-in embedded method discussed in Section 2.3.2.

4.1. Scenario 1: Slow Charging at Off-Peak Load Conditions

In this case, slow battery charging stations operate from 00:00 to 08:00 with the five levels of penetration defined above. The optimization problem stated in Equations (21)–(23) was scripted in Matlab (version R2017, v.9.2) and solved by means of the *fmincon* tool. The parameters of the admittance matrix were taken from OpenDSS simulation tool [30].

When the solution algorithm converges, the state of the system for each level $k = 0, ..., 5$ is given by $x_t^k = [V_{1,t}^1, ..., V_{21,t}^3; \theta_{1,t}^1, ..., \theta_{21,t}^3]$ for $t = 1, ..., 24$. Thereafter, power losses ΔP_t per hour and per level are evaluated by Equation (12) for each state of the system result x_t^k for $k = 0, ..., 5$. Level 0 corresponds to a grid operation with no EV penetration. Levels 1–5 correspond to the EV penetration from 5% to 25% in total daily consumed energy by EVs with respect to overall PQ load consumption.

The 24-h power loss curves by each level for the connection of EV loads under off-peak conditions are depicted in Figure 3. Total real energy losses ΔW to be allocated among network users is evaluated by EV penetration level using Equation (9) and results are depicted in Figure 4. This figure also shows the resultant load and loss factor. Load and loss factors are defined as the ratio between average and maximum values of demands and losses, respectively.

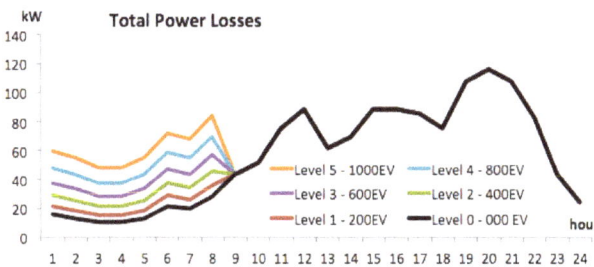

Figure 3. Off-peak load scenario: 24-h power losses by EV penetration level.

Results reveal how the progressive integration of slow charging stations at off-peak load conditions produces a flattening effect of the load curve. The load factor increased from 0.62 at level 0 to 0.77. However, the loss factor also increased from 0.48 to 0.61. This means that 24-h power loss curve is also becoming flat. As result energy losses rose in magnitude from 1.24 MW·h/day, 1.05% (level 0) to 1.70 MW·h/day, 1.08% (level 5). This result is important since despite energy losses grew almost 50% in magnitude, the relative energy losses remains constant around 1.05–1.08%.

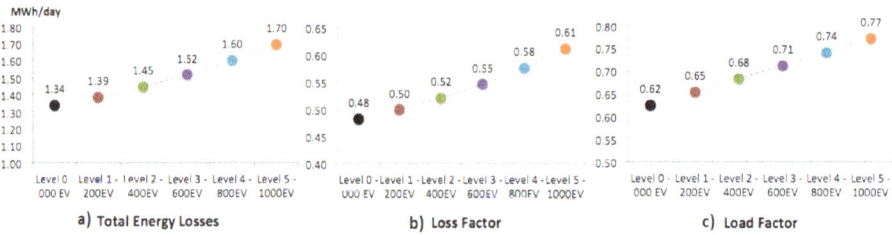

Figure 4. Off-peak load scenario: energy losses, loss and load factors by EV penetration level: (**a**) total energy losses; (**b**) loss factor; (**c**) load factor.

The absolute value of Marginal Loss Coefficients $||MLC_{Di,t}^{p}||$ are directly obtained for at each level k, each bus i, phase p and time t from *fmincon* results via Lagrange multipliers as indicated in Section 2.3.1.

Figure 5 displays the MLCs curves when EVs are charging at off-peak time. The MLCs are applied to agents connected at bus 21 along 24 h. Note that under lower EV penetration (5%, red curve), MLCs observed between 01:00 and 09:00 are significantly lower than ones achieved between 10:00 and 22:00. For a high EV penetration (25%, orange curve), the MLCs obtained between 01:00 and 09:00 are similar to those achieved between 10:00 and 22:00 when no EV is connected (around 0.02–0.03 along the day). Then, under EV charging at off-peak load conditions, the pattern of MLCs is somehow flat, similar to a uniform marginal coefficient. This uniform coefficient produces similar results of a roll-in embedded method applied to recover the power losses.

Figure 6 depicts a complete pattern for calculated MLCs by location and by time for each penetration level. It is worth to note in all phases that MLCs associated to high EV penetration (Level 5, 25%) cover more area (in time and location) than MLCs produced by lower levels.

Table 3 lists the general results for the allocated energy losses for aggregate PQ and EV loads under off-peak condition. The reconciliation factor k_r was around 0.5 in all levels. Equations (18) and (28) were applied for the marginal and pro rata procedure, respectively. Energy losses range from 1339.08 kW·h/day at level 0 to 1696.64 kW·h/day at Level 5. At level 0, EVs do not exist then all losses are assigned to PQ loads. Regarding the allocation results, three facts can be highlighted:

EV charging stations operating under off-peak conditions and marginal loss allocation do not pay for additional energy losses. The marginal procedure assigns lower losses to EV than expected under a pro-rata procedure. This means that EV loads reach a small benefit by their produced losses at off-peak conditions. In fact, PQ loads do no take advantage of the marginal procedure being slightly penalized (they should pay for 1389 kW·h/day with respect to 1372 kW·h/day under the proportional approach).

Pro rata and marginal methods can produce a similar output when EV charging stations are operating under off-peak conditions. The share of energy losses attributable to EV loads (18%) are similar in both approaches: marginal and pro rata. This means that the MLCs are acting as a uniform factor capable to recover the cost of losses.

The EV share of losses is lesser than the EV share of consumption. For instance, at level 5 the ratio between EV and PQ loads consumption is 25%. The EV share of losses is lesser, 18%.

Payment for energy losses by EV location are calculated in a monthly basis using Equations (19) and (29) for marginal and pro rata procedure, respectively. Considering a flat energy price ρ of 0.05 USD/kW·h, left-hand chart of Figure 7 shows how the marginal procedure penalize the slow EV charging stations connected from bus 15 to bus 21. A similar effect is also seen in PQ loads (right-hand chart of Figure 7) connected from bus 15 to bus 21. In this scenario, marginal procedure is applying higher charges to loads (EV and PQ) connected at the end of the line. Figure 7 also indicates that the application of MLCs for loads (EV and PQ) connected near to the origin have a lesser responsibility in the coverage of the entire energy losses.

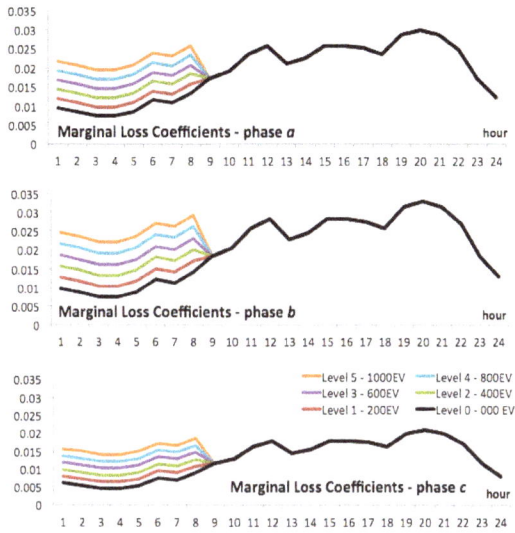

Figure 5. Off-peak load scenario: marginal loss coefficients (MLCs) at Bus 21 by EV penetration level and by time.

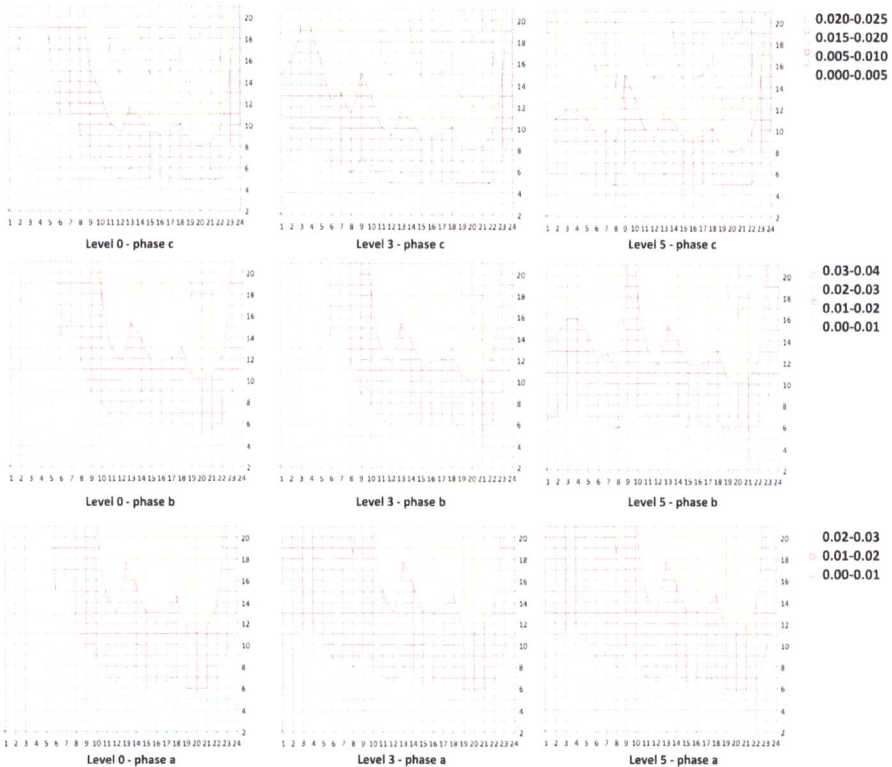

Figure 6. Off-peak load scenario: MLCs pattern by node and by time (Levels 0, 3 and 5).

Table 3. Off-peak load scenario: total energy losses allocated to PQ and EV loads by EV penetration level (kW·h/day).

Level	Pro Rata		Marginal		Energy Losses
	$A_D^{P,PQ}$	$A_D^{P,EV}$	$A_D^{M,PQ}$	$A_D^{M,EV}$	ΔW
Level 0	1339.08, 100%	0.00, 0%	1339.08, 100%	0.00 0,%	1339.08
Level 1	1324.41, 96%	61.93, 4%	1352.52, 98%	33.82, 2%	1386.34
Level 2	1321.66, 91%	124.55, 9%	1363.84, 94%	82.37, 6%	1446.21
Level 3	1329.09, 88%	187.43, 12%	1373.19, 91%	143.32, 9%	1516.51
Level 4	1347.19, 84%	253.33, 16%	1381.59, 86%	218.93, 14%	1600.52
Level 5	1372.92, 81%	323.72, 19%	1389.30, 82%	307.33, 18%	1696.64

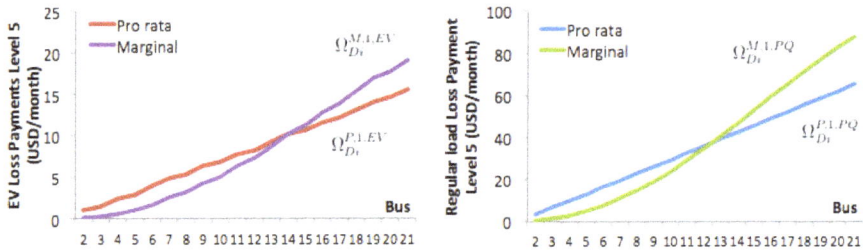

Figure 7. Off-peak load scenario: economic allocation between PQ and EV users connected at Phase 1 (Level 5).

4.2. Scenario 2: Fast Charging at Peak Load Conditions

In this scenario, EV connection was implemented at peak load conditions: from 18:00 to 21:00 with a 7.5 kW charging station considering the same five levels of integration applied in the case of EV charging at peak load conditions (Scenario 1), that is 200, 400, 600, 800 and 1000 units until reach a penetration of 25% of base energy consumption along one day.

The 24-h power loss curves by each level for the connection of EV loads under peak conditions are depicted in Figure 8. It is clear that the load curve becomes more sharp due to the progressive incorporation of slow EV charging stations.

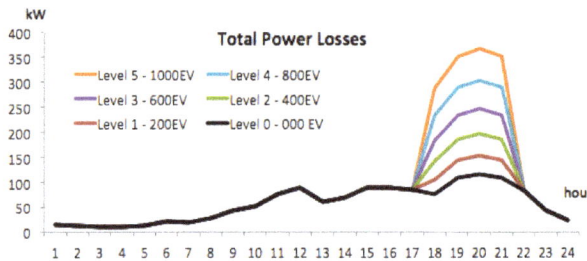

Figure 8. Peak load scenario: 24-h power losses by EV penetration level.

Total real energy losses ΔW to be allocated for each level among network users are indicated in Figure 9. Unlike Scenario 1, results show how the progressive integration of fast charging stations at peak load conditions produces a significative distortion effect of the load curve. EVs are charging only from 17:00 to 22:00. Then, the load factor decreased from 0.62 at Level 0 to 0.42 at Level 5. In this case, average demand does not grow in the same extent than the maximum value. As a result, the load factor falls. The loss factor also fall from 0.48 to 0.26 at Level 5. This means that energy losses drastically rose in magnitude from 1.24 MW·h/day, 1.05% (Level 0) to 2.30 MW·h/day, 1.8% (level 5). In this circumstance, the effects of EV charging stations at peak load condition are too harsh.

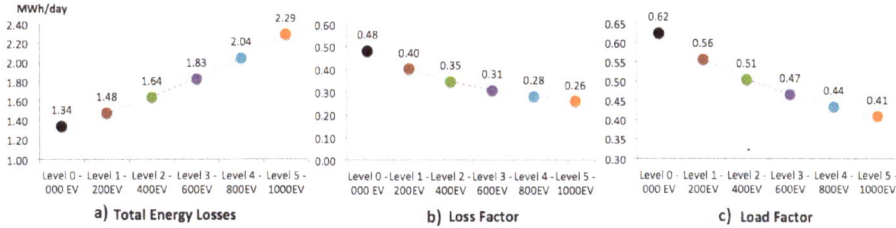

Figure 9. Peak load scenario: energy losses, loss and load factors by EV penetration level: (**a**) total energy losses; (**b**) loss factor; (**c**) load factor.

In Figure 10, the MLCs curves by EV penetration level at bus 21 along 24-h period is presented for the peak load conditions. It should be noted how marginal coefficients are able to reach high values 0.07 at peak time (18:00 and 21:00).

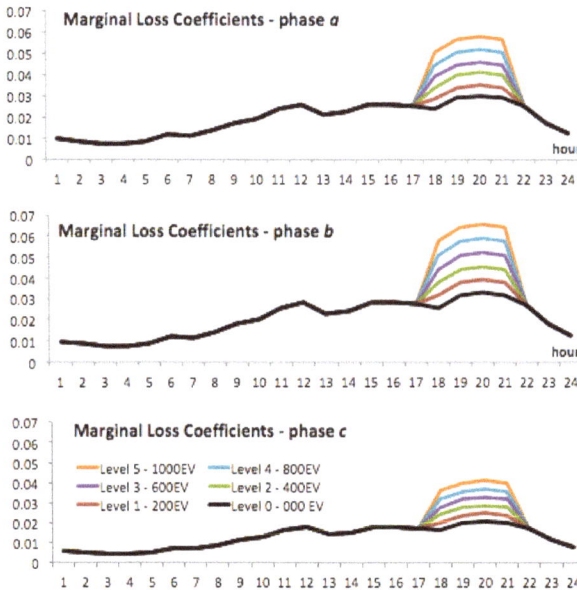

Figure 10. Peak load scenario: MLCs at Bus 21 by EV penetration level and by time.

Figure 11 displays the MLCs curves when EVs are charging at 20:00. The MLCs are applied to agents connected from bus 2 to bus 21. As the load is increasing with the distance, the MLC magnitude at each bus also grows with the distance with respect to the reference bus. Then, closer loads to reference produce lower losses (and lower MLCs) than farther loads and therefore loads connected near to substation pay less for power losses than farthest loads.

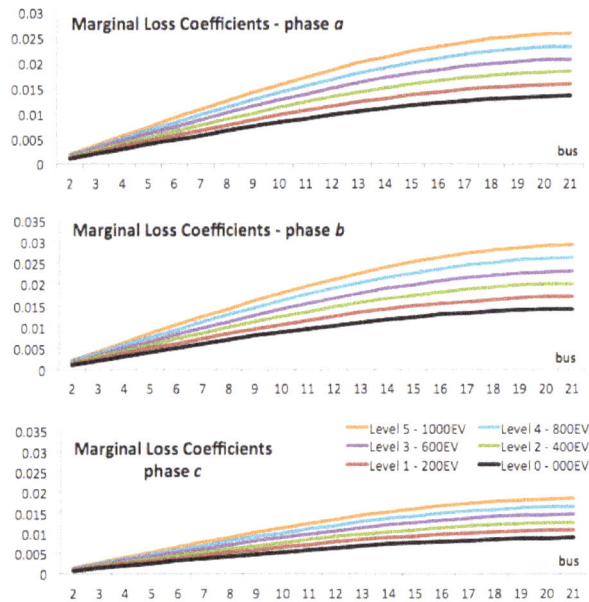

Figure 11. Peak load scenario: MLCs at 20:00 by EV penetration level and by bus.

In Table 4, the energy allocation results for aggregate PQ and EV loads are presented for the peak load condition. The reconciliation factor k_r also fluctuates around 0.5 in all levels. Equations (18) and (28) were applied for the marginal and pro rata procedure, respectively.

Unlike Scenario 1 where aggregate EV loads collect some marginal benefits due to the flattering effect over the load curve, Scenario 2 displays severe charges against EV loads due to energy losses associated with fast charging at peak conditions in all levels. If the marginal procedure is applied, PQ load should assume only a small part of the additional losses (23%). At Level 5, additional loses are 949 kW·h/day and PQ loads have to pay for 224 = 1563−1339 kW·h/day. Otherwise, if the pro rata procedure is applied, PQ loads should cover 46% of the incremental loads observed between Level 0 and Level 5.

Table 4. Peak load scenario: total losses allocated to PQ and EV loads (kW·h/day).

Level	Pro Rata		Marginal		Energy Losses
	$A_D^{P,PQ}$	$A_D^{P,EV}$	$A_D^{M,PQ}$	$A_D^{M,EV}$	ΔW
Level 0	1339.08, 100%	0.00, 0%	1339.08, 100%	0.00, 0%	1339.08
Level 1	1410.22, 96%	66.03, 4%	1382.69, 94%	93.57, 6%	1476.26
Level 2	1499.45, 91%	141.52, 9%	1426.40, 87%	214.56, 13%	1640.96
Level 3	1601.37, 88%	226.19, 12%	1469.93, 80%	357.63, 20%	1827.56
Level 4	1720.74, 84%	324.16, 16%	1516.01, 74%	528.89, 26%	2044.90
Level 5	1851.48, 81%	437.40, 19%	1563.97, 68%	724.91, 32%	2288.88

At Scenario 1 (EVs are charging at peak load conditions) pro rata and marginal allocation results lead to similar pattern. However, at Scenario 2 (EVs are charging at peak load conditions) marginal and pro rata loss allocation produce dissimilar results. EVs must pay for additional energy losses. The marginal procedure assigns higher losses to EV than calculated by the pro-rata procedure. This means that EV loads are duly charged by their produced losses at peak conditions. In this case, PQ loads take advantage of the marginal procedure since they have not to pay for additional losses.

The share of energy losses attributable to EV loads under marginal approach (32%) is significantly higher than the share obtained by the pro rata procedure (19%). If we consider a flat energy price ρ of 0.05 USD/kW·h, the left-hand chart of Figure 12 shows how the marginal procedure strongly penalize EVs connected from the middle to the end of the circuit. Note how EVs connected from bus 9 to bus 21 are facing high charges due to increasing losses. Conversely, the right-hand chart of Figure 12 visualizes how the marginal and pro rata procedures yield in similar charges. This means that there is not significative economical difference for PQ charges but strong incentives to EV loads to perform power loss reduction tasks.

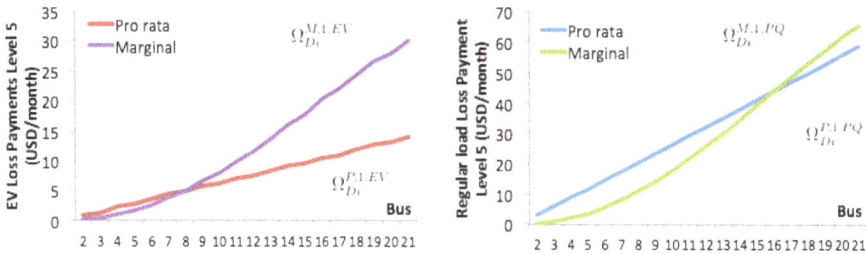

Figure 12. Peak load scenario: economic allocation between PQ and EV users connected at phase 1 (Level 5).

4.3. The Economical Effects in a Single EV Unit Under Off-Peak and Peak Load Conditions

Consider now the perspective of a single EV of 30kW·h capacity when Level 5 is reached (25% of penetration). If a fixed energy price ρ of 0.05 USD/kW·h is considered, the overall charging or energy cost for the EV is 1.5 USD/day. At off-peak and peak conditions, the connection of a single EV unit has different outcomes.

On the one hand, under off-peak load conditions (slow charging from 01:00 to 09:00), when the EV is connected at bus 21, phase 1 (ending node) the payment for losses under marginal procedure is almost 0.43 USD/day. This amount corresponds to 29% of total payment for energy (1.5 USD/day). On the other hand, under peak load conditions (fast charging from 19:00 to 21:00) the payment for losses under marginal procedure is 0.64 USD/day (54% of total payment for energy). This result is important since the best economic solution for the EV is charging under off-peak conditions.

Consider now that the EV is connected at bus 2, phase 1 (very close to substation). In this case, both scenarios show the same result, the EV has to pay only 0.03 USD/day (2% of total payment for energy). This charge is very low when compared with charges applied to loads at the end of the feeder. Then, the incentive is to connect EVs as close as possible to substation since no additional losses are produced.

Economic results for the marginal allocation procedure evidence EV loads connected at farthest loads have to pay important shares due to incremental losses becoming an important incentive (mainly at peak conditions) to provide network support. Under standard pro rata approach, the overall cost is distributed among all loads in a proportional manner and no incentive is provided by time of use and location of the EV charger.

4.4. The Impact of the EV Load Modeling on Loss Allocation Results

All results presented above were obtained assuming a specific EV load parametrization: $a = 0.9537$, $b = 0.0463$, and $\alpha = -2.324$ in Equation (3). To evaluate the effects of the EV load model in the results, we ran the model under peak loading conditions (Level 5) varying α from 0 (PQ load) to -8.0 and b from 0.0 (PQ load) to 0.10. Results of the sensitivity analysis are depicted in Figure 13.

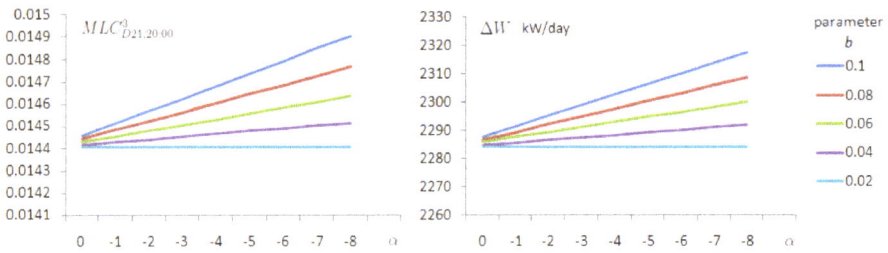

Figure 13. Sensitivity analysis: MLCs and total losses against EV parameters α and b.

If the EV load is regarded as constant PQ ($\alpha = 0$, $b = 0.0$), the marginal loss factor at 20:00, bus 21, Phase 3 ($MLC^3_{D21,20:00}$) is 0.014401769 and total losses to be allocated ΔW is 2284.12 kWh/day. Conversely, if the EV parameters take a non-linear form $\alpha = -8.0$, $b = 0.1$, the marginal loss factor at 20:00, bus 21, Phase 3 ($MLC^3_{D21,20:00}$) is 0.014904362 and total losses to be allocated ΔW increased 2317.50 kW·h/day. These variations on MLCs and energy losses represent 3.5% and 1.5% of the values achieved when loads were assumed as PQ constant, respectively. As a result, we observe significant differences on loss factors and overall energy losses to be allocated among the network users. The adoption of a correct EV load model for economic evaluation of the impacts on losses becomes an important issue to consider to guarantee the fairness of the allocation procedure. As there are several charging protocols of EV batteries [37], future research on economical impacts of EVs on system losses should be devoted to include more detailed models.

5. Conclusions

This paper presents three-phase loss allocation procedure for distribution networks considering widespread connection of non-linear electric vehicle loads. The method was based on the computation of specific marginal loss coefficients (MLCs) per bus and phase. The method was applied in an illustrative unbalanced 12.47 kV feeder with 12,780 residential customers supposing different levels of EV penetration. Two operational cases with two different type of charging stations were considered. Results obtained were also compared with a traditional roll-in embedded method (pro rata).

Depending on the operational scheme adopted, two different situations deserve mention. Firstly, slow EV charging under off-peak demand conditions helps to flatten the load curve yielding moderate MLCs similar to those obtained by means of the pro rata procedure. Secondly, fast EV charging under peak conditions leads to a sharpened load curve with increasing losses and volatile MLCs for EV agents. In this case, marginal loss prices may become a strong incentive to optimize distribution system operation.

Sensitivity analysis results show the influence of the non-linear EV load model in the energy losses allocated. This result highlights the importance of considering an appropriate EV load model to appraise the overall losses to be allocated.

Author Contributions: P.M.D.O.-D.J. conceived the idea behind this research, designed and performed the simulations, performed result analysis and wrote the paper; G.A.R. developed the EV load model, performed result analysis and wrote some parts of the manuscript; M.A.R. developed the simulation case scenarios, performed result analysis and wrote some parts of the manuscript; all the authors analyzed the data and proofread the paper.

Funding: This research received no external funding.

Conflicts of Interest: The authors declare no conflict of interest

Appendix A

```
! Kersting NEV Test system
! W. H. Kersting, A three-phase unbalanced line model with grounded
! neutrals through a resistance, 2008 Ieee Power and Energy Society General
! Meeting, Vols 1-11 (2008) 2651-2652
```

```
! 3 phase approach (kron's reduction) with incremental load
clear
! **** DEFINE SOURCE BUS
new circuit.KersNEV2nThreeP basekV = 12.47 phases = 3  !Define a 3-phase source
~ mvasc3 = 2000000000   mvasc1 = 2100000000
! **** DEFINE DISTRIBUTION LINE
set earthmodel = carson
! **** DEFINE WIRE DATA  STRUCTURE
new wiredata.conductor Runits = mi Rac = 0.306 GMRunits = ft GMRac = 0.0244  Radunits = in Diam = 0.721
new wiredata.neutral   Runits = mi Rac = 0.592 GMRunits = ft GMRac = 0.00814 Radunits = in Diam = 0.563
! **** DEFINE LINE GEOMETRY; REDUCE OUT THE NEUTRAL WITH KRON
new linegeometry.4wire nconds = 4 nphases = 3 reduce = yes
~ cond = 1 wire = conductor units = ft x = -4   h = 28
~ cond = 2 wire = conductor units = ft x = -1.5 h = 28
~ cond = 3 wire = conductor units = ft x = 3    h = 28
~ cond = 4 wire = neutral   units = ft x = 0    h = 24
! **** 12.47 KV LINE!
new line.line1 geometry = 4wire length = 300 units = ft bus1 = sourcebus.1.2.3 bus2 = n1.1.2.3
new line.line2 geometry = 4wire length = 300 units = ft bus1 = n1.1.2.3 bus2 = n2.1.2.3
new line.line3 geometry = 4wire length = 300 units = ft bus1 = n2.1.2.3 bus2 = n3.1.2.3
new line.line4 geometry = 4wire length = 300 units = ft bus1 = n3.1.2.3 bus2 = n4.1.2.3
new line.line5 geometry = 4wire length = 300 units = ft bus1 = n4.1.2.3 bus2 = n5.1.2.3
new line.line6 geometry = 4wire length = 300 units = ft bus1 = n5.1.2.3 bus2 = n6.1.2.3
new line.line7 geometry = 4wire length = 300 units = ft bus1 = n6.1.2.3 bus2 = n7.1.2.3
new line.line8 geometry = 4wire length = 300 units = ft bus1 = n7.1.2.3 bus2 = n8.1.2.3
new line.line9 geometry = 4wire length = 300 units = ft bus1 = n8.1.2.3 bus2 = n9.1.2.3
new line.line10 geometry = 4wire length = 300 units = ft bus1 = n9.1.2.3 bus2 = n10.1.2.3
new line.line11 geometry = 4wire length = 300 units = ft bus1 = n10.1.2.3 bus2 = n11.1.2.3
new line.line12 geometry = 4wire length = 300 units = ft bus1 = n11.1.2.3 bus2 = n12.1.2.3
new line.line13 geometry = 4wire length = 300 units = ft bus1 = n12.1.2.3 bus2 = n13.1.2.3
new line.line14 geometry = 4wire length = 300 units = ft bus1 = n13.1.2.3 bus2 = n14.1.2.3
new line.line15 geometry = 4wire length = 300 units = ft bus1 = n14.1.2.3 bus2 = n15.1.2.3
new line.line16 geometry = 4wire length = 300 units = ft bus1 = n15.1.2.3 bus2 = n16.1.2.3
new line.line17 geometry = 4wire length = 300 units = ft bus1 = n16.1.2.3 bus2 = n17.1.2.3
new line.line18 geometry = 4wire length = 300 units = ft bus1 = n17.1.2.3 bus2 = n18.1.2.3
new line.line19 geometry = 4wire length = 300 units = ft bus1 = n18.1.2.3 bus2 = n19.1.2.3
new line.line20 geometry = 4wire length = 300 units = ft bus1 = n19.1.2.3 bus2 = n20.1.2.3
vsource.source.enabled = no
solve
```

Table A1. EV units connected per bus and per phase at each EV penetration level.

Bus/Phase	Level 1 - 200EV			Level 2 - 400EV			Level 3 - 600EV			Level 4 - 800EV			Level 5 - 1000EV		
	1	2	3	1	2	3	1	2	3	1	2	3	1	2	3
1	0	0	0	0	0	0	0	0	0	0	0	0	0	0	0
2	0	0	0	1	1	1	1	1	1	1	1	1	2	2	1
3	1	1	1	1	1	1	2	2	2	3	3	2	3	4	3
4	1	1	1	2	2	2	3	3	3	4	4	3	5	5	4
5	1	1	1	3	3	2	4	4	3	5	6	4	6	7	6
6	2	2	1	3	4	3	5	5	4	6	7	6	8	9	7
7	2	2	2	4	4	3	6	6	5	8	9	7	10	11	8
8	2	2	2	4	5	4	7	7	6	9	10	8	11	12	10
9	3	3	2	5	6	4	8	9	7	10	11	9	13	14	11
10	3	3	3	6	6	5	9	10	8	11	13	10	14	16	13
11	3	4	3	6	7	6	10	11	8	13	14	11	16	18	14
12	3	4	3	7	8	6	10	12	9	14	16	12	17	20	15
13	4	4	3	8	9	7	11	13	10	15	17	13	19	21	17
14	4	5	4	8	9	7	12	14	11	16	19	15	21	23	18
15	4	5	4	9	10	8	13	15	12	18	20	16	22	25	20
16	5	5	4	10	11	8	14	16	13	19	21	17	24	27	21
17	5	6	4	10	11	9	15	17	13	20	23	18	25	29	22
18	5	6	5	11	12	9	16	18	14	22	24	19	27	30	24
19	6	6	6	11	12	10	17	19	15	23	26	20	29	32	25
20	6	7	5	12	14	11	18	20	16	24	27	21	30	34	27
21	6	7	6	13	14	11	19	21	17	25	29	22	32	33	28
SubTotal	66	74	60	134	149	117	200	223	177	266	300	234	334	372	294
Total	200			400			600			800			1000		

References

1. Rotering, N.; Ilic, M. Optimal charge control of plug-in hybrid electric vehicles in deregulated electricity markets. *IEEE Trans. Power Syst.* **2011**, *26*, 1021–1029. [CrossRef]
2. Deb, S.; Tammi, K.; Kalita, K.; Mahanta, P. Impact of Electric Vehicle Charging Station Load on Distribution Network. *Energies* **2018**, *11*, 178. [CrossRef]
3. Bessa, R.J.; Matos, M.A. Economic and technical management of an aggregation agent for electric vehicles: A literature survey. *Int. Trans. Electr. Energy Syst.* **2012** *22*, 334–350. [CrossRef]
4. Zhang, Z.J.; Nair, N.C. Economic and pricing signals in electricity distribution systems: A bibliographic survey. In Proceedings of the IEEE International Conference on Power System Technology (POWERCON), Auckland, New Zealand, 30 October–2 November 2012; pp. 1–6.
5. Pan, J.; Teklu, Y.; Rahman, S.; Jun, K. Review of usage-based transmission cost allocation methods under open access. *IEEE Trans. Power Syst.* **2000**, *15*, 1218–1224.
6. Carpaneto, E.; Chicco, G.; Akilimali, J.S. Characterization of the loss allocation techniques for radial systems with distributed generation. *Electr. Power Syst. Res.* **2008**, *78*, 1396–1406. [CrossRef]
7. Costa, P.M.; Matos, M.A. Loss allocation in distribution networks with embedded generation. *IEEE Trans. Power Syst.* **2004**, *19*, 384–389. [CrossRef]
8. Bialek, J. Tracing the flow of electricity. *IEE Proc. Gener. Transm. Distrib.* **1996**, *143*, 313–320. [CrossRef]
9. Galiana, F.D.; Conejo, A.J.; Kockar, I. Incremental transmission loss allocation under pool dispatch. *IEEE Trans. Power Syst.* **2002**, *17*, 26–33. [CrossRef]
10. Weckx, S.; Driesen, J.; D'hulst, R. Optimal Real-Time Pricing for UnbalancedDistribution Grids with Network Constraints. In Proceedings of the IEEE Power and Energy SocietyGeneral Meeting (PESGM), Vancouver, BC, Canada, 21–25 July 2013.
11. Heydt, G.T.; Chowdhury, B.H.; Crow, M.L.; Haughton, D.; Kiefer, B.D.; Meng, F.J.; Sathyanarayana, B.R. Pricing and control in the next generation power distribution system. *IEEE Trans. Smart Grid* **2012**, *3*, 907–914 [CrossRef]
12. De Oliveira-De Jesus, P.M.; Castronuovo, E.D.; De Leao, M.P. Reactive power response of wind generators under an incremental network-loss allocation approach. *IEEE Trans. Energy Convers.* **2008**, *23*, 612–621. [CrossRef]
13. Kaur, M.; Ghosh, S. Effective Loss Minimization and Allocation of Unbalanced Distribution Network. *Energies* **2017**, *10*, 1931. [CrossRef]
14. Hong, M. An approximate method for loss sensitivity calculation in unbalanced distribution systems. *IEEE Trans. Power Syst.* **2014**, *29*, 1435–1436. [CrossRef]
15. Li, R.Y.; Wu, Q.W.; Oren, S.S. Distribution locational marginal pricing for optimal electric vehicle charging management. *IEEE Trans. Power Syst.* **2014**, *29*, 203–211 [CrossRef]
16. Kongjeen, Y.; Bhumkittipich, K. Impact of Plug-in Electric Vehicles Integrated into Power Distribution System Based on Voltage-Dependent Power Flow Analysis. *Energies* **2018**, *11*, 1571. [CrossRef]
17. Shirmohammadi, D.; Gorenstin, B.; Pereira, M.V. Some fundamental, technical concepts about cost based transmission pricing. *IEEE Trans. Power Syst.* **1996**, *11*, 1002–1008. [CrossRef]
18. Qian, K.; Zhou, C.; Allan, M.; Yuan, Y. Modeling of load demand due to EV battery charging in distribution systems. *IEEE Trans. Power Syst.* **2011**, *26*, 802–810. [CrossRef]
19. Hernandez, J.C.; Ruiz-Rodriguez, F.J.; Jurado, F. Modelling and assessment of the combined technical impact of electric vehicles and photovoltaic generation in radial distribution systems. *Energy* **2017**, *141*, 316–332 [CrossRef]
20. Kisacikoglu, M.C.; Ozpineci, B.; Tolbert, L.M. EV/PHEV bidirectional charger assessment for V2G reactive power operation. *IEEE Trans. Power Electr.* **2011**, *28*, 5717–5727. [CrossRef]
21. Dharmakeerthi, C.H.; Mithulananthan, N.; Saha, T.K. Impact of electric vehicle fast charging on power system voltage stability. *Int. J. Electr. Power Energy Syst.* **2014**, *57*, 241–249. [CrossRef]
22. Wasley, R.G.; Shlash, M.A. Newton-Raphson algorithm for 3-phase load flow. *Proc. Inst. Electr. Eng.* **1974**, *121*, 630–638. [CrossRef]
23. Arrillaga, J.A.; Arnold, J.P. *Computer Analysis of Power Systems*; John Wiley & Sons, Inc.: Hoboken, NJ, USA, 1990.

24. De Oliveira-De Jesus, P.M.; Alvarez, M.A.; Yusta, J.M. Distribution power flow method based on a real quasi-symmetric matrix. *Electric. Power Syst. Res.* **2013**, *95*, 148–159. [CrossRef]

25. Montenegro, D.; Hernandez, M.; Ramos, G.A. Real time OpenDSS framework for distribution systems simulation and analysis. In Proceedings of the 2012 Sixth IEEE/PES Transmission and Distribution: Latin America Conference and Exposition (T&D-LA), Montevideo, Uruguay, 3–5 September 2012.

26. Mutale, J.; Strbac, G.; Curcic, S.; Jenkins, N. Allocation of losses in distribution systems with embedded generation. *IEE Proc. Gener. Transm. Distrib.* **2000**, *147*, 7–14. [CrossRef]

27. Stoft, S. *Power System Economics*; IEEE Press & Wiley-Interscience: Hoboken, NJ, USA, 2002.

28. Kersting, W.H. A three-phase unbalanced line model with grounded neutrals through a resistance. In Proceedings of the 2008 IEEE Power and Energy Society General Meeting-PESGM, Pittsburgh, PA, USA, 20–24 July 2008; pp. 12651–21652.

29. Gonen, T. *Electric Power Distribution Engineering*, 2nd ed.; CRC Press: Boca Raton, FL, USA, 2008; p. 256

30. Dugan, R.C.; McDermott, T.E. An open source platform for collaborating on smart grid research. In Proceedings of the IEEE Power Energy Society General Meeting, Detroit, MI, USA, 24–29 July 2011; pp. 1–7.

31. Mocci, S.; Natale, N.; Ruggeri, S.; Pilo, F. Multi-agent control system for increasing hosting capacity in active distribution networks with EV. In Proceedings of the IEEE International Energy Conference (ENERGYCON), Cavtat, Croatia, 13–16 May 2014; pp. 1409–1416.

32. Kisacikoglu, M.C.; Erden, F.; Erdogan, N. Distributed control of PEV charging based on energy demand forecast. *IEEE Trans. Ind. Inf.* **2018**, *14*, 332–341. [CrossRef]

33. Ruiz-Rodriguez, F.J.; Hernandez, J.C.; Jurado, F. Voltage behaviour in radial distribution systems under the uncertainties of photovoltaic systems and electric vehicle charging loads. *Int. Trans. Electr. Energy Syst.* **2018**, *28*, 2490 [CrossRef]

34. Munkhammar, J.; Widen, J.; Ryden, J. On a probability distribution model combining household power consumption, electric vehicle home—Charging and photovoltaic power production. *Appl. Energy* **2015**, *142*, 135–143. [CrossRef]

35. Ruiz-Rodriguez, F.J.; Hernandez, J.C.; Jurado, F. Probabilistic Load-Flow Analysis of Biomass-Fuelled Gas Engines with Electrical Vehicles in Distribution Systems. *Energies* **2017**, *10*, 1536. [CrossRef]

36. ElNozahy, M.S.; Salama, M.M. A comprehensive study of the impacts of PHEVS on residential distribution networks. *IEEE Trans. Sustain. Energy* **2014**, *5*, 332–342. [CrossRef]

37. Godina, R.; Paterakis, N.G.; Erdinc, O.; Rodrigues, E.M.G.; Catalão, J.P.S. Impact of EV charging-at-work on an industrial client distribution transformer in a Portuguese Island. In Proceedings of the 2015 Australasian Universities Power Engineering Conference (AUPEC), Wollongong, Australia, 27–30 September 2015; pp. 1–6.

energies

MDPI

Article

Wind Power Monitoring and Control Based on Synchrophasor Measurement Data Mining

Mario Klarić [1], Igor Kuzle [2,*] and Ninoslav Holjevac [2]

[1] Manufacturing and Construction, Dalekovod JSC for Engineering, Zagreb 10000, Croatia; mario.klaric@dalekovod.hr

[2] Faculty of Electrical Engineering and Computing, University of Zagreb, Zagreb 10000, Croatia; ninoslav.holjevac@fer.hr

* Correspondence: igor.kuzle@fer.hr; Tel.: +385-1-6129-875

Received: 31 October 2018; Accepted: 12 December 2018; Published: 18 December 2018

Abstract: More and more countries and utilities are trying to develop smart grid projects to make transformation of their power infrastructure towards future grids with increased share of renewable energy production and near zero emissions. The intermittent nature of solar and wind power can in general cause large problems for power system control. Parallel to this process, the aging of existing infrastructure also imposes requirements to utility budgets in the form of a need for large capital investments in reconstruction or maintenance of key equipment. Synchrophasor and other synchronized measurement technologies are setting themselves as one of the solutions for larger wind power integration. With that aim, in this paper one possible solution for wind power control through data mining algorithms used on a large quantity of data gathered from phasor measurement units (PMU) is described. Developed model and algorithm are tested on an IEEE 14 bus test system as well as on real measurements made on wind power plants currently in operation. One such wind power plant is connected to the distribution grid and the other one to the transmission grid. Results are analyzed and compared.

Keywords: smart grid; wind power; synchronized measurements; PMU; data mining

---

1. Introduction

Many utilities are facing new challenges when trying to develop various kinds of smart grid projects in order to make transformation towards smarter and more sensible [1,2] power grids and utilities. This demanding task becomes even more complex when utilities are facing the aging of existing infrastructure which makes huge demands to public budgets regarding the need for large CAPEX (capital expenditure) amounts in reconstruction and upgrades of power system infrastructure [3–5]. The ageing problems combined can cause severe faults extending along the grid and to cope with the fact that new ways of power system monitoring and control are required [2].

Phasor measurement units (PMU) have already been defined as suitable for many applications of larger renewable energy integration [2,6]. Furthermore, latest developments in the information and telecommunication technologies (ICT) industry creates large possibilities in the areas of data transmission, sensor measurements, energy savings, asset management etc. [7,8] and that provides new opportunities for finding better solutions.

Probably the most suitable installation that imposes itself as an upgrade of the existing system is the usage of simple intelligent electronic devices (IEDs) [9,10]. Together with the increase of the renewable energy share that requires new paradigms and market designs [11], there is also an increasing need for improved monitoring and control possibilities in power systems. Also, operators have a great need for early warning in critical transition situations [12].

In this paper, one possible application for wind power monitoring and control is described which integrates several of these additional services as an example. This kind of project can easily be upgraded, scaled and multiplied for application in the other utilities and transmission systems.

The paper is structured as follows: Section 1 gives an overview and introduction to basic concepts, Section 2 describes the developed big data algorithm and Section 3 depicts test case results. Sections 4 and 5 describe the real system results and provide specifics for different network designs while Section 6 concludes the work and provides guidelines for future work. Fundamental system framework is structured in a way that the integration of additional modules can be arranged around existing basic infrastructure like energy distribution grid, transmission power lines, substations etc. Main groups of additional modules are shown on figure below (Figure 1).

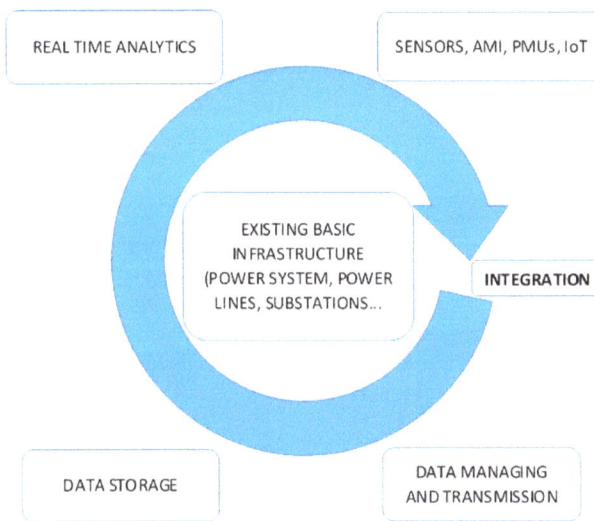

Figure 1. Fundamental smart grid system framework components.

Figure 1 depicts the importance of the integration of all components. Having such a vital role in the whole process the experts working in the field will be required to have a broader insight and understanding of the that process in order to successfully integrate all new technologies. Fundamental basic system architecture is defined as a set of nodes around telecom backbone sending crucial data of the system operation to corresponding servers.

Internet Protocol (IP) multicast [13] is one such technique for one-to-many and many-to-many real-time communication over an IP infrastructure in a network. In smart grid applications, there are different protocols that need to be integrated in a system such as WAMS (Wide Area Monitoring System) that include protocols based on IEEE C37.118, substation automation protocols (IEC 61850) etc. All smart grid services and protocols (WAMS, Smart Metering infrastructure and IEC 61850 based applications) can be efficiently transported over such networks of telecommunication platforms (Figure 2).

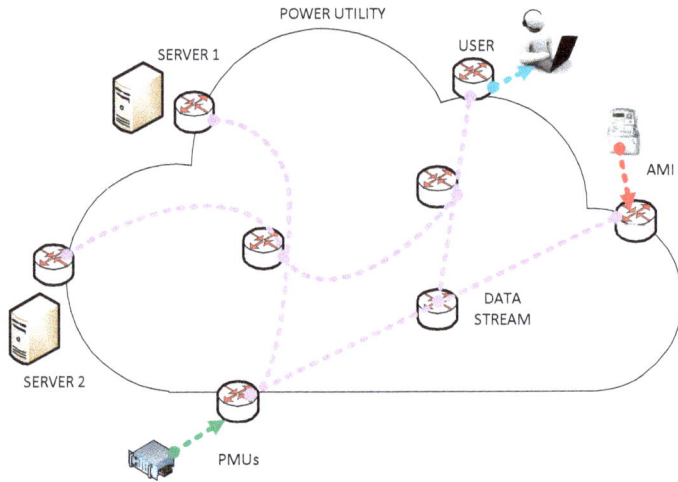

Figure 2. Basic system architecture of IP/MPLS (Multiprotocol Label Switching) techniques.

Having all the data concentrated would enable efficient data storing and processing enhancing the current information stream with the extraction of right information from the big data surrounding. Key characteristic of such future infrastructure enhancements would need to be adaptivity towards existing power grid infrastructure and modularity to allow system's components separation and recombination ("adaptidular" infrastructure). The most important benefits of the new infrastructure (Figure 3) following the adaptidular design paradigm can be described as following:

- Existing capacities and possibilities of existing infrastructure enhancement and upgrading
- Capital expenditures (CAPEX) postponing or abolishing (building of new lines, substations, power infrastructure reconstruction) due to availability of new information in asset management systems, dynamic line rating system, PMU systems etc.
- Maintenance cost cutting through the usage of predictive maintenance enabled through sensor networks and IoT gateways
- Additional services provision: numerous additional services such as meteorological data assessment, air quality mapping, telecom services provision through IP/MPLS etc.

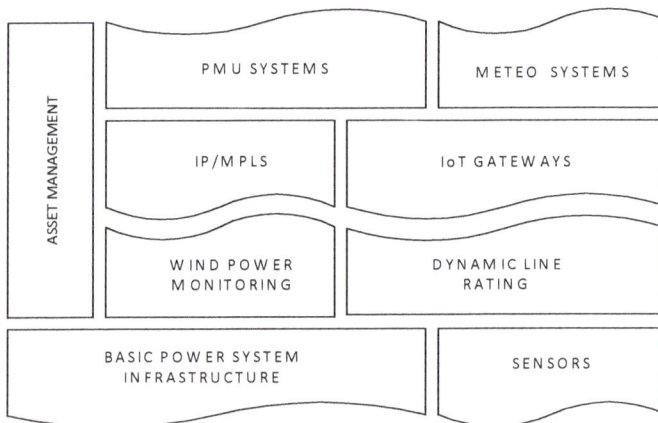

Figure 3. Adaptive and modular hardware and software infrastructure characteristics.

2. Developed Algorithm for Wind Power Monitoring and Control

In order to investigate the possibility and potential of wind power monitoring and control based on big data surrounding an algorithm for monitoring and incorporating synchrophasor measurement was developed. As described earlier, it has all the characteristics of adaptive and modular applications that can easily be installed and commissioned on the existing infrastructure. It also provides ability for later upgrades and integration into large scale applications.

2.1. Big Data Surroundings

The power system infrastructure produces huge amounts of data. The nonlinear nature of this data makes the extraction of useful information complicated [14]. Compared to standard mathematical models, data mining techniques are non-deterministic and provide a feasible and valid solution which is not exact but is simple to obtain, concise, practical and easy to understand. This characteristic is especially suitable when processing the big data streams which are inevitably involved. As mentioned earlier, large wind power capacities are being installed and connected to different voltage levels. Every wind turbine, wind measuring masts inside the wind park transformer substations, etc. represent the source of large quantities of data every second. All these data streams can be further expanded with the installation of new data sensors arrays. These large quantities of data can be deemed unnecessary, but with the usage of different big data algorithms a way to monetize this data can be found.

The most important data that can and should be used in power system data mining algorithms is the data for state estimation and future power system state predictions. These data streams can be classified into three main groups:

1. Phasor values measurements;
2. Loads and production measurements;
3. Other influential variables measurements.

Phasor values like voltages and currents together with belonging phasor angles, can be gathered through PMU measurements and can provide valuable insights into system operation. Also, load and generation data with exact time stamp can easily be measured and collected to afterwards be used for different analyses.

Other influential variables of additional data that are not directly connected to power system monitoring and control are also sometimes highly influential. These include meteorological data from various kinds of measurement systems of which most important are wind speeds and wind directions, air temperature, humidity and pressure, solar irradiance measurements. Together with meteorological data, other measurements such as conductor temperatures, overhead line sags, partial discharges, current transmission line capacity obtained by dynamic line rating (DLR) systems etc. can also be collected [15]. All these data series can be used in wind and solar power system monitoring and control as well as for load forecasting applications and power evacuation possibilities. The prerequisite is to have an efficient solution for data transmission and processing.

2.2. Data Mining Scope

As described earlier, the huge amounts of data inside power creates the big data surroundings. The non-linear nature of the system makes the definition of new models for extraction of useful information from heaps of gathered data even more demanding [16].

Especially demanding is the usage of data from wind power plants since these stochastic sources produce even bigger amounts of data due to dependable variables which influence the output power.

Therefore, good data mining scope thus integrates wide area of variables. This paper defines simplified model which comprises of:

- Wind power plant active and reactive power production (P_{Wind}, Q_{Wind}), at wind power plant point of common coupling (PCC);

- Wind power plant active and reactive power settings (P$_{Settings}$, Q$_{Settings}$), which are operational decisions for the settings of wind power controller placed at wind power (PCC);
- Total system load measurements (P$_L$), expressed in percentage, as a percentage of nominal load;
- Voltage amplitudes and angles (phasors) measurements (V$_i$, δ_i) on selected nodes in the system;
- Line, transformer and generator availability information.

Each operating condition (**OC**) is defined as a mathematical set whose members are the following elements or variables:

$$OC_k = \{V_1, V_2, V_3, \ldots V_i, \delta_1, \delta_2, \delta_3, \ldots \delta i, P_L, P_{Wind}, Q_{Wind}, Z_{th}\} \tag{1}$$

- with i = 1, 2, 3, … *n*; where *n* is the number of nodes in power system with measurements of effective values and voltage angles in the system, and
- with *k* = 1, 2, 3, … *m*; where *m*—total number of input states over which data mining techniques are analyzed.

The abovementioned data can be expanded by defining the finely tuned fractal structures attached to it:

- Wind power total can be divided into wind power of single wind turbine or a cluster of turbines;
- Total system load can be divided into loads on busbar, consumer, or load area level;
- Voltage amplitudes and angles can be enhanced with current amplitudes and angles for each branch as well as Thevenin impedance measurements;
- Wind production is defined with wind speed and can further be detailed with wind direction, air temperature and pressure, solar irradiance and air humidity measurements;
- Line and transformer availability can further be described through breaker status in line bays and transformer bays or through transformer and line monitoring systems.

All this data needs to form large and well-organized databases for further usage in control, planning, asset management and operation and maintenance (O&M) optimization process. Therefore, to take full advantage of the available data efficient algorithms for big data analysis are needed.

2.3. Proposed Algorithm Design

The aim of the developed algorithm is to create a new kind of early warning signal (EWS) and recognize the structure of critical transitions for transmission system and wind power operators in the form of a situational awareness (SA) indicator [17]. These signals should be structured to warn the operators that the alarming operating condition could be reached and that preventive or corrective actions should be done (e.g., wind power curtailment or reactive power support increase) and thus move the system to normal operating state, like described in figure below (Figure 4). Created EWS signal as a situational awareness indicator serves as a main triggering signal for operating decisions in wind power settings in order to change operating condition back to EWS value NORMAL. Therefore, EWS could serve as a first line of defense to reduce the risks of total or partial system blackouts and thus reducing the opportunity costs associated with the costs of electric energy not being delivered.

Commonly used data mining algorithms identified by the IEEE International Conferences on Data Mining (ICDM) are C4.5, k-Means, Support Vector Machine (SVM), Apriori, PageRank, AdaBoost, Neural Networks, Naive Bayes and Classification and regression trees (C&RT). These 10 algorithms cover classification, clustering, statistical learning, association analysis, and link mining, which are all among the most important topics in data mining research and development. In [18] a review on the applications of data mining in power systems is given.

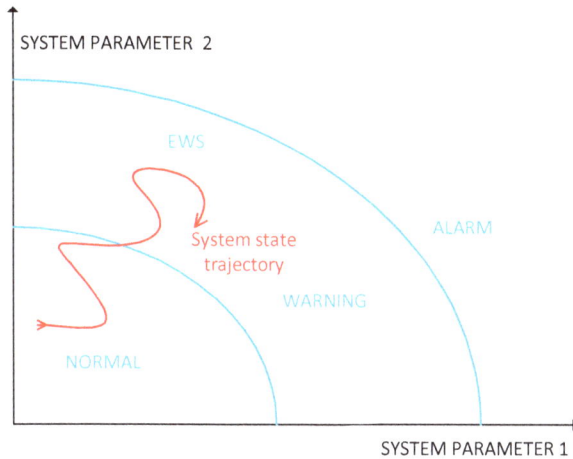

Figure 4. Early warning signal (EWS) concept as a first line of defense inducing preventive and corrective actions after system state change.

The approach described here combines several segments of classification and clustering and statistical learning in one algorithm. Also, it brings combined solution for monitoring and preventive measures operating decisions.

A basic workflow diagram of the proposed algorithm is described on Figure 5. The first step in the algorithm is data management and preparation which consists of time synchronization, format unifying, and ordering of historical raw data from actual power system measurements. Additionally, synthetic data which is produced and gathered from various kinds of simulations based on mathematical models is also included in this step. In this paper DigSilent Power Factory power system analysis software [19] is used as a tool for production of simulation data.

The input data vector in the clustering process is equal to:

$$\mathbf{OC} = [OC_1; OC_2; OC_3; OC_m]^{\mathrm{T}} \tag{2}$$

In this way, mathematically defined power system states are defined as input data in the algorithm. It is important to note that except for the variables defined herein, the input set of system states can be extended to a whole range of additional input signals such as data from various measuring devices for measuring electrical and nonelectric values, meteorological measuring devices, sensors and other devices. The model is therefore adaptive and modular. It is easy to upgrade by simply expanding the operating condition (**OC**) math data set.

The second step is data clustering, with the aim of defining system states on a given database or set of operating conditions. For the algorithm design described in this paper, the analytics software package Statistica [20] was used. Standard variable definition from statistical theory was used where an independent variable (also called experimental or predictor variable), is being manipulated in an experiment to observe the effect on a dependent variable (also called an outcome variable). Total set of operating conditions in this example to be a representative sample needs to be large enough and cover all possible system states and. K-Means algorithm with Euclidian distances was used for clustering of the initial data set in following way:

- Thevenin impedance at bus 8 (Figure 6) was used as dependent variable;
- Thevenin impedance absolute value is used as first dependent variable;
- Thevenin impedance angle was used as second dependent variable;

Clustering was finally made into three clusters which describe normal (NORMAL), transition (WARNING) and problematic (ALARM) conditions. It is important to stress that all three system states should be present in input datasets in order to have a viable solution of this part of the algorithm.

After the clustering of the system states of a particular group or clusters for normal, warning and alarm operating conditions, the same definitions of the target groups serve as inputs for the classification part of the algorithm. With these clustered data, data classes are defined for later analysis of new metric input data:

$$C_A = \{C_{NORMAL}, C_{WARNING}, C_{ALARM}\} \tag{3}$$

C_A—a set of data classes in the algorithm

C_{NORMAL}—data class for normal operating condition

$C_{WARNING}$—the class of data for transition operating condition

C_{ALARM}—data class for normal critical condition

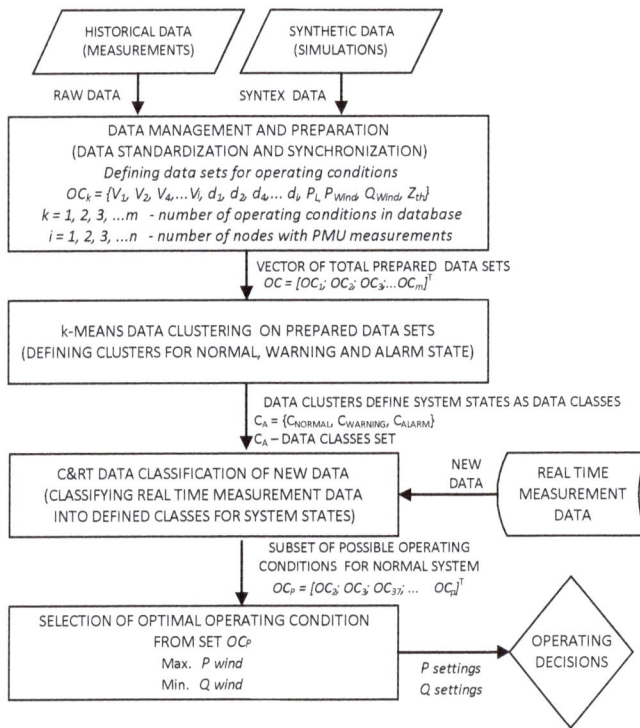

Figure 5. Basic workflow diagram.

The third step consists of data classification of new measurement data and definition of a set of new system operating conditions (OC). Assigned system condition (NORMAL, WARNING and ALARM) were set as independent variables and previously defined variables in data mining scope (P_{Wind}, Q_{Wind}, P_L, V_i, δ_i) as dependent. New measurement data, according to its parameters, in the classification part of the algorithm are classified into predefined groups according to the values of the parameters that are taken as input data. Classification groups are defined as clusters created by earlier clustering of operating conditions.

Classification and regression trees (C&RT) method was used for this classification analysis. For that purpose, software Statistica [20] was used. To assign weight factors to decision making process, misclassification costs [21] were defined heuristically according to table below (Table 1). In columns are predicted variables and in rows are measured variables.

Table 1. Misclassification costs defined for data classification.

EWS	NORMAL	WARNING	ALARM
NORMAL	0	1	5
WARNING	1	0	3
ALARM	1	1	0

To prevent overfitting of the data, a V-fold cross validation is used. 5% of the cases were used as "v-value" [21]. V-fold cross validation where the data set is randomly divided into v equal parts and the learning phase of the algorithm is done on $v - 1$ parts and test on the remaining piece is especially suitable for such situations where a small number of cases is used for classification. Furthermore, pruning on variance that reduces the size of decision trees by removing sections of the tree that provide little power to classify instances was used to get closer look at cost sequence for all calculated classification and regression trees. Cost sequence was calculated for re-substitution and cross-validation costs for all generated C&R (Classification and Regression) trees. In this way, a more simplified decision tree can be chosen according to law of parsimony, anticipating that things are usually connected or behave in the simplest or most economical way, especially with reference to alternative evolutionary paths [22]. To reach a normal system state, as a final result there can be several operating conditions fulfilling the given conditions. This means the output from data classification process will be a set of possible operating conditions (OCs). In the final step, final wind power plant operating decisions are made according to a simple procedure of selecting the best possible solution among the vector of possible operating states (**OC**$_P$) whereby:

$$\mathbf{OC_P} = [OC_2; OC_3; \dots OC_p]^T \tag{4}$$

With the requirement that each element of vector **OC**$_P$ is also an element of the class C_{NORMAL}.

A final operating decision still needs to be made, meaning settings of wind power plant controller ($P_{setting}$ and $Q_{setting}$) at the point of common coupling need to be defined. Variable $P_{setting}$ is defined as setting of for output active power. If this setting is lower than available wind power, the result will be wind power curtailment. This variable is defined as a continuous variable. Variable $Q_{setting}$ is defined as setting of regime for reactive power regulation. This variable is defined as categorical variable (of total output Q or cos φ) meaning one setting represents one possible category (e.g., cos φ = 0.9 lagging or Q equal to 0.5 p.u.). This way reactive power control variable is discretized. Final operating decisions for wind power plants are made according to simple process of selecting the best possible solution among the set of possible operating conditions (OCs). Final operating condition is chosen to minimize the opportunity costs of wind energy export and thus maximizing the produced energy. Also, according to [23], to prolong the lifetime of wind turbines it is necessary to lower reactive power production and its influence on power electronics in turbine converters. In harmony with the availability of wider range of PMU measurements the operation can be optimized with both available measurement and analysis results [24]. Therefore, final decisions can be summarized as maximization of output active power and minimization of reactive power (Equations (5) and (6)).

$$\max \{P_{wind}\} \tag{5}$$

$$\min \{Q_{wind}\} \tag{6}$$

Power transformers at point of common coupling (PCC) have limited capacity. Therefore, additional condition needs to be fulfilled in order not to endanger operational limits (Equation (7)) where S_{TR} is the power transformer capacity (MVA).

$$S_{TR} \leq (P_{Wind}^2 + Q_{Wind}^2)^{1/2} \tag{7}$$

3. Test System Example

3.1. Test System Description

IEEE 14 bus test system was used as a first test case for the application of the proposed algorithm. Instead of synchronous compensator that is originally included in the IEEE 14 test system connected to bus 8, a wind power plant on that given bus was defined with rated power of 20 MW, which can be seen on Figure 6.

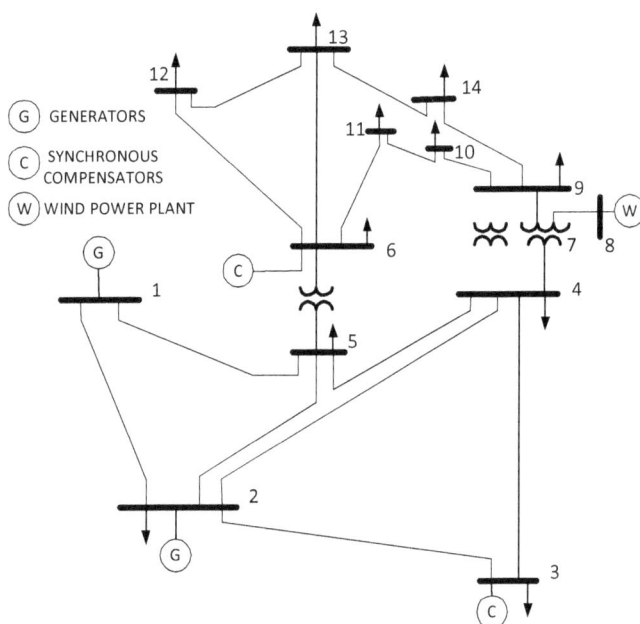

Figure 6. IEEE 14 bus test system with addition of a win power plant on bus 8.

3.2. Operating Conditions

In this numerical example, only synthetic data was produced and analyzed. Power system calculations were made in DigSilent Powerfactory software (DIgSILENT GmbH, Gomaringen, Germany) as was mentioned earlier with the series of power flow simulations for different network conditions.

Operating conditions (OCs) were calculated for a wide range of different simulation scenarios including:

- Variable wind power plant active power production change in an interval from 0–100% of rated power in discrete steps of 25%;
- Variable wind power plant reactive power settings definition in three different modes:

  - power factor regulation (setup point change from 0.9 p.u. lagging to 0.9 p.u. leading in steps of 0.05 p.u.);

- reactive power regulation (setup point change from 1.0 p.u lagging to 1.0 p.u. leading in steps of 0.05 p.u.);
- voltage regulation (setup point change from 0.9 p.u. lagging to 0.9 p.u. leading in steps of 0.05 p.u.).

- Power system load change in an interval from 1 p.u. to 1.6 p.u. (100–160%) in steps of 0.2 p.u. (20%);
- Line, transformer and generator availability status change switching off in different combinations for different OCs.

This way, a total of 396 OCs was created with most of the variables defined as continuous variables. Thus, large database covering a large number of possible network situations was created and further analyses were enabled.

3.3. Test Model Results

Large number of different analyses made in Statistica software, (Tibco Software, Palo Alto, CA, USA) as mentioned earlier. For example, voltage isolines for bus 8 at subject test system in various system load conditions and wind farm production (wind farm working in cos φ regulation mode with cos φ = 1) are given in following figure (Figure 7).

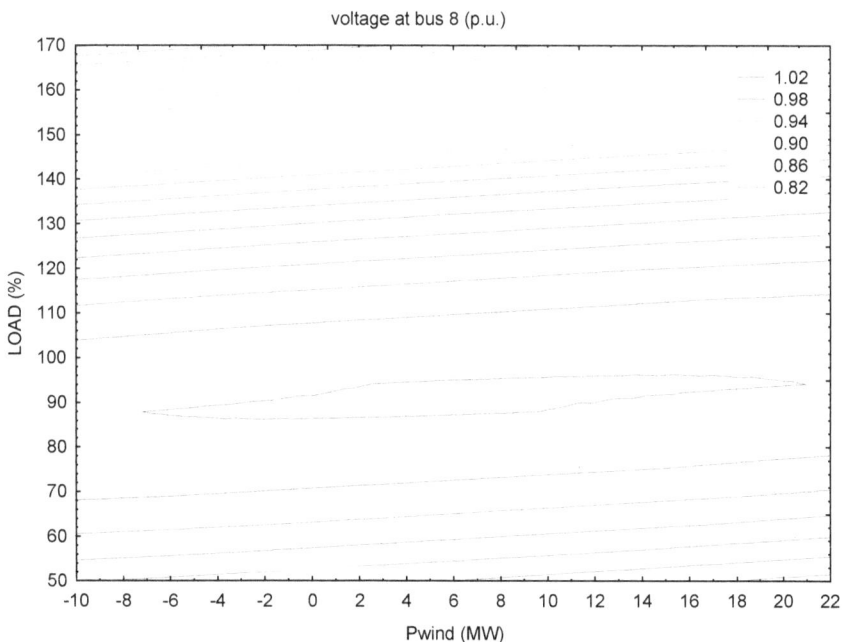

Figure 7. Voltage profiles for bus 8 of the IEEE 14 test system for different load conditions and wind power production with reactive power mode cos φ = 1.

From Figure 7 it can be seen how with the increase of load and/or wind production, voltage isolines become denser, which is explained through larger voltage sensitivity in these operating regions and conditions.

In this series of calculations, only simulation data was used. Simulation data was generated on an IEEE 14 bus system. Thus, data preparation step was simplified and there was no need for data formatting/unifying and time synchronization.

After data clustering, resulting centroids of Thevenin equivalent, for k-means clustering based on a total of 396 training cases, are given in following table (Table 2):

Table 2. Centroids of Thevenin equivalent for the k-means clustering—IEEE 14 bus test model.

EWS	$Z_{th\,abs}$	$Z_{th\,arg}$	No. of Cases
NORMAL	0.438743	87.91876	295
WARNING	0.554339	80.64339	85
ALARM	2.179344	69.31481	16

Thevenin impedances can be used for a wide range of protection applications [25].

According to EWS centroids given in Table 2 and EWS clustering depicted in Figure 8, it can be seen how power system changes its impedance to higher absolute and more resistive values during warning and alarm operating conditions. This represents an expected behavior that can be detected and further actions can be planned accordingly.

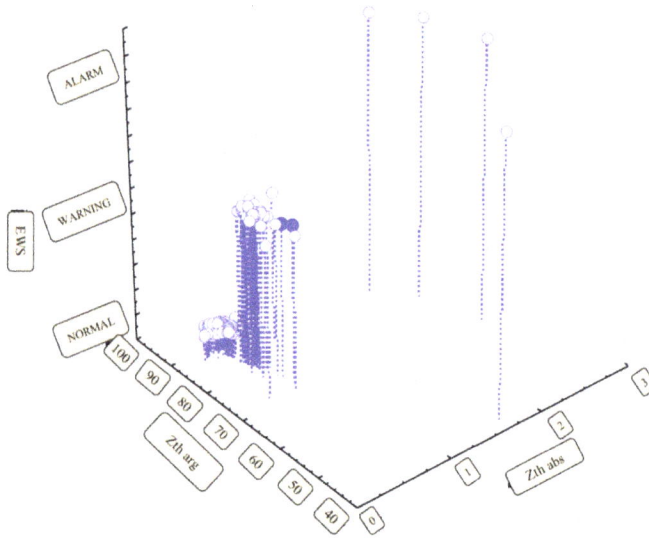

Figure 8. EWS clusters based on Thevenin equivalent at bus 8.

Further validation of such clustering was made using Jacobi matrix eigenvalues. Since Jacobi matrix for IEEE 14 bus test system has 14 eigenvalues, root mean square value of all 14 values was used as leveled variable which is labeled as lambda. In Figure 9 it can be seen how lambda values are clustered for normal, warning and alarm operating conditions. In this way, the early warning signal is verified with Jacobi matrix eigenvalues.

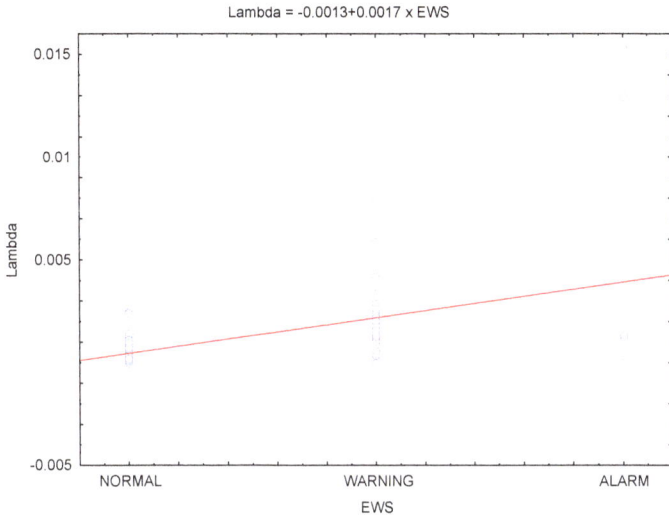

Figure 9. EWS clusters based on Jacobi matrix eigenvalues with least square method fitting.

Calculated clusters are used for further classification of cases of OCs. Classification was made with the usage of clustered EWS in OCs as dependent variables. Independent variables are voltage values and voltage angles on all buses as well as total system load, wind power active power and wind power reactive power settings.

As a stopping rule during classification tree calculation, prune on misclassification error criteria was used as described in Breiman et al. [21]. The pruning process results in a sequence of optimally pruned trees and a criterion to select the "right-sized" tree is applied afterwards. Trees with smallest misclassification costs often have hidden over-fitting. Thus, it is needed to make automatic tree selection procedure to avoid "over fitting" and "under fitting" of the data. To distinguish calculated trees, re-substitution costs and cross-validation costs are calculated according to [21] for all trees and cost sequence is shown on the figure below (Figure 10).

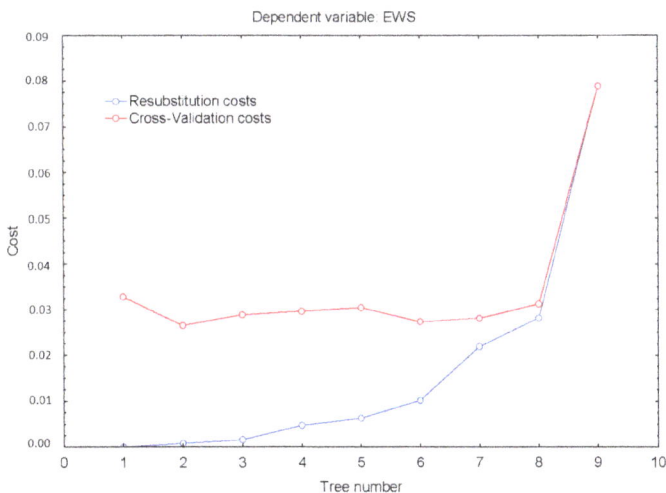

Figure 10. Early warning signal (EWS) cost sequence for different classification and regression trees.

For making the selections "1 Standard error rule" proposed by Breiman et al. [21] proposed a "1, of the "right-sized" tree. In this way, tree number 2 is selected to avoid "over fitting" and "under fitting" of the data (Figure 10). Also, through this classification analysis, it is possible to calculate importance for all independent variables according to the definition of variable importance given in [21]. Importance for all dependent variables is given on a Figure 11.

Practically all voltage and angle measurements from PMU devices have very high importance, except measurements from node 1 which was chosen as the slack busbar. Voltage angles (d1, d2, to d14) have very high importance which is understandable since they represent active power flows. Wind power active power production and reactive power settings have lower importance which can be described through moderate installed power of wind power plant compared to the network size.

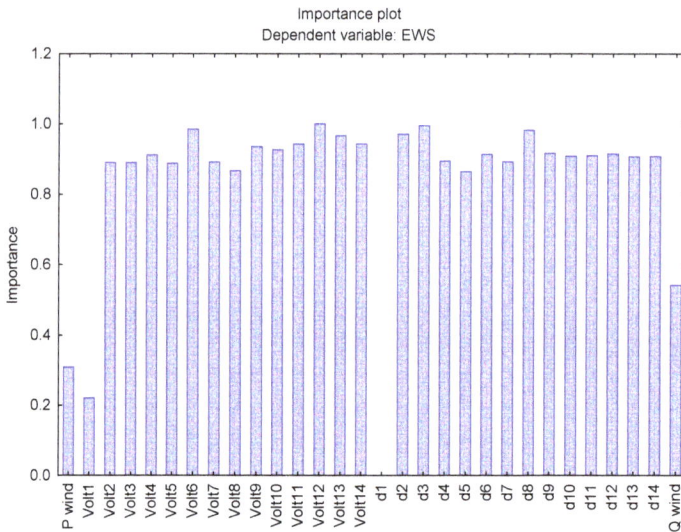

Figure 11. EWS importance plot for different dependent variables.

4. Radial Distribution System Example

4.1. Distribution System Description and Operating Conditions

Wind power plant (WPP) VE ZD 6 is 9.2 MW renewable energy source connected to 30 km long radial transmission 35 kV voltage level line in the middle of its length (Figure 12). Subject 35 kV power line is part of distribution system operator network in central Croatia. General characteristics of this network are low loadings especially during the night time. Therefore, large voltage deviations can occur in this relatively long and lightly loaded medium voltage line. Since wind power plant is connected through a T-junction situated in the middle of its length, it can serve as a voltage controller and provide preventive measures for system support.

In this numerical example, synthetic data was used to perform data mining calculations and real operation measurement data was used to verify the model. Measurements were made on the point of common coupling (PCC) using PMU device Arbiter 1133a and metering device ION 7660.

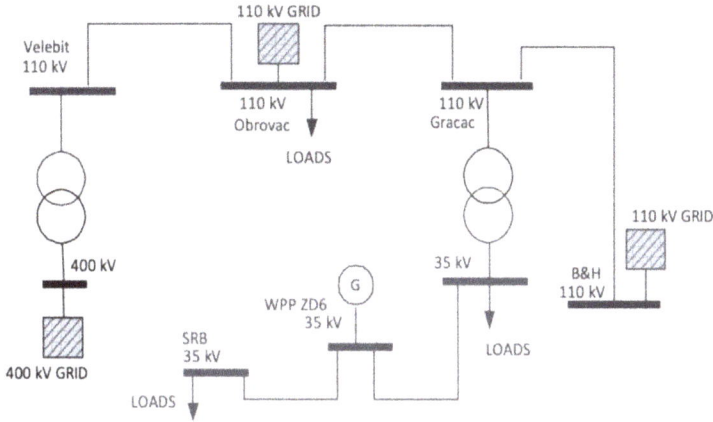

Figure 12. Wind power plant VE ZD 6 connected to radial 35 kV line.

4.2. Distribution System Results

After data clustering, resulting centroids of Thevenin equivalent, for k-means clustering based on a total of 135 training cases, are given in following table (Table 3).

Table 3. Centroids of Thevenin equivalent in PCC for k-means clustering—WPP VE ZD 6 model.

EWS	$Z_{th\ abs}$	$Z_{th\ arg}$	No. of Cases
NORMAL	4.090202	74.98206	41
WARNING	3.781597	78.12948	71
ALARM	3.228652	80.60002	23

In accordance to the EWS centroids given in Table 3, EWS clustering was made and the results are shown on Figure 13.

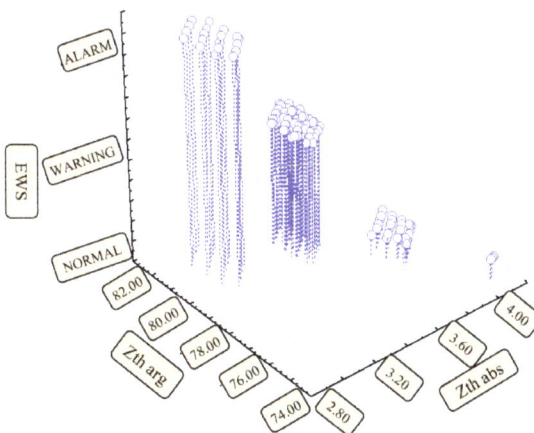

Figure 13. Early warning signal (EWS) clusters based on Thevenin equivalent at wind power plant VE ZD 6.

After data classification, resulting C&RT is given in Figure 14. Operating conditions can easily be recognized by monitoring voltage angle at transformer substation 110/35 kV Obrovac.

In these figures (Figures 14 and 15) ID represents ordinal number of the decision tree leaf, and N the number of cases on that leaf. Additional classification was made on a reduced set of parameters for a substation 110/35 kV Obrovac, where set of input parameters was built from available measurements since at the testing period real operation measurements were not available. Those measurements include voltage amplitudes and voltage angles, as well as active and reactive power at PCC of WPP VE ZD 6. Voltage angle was measured with reference angle at transformer substation 400/110 kV Velebit. This high voltage node was also defined as slack node in the calculations of operating conditions.

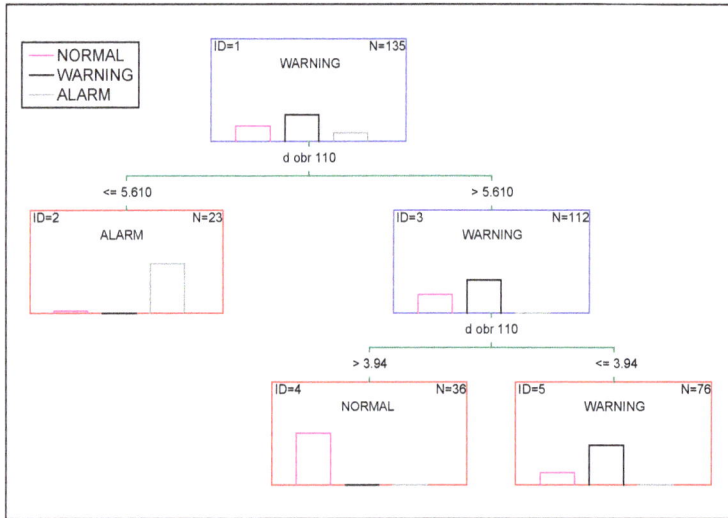

Figure 14. Resulting C&RT for a complete set of all possible parameter variations (VE ZD 6).

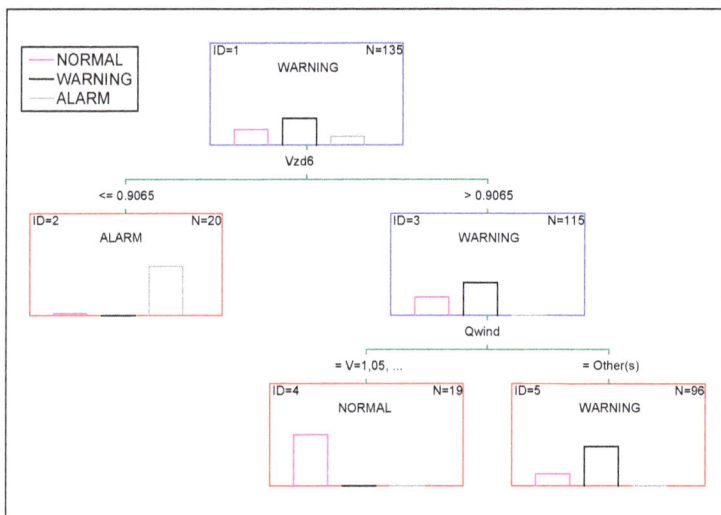

Figure 15. Resulting C&RT for reduced set of possible parameter variations (VE ZD 6).

With such reduced set of input parameters, it is evident that there are only two variables in the classification tree of decision-making process: effective values of voltage at the PCC and the settings of reactive power control. Therefore, in defining the conditions in the local network around the WPP, it is enough to monitor the effective voltage values at that PCC and base decision making on that variable.

Decision-making in this case refers to the change of the wind power for reactive power production to aid the system conditions. The peak power of the WPP is 9.2 MW which can easily be evacuated through the corresponding 35 kV transmission line (approximately 16 km long). The problem occurs in regard to the voltage security of the power system. Final operating decision making thus results in $P_{setting} = \max \{P_{WIND}\} = 9.2$ MW and $Q_{setting} = \min \{Q_{WIND}\} = $ Const. $\varphi = 1$ for all operating conditions with voltages at PCC above 31.7275 kV (0.9065 p.u). 31.7275 kV (0.9065 p.u) represents the critical voltage value that was obtained through the analysis process.

5. Meshed Transmission System Example

5.1. Transmission System Description and Operating Conditions

Wind power plant (WPP) VE ZD 2&ZD 3 has a capacity of 36.8 MW and is connected to meshed 110 kV grid (Figure 16). Subject 110 kV network is part of transmission system operator network in southern Croatia. General characteristic of this network are also low loadings, especially during night. Therefore, large voltage deviations occur. Furthermore, the PCC of the wind power plant is represented through the power transformer with rated power of 40 MVA.

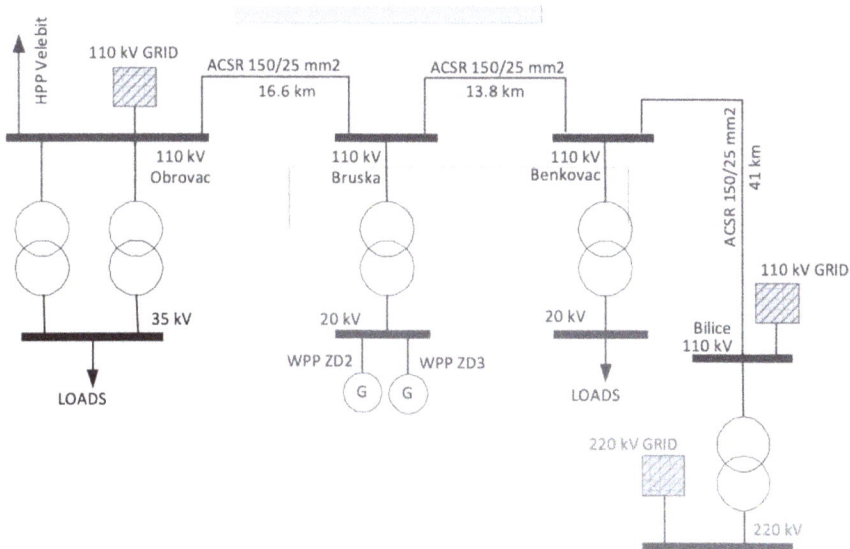

Figure 16. Wind power plant VE ZD 2 & ZD 3 connected to a meshed 110 kV grid.

In this numerical example, synthetic data was used in order to perform data mining calculations and measurement data was used to verify the model. Measurements were made on point of common coupling (PCC) using PMU device Arbiter 1133a and metering device ION 8800.

5.2. Transmission System Results

After data clustering, resulting centroids of Thevenin equivalent, for k-means clustering based on a total of 150 training cases, are given in following table (Table 4).

Table 4. Centroids of Thevenin equivalent in PCC for k-means clustering—WPP VE ZD 2&ZD 3.

EWS	$Z_{th\ abs}$	$Z_{th\ arg}$	No. of Cases
NORMAL	11.98351	83.80223	130
WARNING	10.42744	81.64916	14
ALARM	7.33635	71.44346	6

According to EWS centroids given in Table 4, EWS clustering was made and the results are given on Figure 17.

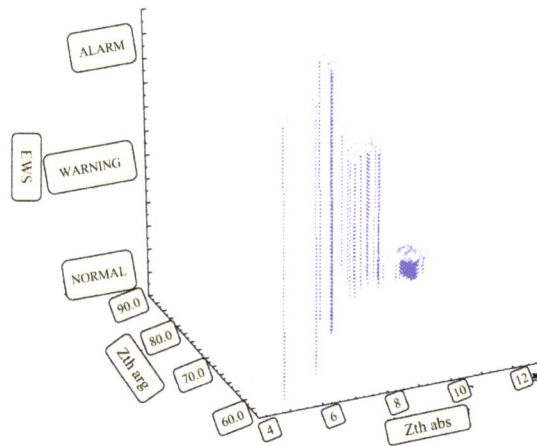

Figure 17. EWS clusters based on Thevenin equivalent at WPP VE ZD 2&ZD 3.

After data classification, resulting C&RT tree is given on figure below (Figure 18). Operating conditions can easily be recognized and predicted by monitoring of voltage amplitudes and voltage angle at 4 influential nodes of the observed segment of the transmission network.

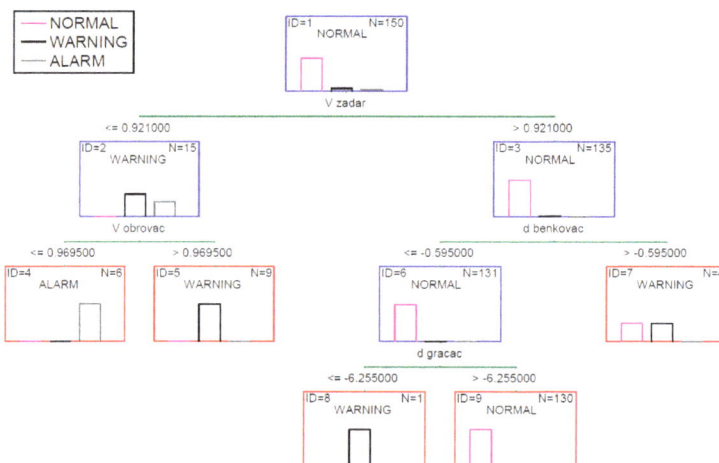

Figure 18. Resulting C&RT for a complete set of all possible parameter variations—WPP VE ZD 2 & ZD 3.

During testing, only measurements at substation 110/20 kV Bruska were made. Therefore, additional classification is made on a reduced set of parameters, where set of input parameters was built from available measurements (Figure 19). Those measurements include voltage amplitudes and voltage angles, as well as active and reactive power production measured on PCC of WPP VE ZD 2 & ZD 3. Voltage angle was measured with reference to angle at substation 400/110 kV Velebit. This node was also defined as slack node in the calculations of operating conditions.

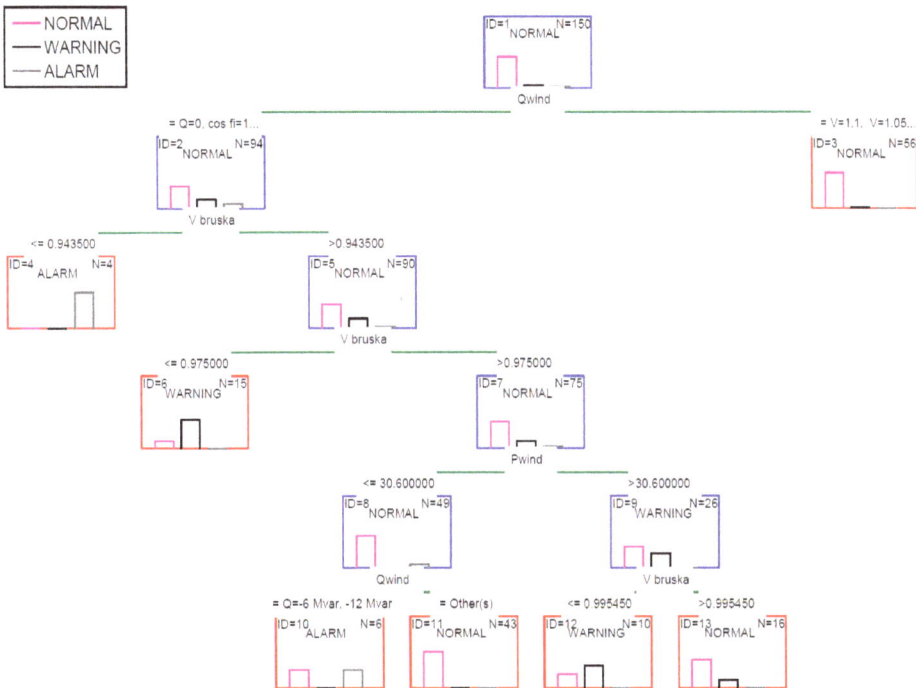

Figure 19. Resulting C&RT for reduced set of possible parameters—WPP VE ZD 2 & ZD 3 model.

It is evident that there are several variables in the classification tree of the decision-making process that have significant importance:

- WPP reactive power production setting;
- Voltage amplitudes at substation 110/20 kV Bruska (PCC of WPP VE ZD 2&ZD 3);
- Curtailment amount of the active power generation of the wind power plant.

Thus, to define the opportunities in a local network around the wind power plant, it is enough to monitor voltage values at that measuring point and reach decisions based on that variable. As a critical voltage value threshold, the value of 0.9435 p.u. i.e., 103.785 kV is defined. When voltage drops below the above-mentioned value at the WPP VE ZD 2 & ZD 3 or at substation 110/20 kV Bruska, critical conditions can be expected in the system.

Decision-making primarily relates to the change of wind power operational regime in regard to the reactive power production. The results are also understandable from the power flows point of view since the installed active power of the wind power plant is 36.8 MW, which is not a problem in the observed segment of the transmission system. The evacuation of installed rated power over a distance of about 16 km over the 110 kV transmission line towards substation 110/35 kV Obrovac and 110/35 kV Benkovac is done without any problems. Therefore, the only problem is the problem of voltage control to avoid out-of-limit voltages through reactive power production regulation.

5.3. Transmission System Model Validation

Transmission system model validation was done using real operation measurements from the same WPP and the following tests were analyzed:

1. Active power curtailment test;
2. Tripping of 110 kV transmission line Obrovac—Bruska test;
3. Change of reactive power regulation regime test.

Results are given in following figures (Figure 20—curtailment test; Figure 21—tripping test; Figure 22—reactive power change test).

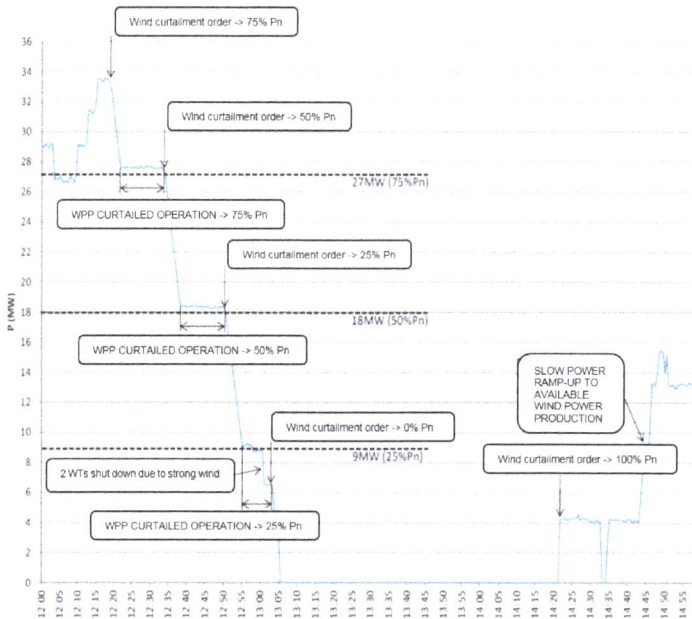

Figure 20. Validation of resulting C&RT—WPP VE ZD 2 & ZD 3 power measurements—wind curtailment test.

During all these curtailment cases, voltages in the network, as well as other parameters from C&RT process were classified as normal operating conditions.

Next, the test of tripping transmission line 110 kV Obrovac—Bruska was measured after circuit breaker (CB) tripping in line bay in substation 110/20 kV Bruska. Subject transmission line was reconnected after several minutes by circuit breaker in the same line bay.

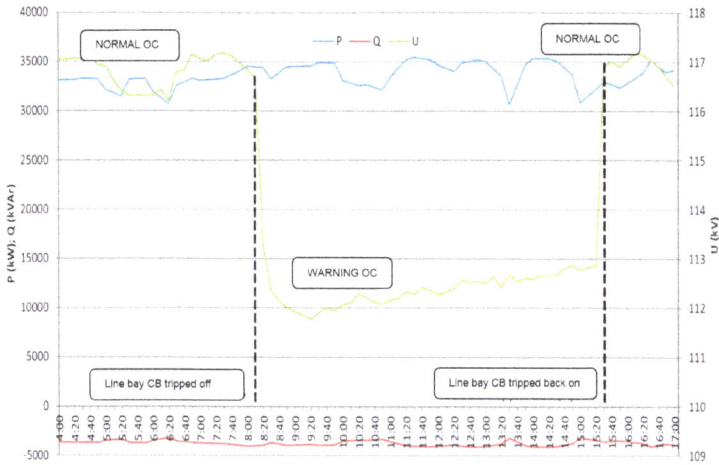

Figure 21. Validation of resulting C&RT—WPP VE ZD 2 & ZD 3 power and voltage measurements—110 kV transmission line Obrovac—Bruska tripping case.

During these line tripping cases, voltages in the network, as well as other parameters from C&RT process were classified as normal and warning operating conditions. Warning EWS is generated right after circuit breaker tripping. This scenario can be expected as there are lines going from WPP towards the rest of the meshed grid. With recognition of such circumstances the message to the operator can be sent to perform corrective actions if other circumstances in the grid worsen. Possible actions include change of the regulation of reactive power regime which was tested and depicted in the following figure (Figure 22). Warning EWS is generated after reactive power was set to constant reactive power −15 Mvar lagging.

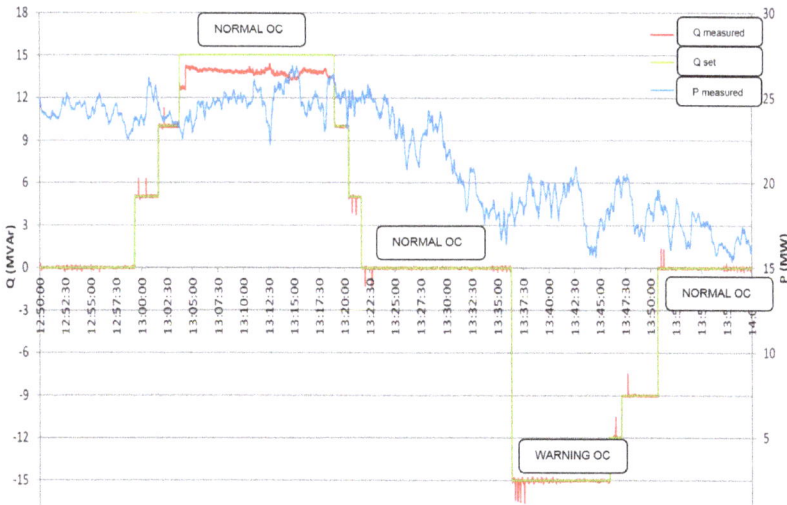

Figure 22. Validation of resulting C&RT—WPP VE ZD 2 & ZD 3 active and reactive power measurements—reactive power production regime change test.

171

It has been shown that following several voltage angles and amplitudes can give clear early warning signal about the situation in the grid and clear classification of the operating conditions can be done. The proposed method interprets and analyzes the measurement data (PMU measurements included) and provides the foundation to determine which measurements are necessary and crucial to avoid need to constantly display large quantities of data which can lead to inefficient recognition of important event.

6. Conclusions

In this paper, one solution for wind power monitoring and control is described through the enhancement of power system infrastructure with various ICT elements. Adaptive and modular (adaptidular) ICT installation were suggested and designed for the subject upgrade of power systems providing the foundations for the smart grid features.

Wind power monitoring and control was done through data mining algorithm using large streams of data coming from various devices. Most important devices used are PMUs measuring phasor values of all electrical variables in the system.

Input data defined for the data mining algorithm comprises of wind power plant active power production, wind power plant reactive power production setting, total system load, voltage amplitudes and voltage angles (phasors) and line, transformer and generator availability. All these data streams create a large database for further usage in control, planning, asset management and operation and maintenance processes.

Developed data mining algorithm for wind power monitoring and control is defined through three main components: the first component is data management and preparation; the second one is data clustering; and the third is data classification.

In the described approach, data management and preparation are used for data formatting and synchronization. Data clustering is done with k-means algorithm in order to define three clusters of data: NORMAL, WARNING and ALARM operating conditions. Thus, defined clusters are used as dependent variables during classification with the usage of C&RT algorithm. Independent variables in the classification algorithm are voltage phasor values (amplitudes and angles), system load, wind power plant active power production and wind power plant reactive power setting. The resulting classification tree can serve as decision tree which has combined decisions and measurements in its leaves. In this way, early warning signal for situational awareness is created with several actions in the decision tree that can serve as preventive measures and influence the system operating condition. The possible necessary decisions are in this model related to wind power plant and its active power production or reactive power setting regime change. The result of such decisions is the transition of the system state trajectory from WARNING and ALARM operating condition towards NORMAL system operating condition.

From the resulting C&RT tree as well as from the importance plot, most important variables can be detected. From these observations, the optimization of PMU infrastructure can be made by installing only devices necessary for the measurements of variables present in decision tree and/or importance plot, to provide the possibility for this algorithm to enable automated decision making.

Improvements of the proposed algorithm can be done with the definition of additional operating conditions either through recorded real system states measurement or through synthetized data from various kinds of simulations. Although mathematical techniques accurately simulate the synthetic data, they usually fail to handle the irregularities that exist in real data. It can also be concluded that in the future, more data will be generated from various measurement devices and efficient algorithms, such as one proposed here, will be needed to deal with such big data structures.

Author Contributions: M.K. and I.K. conceived and designed the experiments; M.K. performed the experiments; M.K. and I.K. analyzed the data; N.H. contributed to analysis tools design; M.K., I.K. and N.H. wrote the paper.

Acknowledgments: The work of the authors is a part of the H2020 project CROSSBOW—CROSS Border management of variable renewable energies and storage units enabling a transnational Wholesale market

Energies **2018**, *11*, 3525

(Grant No. 773430). This document has been produced with the financial assistance of the European Union. The contents of this document are the sole responsibility of authors and can under no circumstances be regarded as reflecting the position of the European Union. This work has been supported in part by the Croatian Science Foundation under the project WINDLIPS—Wind Energy integration in Low Inertia Power System (Grant No. PAR-02-2017-03).

Conflicts of Interest: The authors declare no conflict of interest.

References

1. Marinakis, A.; Larsson, M. Survey of Measurement and Intelligence-based Applications in Power Systems. In Proceedings of the IEEE PowerTech Conference 2013, Grenoble, France, 16–20 June 2011; pp. 1–6.
2. Ranganathan, P.; Nygard, K. Smart Grid Data Analytics for Decision Support. In Proceedings of the IEEE Electrical Power and Energy Conference (EPEC) 2011, Winnipeg, MB, Canada, 3–5 October 2011; pp. 315–321.
3. Panciatici, P.; Bareux, G.; Wehenkel, L. Operating in the fog. *IEEE Power Energy Mag.* **2012**, *10*, 40–49. [CrossRef]
4. Kamwa, I.; Samantaray, S.R.; Joos, G. Development of Rule based Classifiers for Rapid Stability Assesment of Wide Area Post-Disturbance Records. *IEEE Trans. Power Syst.* **2009**, *24*, 258–270. [CrossRef]
5. He, M.; Zhang, J.; Vittal, V. Robust Online Dynamic Security Assessment Using Adaptive Ensemble Decision-Tree Learning. *IEEE Trans. Power Syst.* **2009**, *28*, 4089–4098. [CrossRef]
6. Kezunovic, M.; Xie, L.; Grijalva, S. The role of big data in improving power system operation and protection. In Proceedings of the 2013 IREP Symposium, Bulk Power System Dynamics and Control—IX Optimization, Security and Control of the Emerging Power Grid (IREP), Rethymno, Greece, 25–30 August 2013; pp. 1–9.
7. Cleveland, F. Use of Wireless Data Communications in Power System Operations. In Proceedings of the 2006 Power System Conference and Exposition, Atlanta, GA, USA, 29 October–2 November 2006; pp. 631–640.
8. Yorozu, Y.; Hirano, M.; Oka, K.; Tagawa, Y. Low Throughput Networks for the IoT: Lessons learned from industrial implementations. In Proceedings of the 2015 IEEE 2nd World Forum on Internet of Things (WF-IoT), Milan, Italy, 14–16 December 2015; pp. 181–186.
9. Ivanković, I.; Kuzle, I.; Holjevac, N. Wide Area Information-Based Transmission System Centralized Out-of-Step Protection Scheme. *Energies* **2017**, *10*, 633. [CrossRef]
10. Ivanković, I.; Rubeša, R.; Kuzle, I. Modeling 400 kV Transmission Grid with System Protection and Disturbance Analysis. In Proceedings of the IEEE International Energy Conference, Leuven, Belgium, 4–8 April 2016; pp. 1–7.
11. Ilak, P.; Krajcar, S.; Rajšl, I.; Delimar, M. Pricing Energy and Ancillary Services in a Day-Ahead Market for a Price-Taker Hydro Generating Company Using a Risk-Constrained Approach. *Energies* **2014**, *7*, 2317–2342. [CrossRef]
12. Ivanković, I.; Kuzle, I.; Holjevac, N. Algorithm for Fast and Efficient Detection and Reaction to Angle Instability Conditions Using Phasor Measurement Unit Data. *Energies* **2018**, *11*, 681. [CrossRef]
13. CISCO. PMU Networking with IP Multicast. Document ID 1458164583372983. Available online: www.cisco.com (accessed on 17 March 2014).
14. Li, Z.; Wu, W. Phasor Measurement Aided Decision Trees for Power System Security Assessment. In Proceedings of the 2nd International Conference on Information and Computing Science 2009, Manchester, UK, 21–22 May 2009; pp. 358–361.
15. Wydra, M. Performance and Accuracy Investigation of the Two-Step Algorithm for Power System State and Line Temperature Estimation. *Energies* **2018**, *11*, 1005. [CrossRef]
16. Pérez-Chacón, A.R.; Luna-Romera, J.M.; Troncoso, A.; Martínez-Álvarez, F.; Riquelme, J.C. Big Data Analytics for Discovering Electricity Consumption Patterns in Smart Cities. *Energies* **2018**, *11*, 683. [CrossRef]
17. Cotilla-Sanchez, E.; Hines, P.; Danforth, C. Predicting critical transitions from time series synchrophasor data. In Proceedings of the 2013 IEEE Power and Energy Society General Meeting, Vancouver, BC, Canada, 21–25 July 2013.
18. Kazerooni, M.; Zhu, H.; Overbye, T. Literature review on the applications of data mining in power systems. In Proceedings of the Power and Energy Conference, Chicago, IL, USA, 14–17 April 2014.
19. Power System Software & Engineering. Available online: https://www.digsilent.de/en/ (accessed on 15 October 2018).
20. TIBCO Statistica is Now Part of TIBCO Data Science. Available online: https://www.tibco.com/products/tibco-statistica (accessed on 15 October 2018).

21. Breiman, L.; Friedman, J.H.; Olshen, R.A.; Stone, C.J. *Classification and Regression Trees*; Wadsworth Inc.: Pacific Grave, CA, USA, 1984.
22. Han, J.; Kamber, M.; Pei, J. *Data Mining Concepts and Techniques*, 3rd ed.; Morgan Kaufmann: San Francisco, CA, USA, 2012.
23. Zhou, D.; Blaajberg, F. Reactive Power Impact on Lifetime Prediction of Two-level Wind Power Converter. In Proceedings of the PCIM Europe 2013, Nuremberg, Germany, 14–16 May 2013; pp. 564–571.
24. Ivanković, I.; Kuzle, I.; Holjevac, N. Multifunctional WAMPAC system concept for out-of-step protection based on synchrophasor measurements. *Int. J. Electr. Power Energy Syst.* **2017**, *87*, 77–88. [CrossRef]
25. Abdelkaber, S. Online Tracking of Thévenin Equivalent Parameters Using PMU Measurements. *IEEE Trans. Power Syst.* **2012**, *27*, 975–983. [CrossRef]

Article

On Conceptual Structuration and Coupling Methods of Co-Simulation Frameworks in Cyber-Physical Energy System Validation

Van Hoa Nguyen [1,*], **Yvon Besanger** [1], **Quoc Tuan Tran** [2] and **Tung Lam Nguyen** [1,2]

1   Univ. Grenoble Alpes, CNRS, Grenoble INP, G2Elab, F-38000 Grenoble, France;
    Yvon.besanger@g2elab.grenoble-inp.fr (Y.B.); tung-lam.nguyen@g2elab.grenoble-inp.fr (T.L.N.)
2   Alternative Energies and Atomic Energy Commission (CEA), National Institute for Solar Energy (INES),
    50 Avenue du Lac Léman, F-73375 Le Bourget-du-Lac, France; Quoctuan.tran@cea.fr
*   Correspondence: van-hoa.nguyen@grenoble-inp.fr; Tel.: +33-4-7692-7187

Received: 31 October 2017; Accepted: 22 November 2017; Published: 29 November 2017

Abstract: Co-simulation is an emerging method for cyber-physical energy system (CPES) assessment and validation. Combining simulators of different domains into a joint experiment, co-simulation provides a holistic framework to consider the whole CPES at system level. In this paper, we present a systematic structuration of co-simulation based on a conceptual point of view. A co-simulation framework is then considered in its conceptual, semantic, syntactic, dynamic and technical layers. Coupling methods are investigated and classified according to these layers. This paper would serve as a solid theoretical base for specification of future applications of co-simulation and selection of coupling methods in CPES assessment and validation.

Keywords: cyber-physical energy system; co-simulation; conceptual structuration; coupling method

1. Introduction

Moving towards a decarbonized scenario, the power grid is expecting a high penetration of distributed and renewable energy resources and advanced Information and Communication Technologies (ICT) [1,2], which has a strong impact on the system architecture and is transforming the classical grid into a cyber-physical energy system (CPES)—Smart Grid [3]. The traditional design and validation methods, which focus in single domain, do not quite keep up with the changes [4,5]. On the other hand, emerging issues such as cyber-security requires also new tools and methods for assessment. It is, therefore, necessary to develop an integrated approach for such complex system in a holistic manner, taking into account the interaction and inter-dependencies among domains [6].

Power systems and communication networks, however, are very different in term of dynamic behavior and hierarchy. The simulation of these systems require therefore different model of computation and solvers (i.e., a power system is often simulated as continuous system with capability of generating discrete events; ICT system is in general simulated as discrete event simulation) [7]. A holistic approach requires essentially a consistent semantics for specification of the complete CPES, across multiple domains, which is rich enough to support heterogeneous design. While such a unified approach is not easy to achieve, researchers often employ the co-simulation approach: creating joint simulation of the already well-established tools and semantics; to consider the impact of ICT solutions to the power system when they are simulated with their suitable solvers [8]. Co-simulation offers a flexible solution which allows consideration of network behavior and physical energy system state at the same time. As the calculation load is shared among simulators, co-simulation also enables the possibility of large scale system assessment.

Moreover, the CPES requires the physical infrastructure and computational cyber-infrastructure to holistically and consistently coordinate to ensure its efficient and reliable functionality. It introduces a huge data influx for which big data analytic and applications are therefore of paramount importance [9]. Co-simulation framework, gathering data and analysis from multiple sources and domains, is potentially the solution to experiment such big data applications over large scale systems. On the other hand, distributed simulation with appropriate time scale, besides the benefit of computational load sharing, will avoid overloading slower simulation with unnecessary data influx from faster applications, and thus increases the performance.

Recent developments of co-simulation have led to an important portfolio of experimental and demonstration platforms. Some well-detailed, but application-oriented reviews on co-simulation frameworks [4,5,8,10], in particular, and test-bed for CPES validation in general [11] exist in the literature. In the context of this paper, we aim to providing a more systematic point of view from a higher abstraction level by offering a conceptual structuration of co-simulation framework in CPES assessment as well as the associated problems and a detailed review on coupling methods used and can potentially be used in this framework. It would serve as a solid theoretical base for specification of future applications of co-simulation and selection of coupling methods in CPES assessment and validation.

The paper is organized as follows: Section 2 presents the structuration and the different layers of a co-simulation framework in CPES context; Section 3 provides a systematic review on coupling methods according to the different layers, with a particular emphasis on usage and requirement of operational integration and formal integration, in the vision of helping users to establish their own co-simulation framework. The applications and associated abstraction levels of the coupling methods are finally considered in Section 4, along with some further perspectives.

2. Conceptual Structuration of Co-Simulation Framework

Establishing a standardizable holistic framework for CPES using co-simulation is a difficult and complex task because it requires a strong interoperability among the participating elements, especially in case of multiple partner involvement. This implies necessary efforts on harmonization, adaptation and eventually changes of actual employed standards and protocols in individual models to be able to integrate into holistic experiments.

Coupling different simulators introduces several new issues with respect to the classical modelling and simulation approaches. A generic layered structuration of co-simulation framework is therefore necessary to improve the interoperability of simulators as well as to highlight the intersection of domains and the issues that need to be solved in the process of designing a co-simulation framework. Based on existing models for generic interoperability [12] and multi-modeling [13,14], a generic five layers structuration of a co-simulation framework can be proposed (cf. Table 1).

2.1. Conceptual Layer: Architecture

As aforementioned in Table 1, the conceptual layer is highest layer of a co-simulation framework. It involves the meta-modeling process and the topology of the framework (e.g., in Figure 1 the CPES can be considered in a system of systems approach or the in layered approach: market-driven ICT network governing the power system). In this layer, the chosen scenario is analyzed around the three main points: system configuration according to the scenario, purpose of investigation (deduced from the scenario and the desired research/contribution) and use-case/test-case definition; as the scenario and the system configuration influences strongly the conceptual design. Moreover, the questions of abstraction level, interaction to environment and complex system representation manner (e.g., System of systems, coupled systems or multi-agent system) are subjects of interest in this layers.

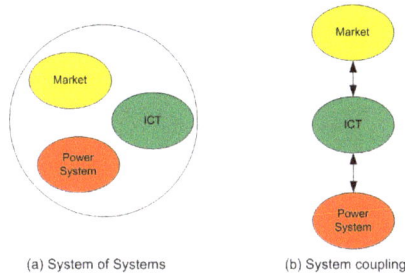

(a) System of Systems   (b) System coupling

Figure 1. Illustration of different possible topologies in conceptual layer of a CPES co-simulation framework.

Table 1. Generic layers of a co-simulation framework.

Layer	Description	Associated Problems
Conceptual	Highest level where the models are considered as "black boxes" and the level concerns the co-simulation framework representation.	Generic structure of the framework; Meta-Modeling of the components.
Semantic	The level concerns the signification and the role of the co-simulation framework with respect to the open questions of the investigated CPES and studied phenomenon.	Signification of individual models; Interaction graph among the models; Signification of each interaction.
Syntactic	The level concerns the formalization of the co-simulation framework.	Formalization of individual models in the respective domains; Specification and handling the difference among the formalism.
Dynamic	The level concerns the execution of the co-simulation framework, the synchronization techniques and harmonization of different models of computation.	Order of execution and causality of models; Harmonization of different models of computation; Resolution for potential conflict in simultaneity of actions.
Technical	The level concerns the implementation details and evaluation of simulation.	Distributed or centralized implementation; Robustness of the simulation; Reliability and efficiency of the simulation.

The conceptual layer should define the structure on which the co-simulation framework is developed and the formal semantic relations/syntactic formulation are integrated.

2.2. Semantic Layer: Formal integration

The semantic layer concerns the signification and role of individual models in the general framework, as well as their in-between interactions. It is necessary to note that the models may be represented at different spatial and temporal scales, with possible intersection among their abstraction domains. In that boundary, one needs to consider and specify how information in a model can be perceived in another one (cf. Figure 2). The interaction among models has to provide a semantic coherence throughout the whole system.

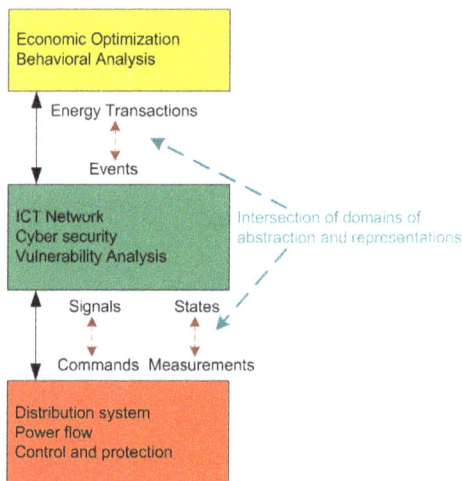

Figure 2. Illustration of semantic layer of a CPES co-simulation framework.

It is necessary to clarify that the notion of semantic layer here is addressed towards the co-simulation framework, and is not necessarily the semantic model of the energy system, although there is an overlapping area between the two concepts. More specifically, the semantic layer of a co-simulation framework explains the behavior and interaction of involved individual simulators and models, and not necessarily the semantic relations among their representing counterparts. It can be considered as an abstract layer specifying the experiments, and in some ways, similar to the concept of abstract test suite in TTCN-3 standard [15]. The dichotomies and communication among entities defined in this layer will decide how the experiment works and will be governed by the master algorithm.

The required interoperability in semantic layer and in the following syntactic layer needs to be done via applying a common information model to the framework. However, taking into account that a co-simulation framework often includes multiple domains (e.g., power system, ICT, thermal), it may be necessary to include several existing information models and thus to resolve interfacing issues. It is recommended to employ standardized information models to promote reusability and the possibility of further integration of new elements into the framework.

2.3. Syntactic Layer

This layer of a co-simulation framework influences strongly the reusability of models and interoperability of the framework. In a co-simulation framework, each model comes from different scientific domains, and it is very likely that the employed formalisms are different. A consistent co-simulation framework has to manage and harmonize these differences. Not only it has to provides a meaningful and rigorous translation of inter-formalism information, but also, the differences in term of representation, underlying model of computation and spatial-temporal scale need to be handled (cf. Figure 3).

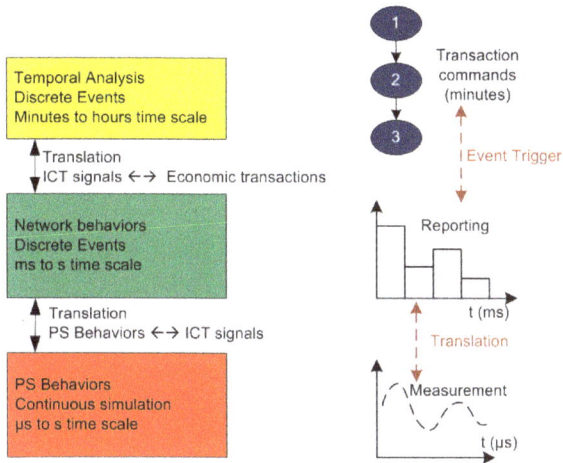

Figure 3. Illustration of syntactic layer of a CPES co-simulation framework.

In order to improve the reusability of models, it is required that the models are syntactic coherent. In general, *syntactic interoperability* of a co-simulation framework often requires the utilization of ontologies (not only for modeling, but also for data exchange). As for power system, we acknowledge several information model standards issued by the International Electrotechnical Commission (IEC) that can be considered as candidates: IEC 61850 [16], Multispeak (http://multispeak.org) and Common Information Model (CIM) [17,18]. Object Linking and Embedding for Process Control - Unified Architecture (OPC UA) [19] can also be implemented as a semi-information model when it is necessary to involve Supervisory Control and Data Acquisition (SCADA) system (i.e., hardware-in-the-loop situation). Several works are underway to harmonize the difference and to bring these standards together [20–22]. There exist a few semiautomatic converters between their models [23–25]. The mapping of ontological models to physical power system components and use-cases is classically implemented using hard-coding. One of the challenges is to enable auto-configuration of the topology of the considered CPES on-the-go and to update the database accordingly. In the case with CIM, an adaptive CIM compliant approach to resolve this challenge is proposed in [26]. The utilization of ontology in CPES is mainly for exchanging data between applications and encapsulating entire power system models in a standardized manner. As aforementioned, the concept of syntactic layer in a co-simulation framework is not always the same and is often larger than just using an ontology. Not only should it include ontological information on individual parts of the system and their interconnections, but also specify the format, syntax of the interconnections among simulators and models, as well as specifications of interfacing ontologies of different domains or transition among different time scale, which are not covered by current existing ontologies.

While being very similar to semantic layer, the syntactic layer can be simply imagined as the manner in which semantic models and interactions are represented. The notions of semantic and syntactic layers here, even though involving the utilization of energy system ontologies (e.g., IEC 61850 or CIM), are larger and require also an experiment description ontology, e.g., TTCN-3 [15]. These ontologies, if employed, can be considered as a common (or transitive) zone alongside the abstraction layers.

2.4. Dynamic Layer: Execution and Synchronization

The dynamic layer concerns executions aspects of a co-simulation framework. One needs to define the order of execution and causality of models as well as the resolution for potential conflict (i.e., cyclic dependency, deadlock) in simultaneity of actions. Synchronization techniques are necessary to ensure the consistency of the co-simulation outcome, especially when the framework consists of models with different model of computation (i.e., Continuous simulation and Discrete Event Simulation) or when the simulation is distributed (cf. Figure 4).The necessary message payloads model should also be defined.

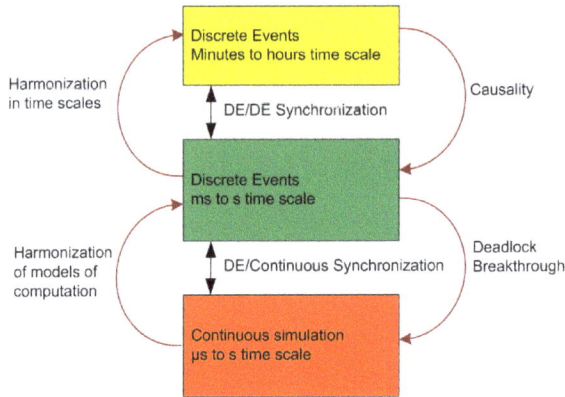

Figure 4. Illustration of dynamic layer of a CPES co-simulation framework.

Even though it is not directly related to semantic layer, it is demonstrated that different choices in dynamic layer may lead to differences in the result of simulation [27,28].

In general, for the dynamic layer, the two main issues influencing the correctness and the reliability of a co-simulation framework are harmonization of models of computation and synchronization of simulators. Harmonization of different models of computation is also a critical problem because CPES often involves a juxtaposition of various domains, where their dynamic behavior requires different solvers and consequently various models of computation. On the other hand, one can ensure, via synchronization, that operations occur in the logically correct order, whether they proceed concurrently or they must obey causality. As the size of system increases (physically or by increasing the number of simulators) or as the speed of operation increases, synchronization plays an increasingly dominant role in the stability of the framework.

Models of computation can be classified as of [29] : Imperative (e.g., Emulators), Finite State Machine (e.g., a set of states, rule-based control), Dataflow (e.g., ODEs, DEAs), Discrete Event (e.g., communication, zero-crossing), etc. Discrete Event simulation can be further broken down into: event scheduling, process interaction and activity scanning [30].

Power system models are, in general, represented by continuous models which is also capable to produce events (e.g., zero crossing, switching, etc.). The continuous models often use the *imperative* or *dataflow* model of computation, in which the model react to the availability of data at their inputs by performing some computation, via mean of according solvers, and producing data on their outputs. Dataflow is concurrent with no notion of time. In general case of power system, the model is based on a set of differential equations defining the peculiarity of the state variables and the environment factors of a system (e.g., steady-state simulations, electromagnetic transients and circuit simulations, or electromechanical phenomena). ICT and market models are often represented in discrete events chaining over a discrete set of points in time (thus associated to discrete events model of computation). More specifically, *discrete event* models are discrete, dynamic and stochastic in nature and are "run" whereas continuous model can be "run" or "solved" according to their models of computation [31]. General speaking, the two most encountered MoC for CPES assessment are Dataflow and Discrete Event (Figure 5), and they are most of the time, the subjects of harmonization.

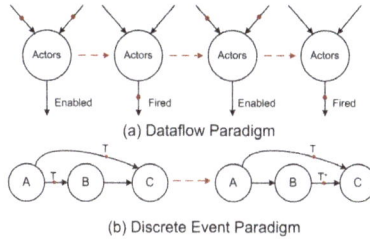

(a) Dataflow Paradigm

(b) Discrete Event Paradigm

Figure 5. Examples of (**a**) Dataflow process and (**b**) Discrete Event process.

As for synchronization techniques, two well established approaches are called conservative processing and optimistic processing. In general, the conservative approach requires all the participating simulators to wait for each other to finish their step before advancing to the next step. The synchronization is checked with time stamps, via Null Message algorithm [32] or Global Synchronization approaches, such as: Bounded Lag algorithm [33], Time Buckets [34] or Composite Synchronization [35]. On the other hand, the optimistic approach allows the individual simulators to advance on their own events. When a conflict is detected, then the simulators must perform a leap backwards (time warp [36]) and discard the all the results from the moment in question.

2.5. Technical Layer: Implementation and Evaluation

This layer involves the choices of techniques for implementation and the evaluation of results. Various practical issues need to be considered: the choice of distributed or centralized simulation, global model of computation, technical implementation of interface or latency assessment. As co-simulation requires many coupling of simulators, it is necessary to consider in the end, the reliability, efficiency of the coupling, the stability of co-simulation framework (especially in cases involving hardware-in-the-loop), as well as to evaluate the robustness and accuracy of the results (cf. Figure 6). An in-depth discussion on the technical aspects of co-simulation is however further than the scope of this paper (i.e., conceptual structuration).

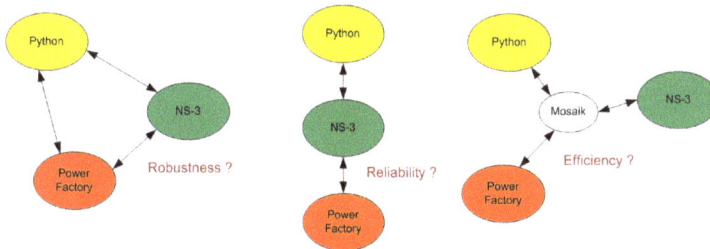

Figure 6. Illustration of technical layer of a CPES co-simulation framework.

The structuration proposed in this section is important to classify different models in a co-simulation framework into their according level of abstraction. It provides insights to identify the potential issues as well as to search for solution for coupling across different scale and abstraction level. As a consequence, the modularity, scalability and interoperability of a co-simulation framework might be improved.

In the context of this paper, we are interested also, in next sections, different existing coupling methods and synchronization techniques for co-simulation as well as their position with respect to the proposed structuration. It would provide a methodological and structural review on the state of art in co-simulation for CPES assessment.

3. Review on Coupling Methods

In general, a co-simulation framework requires the joint and simultaneous executions of models in different tools, via mean of information exchange during the execution of the simulations. Information is exchanged through either ad-hoc interfaces (Figure 7a) or via intermediate buffer governed by a master algorithm (Figure 7b). Master algorithm (where exists) is responsible for instantiating the simulators and for orchestrating the information exchange (simulator-simulator or simulator-orchestrator).

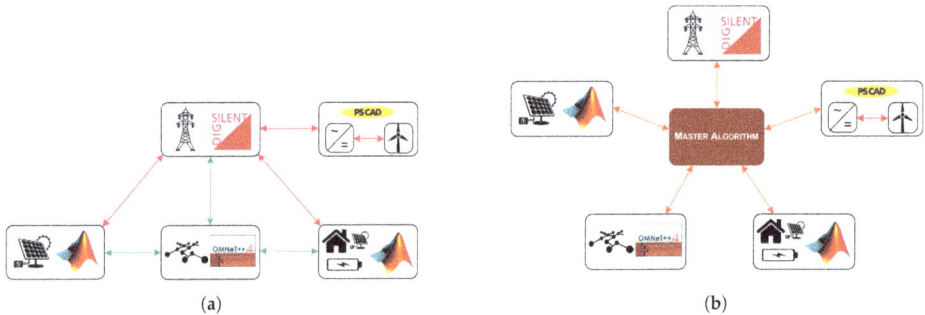

(a)   (b)

Figure 7. (**a**) Ad-hoc Co-Simulation; (**b**) Co-Simulation with an orchestrator.

The existing methods for simulator coupling in co-simulation can be classified into operational integration and formal integration. The operational coupling technique is used for specific problematic and is often not generic. In general, the operational integration aims for co-simulation at dynamic and technical layers while semantic and syntactic interrelation among the simulator are not properly addressed, thus does not provide reusability and modularity. On the other hand, formal integration aims at solving the problem at semantic and syntactic level (via either model coupling or simulator coupling). The semantic coherency among entities is ensured by encapsulating the models in a more generic layer or by orchestrating the simulators via mean of a master algorithm. The semantic and syntactic of the interaction is then governed by the said master federate.

3.1. Operational Integration

Operational integration approach most of the time case-specific and provides a good flexibility in practice. In this approach, the questions in semantic layer, which are directly related to the objectives of the test, as well as experiment behaviors are considered and interpreted outside of the simulation framework. The co-simulation focuses on engineering solutions at syntactic, dynamic and technical layers. The advantage of this approach relies in the simplicity of set-up which correctly answers to the user's need. However, the particular specifications make it hard for others to reproduce and compare the experiment result. In a scientific and industrial context, it may sometimes limit the valorization of the research.

Two most popular approaches of operational integration for co-simulation are: Model exchange (i.e., coupling the models together into one simulator) and ad-hoc coupling.

3.1.1. Model Exchange—Functional Mock-up Interface

In this approach, each model is built in its own formalism, then implanted to the same tool. The syntactic level in this case is completely hidden as the differences among the formalism are handled vie the implanting tool, i.e., technical layer. The models are forced to be simulated in the same rate, governed by the simulator. Therefore, this approach present a lot of limit in term of coupling models with different time scale.

In term of application, the approach is widely used in ICT domain simulation. The platforms such as NS-2 and NS-3, OMNet++ [37] or GTNetS [38] provides the possibility of using various models representing different aspects of the network at the same time. In CPES system assessment, it is not uncommon either to encapsulate a model and integrate it to another one; as an alternative when online co-simulation is hard to achieve.

Functional Mock-up Interface (FMI) is a standard for co-simulation issued by the consortium of the European ITEA2 Modelisar project in 2010 and in revised version in 2014. The basic component of FMI is the Functional Mock-up Unit (FMU), which gives access to the equations of the model (mode Model Exchange) or implements a solver manipulating the equations of the model (mode Co-simulation). The main purpose of FMI is providing a common interface by defining abstract functions to be implemented by every simulation component and defining how to export that said component as a shared library. FMU for model exchange can be considered as an operational integration while FMU for co-simulation can be applied to either operational integration or formal integration, depending on the level of semantic coherency among the models. It allows the integration of simulation models in a distributed and parallel way, thus possibility to apply to large-scale system [39,40].

FMI has been widely applied to the domain of mecatronics and automobile. Until recently, researchers have investigated application of FMI to CPES assessment in an effort to improve interoperability and reusability of models [41,42]. In [43], a new method of exploiting FMU via web-service and Service Oriented Architectures (SOA) is proposed. The architecture is promising and provides the framework for a potential "co-simulation as a service".

3.1.2. Ad-Hoc Coupling of Simulator

The most common architecture for simulator coupling in an operational integration is linking and integrating existing independent domain specific tools. In the case of CPES, most of the time, a power system simulator is coupled with the communication system simulator. This configuration is intuitive, concise and efficient and is often realized via sockets or direct integration to ICT simulator, as of survey result in [44].

Some co-simulation frameworks using ad-hoc connections are presented in Table 2.

Table 2. Some co-simulation frameworks using ad-hoc connections in the domain of CPES.

Co-Simulation Framework	Power System Simulator	Network Simulator	Synchronization Strategy	Applications	Time Scale
GECO [45,46]	PSLF	NS-2	Global Event-Driven	PMU-based WAMPAC	ms to s
ADEVS [47]	ADEVS	NS-2	DEVS	WAMPAC	ms to s
VPNET [48]	VTB	OPNET	Master-slave	WAMPAC	ms to s
GridSim [49]	PowerTech TSAT	GridStat	Time stepped	WAMPAC	ms to s
PowerNet [50]	Modelica	NS-2	Master - Slave	Control	Unknown
Bergmann et al. [51]	NETOMAC	NS-2	Time stepped	Evaluation of DERs	μs to ms
Babazadeh et al. [52,53]	RTlab	OPNET SITL	Asynchronous	WAMC, HVDC	μs to ms
Godfrey et al. [54]	OpenDSS	NS-2	Unknown	DER Integration	ms to s
TASSCS [55]	PowerWorld	OPNET	Unknown	Cyber-security of SCADA	μs to ms
Greenbench [56]	PSCAD	OMNet++	Global Event-Driven	Cyber-security LV Grid	ms to s

In the ad-hoc approach, the interfaces are specifically configured for to be implemented only within the simulators, with possibly high requirements of adaptations. In this set-up, the inter-dependency of involved simulators is not observed and assessed by an orchestrator, therefore, time synchronization is difficult when more than three simulators are involved in the experiment. If not properly configured, the inter-dependency of simulators may lead to deadlock and inaccurate results, especially in case of presence of communication delay among them. Scalability and stability of the experiments are therefore questionable and most of the time, ad-hoc co-simulation approach is not suitable for large scale system.

Moreover, most of the proprietary simulation tools do not provide direct interfaces to advances simulation tools and co-simulation. The interfaces developed for ad-hoc co-simulation depend on the simulators it connects; therefore, the re-usability of models and interfaces is not always evident.

Most of the time, re-utilization of the simulator in another co-simulation framework requires heavy adaptation efforts. As aforementioned, one can resort to the Functional Mock-up Interface (FMI) standard to remedy the issue. As an interface, FMI can be used in both ad-hoc and with master algorithm mode.

In general, operational integration responds quickly to the needs of a specific study, however, these approaches do not satisfy the requirements in conceptual, semantic and syntactic levels. As a consequence of this lack of generality and information, the operational models are strongly related to the technical implementation and are very limited in term of interoperability and reusability. On the other hand, operational integration demonstrates also difficulty in handling time scale and abstraction level transition among models.

3.2. Formal Integration

Formal integration is a more generic approach than operational integration as the co-simulation framework needs to ensure correct inter-relation among models in semantic and conceptual layers. In order to provide and handle, it is generally required to be orchestrated by a master algorithm. Analogy to the operational integration approach, the aforementioned master algorithm can be implemented in a single simulator where all the models are encapsulated into, or can be implemented as an orchestrator of different simulators. The principle difference between the two methods is that the models in this case are required to be semantically coherent and the exchanged signals are syntactically ready to be processed, without any further interpretation at higher layers.

This approach can be implemented in several forms:

- Implantation the elementary model into a more generic formalism. This requires however a lot of efforts and collaborations, especially in case of multi-domain experiments, where the behaviors of elements in one domain are not always evident to the experts of the other domain.
- Building and integrating an interface for translating from a particular formalism of a model to a more generic one.
- Simulator coupling handled and governed by an orchestrator.

The first two approaches requires the existence of a formalism generic enough to correctly and completely encapsulate the other involved models, without losing their semantic and conceptual meaning. This requirement appears to be a challenge. In ICT network simulation, Discrete Event System Specification (DEVS) and its derivations are generally accepted to satisfy such conditions.

3.2.1. Encapsulation of Formalism—Discrete Event System Specification (DEVS)

In general, DEVS is a generic formalism that allows encapsulating other Discrete-event models and discrete-time continuous models [57]. In DEVS, a model is represented by a 7-tuple $(I, O, S, \delta_{ext}, \delta_{int}, \lambda, t_a)$, where I and O represent the inputs and outputs of the model, S is the vector of state variables, t_a represents time advancement in simulation, δ_{int} is the evolution of internal states of model while δ_{ext} represent the influence of external events to model's internal states. Model coupling in DEVS consists of encapsulating the individual models into a big global DEVS multi-model [58–60].

The approach of encapsulation of formalism provides a strong syntactic coherence for the individual models as well as the link among them. Moreover, there are various algorithms to simulate in a distributed manner the DEVS-based model [57,60], leading to a good flexibility in term of management of simulation-related constraints.

On the other hand, the approach requires a considerable effort to encapsulate the models into DEVS, thus limits the reusability of existing models. Moreover, it is arguable that encapsulation of formalism could provide semantic and conceptual coherence and consistency [14].

3.2.2. Waveform Relaxation Method

Waveform relaxation method (WRM) [61] is a family of techniques popularly used in solving large systems of nonlinear ordinary differential equations (ODE) [62] or Differential Algebraic Equations (DAE) [63]. The basic idea of WRM is replacing the problem of solving a differential in multiple variables by one of solving a sequence of differential equations in one variable, in which the waveform of other variables are predefined. The solutions obtained from these equations are then substituted into the others one dimensional differential equations, which are then re-solved using the new waveform. The procedure is repeated until convergence condition is reached. Consider a first order two dimensional differential equations in $x(t) \in \mathbb{R}^2$ on $t \in [0, T]$:

$$
\begin{cases}
(\mathbf{A}) & \dot{x}_1 = f_1(x_1, x_2, t), \quad x_1(0) = x_{10} \\
(\mathbf{B}) & \dot{x}_2 = f_2(x_1, x_2, t), \quad x_2(0) = x_{20}
\end{cases}
$$

Then the iterative algorithm of WRM can be applied to solve this system of ODEs as illustrated on Figure 8. Similar to the case of nonlinear algebraic equations, Gauss-Seidel (or Serial Sequence) [62] and Jacobi (or Parallel Sequence) [64] techniques can be applied.

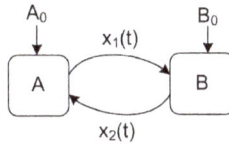

Figure 8. Illustration of the waveform relaxation method.

Originally aimed to solve complex ODEs, the WRM techniques recently draws attentions of the researchers for its applications to co-simulation, in particular, the coupling of continuous simulation [43]. In this approach, an individual simulator is represented by complete I/O waveform at the coupling points, to the other simulators. The WRM is executed until the convergence conditions are reached (Figure 9). Therefore, WRM is a method of co-simulation of semantic, syntactic and dynamic levels.

The WRM technique is of great interest in case of strong coupling of simulators, especially in the presence of communication latency. In general, WRM requires multiple executions of simulators until reaching the convergence conditions, which is the main source of execution time of the co-simulation framework. On the other hand, the other methods of coupling require one single execution of simulators, with multiple exchanges in one run instance. In the case without presence of latency or with very small latency (e.g., co-simulation of different simulators in one computer), WRM has in general longer execution time. However, when communication latency is not negligible, WRM may reach convergent point way much faster than the other methods [43]. In theory, the performance of WRM for differential systems can be improved with parallel WRM [65] and the accuracy/convergence rate can be improved with two-stage WRM [66]. However, application of these methods to co-simulation is not yet investigated.

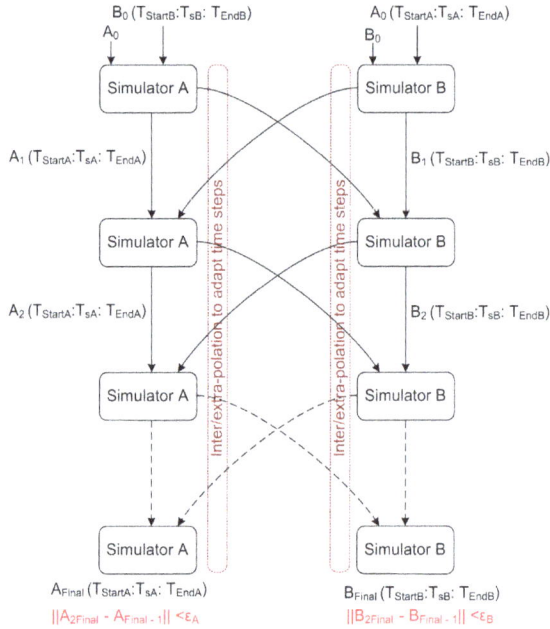

Figure 9. WRM method applied to Co-simulation.

Three important issues need to be considered in the implementation of a co-simulation framework based on WRM techniques:

- Until now, the method is exclusively applied for coupling of continuous simulation, particularly of dataflow MoC type. Discretized WRM and applications of WRM to discrete systems [67,68] is still a new research subject.
- In the implementation of WRM, the convergence condition needs to be properly addressed. In literature, there exist works on convergence conditions of WRM for ODE [69] and DAE [70,71]. A convergence analysis for Discrete-time WRM is also proposed in [72].
- in case of coupling models with asynchronous time steps, it is necessary to use an inter/extra-polation algorithm to harmonize the difference.

To conclude, WRM is an interesting approach to co-simulation, with a well-established mathematical foundation. However, the applicability of the method to discrete event simulator needs further investigation.

3.2.3. Federation of Simulation—Simulator Coupling with Orchestrator

The first two approaches of formal integration ensure semantic coherence by converting the models into a more generic formalism (i.e., DEVS and waveform), in this section, we consider formal integration by simulator coupling. While ad-hoc co-simulation is concise and efficient, for semantic and conceptual levels, it is necessary to implement a coordinator/master algorithm to orchestrate the framework. In this approach, the master algorithm is responsible for instantiating, interfacing and coordinating the involved simulators. The simulators can communicate via a common buffer/message bus governed by the master algorithm or directly among each other's (in which the master algorithm define the routing). Beside the benefit of semantic coupling, co-simulation with a master algorithm also helps to simplify the framework architecture (i.e., N interfaces for a N domains system, rather than N^2 interfaces that would be required if each domain had to explicitly interface to each other domain).

As of today, the master algorithm is most of the time implemented manually via self-developed (e.g., Java, Python, Mathlab) or third party software such as Ptolemy (http://ptolemy.eecs.berkeley.edu/) (used in [73–75], Simantics (https://www.simantics.org/) or Mosaik (http://mosaik.offis.de/) (e.g., [45,76,77]. The IEEE 1516 High level Architecture (HLA) is one of a few standards for co-simulation coordinator [78]. HLA is developed by the US M&S Co, mainly driven by military research in enabling joint simulation training. The first civil version is published by IEEE in 2000 and the evolved HLA is published in 2010. A HLA-based simulation consists of federates (participating simulators) and Run-Time Infrastructure (RTI), which provides an API for bidirectional communication from and to federates. The advantage of HLA is the possibility of highly paralleled simulations for large-scale systems; whereas its disadvantage is the introduction of additional time-synchronization issues [44]. Thus, we can register various researches focusing on the optimization aspect of the synchronization techniques in HLA [79,80].

Table 3 provides a non-exhaustive list of examples of co-simulation frameworks orchestrated by a master algorithm.

Table 3. Some co-simulation frameworks with master orchestrator in the domain of CPES.

Co-Simulation Framework	Power System Simulator	Network Simulator	Master Algorithm	Applications	Time Scale
EPOCHS [81]	PSCAD/EMTDC,PSLF	NS-2	HLA	Protection and Control	ms to minutes
INSPIRE [82]	PowerFactory	OPNET	HLA	WAMPAC	ms to minutes
SINARI [83]	PSCAD	NS-2	JAVA	vulnerability analysis	ms to s
VIRGIL [84]	PowerFactory	OMNET++	Ptolemy 2	Control and optimization	ms to s

DEVS and WRM are specific for discrete event and continuous simulation with possible adaptations to cover the other types of simulations. On the other hand, simulator coupling with an orchestrator provides more flexibility in term of coherence level. In general, depending on the specific set up and the master algorithm, co-simulation in this approach requires consistency in from dynamic to conceptual level.

3.2.4. Multi-Agent Approach for Co-Simulation

Beside the aforementioned approaches, we can also mention the multi-agent simulation approach for co-simulation at semantic and conceptual layers, which is interaction-oriented and can be considered as an overlapping of operational integration and formal integration. While offering conceptual and semantic interoperability, multi-agent simulation approach provides an operational integration of models and the simulators are executed via a well specified master algorithm. As a consequence, this approach does not always offer reusability and modularity for existing simulators.

The principle idea of the approach is considering the co-simulation framework as a multi-agent system. The most notable framework using this approach is *Agents and Artifact for multi-modeling* (AA4MM). The meta-model AA4MM considers the framework as interactive separated layers [85,86]. In semantic and syntactic levels, it defines a set of concepts representing the framework as a junction of interacting autonomous and heterogeneous models. In dynamic layer, the execution of the framework is considered to be a set of interacting autonomous and heterogeneous simulations. In addition, at last, in technical layer, the meta-model specifies an environment of interacting autonomous and heterogeneous simulators. An agent is associated to its models, simulations and simulators. The co-simulation framework is built upon the interaction among agents and the management of shared resources and environment (artifacts) (Figure 10). In general, AA4MM offers a homogeneous view on aspects of a co-simulation framework at semantic, syntactic, dynamic and technical level [14].

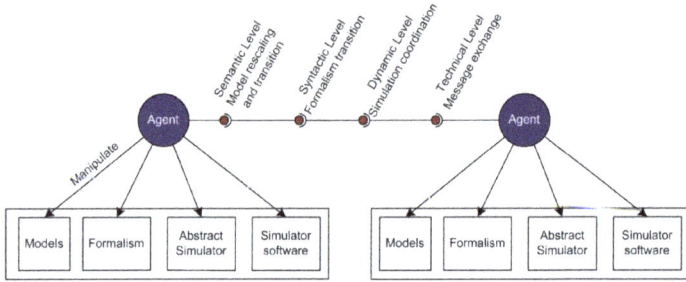

Figure 10. Interaction among Agents in the AA4MM meta-model (adapted from [14]).

Providing a great synergy and strong coupling capability, the implementation of multi-agent approach to CPES assessment is however still in the very first stage.

4. Conclusions

As the smart grid evolves into a cyber-physical energy system with inter-dependent domains, it is necessary to consider the design and validation problems at system level. That leads to the necessity of a holistic approach combining multi-domain validation, including hardware testing and simulation. While it requires a lot of effort to establish a general framework to cover all the domains in a CPES system, researchers have adopted the co-simulation approaches to consider the inter-dependency among domains as well as to validate the system as a whole. Flexible and adaptive, co-simulation is considered suitable for CPES assessment. However, in order to build a co-simulation framework, it is imperative to take into account various problems in different abstract levels, which is confusing and not always visible if one does not have a clear structuration of the framework.

In this paper, a conceptual structuration of co-simulation framework in CPES assessment is presented. A detailed review on coupling methods used and can potentially be used in co-simulation framework, along with their abstraction levels in the structuration, is then followed. It is supposed to provide the readers with a systematic view of the state of the art. This classification can be represented in a simple illustrative diagram (Figure 11).

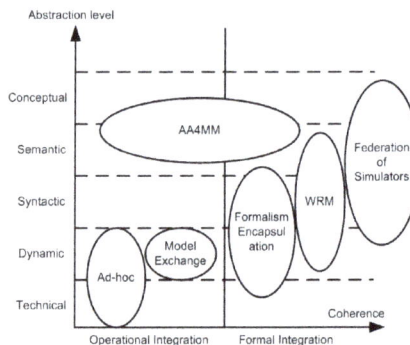

Figure 11. Reviewed coupling methods and their associated abstraction levels.

The paper highlighted multiple problems in different abstraction levels of a co-simulation framework. It is an important contribution to establish a solid theoretical base for specification of future applications of co-simulation and selection of coupling methods in CPES assessment and validation.

It is, however, necessary to take a closer look at the dynamic level where the synchronization among simulators happens. Several technical challenges persist at this level, such as the difficulty in implementing simultaneously continuous and discrete system, the lack of formalism to efficiently cover continuous simulation and discrete event simulation, harmonization of different time scales, cyclic dependency in conservative approach or time warp mechanism in optimistic approach, performance evaluation and stability analysis of different synchronization techniques, etc. These issues are interesting perspectives for further investigation.

Acknowledgments: This work is supported by the European Community's Horizon 2020 Program (H2020/2014-2020) under project "ERIGrid" (Grant Agreement No. 654113, www.erigrid.eu), which provides also funding for its open access. The participation of G2Elab and CEA-INES is also partially supported by the Carnot Institute "Energies du Futur" under the PPInterop II project (www.energiesdufutur.eu). The authors would like to express their gratitude towards the anonymous reviewers for helping to improve the paper with their valuable comments.

Author Contributions: V.H. Nguyen initiated the study on structuration and abstraction layers of co-simulation framework in CPES assessment. Y. Besanger and Q.T. Tran proposed the classification of coupling methods. T.L. Nguyen contributed to the review of existing works in operational and formal integration. V.H. Nguyen wrote the paper.

Conflicts of Interest: The authors declare no conflict of interest. The founding sponsors had no role in the study, in the writing of the manuscript and in the decision to publish the results.

Abbreviations

The following abbreviations are used in this manuscript:

CHIL	Controller-Hardware-in-the-loop
CPES	Cyber-Physical Energy System
DAE	Differential Algebraic Equation
DEVS	Discrete Event System Specification
DRES	Distributed Renewable Energy Resources
DRTS	Digital Real-Time Simulation
ICT	Information and Communication Technologies
IEC	International Electrotechnical Commission
FMI	Functional Mock-up Interface
FMU	Functional Mock-up Unit
HIL	Hardware-in-the-loop
HLA	High Level Architecture
HUT	Hardware Under Test
ODE	Ordinary Differential Equation
OLE	Object Linking and Embedding
OPC	OLE for Process Control
OPC UA	OPC Unified Architecture
PDES	Parallel Discrete-Event Simulation
PHIL	Power-Hardware-in-the-loop
SCADA	Supervisory Control And Data Acquisition
WAMPAC	Wide Area Monitoring Protection And Control
WRM	Waveform Relaxation Method

References

1. International Energy Agency. *Smart Grids in Distribution Network-Roadmap Development and Implementation*; IEA: Paris, France, 2015.
2. European Commission. *Energy Roadmap 2050*; Communication from the Commission to the European Parliament, the Council, the European Economic And Social Committee and the Committee of the Regions, Smart Grid Task Force; European Commission: Brussels, Belgium, 2011.
3. Farhangi, H. The path of the smart grid. *IEEE Power Energy Mag.* **2010**, *8*, 18–28.

4. Mets, K.; Ojea, J.A.; Develder, C. Combining Power and Communication Network Simulation for Cost-Effective Smart Grid Analysis. *IEEE Commun. Surv. Tutor.* **2014**, *16*, 1771–1796.

5. Palensky, P.; van der Meer, A.A.; López, C.D.; Jozeph, A.; Pan, K. Co-Simulation of Intelligent Power Systems—Fundamentals, software architecture, numerics, and coupling. *IEEE Ind. Electron. Mag.* **2017**, *11*, 34–50.

6. Strasser, T.; Andrén, F.P.; Lauss, G.; Bründlinger, R.; Brunner, H.; Moyo, C.; Seitl, C.; Rohjans, S.; Lehnhoff, S.; Palensky, P.; et al. Towards holistic power distribution system validation and testing—An overview and discussion of different possibilities. *Elektrotech. Informationstech.* **2017**, *134*, 71–77.

7. Nguyen, V.H.; Besanger, Y.; Tran, Q.T.; Nguyen, T.L.; Boudinet, C.; Brandl, R.; Strasser, T. Using Power-Hardware-in-the-loop Experiments together with Co-simulation in a holistic approach for cyber physical energy system validation. In Proceedings of the IEEE PES International Conference on Innovative Smart Grid Technologies IEEE ISGT Europe 2017, Turin, Italy, 26–29 September 2017.

8. Palensky, P.; van der Meer, A.A.; López, C.D.; Jozeph, A.; Pan, K. Applied co-simulation of intelligent power systems: Implementation, usage, examples. *IEEE Ind. Electron. Mag.* **2017**, *11*, 6–21.

9. Dang-Ha, T.H.; Olsson, R.; Wang, H. The Role of Big Data on Smart Grid Transition. In Proceedings of the IEEE International Conference on Smart City (SmartCity), Chengdu, China, 19–21 December 2015; pp. 33–39.

10. Li, W.; Ferdowsi, M.; Stevic, M.; Monti, A.; Ponci, F. Co-simulation for smart grid communications. *IEEE Trans. Ind. Inform.* **2014**, *10*, 2374–2384.

11. Cintuglu, M.; Mohamed, O.; Akkaya, K.; Uluagac, A. A Survey on Smart Grid Cyber-Physical System Testbeds. *IEEE Commun. Surv. Tutor.* **2017**, *19*, 446–464.

12. Tolk, A.; Muguira, J. The levels of conceptual interoperability model. In Proceedings of the 2003 Fall Simulation Interoperability Workshop, Orlando, FL, USA, 14–19 September 2003.

13. Cantot, P.; Igarza, J.L.; Luzeaux, D.; Rabeau, R. *Simulation et ModéLisation des SystèMes de SystèMes-IC2 Traité Informatique et Systèmes d'Information*; Hermes, Lavoisier: Paris, France, 2009.

14. Siebert, J. Approche Multi-Agent pour la Multi-ModéLisation et le Couplage de Simulations. Application à L'éTude des Influences Entre le Fonctionnement des RéSeaux Ambiants et le Comportement de Leurs Utilisateurs. Ph.D. Thesis, Université Henri Poincaré—Nancy I, Nancy, France, 2011.

15. Centre for Testing and Interoperability. *TTCN-3 Tutorial*; Technical Report; ESTI: Valbonne, France, 2013.

16. International Electrotechnical Commission. *TC IEC61850—Power Utility Automation*; Technical Report; 57—Power System Management and Associated Information Exchange; IEC: Geneva, Switzerland, 2003.

17. International Electrotechnical Commission. *Energy Management System Application Program Interface (EMS-API)—Part 301: Common Information Model (CIM) Base*; Technical Report; TC 57—Power System Management and Associated Information Exchange; IEC: Geneva, Switzerland, 2014.

18. International Electrotechnical Commission. *Application Integration at Electric Utilities—System Interfaces for Distribution Management—Part 11: Common Information Model (CIM) Extensions for Distribution*; Technical Report; TC 57—Power System Management and Associated Information Exchange; IEC: Geneva, Switzerland, 2013.

19. Electrical Power Research Institute. *OPC Unified Architecture—Part 1: Overview and Concepts*; Technical Report; TC 65/SC 65E–TR 62541-1:2010; Electrical Power Research Institute: Palo Alto, CA, USA, 2010.

20. International Electrotechnical Commission. *Application Integration at Electric Utilities—System Interfaces for Distribution Management—Part 14: Multispeak-CIM Harmonization*; Technical Report; TC 57—Power System Management and Associated Information Exchange; IEC: Geneva, Switzerland, 2013.

21. Gary, A.; Greg, R.; Gerald, R. MultiSpeak and IEC 61968 CIM: Moving Toward Interoperability. In Proceedings of the Grid-Interop Forum, Atlanta, GA, USA, 11–13 November 2008.

22. Nguyen, V.H.; Besanger, Y.; Tran, Q.T. CIM and OPC UA for interoperability of micro-grid platforms. In Proceedings of the IEEE PES Innovative Smart Grid Technologies Conference ISGT North America 2016, Minneapolis, MN, USA, 6–9 September 2016.

23. Kostic, T.; Preiss, O.; Frei, C. Towards the formal integration of two upcoming standards: IEC 61970 and IEC 61850. In Proceedings of the Large Engineering Systems Conference on Power Engineering, Montreal, QC, Canada, 7–9 May 2003.

24. Electrical Power Research Institute. *Harmonizing the International Electrotechnical Commission Common Information Model (CIM) and 61850—Key to Achieve Smart Grid Interoperability Objectives*; Technical Report; EPRI1020098; EPRI: Palo Alto, CA, USA, 2010.

25. Rohjans, S.; Uslar, M.; Appekath, J. OPC UA and CIM: Semantics for the smart grid. In Proceedings of the IEEE PES Transmission and Distribution Conference and Exposition, New Orleans, LA, USA, 19–22 April 2010.

26. Le, M.T.; Nguyen, V.H.; Besanger, Y.; Tran, Q.T. CIM compliant multiplatform approach for cyber-physical energy system assessment. In Proceedings of the IEEE PES Innovative Smart Grid Technologies Conference ISGT Asia, Auckland, New Zealand, 4–7 December 2017.

27. Michel, F. Formalisme, Outils et éLéMents MéThodologiques pour la Modélisation et la Simulation Multi-Agents. Ph.D. Thesis, Université des Sciences et Techniques du Languedoc, Montpellier, France, 2004.

28. Chevrier, V.; Fates, N. How important are updating schemes in multi-agent systems? An illustration on a multi-turmite model. In Proceedings of the 9th International Conference on Autonomous Agents and MultiAgent System AAMAS 2010, Toronto, ON, Canada, 9–14 May 2010.

29. Chang, W.T.; Ha, S.; Lee, E.A. Heterogeneous Simulation–Mixing Discrete-Event Models with Dataflow. *J. VLSI Signal Process.* **1997**, *15*, 127–144.

30. Kesaraju, V.; Ciarallo, F. Integrated simulation combining process-driven and event-driven models. *J. Simul.* **2012**, *6*, 9–20.

31. Banks, J.; Carson, I.I.; Nelson, B.L.; Nicol, D. *Discrete Event System Simulation*, 4th ed.; Prentice-Hall International: New York, NY, USA, 2005.

32. Chandy, K.; Misra, J. Distributed Simulation: A case study in design and verification of distributed program. *IEEE Trans. Softw. Eng.* **1979**, *24*, 440–452.

33. Lubachevsky, B.; Shwartz, D.; Weiss, A. An analysis of rollback-based simulation. *ACM Trans. Model. Comput. Simul.* **1991**, *1*, 154–193.

34. Steinman, J. Breathing Time Warp. In Proceedings of the Seventh Workshop on Parallel and Distributed Simulation, San Diego, CA, USA, 16–19 May 1993.

35. Nicol, D.; Liu, J. Composite synchronization in parallel discrete-event simulation. *IEEE Trans. Parallel Distrib. Syst.* **2002**, *13*, 433–446.

36. Jefferson, D. Virtual time. *ACM Trans. Program. Lang. Syst.* **1985**, *7*, 404–425.

37. Pongor, G. OMNET: Objective modular network testbed. In Proceedings of the MASCOTS93 International Workshop on Modeling, Analysis and Simulation on Computer and Telecommunication Systems, San Diego, CA, USA, 17–20 January 1993.

38. Riley, G. The Georgie Tech Network Simulator. In Proceedings of the ACM SIGCOMM Workshop on Models, Methods and Tools for Reproductible Network Research, Bolton Landing, NY, USA, 19–22 October 2003.

39. Awais, M.; Gawlik, W.; DeCillia, G.; Palensky, P. Hybrid Simulation using SAHISim Framework. In Proceedings of the 8th EAI International Conference on Simulation Tools and Techniques, Athens, Greece, 24–26 August 2015.

40. Galtier, V.; Vialle, S.; Dad, C.; Tavella, J.P.; Lam, Y.M.; Plessis, G. FMI-based Distributed Multi-Simulation with DACCOSIM. In Proceedings of the Spring Simulation Multi-Conference, Alexandria, VA, USA, 12–15 April 2015.

41. Widl, E.; Delinchant, B.; Kubler, S.; Li, D.; Muller, W.; Norrefeldt, V.; Nouidui, T.S.; Stratbucker, S.; Wetter, M.; Wurtz, F.; et al. Novel simulation concepts for buildings and community energy systems based on the Functional Mock-up Interface specification. In Proceedings of the Workshop on Modeling and Simulation of Cyber-Physical Energy Systems (MSCPES), Berlin, Germany, 14 April 2014.

42. Nguyen, T.L.; Tran, Q.T.; Caire, R.; Besanger, Y.; Hoang, T.T.; Nguyen, V.H. FMI Compliant Approach to Investigate the Impact of Communication to Islanded Microgrid Secondary Control. In Proceedings of the IEEE PES International Conference on Innovative Smart Grid Technologies (ISGT Asia 2017), Auckland, New Zealand, 4–7 December 2017.

43. Raad, A.; Reinbold, V.; Delinchant, B.; Wurtz, F. Energy building co-simulation based on the WRM algorithm for efficient simulation over FMU components of Web Service. In Proceedings of the BS'15 Building Simulation Conference, Hyderabad, India, 7–9 December 2015.

44. Müller, S.C.; Georg, H.; Nutaro, J.J.; Widl, E.; Deng, Y.; Palensky, P.; Awais, M.U.; Chenine, M.; Kuch, M.; Stifter, M.; et al. Interfacing Power System and ICT Simulators: Challenges, State-of-the-Art and Case Studies. *IEEE Trans. Smart Grid* **2016**, *PP*, doi:10.1109/TSG.2016.2542824.

45. Lin, H.; Sambamoorthy, S.; Shukl, S.; Thorp, J.; Mili, L. A study of communication and power system infrastructure interdependence on PMU-based wide area monitoring and protection. In Proceedings of the Power and Energy Society General Meeting, San Diego, CA, USA, 22–26 July 2012.

46. Lin, H.; Veda, S.; Shukla, S.; Mili, L.; Throp, J. GECO: Global Event-Driven Co-Simulation Framework for Interconnected Power System and Communication Network. *IEEE Trans. Smart Grid* **2012**, *3*, 1444–1459.

47. Nutaro, J.; Kuruganti, P.; Miller, L.; Mullen, S.; Shankar, M. Integrated hybrid-simulation of electric power and communication systems. In Proceedings of the IEEE PES General Meeting 2007, Tampa, FL, USA, 24–28 June 2007.

48. Li, W.; Monti, A.; Luo, M.; Dougal, R. VPNET: A co-simulation framework for analyzing communication channel effects on power systems. In Proceedings of the IEEE Electric Ship Technologies Symposium, Alexandria, VA, USA, 10–13 April 2011; pp. 143–149.

49. Anderson, D.; Zhao, C.; Hauser, C.; Venkatasubramanian, V.; Bakken, D.; Bose, A. A virtual smart grid–real-time simulation for smart grid control and communication design. *IEEE Power Energy Mag.* **2012**, *10*, 49–57.

50. Liberatoire, V.; Al-Hammouri, A. Smart Grid Communication and Co-Simulation. In Proceedings of the IEEE Energytech, Cleveland, OH, USA, 25–26 May 2011; pp. 1–5.

51. Bergmann, J.; Glomb, C.; Götz, J.; Heuer, J. Scalability of Smart Grid Protocols: Protocols and Their Simulative Evaluation for Massively Distributed DERs. In Proceedings of the 1st IEEE International Conference on Smart Grid Communication, Gaithersburg, MD, USA, 4–6 October 2010; pp. 131–136.

52. Babazadeh, D.; Chemine, M.; Kun, M.; Al-Hammouri, A.; Nordstrom, L. A platform for Wide Area Monitoring and Control System ICT analysis and Development. In Proceedings of the IEEE PowerTech, Grenoble, France, 16–20 June 2013; pp. 1–7.

53. Babazadeh, D.; Nordstrom, L. Angent-based control of VSC-HVDC Transmission Grid—A Cyber Physical System Perspective. In Proceedings of the IEEE MSCPES 2014, Berlin, Germany, 14 April 2014.

54. Godfrey, T.; Mullen, S.; Dugan, R.; Rodine, C.; Griffith, D.; Golmie, N. Modeling smart grid applications with co-simulation. In Proceedings of the 1st IEEE International Conference on Smart Grid Communication, Gaithersburg, MD, USA, 4–6 October 2010; pp. 291–296.

55. Mallouhi, M.; Al-Nashif, Y.; Cox, D.; Chadaga, T.; Hariri, S. A testbed for analyzing security of SCADA control systems (TASSCS). In Proceedings of the 2011 IEEE PES Innovative Smart Grid Technologies Conference, Anaheim, CA, USA, 17–19 January 2011.

56. Wei, M.; Wang, W. Greenbench: A benchmark for observing power grid vulnerability under data-centric threats. In Proceedings of the IEEE INFOCOM 2014—IEEE Conference on Computer Communications, Toronto, ON, Canada, 27 April–2 May 2014; pp. 2625–2633.

57. Zeigler, B.; Praehofer, H.; Kim, T. *Theory of Modeling and Simulation*; Academic Press: Cambridge, MA, USA, 2000.

58. Himmelspach, J.; Uhrmacher, A. Plug'n Simulate. In Proceedings of the ANSS07 40th Annual Simulation Symposium, Norfolk, VA, USA, 26–28 March 2007.

59. Himmelspach, J.; Rohl, M.; Uhrmacher, A. Component based models and simulation experiments for multi-agent systems in JAMES II. In Proceedings of the International Workshop on Agent Theory to Agent Implementation, Estoril, Portugal, 12–13 May 2008.

60. Quesnel, G.; Duboz, R.; Ramat, E. The virtual laboratory environment—An operational framework for multi-modeling, simulation and analysis of complex dynamical systems. *Simul. Model. Pract. Theory* **2008**, *17*, 641–653.

61. Lelarasmee, E.; Ruehli, A.; Sangiovanni-Vincentelli, A. *The Wave form Relaxation Method for Time-Domain Analysis of Large Scale Integrated Circuits*; University of California: Oakland, CA, USA, 1982; Volume 1.

62. White, J.; Vincentelli, A.; Odeh, F.; Ruehli, A. Waveform Relaxation: Theory and Practice. *Trans. Soc. Comput. Simul.* **1985**, *2*, 95–133.

63. Crow, M.; Ilic, M. The waveform relaxation method for systems of differential/Algebraic Equations. *Math. Comput. Model.* **1994**, *19*, 67–84.

64. Sand, J.; Burrage, K. A Jacobi Waveform relaxation method for ODE. *SIAM J. Sci. Comput.* **1998**, *20*, 534–552.

65. Burrage, K.; Dyke, C.; Pohl, B. On the performance of parallel waveform relaxations for differential systems. *Appl. Numer. Math.* **1996**, *20*, 39–55.

66. Garrappa, R. An analysis of convergence for two-stage waveform relaxation methods. *J. Comput. Appl. Math.* **2004**, *169*, 377–392.

67. Sun, W.; Fan, W.G.; Wang, T. A discretized waveform relaxation method in flexible holonomic systems. In Proceedings of the 2014 International conference on Information Science, Electronics and Electrical Engineering, Sapporo, Japan, 26–28 April 2014.

68. Liu, H.; Fu, Y.; Li, B. Discrete waveform relaxation method for linear fractional delay differential-algebraic equations. *Discrete Dyn. Nat. Soc.* **2017**, *2017*, 6306570.
69. Hout, K. On the convergence of waveform relaxation methods for stiff nonlinear ordinary differential equations. *Appl. Numer. Math.* **1995**, *18*, 175–190.
70. Bartoszewski, Z.; Kwapisz, M. On the convergence of waveform relaxation methods for differential-Functional System of Equations. *J. Math. Anal. Appl.* **1999**, *235*, 478–496.
71. Jiang, Y.L.; Wing, O. A note on convergence conditions of waveform relaxation algorithms for nonlinear differential-algebraic equations. *Appl. Numer. Math.* **2001**, *36*, 281–297.
72. Crisci, M.; Russo, E.; Vecchio, A. Discrete-time waveform relaxation Voltera-Runge-Kutta methods: Convergence analysis. *J. Comput. Appl. Math.* **1997**, *86*, 359–374.
73. Lee, E.; Zheng, H. Operational Semantics of Hybrid Systems. In *Hybrid Systems: Computation and Control, Proceedings of the 8th International Workshop, HSCC 2005, Zurich, Switzerland, 9–11 March 2005*; Springer Science Business Media: Berlin, Germany, 2005; pp. 25–53.
74. Wetter, M. Co-simulation of building energy and control systems with the building controls virtual test bed. *J. Build. Perform. Simul.* **2011**, *3*, 185–203.
75. Chatzivasileiadis, S.; Bonvini, M.; Matanza, J.; Yin, R.; Liu, Z.; Nouidui, T.; Kara, E.; Parmar, R.; Lorenzetti, D.; Wetter, M.; et al. Cyber physical modeling of distributed resources for distribution system operations. *Proc. IEEE* **2015**, *104*, 789–806.
76. Georg, H.; Wietfeld, C.; Muller, S.; Rehtanz, C. A HLA based simulator architecture for co-simulating ICT based power system control and protection systems. In Proceedings of the 3rd Smart Grid communications (SmartGridComm), Tainan City, Taiwan, 5–8 November 2012.
77. Galus, M.D.; Waraich, R.A.; Noembrini, F.; Steurs, K.; Georges, G.; Boulouchos, K.; Axhausen, K.W.; Andersson, G. Integrating Power systems, transport systems and Vehicle Technology for Electric Mobility Impact assessment and efficient control. *IEEE Trans. Smart Grid* **2012**, *3*, 934–949.
78. IEEE. *IEEE Standard for Modeling and Simulation (M & S) High Level Architecture (HLA)—Framework and Rules*; Technical Report; IEEE SA-1516-2010; IEEE: Piscataway, NJ, USA, 2010.
79. Hemingway, G.; Neema, H.; Nine, H.; Sztipanovits, J.; Karsai, G. Rapid synthesis of high-level architecture-based heterogeneous simulation: A model-based integration approach. *J. Simul.* **2012**, *88*, 217–232.
80. Wang, X.; Turner, S.J.; Yoke Hean Low, M.; Gan, P.B. Optimistic Synchronization in HLA-Based Distributed Simulation. *J. Simul.* **2005**, *81*, 279–291.
81. Hopkinson, K.; Wang, X.; Giovanini, R.; Thorp, J.; Birman, K.; Coury, D. EPOCHS: A platform for Agent-Based Electric Power and Communication Simulation Built from Commercial Off-The-Shelf Components. *IEEE Trans. Power Syst.* **2006**, *21*, 548–558.
82. Georg, H.; Muller, S.; Rehtanz, C.; Wiefeld, C. Analyzing Cyber-Physical Energy Systems: The INSPIRE Co-Simulation of Power and ICT system using HLA. *IEEE Trans. Ind. Inform.* **2014**, *10*, 2364–2373.
83. Caire, R.; Sanchez, J.; Hadjsaid, N. Vulnerability analysis of coupled heterogeneous critical infrastructures: A Co-simulation approach with a testbed validation. In Proceedings of the IEEE ISGT Europe 2013, Lyngby, Denmark, 6–9 October 2013.
84. Chatzivasileiadis, S.; Marco, B.; Javier, M.; Rongxin, Y.; Thierry, S.N.; Emre, C.K.; Rajiv, P. Cyber–Physical Modeling of Distributed Resources for Distribution System Operations. *Proc. IEEE* **2016**, *104*, 789–806.
85. Ricci, A.; Viroli, M.; Omicini, A. Give Agents their artifacts: The A & A approach for engineering working environments in MAS. In Proceedings of the AAMAS07 6th International Joint Conference on Autonomous Agents and Multi-Agent Systems, Honolulu, HI, USA, 14–18 May 2007.
86. Omicini, A.; Ricci, A.; Viroli, M. Artifacts in the A & A meta-model for multi-agent systems. *Auton. Agent Multi-Agent Syst.* **2008**, *17*, 432–456.

Review

Operational Range of Several Interface Algorithms for Different Power Hardware-In-The-Loop Setups

Ron Brandl

Fraunhofer Institute of Wind Energy and Energy System Technology, 34121 Kassel, Germany;
ron.brandl@iwes.fraunhofer.de; Tel.: +49-561-7294-103

Received: 16 October 2017; Accepted: 17 November 2017; Published: 23 November 2017

Abstract: The importance of Power Hardware-in-the-Loop (PHIL) experiments is rising more and more over the last decade in the field of power system and components testing. Due to the bidirectional exchange between virtual and physical systems, a true-to-reality interface is essential; however, linking several dynamic systems, stability issues can challenge the experiments, the components under test, and the individuals performing the experiments. Over the time, several interface algorithms (IA) have been developed and analyzed, each having different advantages and disadvantages in view of combining virtual simulations with physical power systems. Finally, IA are very specific to the kind of PHIL experiment. This paper investigates the operational range of several IA for specific PHIL setups by calculations, simulations, and measurements. Therefore, a selection of the mainly used respectively optimized IA is mathematically described. The operational range is verified in a PHIL system testing environment. Furthermore, in order to study the influence of different PHIL setups, according to software and hardware impedance, different tests using linear and switching amplifiers are performed.

Keywords: Power Hardware-in-the-Loop (PHIL); interface algorithm (IA); operational range of PHIL; linear/switching amplifier

1. Introduction

Challenging global goals for decentralized energy resources integration [1] lead on the one hand to the improvement of energy efficiency and the decrease of CO_2 emissions and, on the other hand, to the reduction of development costs and time for new smart grid components. Power Hardware-in-the-Loop (PHIL) technologies are present in the field of component testing, in power system stability studies as risk-free alternative for field tests (or preliminary testing), and are supporting the verification of new control strategies [2,3].

Nonetheless, by running realistic and stable experiments by bidirectional connection of a virtual simulated system (VSS) with a physical power system (PPS), an interface algorithm (IA) has to be used to eliminate inaccuracies and disturbances that result from the necessary coupling of devices like power amplifiers and measurement probes [4,5].

For the interaction between physical and virtual power domains, a power adaptive coupling element is needed. Considering this aspect, this paper will concentrate on the use of linear or switching amplifiers, which adapt control signals from the VSS to the necessary level of the PPS and is furthermore acting as a power sink or source.

In addition to physical connections via amplifiers in a testing environment, further software adaptions need to be made in order to perform secure and stable experiments. For this purpose, different IA are analyzed and compared.

The paper is structured as an overview of already existing IA studies on different laboratory setups. The results can be used to simplify preliminary stability studies of new PHIL testbeds and planned PHIL experiments by the choice of the most suitable IA.

2. Interface Algorithms of Power Hardware-In-The-Loop Systems

In general, a PHIL system architecture can be illustrated as a Thévenin equivalent, as depicted in Figure 1.

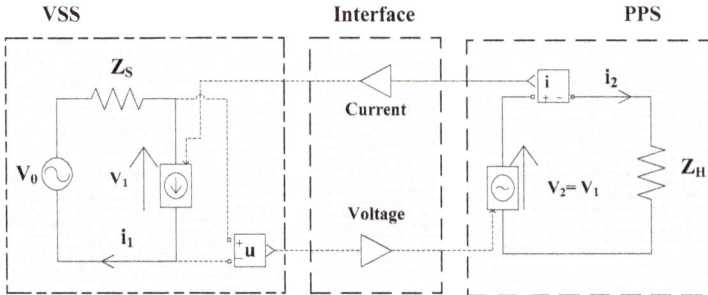

Figure 1. General Power Hardware-in-the-Loop (PHIL) system scheme showing the virtual and physical domains and their interfaces.

The VSS, represented by the voltage source V_0 and the internal simulation impedance Z_S, is used to run power system simulations mainly for power system stability studies.

The PPS, represented by the voltage or power source V_2, the measurement i_2, and the hardware impedance Z_H, contains a power amplifier, the hardware-under-test (HUT), and the measurement probes.

A PHIL system needs an interface to pass voltage or current from the VSS to the PPS and feed the dynamic behavior measured from the PPS as current and/or voltage flow back to the VSS.

The use of a power amplifier, as well as of probes, is essential for PHIL systems, but they influence the system dynamics and can introduce disturbances into the system [6,7]. Several algorithms were proposed in the last decade to ensure a stable operation of such a testing setup; however, the use of damping or filtering methods to ensure safe and stable experimental conditions influences the accuracy of the dynamic PHIL setup behavior. Finally, a compromise has to be found between more realistic test cases and larger stable operational ranges.

In the following, a comparison of the stable operational ranges of several IA and how to investigate them is presented.

2.1. Ideal Transformer Method

The Ideal Transformer Method (ITM) presents the most general and straightforward method for linking the VSS with the PPS [5]. Its scheme is depicted in Figure 2 and can be described with the following open loop transfer function by Equation (1):

$$-F_0(s) = e^{-sT_D} \cdot T_{PA}(s) \cdot T_M(s) \cdot Z_S(s)/Z_H(s) \tag{1}$$

where T_D is the total time delay produced by the measurement probe T_{D2} and the power amplifier T_{D1}, and T_{PA} and T_M represent the dynamic behavior of the power amplifier and the measurement probe, respectively.

The ratio between the simulated impedance Z_S and hardware impedance Z_H is critical for the stable operational range of PHIL experiments, since its values will be changing during the test. Therefore, Nyquist calculations of the stable operational range of the ITM were made by using several impedance ratios.

Figure 3 shows the stable range of the ITM; everything that is below the added surface at the impedance ratio 1 will be a stable system. The intersection at $Z_S/Z_H = 1$ depicts the range until the system behaves in stable conditions. Above this, the system will not be stable according to the Nyquist criteria.

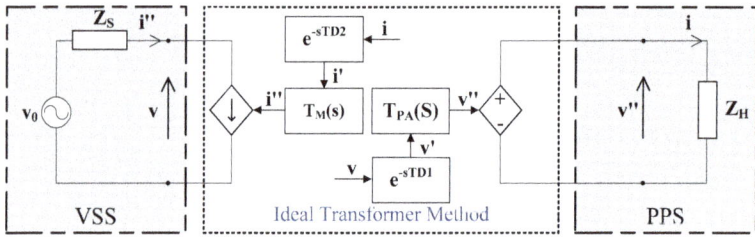

Figure 2. Scheme of the Ideal Transformer Method (ITM) interface.

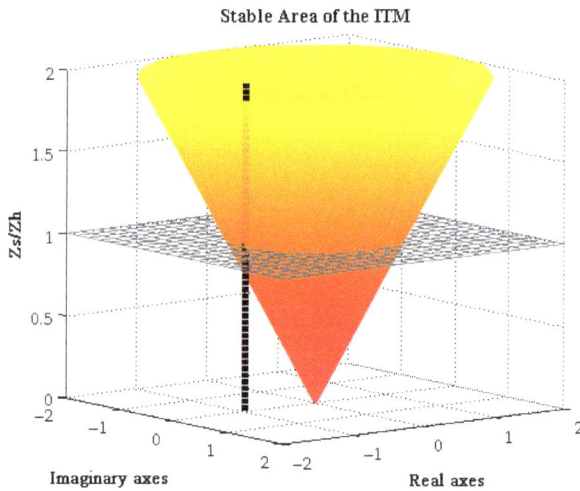

Figure 3. Stable operational ranges of the ITM.

2.2. Advanced Ideal Transformer Method

To optimize the stable ranges of the ITM [8], an improvement of the method to an Advanced Ideal Transformer Method (AITM) is proposed by adding an extra compensation impedance Z_C in the VSS, as shown in Figure 4. The mathematical explanation of the AITM is given by Equation (2). Note that for the AITM, Z_C needs to be greater than 0 to avoid short-circuit conditions of the injecting current source given the "i''".

$$- F_0(s) = e(-sT_D) \cdot T_{LV}(s) \cdot T_M(s) \cdot Z_S(s) / (Z_C(s) + Z_H(s)) \tag{2}$$

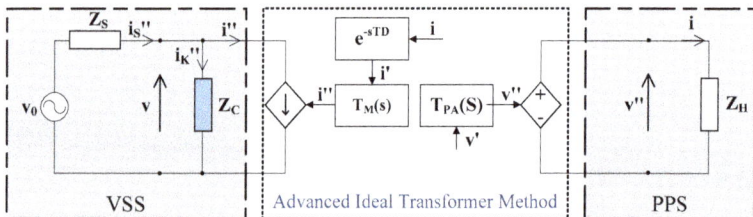

Figure 4. Scheme of the Advanced Ideal Transformer Method (AITM) interface.

Figure 5 shows the stable ranges of the AITM for different compensation impedances. It can be seen that only for Z_C of 0.5 Ω will the system go into an unstable state, where the ratio of Z_S/Z_H is over 1.5 (red-yellow surface).

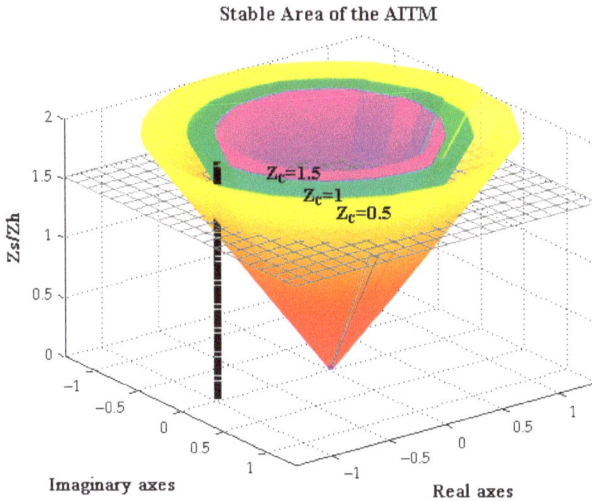

Figure 5. Stable operational ranges of the AITM.

Nonetheless, adding additional components to the system (in software and hardware) can affect the accuracy of the results. Therefore, Z_C has to be as small as possible. This is shown below in the Bode diagram (Figure 6), where the ITM is compared to the AITM with a resistance of exemplary R_C = 0.5 Ω. By inserting a resistance, the magnitude of the system will be affected.

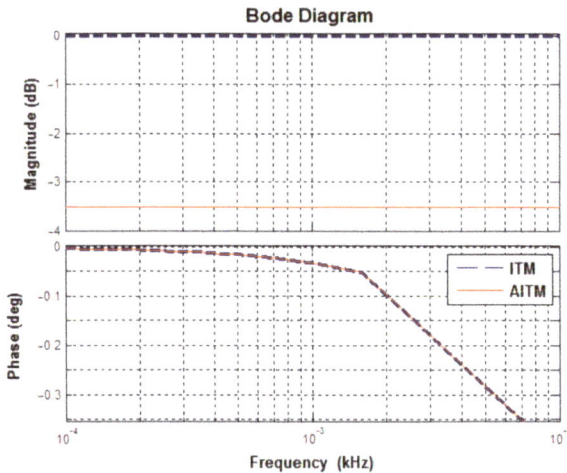

Figure 6. Comparison of ITM and AITM.

2.3. Partial Circuit Duplication

The method of the Partial Circuit Duplication (PCD) is presented in [9,10], and consists of additional coupling impedances Z_{SH} in the VSS and PPS (Figure 7). This method can be expressed as

Equation (3). Note that the influence of the power amplifier and the measurement probe is omitted in Equation (3).

$$- F_0(s) \approx Z_S(s) \cdot Z_H(s) / ((Z_S(s) + Z_{SH}(s)) \cdot (Z_H(s) + Z_{SH}(s))) \tag{3}$$

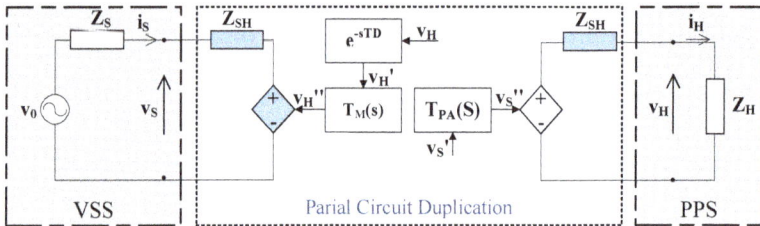

Figure 7. Scheme of the Partial Circuit Duplication (PCD) interface.

Figure 8 shows that even for small values of the coupling impedance Z_{SH}, the system will still operate in a stable condition. However, adding additional impedances in the VSS and the PPS means also a higher influence on the PHIL results due to the power consumption of the added components.

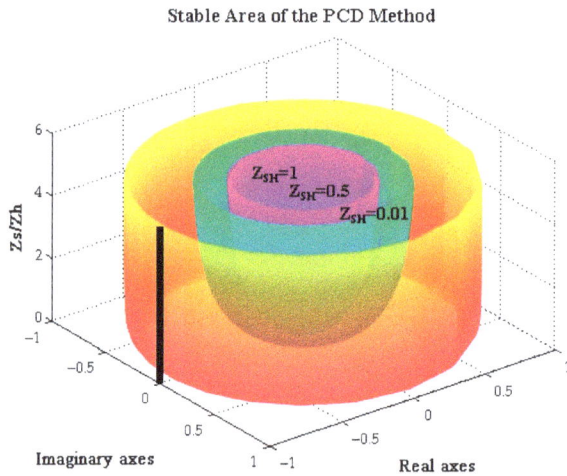

Figure 8. Stable operational ranges of the PCD method.

2.4. Damping Impedance Method

The Damping Impedance Method (DIM) combines the methods of the ITM and the PCD [5,11,12]. Its scheme is presented in Figure 9 and Equation (4) describe its dynamics. The DIM consists of an additional damping impedance Z^* and will ensure absolute stability when Z^* matches Z_H. Note that the influence of the power amplifier and the measurement probe are omitted.

$$- F_0(s) \approx Z_S(s) \cdot (Z_H(s) - Z^*(s)) / ((Z_H(s) + Z_{SH}(s))(Z_S(s) + Z_{SH}(s) + Z^*(s))) \tag{4}$$

Compared to the calculations in the sections above, Figure 10 presents a wider stable operational range than any other method. For the case of $Z^* = Z_H$ there is no increase of stable conditions, only for $Z^* \gg Z_H$ or $Z^* \ll Z_H$ can the system be unstable.

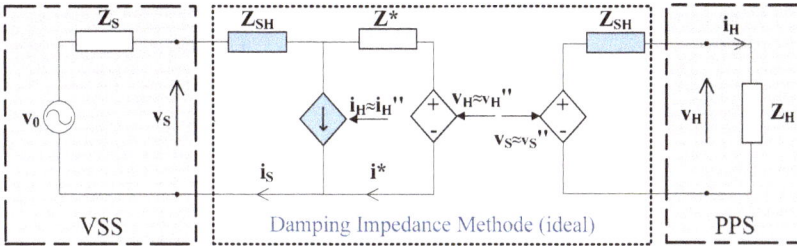

Figure 9. Scheme of the Damping Impedance Method (DIM) interface.

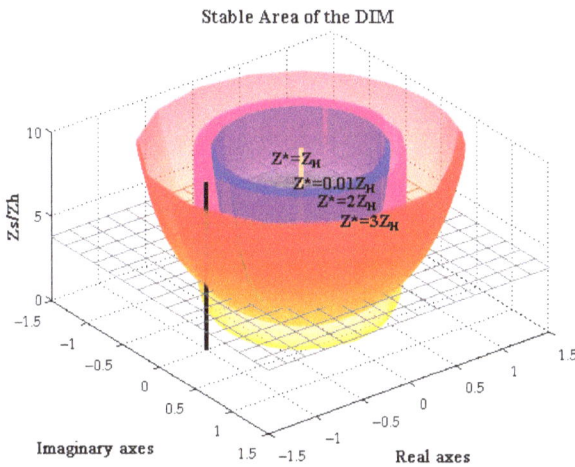

Figure 10. Stable operational ranges of the DIM for several damping impedances.

To ensure an absolute stable case during the experiment, the damping impedance should be adapted during the test to match Z_H in any case [12,13].

The DIM needs high implementation efforts and costs for the additional hardware impedance Z_{SH}.

2.5. IA Extendible Feedback Current Filter

A Feedback Current Filter (FCF) can be used to extend the stable operational ranges of the discussed IA [4]. The feedback current or voltage is filtered by a band pass or low pass filter to cut undesired harmonics and noise. Figure 11 shows an example of the ITM with additional FCF. The mathematical expression is given by Equation (5). Note that the influence of the power amplifier and the measurement probe are omitted in Figure 11.

$$- F_0(s) = e(-sT_D) \cdot T_{LV}(s) \cdot T_M(s) \cdot T_{FCF}(s) \cdot Z_S(s)/Z_H(s) \tag{5}$$

T_{FCF} represents the dynamic influence of the FCF in the system. For the use of a low pass filter with different cutting frequencies f_C, the following calculations were made.

As Figure 12 depicts, the ITM can be improved by adding a FCF. Without an additional filter, the stable area of the ITM was trespassed at a ratio higher than 1. With a FCF of $f_C = 1000$ Hz, the ratio can be increased to 3.13. The advantage of the FCF is that every IA can be improved with it, but contrary to the increased operational range, the FCF can affect the accuracy of exchanged signals between the VSS and PPS depending on the chosen cut-off frequency f_C, as it can be seen in the loss of magnitude in Figure 13.

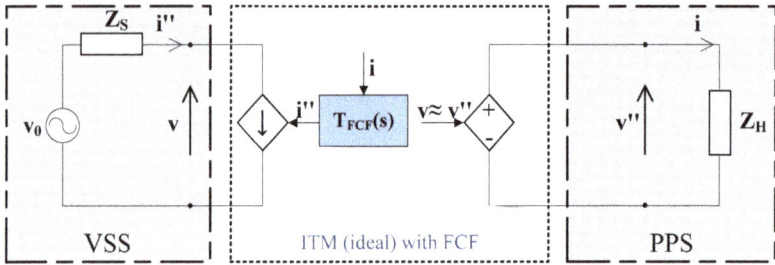

Figure 11. Scheme of an ideal ITM interface with Feedback Current Filter (FCF).

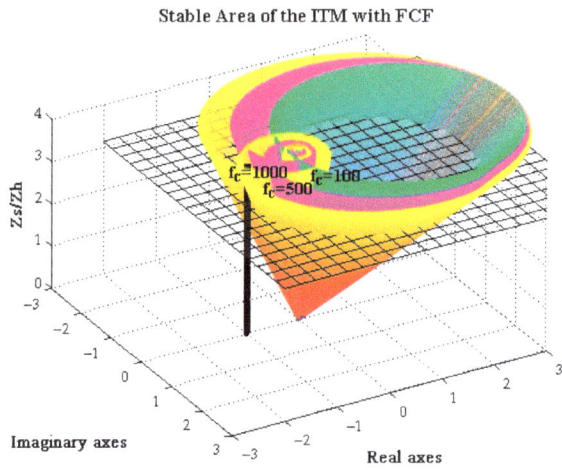

Figure 12. Stable operational ranges of the ITM with FCF.

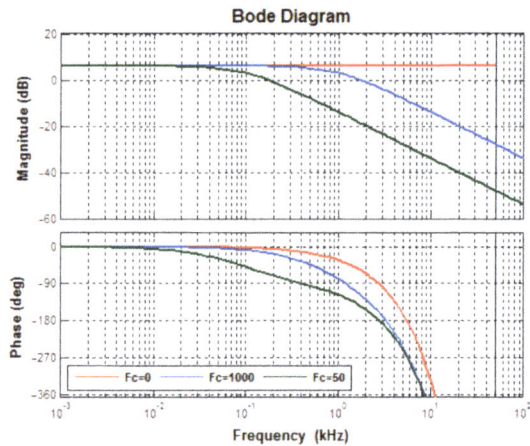

Figure 13. Influence of accuracy using the ITM with FCF.

2.6. Summary

Table 1 summarizes the mathematical explanations and gives an overview of the advantages and issues of each presented IA.

Table 1. Comparison of interface algorithms (IA).

Interface Algorithm	Mathematical Expression	Pro and Contra
Ideal Transformer Method	$Z_S(s)/Z_H(s)$	+ best accuracy for PHIL + easy implementation − low stability
Advanced Ideal Transformer Method	$Z_S(s)/(Z_C(s) + Z_H(s))$	+ high accuracy + easy implementation + good stability
Partial Circuit Duplication	$Z_S(s) \cdot Z_H(s)/$ $((Z_S(s) + Z_{SH}(s)) \cdot (Z_H(s) + Z_{SH}(s)))$	+ extreme high stability − additional hardware required − low accuracy
Damping Impedance Method	$Z_S(s) \cdot (Z_H(s) - Z^*(s))/$ $((Z_S(s) + Z_{SH}(s))(Z_H(s) + Z_{SH}(s) + Z^*(s)))$	+ great stability + good accuracy − additional hardware required
Feedback Current Filter	$T_{FCF}(s) \cdot Z_S(s)/Z_H(s)$	+ Extendible feature for IA + easy implementation − accuracy depending on f_C

3. Comparison of Power Amplifier with Power Hardware-in-the-Loop Systems

Contrary to Controller Hardware-in-the-Loop systems, PHIL Systems need additional voltage and power sinks or sources to adapt low-level signals from the Real-Time Simulator to the level of the HUT [6,14]. For carrying out PHIL experiments [4], the dynamic behavior of power amplifiers has to be considered, due to additional delays and internal filters that they introduce.

For upcoming investigations, two different kinds of amplifier are planned to be used to compare the behavior of PHIL systems: the switching amplifier and linear amplifier.

3.1. Switching Amplifier

Switching amplifiers are alternating/direct current converters and consist of a rectifier and an inverter. The advantages of the switching amplifiers are their efficiency in lower power ranges, but they can be built up to megawatt-ranges in compact and cost-reduced ways. Because of their internal pulsing of the output voltage coming from switching components, they are susceptible to harmonics and flickers. Furthermore, the reaction of the switching compared to linear amplifiers are slower [6] (see swell rate in Table 2).

Table 2. Parameters of the amplifiers and measurement probes.

System Parameter	Linear Amplifier	Switching Amplifier	Current Probe	Voltage Probe
Model	S & S PAS 90000	Ametek RS 270	LEM HTA 1000	LEM CV 3–100 V
Power	3×30 kVA	$3 \times 3 \times 30$ kVA	1000 A	1000 V
Bandwidth	DC ... 5 kHz	DC ... 2 kHz	DC ... 50 kHz	DC ... 800 kHz
Swell rate	>52 V/µs	>0.5 V/µs	>50 A/µs	0.4 µs to 90% V_N

3.2. Linear Amplifier

The linear amplifier provides, besides the functionalities of the switching amplifier, the possibility to operate it in a linear range, which implies faster response and adaptation of the low-level signals from the VSS; on the contrary, disturbances and transient effects are also amplified which switching amplifiers may damp. The disadvantage of the linear amplifier is the limited operational range compared to switching amplifiers. Therefore, the costs and dimensions are higher for the same power levels [15].

3.3. Comparison of Power Amplifiers

Table 2 provides the parameter of the used amplifiers.

To analyze dynamic behaviors of power amplifiers, a step signal can be input to investigate their response and related characteristics. The step response and mathematical expression of the behavior of the used amplifiers are given in Figures 14 and 15, as well as in Equation (6) for the switching amplifier and in Equation (7) for the linear amplifier.

$$-F_0(s)_{\text{swichted}} = e^{-90\mu s} \cdot \left(-5849s^2 + 2.2 \times 10^9 s + 2.2 \times 10^{12}\right) \Big/ \left(s^3 + 181 \times 10^3 s^2 + 2.5 \times 10^9 s + 2.2 \times 10^{12}\right) \tag{6}$$

$$-F_0(s)_{\text{linear}} = e^{-30\mu s} \times 5.5 \times 10^{12} \Big/ \left(s^2 + 12 \times 10^9 s + 1.4 \times 10^{12}\right) \tag{7}$$

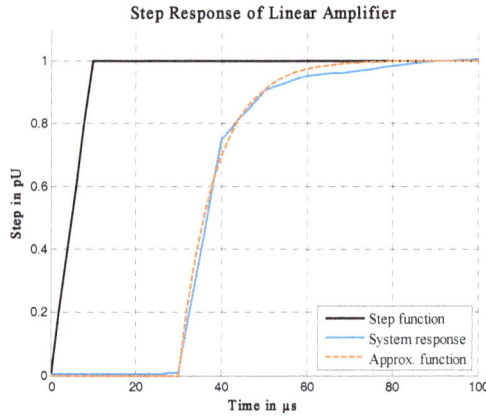

Figure 14. Step response of a linear amplifier.

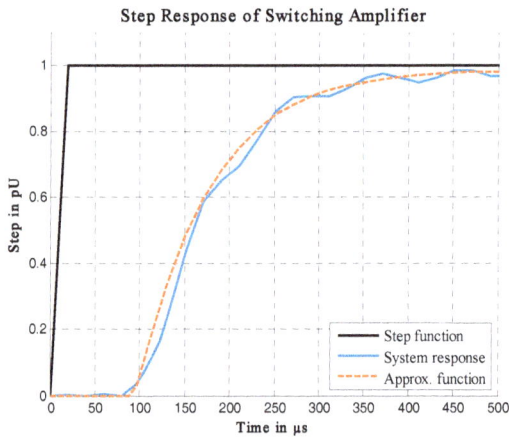

Figure 15. Step response of a switching amplifier.

The above diagrams (Figure 16) show that the linear amplifier is three times faster than the switching amplifier and has a smoother rise of the signal.

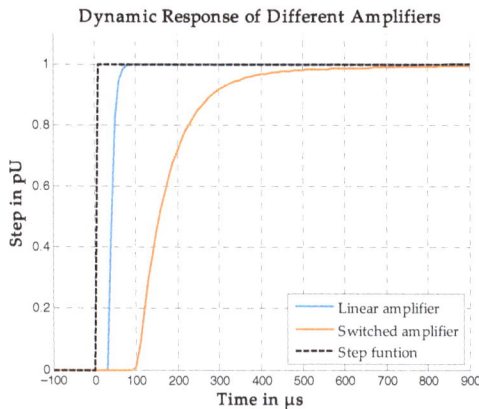

Figure 16. Comparison of step response.

According to the step response test, the system with a linear amplifier delays the response with $T_{D,linear} = 30$ µs and the switched amplifier with $T_{D,switched} = 90$ µs. Both determined delays, including delays of input and output cards of the simulator, as well as delays generated by the measurement probes.

4. Preliminary Simulations of Power Hardware-in-the-Loop Systems

For a preliminary assessment of the possible operation ranges of PHIL experiments, the PHIL system was studied in pure simulation first. Therefore, the tested power system model, simplified as Thévenin circuit is connected to a chosen IA with an impedance model of the PPS as well as the power amplifiers' dynamics analyzed in the previous section.

Figure 17 depicts the simulations model, where several IA are added to verify their operational ranges. During the simulation, the PPS was connected at 0.2 s to the VSS. Depending on the IA and the ratio of Z_S and Z_H, the simulation will be stable or not. Section 4.4 summarizes the conclusion of the investigated simulations.

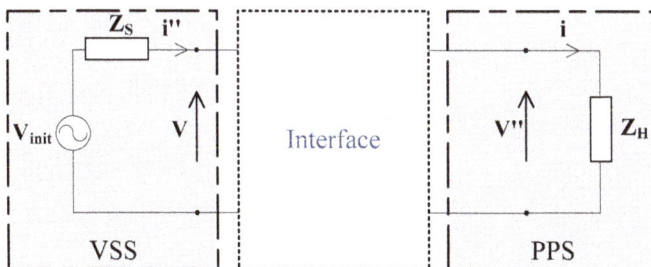

Figure 17. Simulation model for IA studies.

4.1. Simulation of the Ideal Transformer Method

Table 3 lists the used parameter for the preliminary ITM simulation studies. Figure 18 shows the results.

Table 3. Parameters of ITM simulation.

ITM	System A	System B
VSS impedance R_S	$1\,\Omega$	$2\,\Omega$
PPS impedance R_H	$1\,\Omega$	$1\,\Omega$
Transfer function	$e^{-T_D} \cdot T_{System} \cdot Z_S/Z_H$	
Result	Stable	Unstable

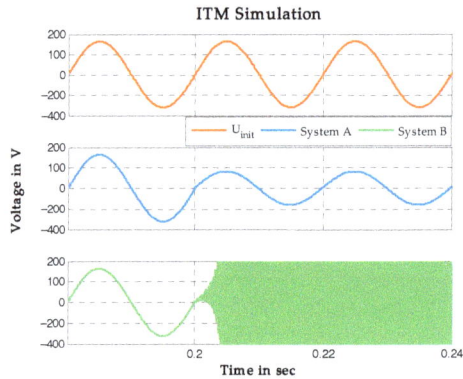

Figure 18. Results of the ITM studies.

4.2. Simulation of the Advanced Ideal Transformer Method

Table 4 lists the used parameter for the preliminary AITM simulation studies. Figure 19 shows the results.

Table 4. Parameters of AITM simulation.

AITM	System A	System B
VSS impedance R_S	$3\,\Omega$	$4\,\Omega$
PPS impedance R_H	$1\,\Omega$	$1\,\Omega$
Coupling imped. R_{SH}	$3.5\,\Omega$	$3.5\,\Omega$
Transfer function	$e^{-T_D} \cdot T_{System} \cdot Z_S/(Z_K + Z_H)$	
Result	Stable	Unstable

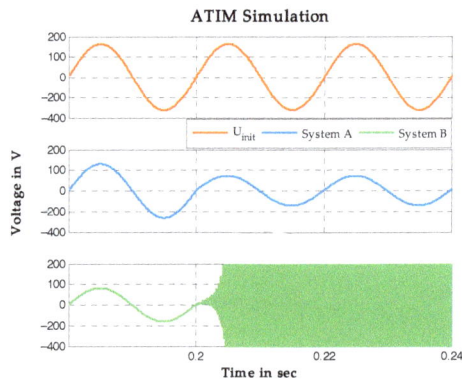

Figure 19. Results of the AITM studies.

4.3. Simulation of the Damping Impedance Method

Table 5 lists the used parameters for the preliminary DIM simulation studies. The damping impedance is designed with a higher value as the hardware impedance to simulate variable conditions. Figure 20 shows the results.

Table 5. Parameters of DIM simulation.

DIM	System A	System B	System C
Virtual impedance R_S	$2\,\Omega$	$5\,\Omega$	$6\,\Omega$
Coupling impedance R_{SH}	$0.1\,\Omega$	$0.1\,\Omega$	$0.1\,\Omega$
Damping impedance R^*	$3\,\Omega$	$3\,\Omega$	$3\,\Omega$
Hardware impedance R_H	$1\,\Omega$	$1\,\Omega$	$1\,\Omega$
Transfer function	$e^{-T_D} \cdot T_{System} \cdot Z_S(s)(Z_H(s) - Z^*(s))/((Z_S(s) + Z_{SH}(s))(Z_H(s) + Z_{SH}(s) + Z^*(s)))$		
Simulation	Stable	Stable	Unstable

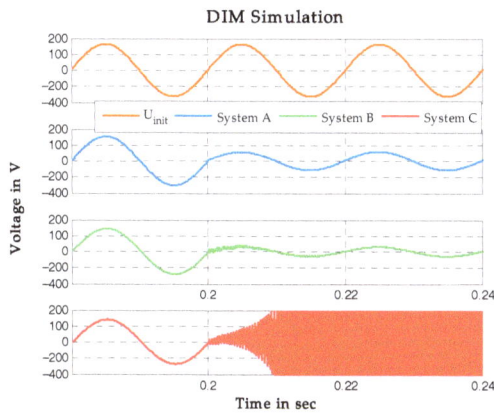

Figure 20. Results of the DIM studies.

4.4. Conclusion of the IA Simulation Studies

Table 6 gives a comparison of the investigations' results, at which impedance ratios of the simulation remain stable.

Table 6. Stable cases of the IA simulations.

Stable Case	ITM	AITM	DIM
Virtual impedance R_S	$1\,\Omega$	$5\,\Omega$	$6\,\Omega$
Hardware impedance R_H	$1\,\Omega$	$1\,\Omega$	$1\,\Omega$
Ratio R_S/R_H	1	5	6

The differences between the calculated and simulations' results are due to the added dynamics of the switching amplifier.

Table 6 shows which ratio of the impedances from VSS and PPS different IAs can handle. Furthermore, according to the impedance ratio, it can be stated that the DIM shows a higher stability than the AITM, which shows a higher stability than the ITM.

These investigations are not sufficient to set up PHIL experiments. In reality, several disturbances occur due to additional delays, amplifications errors, and electromagnetic compatibility (EMC) of wires and elements, which cannot be easily investigated by calculations or simulations. Therefore, real tests for stable operational ranges of the PHIL system itself have to be made to ensure a stable and safe experiment [11,16], and are shown in Section 5.

Nevertheless, the carried out preliminary studies provide a good overview concerning the choices of which IA can be used for different test cases.

5. Verification of IA in Real PHIL Systems

As mentioned above, running PHIL experiments can include additional disturbances which cannot be calculated or simulated. The following list will give a summary of what aspects have to be considered when setting up a PHIL system.

- Reduce electromagnetic influences by using screened and short wires, especially for the low-level signals;
- Reduce delays by using fast components and short connections;
- All devices have to be in the same emergency circuit;
- Integrate error detectors and protection devices (i.e., in the real-time simulator and power amplifier [17]);
- Ensure the safety of the experiment setup, especially when using hardware like batteries and rotating machines.

The uses of power amplifiers with internal filters and resistances or inductances can increase the testing system stability due to the intrinsic behavior of the different components.

Two of the mentioned IAs (ITM and DIM) have been verified in a real PHIL system. The easy implementation and high accuracy makes the ITM the best choice for linear and not so complex cases. In addition, the DIM with its higher stable operational range and good accuracy is mostly used to investigate non-linear cases with a high range of different or variable operational points. Therefore, these two methods will be compared in the following by real lab-based experiments using different power amplifiers.

5.1. Testing the Ideal Transformer Method

5.1.1. Ideal Transformer Method with Linear Amplifier

The ITM has been tested while using a linear amplifier. Table 7 gives the used parameter of the experiment. The results are shown in Figure 21.

Figure 21 depicts the stable areas of the used parameters. The integral of the several curves represents the range of the stable operation points.

Table 7. Parameters of ITM experiment.

Parameter of ITM Experiment with Linear Amplifier	
Voltage U_{init}	230 V at 50 Hz
Virtual impedance R_S	variable
Ratio of R_S/R_H	(0.9:0.1:1.8)
Physical impedance R_H	105.90 Ω
FCF f_C	(1:1:10) kHz
Additional delay T_D	(0:25:400) µs
Method	ITM with FCF
Amplification system	Linear amplifier
Real-time simulator	OP5600 from OPAL-RT

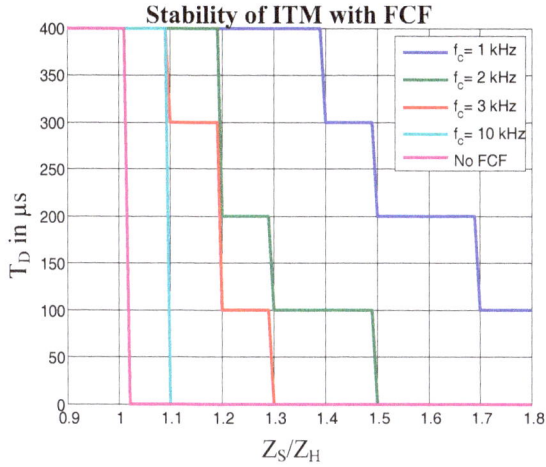

Figure 21. Results of the ITM with linear amplifier.

5.1.2. Ideal Transformer Method with Switching Amplifier

The ITM has been tested while using a switching amplifier. Table 8 gives the used parameters of the experiment. The results are shown in Figure 22.

Table 8. Parameters of ITM experiment.

Parameters of ITM Test with Switching Amplifier	
Voltage U_{init}	230 V at 50 Hz
Virtual impedance R_S	variable
Physical impedance R_H	31.8 Ω
FCF f_C	(1:1:10) kHz
Additional delay T_D	(0; 50; 500; 1000) μs
Method	ITM with FCF
Amplification system	Switched amplifier
Real-time simulator	OP5600 from OPAL-RT

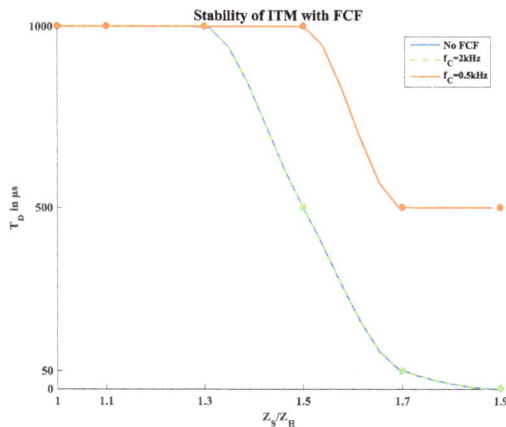

Figure 22. Results of the ITM with switching amplifier.

Figure 22 depicts the stable areas of the used parameters. It can be seen that there is no change about using a cut-off frequency higher than 2 kHz compared to no FCF. According to the bandwidth of the amplifier (see Table 2), the amplifier cuts off the frequency over 2 kHz by its internal filters. This means that a FCF with $f_C \geq 2$ kHz won't affect the results.

5.1.3. Conclusion of the ITM Experiments

The results of the ITM testing with different amplifiers lead to the conclusion that the internal filters of the switching amplifier have a positive effect on the system's stability range. This implies that for the ITM with lower stability, a linear amplifier leads faster to unstable conditions. The use of additional FCF can stabilize the system and, by improving the method to the AITM, can increase the operational ranges of a PHIL experiment as well.

It can be seen in Figure 21 that a FCF over the bandwidth limit of 5 kHz (see Table 2) of the linear amplifier still affect the results. This comes from the used measurement probes in this case, as they had a low accuracy and created higher delays. Therefore, more accurate probes were used for the PHIL system experiments with the switching amplifier.

5.2. Testing the Damping Impedance Method

5.2.1. Damping Impedance Method with Linear Amplifier

The DIM has been tested while using a linear amplifier. Table 9 gives the used parameter of the experiment. The results are shown in Figure 23.

Table 9. Parameters of DIM experiment.

Paramteres of DIM Experiment with Linear Amplifier	
Voltage U_{init}	230 V at 50 Hz
Virtual impedance R_S	variable
Damping impedance R^*	variable
Coupling impedance R_{SH}	variable
Physical impedance R_H	105.90 Ω
Additional delay T_D	(220) μs
Method	DIM
Amplification system	Linear amplifier
Real-time simulator	OP5600 from OPAL-RT

Figure 23 depicts the stable areas of the used parameters.

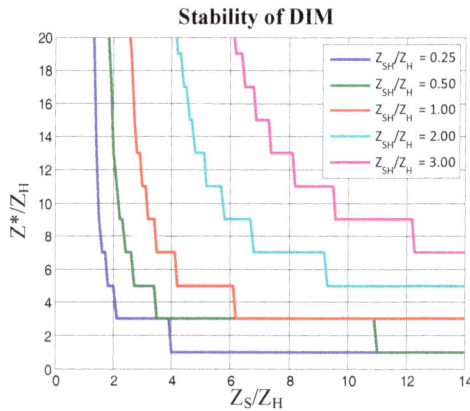

Figure 23. Results of the DIM with linear amplifier.

5.2.2. Damping Impedance Method with Switching Amplifier

The DIM has been tested while using a switching amplifier. Table 10 gives the used parameters of the experiment. The results are shown in Figure 24.

Table 10. Parameters of DIM experiment.

Parameters of DIM Experiment With Switching Amplifier	
Voltage U_{init}	230 V at 50 Hz
Virtual impedance R_S	variable
Damping impedance R^*	variable
Coupling impedance R_{SH}	0.23 Ω
Physical impedance R_H	31.8 Ω
Additional delay T_D	variable
Amplification system	Switched amplifier
Real-time simulator	OP5600 from OPAL-RT

Figure 24 depicts the stable areas for different cases of the experiment.

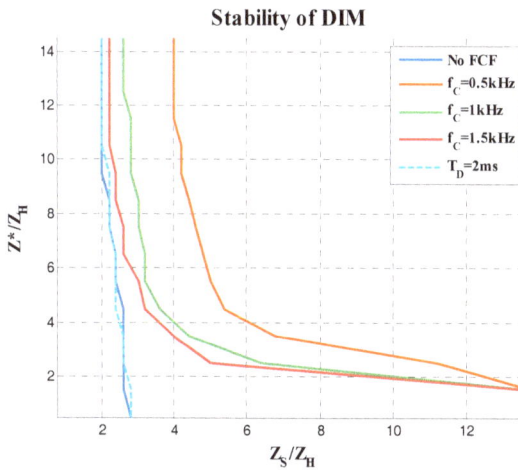

Figure 24. Results of the DIM with switching amplifier.

5.2.3. Conclusion of the DIM Experiments

Using methods with additional stability functionality, like the tested DIM, results in a much higher operational range for the system. It is also seen that if the impedances between the VSS and PPS are matching, respectively $Z_S/Z_H \leq 1$, the highest stability is given.

Furthermore, additional delays do not affect the DIM (Figure 24 dashed blue line) as they affect the ITM. Adding an extra FCF can increase the stability range of the DIM.

Moreover, it was shown that the DIM has a wider stable operational range than the ITM.

6. Experimental Investigation of ITM and DIM

The experimental studies serve as proof of concept for the undertaken analysis of the ITM and DIM in a laboratory setup by using ohmic load. For this purpose, the VSS or PPS changed continuously to test the limits of the stability of the entire system.

Table 11 presents the performed test cases.

Table 11. Test cases of the ITM and DIM.

Test Case	1	2
Description	Review of the ITM interface with resistive physical load.	Review the DIM interface with resistive physical load.
Scenario	(a) Variation of virtual impedance (b) Variation of physical impedance	(a) Variation of virtual impedance (b) Variation of physical impedance

6.1. Test Case 1: ITM

In scenario (a), the virtual impedance is variable in the range of $0 < R_s < 1$ kΩ. Due to laboratory setup limitation, the physical impedance was chosen with a constant value of $R_H = 53$ Ω during the entire test. A FCF set to $f_G = 2$ kHz was chosen. In both scenarios of test case 1, the ITM has been used.

In scenario (b), a constant value for the virtual impedance RS 100 $= \Omega$ was set and the physical impedance varies.

Figure 25 depicts the scheme of the performed experimental setup.

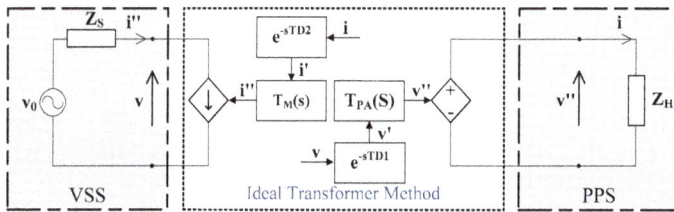

Figure 25. Experimental setup of the ITM.

A shutdown of the HUT occurred for both scenarios (see Figures 26 and 27), due to a detected overshoot of the selected 20% of total harmonic distortion (THD) threshold. Only stable operations of $R_S/R_H < 1.5$ could be achieved. This finding validates the stability analysis of the linear amplifier at $f_G = 20$ kHz presented in Section 5 (see Figure 21).

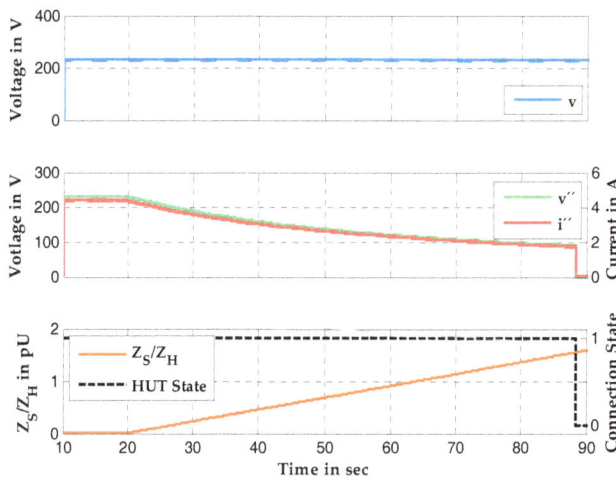

Figure 26. ITM studies with variable software impedance Z_S.

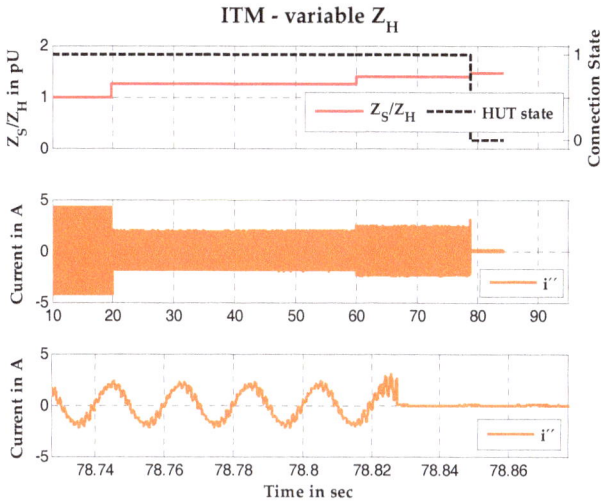

Figure 27. ITM studies with variable hardware impedance Z_H.

6.2. Test Case 2: DIM

Same investigation scenarios have been undertaken for the DIM as in test case 1. Due to laboratory setup limitation the value of the coupling impedance was selected with $R_{SH} = 104.9 \, \Omega$.

Contrary to the performed studies of the ITM, the DIM experiment does not show any unstable operations during the entire experiment. It should be noted that the large coupling impedance has a great effect on the values, therefore it should be wisely chosen according to the envisaged investigations.

Figure 28 depicts the scheme of the performed experimental setup.

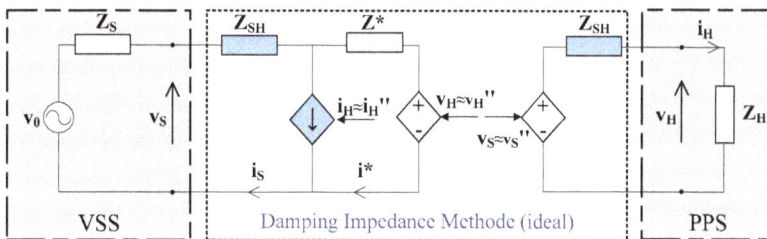

Figure 28. Experimental setup of the DIM.

The experiments (shown in Figures 29 and 30) confirm the conclusions of the analytical investigations performed for checking the stability limits of the ITM and DIM. It can be noted that the DIM has proven a wider range of stability under the same testing scenarios and conditions.

DIM - variable Z$_S$

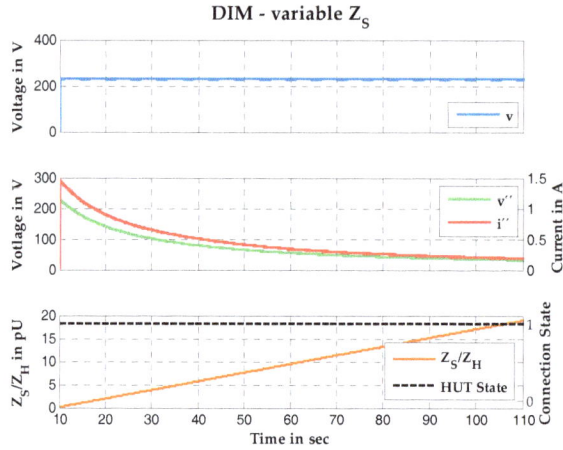

Figure 29. DIM studies with variable software impedance Z$_S$.

DIM - variable Z$_S$

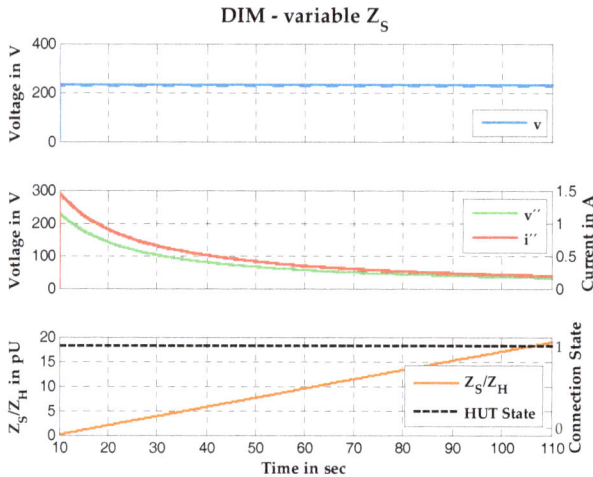

Figure 30. DIM studies with variable hardware impedance Z$_H$.

7. Conclusions

The paper presented an overview of interfaces for PHIL systems and compared the stable operational ranges of different interface algorithms, like the ITM, AITM, etc.

It has presented a way to investigate the functionality of different IA by using their mathematical equations, simulations—including physical component behaviors—and real laboratory experiments in Section 2.

According to the analyzed behavior of different IAs, the achieved results from the IA performance can be assessed.

Pure simulations including the characteristic of the PHIL system should be made to analyze which IA can be used (see Section 4).

It is important to investigate the capabilities of the PHIL system by performing preliminary stability tests (see Section 3). As shown in Section 5, good knowledge about the used components in a PHIL system is essential.

Energies **2017**, *10*, 1946

As a conclusion of the performed investigations, it can be stated that it is reasonable to start tests with high-accuracy IA and then, if necessary, gradually migrate towards IAs with wider operational ranges.

After investigating the combination of real-time simulations and power interfaces in a linear manner, as a next step to provide a generic stability description of PHIL systems, non-linear HUT is planned to be included in the initial studies of PHIL setups (e.g., inverters, active loads, etc.).

Acknowledgments: I acknowledge the support of our work by the German Ministry for the Environment, Nature and Nuclear Safety, and the Projektträger Jülich within the project "DEA-Stabil: Beitrag der Windenergie und Photovoltaik im Verteilungsnetz zur Stabilität des deutschen Verbundnetzes" (FKZ 0325585). Furthermore, this work is partly supported by the European Community's Horizon 2020 Program (H2020/2014-2020) under project "ERIGrid: European Research Infrastructure supporting Smart Grid Systems Technology Development, Validation and Roll Out" (Grant Agreement No. 654113).

Conflicts of Interest: The authors declare no conflicts of interest. The founding sponsors had no role in the design of the study; in the collection, analyses, or interpretation of data; in the writing of the manuscript, and in the decision to publish the results.

References

1. Bundesministerium für Wirtschaft und Energie. BMWi—EEG 2014. 2016. Available online: http://www.bmwi.de/DE/Themen/Energie/Erneuerbare-Energien/eeg-2014.html (accessed on 11 July 2016).
2. Dufour, C.; Andrade, C.; Bélanger, J. Real-Time Simulation Technologies in Education: A Link to Modern Engineering Methods and Practices. In Proceedings of the 11th International Conference on Engineering and Technology Education INTERTECH 2010 Ilhéus, Bahia, Brazil, 7–10 March 2010.
3. Menghal, P.M.; Jaya Laxmi, A. Real time simulation: A novel approach in engineering education. In Proceedings of the 2011 3rd International Conference on Electronics Computer Technology (ICECT), Kanyakumari, India, 8–10 April 2011; pp. 215–219.
4. Viehweider, A.; Lauss, G.; Lehfuß, F. Stabilization of Power Hardware-in-the-Loop simulations of electric energy systems. *Simul. Model. Pract. Theory* **2011**, *19*, 1699–1708. [CrossRef]
5. Ren, W.; Steurer, M.; Baldwin, T.L. Improve the Stability and the Accuracy of Power Hardware-in-the-Loop Simulation by Selecting Appropriate Interface Algorithms. In Proceedings of the 2007 IEEE/IAS Industrial & Commercial, Edmonton, AB, Canada, 6–11 May 2007; pp. 1–7.
6. Lehfuss, F.; Lauss, G.; Kotsampopoulos, P.; Hatziargyriou, N.; Crolla, P.; Roscoe, A. Comparison of multiple power amplification types for power Hardware-in-the-Loop applications. In Proceedings of the Complexity in Engineering (COMPENG), Aachen, Germany, 11–13 June 2012; pp. 1–6.
7. Lauss, G.; Faruque, M.O.; Schoder, K.; Dufour, C.; Viehweider, A.; Langston, J. Characteristics and Design of Power Hardware-in-the-Loop Simulations for Electrical Power Systems. *IEEE Trans. Ind. Electron.* **2016**, *63*, 406–417. [CrossRef]
8. Zhang, Z.; Fickert, L.; Zhang, Y. Power hardware-in-the-loop test for cyber physical renewable energy infeed: Retroactive effects and an optimized power Hardware-in-the-Loop interface algorithm. In Proceedings of the 2016 17th International Scientific Conference on Electric Power Engineering (EPE), Prague, Czech, 16–18 May 2016; pp. 1–6.
9. Steurer, M.; Ren, W.; Bogdan, F.; Sloderbeck, M.; Woodruff, S. Controller and Power Hardware-In-Loop Methods for Accelerating Renewable Energy Integration. In Proceedings of the 2007 IEEE Power Engineering Society General Meeting, Tampa, FL, USA, 24–28 June 2007.
10. Kuffel, R.; Wierckx, R.P.; Duchen, H.; Lagerkvist, M.; Wang, X.; Forsyth, P.; Holmberg, P. Expanding an Analogue HVDC Simulator's Modelling Capability Using a Real-Time Digital Simulator (RTDS). In Proceedings of the (ICDS '95) First International Conference on Digital Power System Simulators, College Station, TX, USA, 5–7 April 1995; p. 199.
11. Brandl, R.; Hernández, H.; Geibel, D. HIL-Methods Supporting the Development Process from Simulatins to Real Environment Testing. *DER J.* **2012**, *8*, 343–356.
12. Lehfuß, F.; Lauss, G. Power Hardware-in-the-Loop Simulations for Distributed Generation. In Proceedings of the 21st Conference on Electricity Distribution, Frankfurt, Germany, 6–9 June 2011.

13. Paran, S.; Edrington, C.S. Improved power hardware in the loop interface methods via impedance matching. In Proceedings of the 2013 IEEE Electric Ship Technologies Symposium (ESTS 2013), Arlington, VA, USA, 22–24 April 2013; pp. 342–346.

14. Roscoe, A.J.; Mackay, A.; Burt, G.M.; McDonald, J.R. Architecture of a Network-in-the-Loop Environment for Characterizing AC Power-System Behavior. *IEEE Trans. Ind. Electron.* **2010**, *57*, 1245–1253. [CrossRef]

15. De Jong, E.; de Graaff, R.; Vaessen, P.; Crolla, P.; Roscoe, A.; Lehfuß, F.; Lauss, G.; Gafaro, F. (Eds.) *European White Book on Real-Time Powerhardware-in-the-Loop Testing*; University of Strathclyde: Glasgow, UK, 2012.

16. Craciun, B.-I.; Kerekes, T.; Sera, D.; Teodorescu, R.; Brandl, R.; Degner, T.; Geibel, D.; Hernandez, H. Grid integration of PV power based on PHIL testing using different interface algorithms. In Proceedings of the IECON 2013—39th Annual Conference of the IEEE Industrial Electronics Society, Vienna, Austria, 10–13 November 2013; pp. 5380–5385.

17. Lauss, G.; Lehfus, F. Safety issues for power-harware-in-the-loop simulations. In Proceedings of the 2013 IEEE Grenoble PowerTech, Grenoble, France, 16–20 June 2013; pp. 1–6.

energies

MDPI

Article

Co-Simulation of Smart Distribution Network Fault Management and Reconfiguration with LTE Communication

Michele Garau [1,2], Emilio Ghiani [1,*], Gianni Celli [1], Fabrizio Pilo [1] and Sergio Corti [3]

[1] Department of Electrical and Electronic Engineering, University of Cagliari, Piazza d'Armi, 09123 Cagliari, Italy; celli@diee.unica.it (G.C.); pilo@diee.unica.it (F.P.)

[2] Department of Information Security and Communication Technology, NTNU—Norwegian University of Science and Technology, O.S. Bragstads plass 2B, 7491 Trondheim, Norway; michele.garau@ntnu.no

[3] Ricerca sul Sistema Energetico—RSE S.p.A., via R. Rubattino 54, 20134 Milano, Italy; sergio.corti@rse-web.it

* Correspondence: emilio.ghiani@diee.unica.it; Tel.: +39-070-675-5872

Received: 17 April 2018; Accepted: 21 May 2018; Published: 23 May 2018

Abstract: Transition towards a smart grid requires network modernization based on the deployment of information and communication technologies for managing network operation and coordinating distributed energy resources in distribution systems. The success of the most advanced smart grid functionalities depends on the availability and quality of communication systems. Amongst the most demanding functionalities, those related to fault isolation, location and system restoration (FLISR) to obtain a self-healing smart grid are critical and require low latency communication systems, particularly in case of application to weakly-meshed operated networks. Simulation tools capable of capturing the interaction between communication and electrical systems are of outmost utility to check proper functioning of FLISR under different utilization conditions, to assess the expected improvements of Quality of Service, and to define minimum requirements of the communication system. In this context, this paper investigates the use of public mobile telecommunication system 4G Long Term Evolution (LTE) for FLISR applications in both radially and weakly-meshed medium voltage (MV) distribution networks. This study makes use of a co-simulation software platform capable to consider power system dynamics. The results demonstrate that LTE can be used as communication medium for advanced fault location, extinction, and network reconfiguration in distribution networks. Furthermore, this paper shows that the reduction of performances with mobile background usage does not affect the system and does not cause delays higher than 100 ms, which is the maximum allowable for power system protections.

Keywords: smart grid; cyber physical co-simulation; information and communication technology; 4G Long Term Evolution—LTE; network reconfiguration; fault management

1. Introduction

1.1. Motivation

The development of future energy systems in accordance with the smart grid (SG) paradigm requires a radical change in the management of the electricity distribution network, which needs to become intelligent and adaptive. Smart distribution networks (SDN) have systems in place to control a combination of distributed energy resources (DERs). Distribution system operators (DSOs) have the possibility of managing electricity flows using a flexible network topology [1,2]. The transition towards SDN involves software, automation, and controls to ensure that the power distribution network, not only remains within its operating limits (e.g., node voltages and branch currents within acceptable limits), but is also operated in an optimal way. In the SDN context, therefore, Information and Communication Technologies (ICT) are not a simple add-on to the electrical system, but their availability and efficiency are essential for operating the entire power distribution system. In fact, the electric system is managed and controlled through ICT network, which allows a bidirectional exchange of large amounts of data, creating a keen interdependence between electric system and ICT system. In the ICT system, the communication between the SDN components is characterized by non-idealities such as latencies and packet losses that may reflect upon the power system operation; furthermore, components such as antennas, routers, modems, etc. are subjected to faults and malfunctioning that may cause system reliability reduction or service interruption [3].

In this context, this article aims at providing—by means of a co-simulation-based assessment method—an evaluation of the performance of LTE as communication technology for smart grid application, considering a highly time-critical application like fault location, isolation and system reconfiguration (FLISR).

1.2. Literature Review on Simulation of Communication Systems for Smart Grids

With recent enhancements in wireless solutions, which guarantee a reliable low-cost communication, a strong interest is upon the possibility of exploiting last generation communication systems for supporting the transition of distribution network towards a Smart Grid scenario. However, the best option for communication technology solution to fit SG applications is still not clear, even though LTE technology is considered one of the most promising. LTE, with its widespread distribution, broad coverage, high throughput, device-to-device (D2D) capability, despite not being originally designed for smart grid applications, represents a valuable candidate for usage in a SG communication system [4,5]. A comprehensive analysis of an LTE-based smart grid operation analysis with a co-simulation approach is still missing in literature. In [6], the communication challenges when choosing a technology supporting distribution automation applications was investigated with the communication software OPNET. LTE performances were analyzed in terms of coverage, delay and reliability with variable real-world deployment constraints, but the impact on distribution network was only analyzed in terms of requirements, and no interrelation between communication network and distribution network was analyzed in a joint way. A similar approach was adopted in Reference [7], where OPNET simulated using LTE for transmitting Phase Measurement Unit (PMU) packets in a fault monitoring system. Performances in terms of latencies, channel utilization, and response with variable load were examined. An analogous methodology was applied in [8], where LTE was analyzed in an OPNET environment to investigate the impact of SG communication on public shared LTE networks. Finally, in [9], LTE latencies were theoretically investigated based on requirement documents released by the National Institute of Standards and Technology (NIST), and the traffic distribution of smart grid distribution automation considering a smart grid application reserved bandwidth.

All the mentioned publications miss catching the cyber-physical behavior of smart grid, where electric and communication systems are strictly interdependent. A simulation platform where both domains are jointly simulated is fundamental in order to correctly analyze the smart grid behavior providing test platforms for smart grid applications that can be used for engineering smart grids from use case design to field deployment [10].

Smart grid simulators may be classified according to their modeling capabilities of power and communication systems. Three alternative approaches have been proposed in literature to tackle this kind of studies: Co-simulation, comprehensive simulation, and hardware-in-the-loop.

Co-simulation usually involves the integration of two or more simulators to capture cyber physical interdependency of a process or system. By co-simulating conventional power system simulation with communication and automation systems, the impact and dependencies of communication on the system can be investigated [11,12]. In co-simulation, each system is analyzed by its own dedicated simulator, and all simulators are executed simultaneously by appropriately designed run time interfaces (RTI) and coordinated simulation management. Various solutions for realizing a co-simulation tool, that differ in the targeted field of researching smart grids, and consequently in architectural choices, e.g., software components, time synchronization strategy, and scalability, can be found in the literature. Among them, for instance, EPOCHS is recognized to be one of the first co-simulation tools for power systems [13]. It was developed integrating three different commercial software: PSCAD/EMTDC and GE Power Systems Load Flow Software (PSLF) simulating the power grid, and ns-2 simulating the telecommunication network. PSCAD/EMTDC is dedicated to simulate electromagnetic transients, whilst PSLFs simulates the electrical system for long-term scenarios. Another important pioneer platform for co-simulation is GECO [14]. It exploits the event-driven method for synchronizing the simulation of the power system (with PSLF) and the communication network (modeled with ns-2). In this tool, each iteration of the numerical solution of the power flow is an event. All events are integrated in the event scheduler of ns-2, allowing a perfectly integrated simulation and minimization of synchronization errors. If compared to time step synchronization, event driven synchronization permits reducing simulation time and simulating large power systems with reduced computational burden. An alternative approach is comprehensive simulation, that combines power system and communication network simulation in one environment. In this case, the main concept is to bring together both system models and solving routines which leads either to integrate power systems simulation techniques into a communication network simulator or vice versa. A comprehensive simulation approach has been adopted for instance in Reference [15], where the authors presented a modular simulation environment based on OMNeT++, exploiting existing models for the communication network but purposely developing extra models for the electrical network. Finally, co-simulation could be realized with hardware in the loop (HIL) with software simulators and hardware components integrated in a real test bed, often used for testing control and protection systems in power systems [16]. HIL approach allows a perfect correspondence with a real system but with higher investment costs. A detailed state-of-the-art review of appropriate tools for simulating both domains of power system and ICT processes in the evolution of smart grids was presented in Reference [17].

The authors of Reference [18] proposed a classification of different fields of application of the co-simulation/HIL approach for smart grid analysis. Three macro-areas were identified:

- wide area monitoring and control (WAMC);
- optimization and control in distribution networks;
- integration of distributed generation.

In these fields of application, co-simulation approach allows emphasizing several critical aspects related to the interaction between the electrical system and ICT for smart grid operation, in particular:

- impact of latencies on correct operation of the electrical system [19,20];
- use of artificial intelligence in the management of smart grid [21,22];
- effect of cyber attack on smart grid management algorithms [23–26].

1.3. Contributions of This Paper

The objective of this paper is to demonstrate that LTE may provide appropriate performance for supporting data communication required to perform fault location, extinction, and a subsequent network reconfiguration in smart power distribution networks. For this reason, the co-simulation tool adopted has been purposely developed to simulate the highly time-critical smart grid application of fault management and network reconfiguration and permits reproducing and evaluating the behavior of the public mobile telecommunication system 4G LTE as communication technology for smart grid applications. In particular, this study focuses on the impact of LTE performances on network operation during fault management and reconfiguration. The architecture for co-simulation proposed in this paper coordinates two software packages, i.e., OMNeT++ for the ICT system and DIgSILENT PowerFactory with a Python script for the power system. A MATLAB Graphical User Interface (GUI) which allows the user to personalize the input data and to interact with the simulation as shown in Figure 1 was developed.

The co-simulator uses electromagnetic transient analysis capabilities of PowerFactory and the wide choice of libraries for communication systems analysis that are offered in the OMNeT++ open source environment. Specifically, in this paper, a system-level simulator for LTE and LTE-Advanced networks (SimuLTE) was used [27].

The Python script coordinates both dynamic simulations and allows data exchange between software packages through dedicated interfaces. PowerFactory provides a Python API that allows accessing software functionalities, element parameters, simulation results, etc.

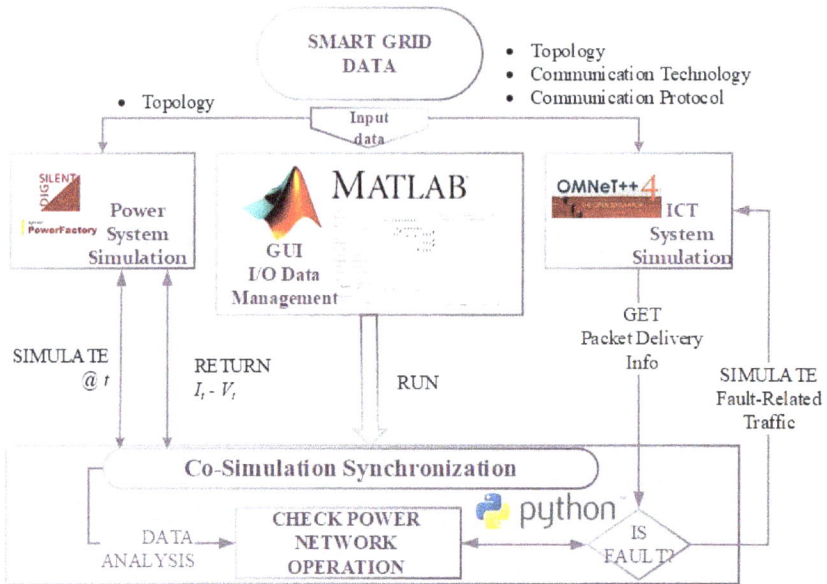

Figure 1. Schematic representation of the co-simulation tool proposed.

The integration of OMNeT++ in the co-simulation framework was obtained with a TCP socket connection programmed in C++. The Python script is the Run Time Interface (RTI) of the co-simulation tool. For each time step Δt, the script calls PowerFactory for solving the differential algebraic equations that describe the electric network analyzed during the time interval, and contemporarily calls OMNeT++ that executes the simulation during the subsequent time step. The scheduler is the heart of the simulation in the OMNeT++ environment, as its purpose is to handle the event list and run the scheduled event for the next instance. A customized scheduler was purposely developed in RTI to properly coordinate the OMNeT++ simulation. In the proposed application, when a short circuit condition is detected in the electric network, a new event is scheduled in OMNeT++ simulating the communication among the distributed devices involved in FLISR.

2. Protection and Reconfiguration of Smart Distribution Networks

This section describes the smart distribution network reference scenario, the innovative protection schemes and the fault management approach co-simulated.

2.1. Smart Distribution Network Structure

In distribution systems, supervised control and data acquisition (SCADA) is typically positioned at feeder level, and the majority of secondary distribution substations are not extensively monitored or controlled. Each secondary substation is equipped with manual or automatic sectionalizer (AS) or load-break switches used in conjunction with source-side circuit breakers, such as reclosers or circuit breakers, positioned at the origin of MV distribution feeders or in critical points, to automatically isolate faulted sections of electrical distribution systems with support of SCADA systems. The power to operate the control circuitry and the mechanism is obtained from the line through sensing-current transformers (Figure 2a). No auxiliary power supply, external connections, or external equipment is required. The AS permits disconnecting a portion of the distribution system or a single MV user (typically passive) when the source-side circuit breaker opens to de-energize the circuit.

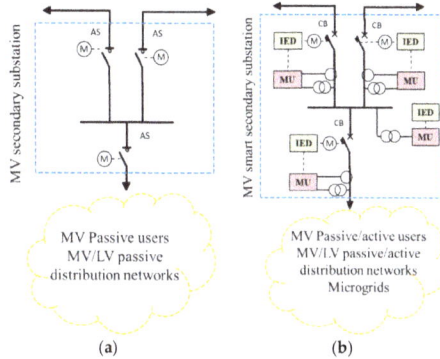

(a)   (b)

Figure 2. Schematic representation of (**a**) conventional secondary substation, (**b**) smart secondary substation.

In future distribution networks, secondary substations will be transformed into smart secondary substations (SSS) with a pervasive use of digital communication and intelligent electronic devices (IEDs) to enable local and/or remote sensing and control of substation equipment [28] (Figure 2b). IEDs, microprocessor-based controllers of power system equipment such as circuit breakers, transformers, and distributed generation, can be used for protection purposes, power quality analysis, network monitoring, energy metering, and so on. Real-time control of each network component is required and the network is equipped with smart meters and communication devices as well as faster protection devices and controls for power flow monitoring, distributed generation management, and network automation [29]. Distribution system operators (DSO) are already developing a significant refurbishment activity of secondary substations with new solutions for technological improvement of MV and low voltage (LV) equipment, MV/LV transformers, protection system, remote control devices and auxiliary components [30,31], in order to create SSS. SSS is equipped with reliable power components, high performance protection schemes, efficient flow monitoring system and reliable communication infrastructures, organized in order to:

- manage energy flows;
- contribute to voltage regulation;
- ensure fast reconfiguration after a failure;
- identify and pursue efficiency opportunities.

With smart distribution networks, radial operation of the network could be abandoned with significant benefits. Indeed, with a closed-loop or weakly meshed network, reduction of power losses, improvement of voltage profile, and a greater flexibility with reconfiguration, as well as superior ability to cope with load/generation growth with less need of network upgrades [32–34]. SDN allows changing between radial and meshed operation enabling exploitation of the advantages of both schemes.

2.2. Fault Detection and Reconfiguration Scheme for Smart Distribution Networks

In distribution systems, network automation and protection systems are designed to minimize the number of power interruptions and to limit outage duration. With smart grid enhancement, the number and outage time of interruptions is expected to be further reduced compared to the current situation.

The operation of circuit breakers is highly time critical since it is necessary to guarantee an instantaneous trigger on breakers to assure an efficient intervention during or after a fault extinction. The implementation of such systems requires a smart grid infrastructure that allows fast location of the fault's area, interruption of the short circuit current, as well as automation systems to reduce outage duration with automatic reconfiguration. It strongly relies on the performance of the communication system. Compared to wired solutions, such as power line carrier (PLC) or Fiber Optics, this paper investigates the use of wireless technologies for smart grid applications. In fact, they may provide communication abilities with lower cost of equipment and installation, quicker deployment, wide access and flexibility [35].

In this paper, the analyzed communication system was the LTE architecture used in public communication networks. SSS were connected to the communication network through LTE user equipment (UE). A distribution management system (DMS) with supervision/protection/reconfiguration capabilities was also used on the same communication system. Under the proposed protection scheme, each SSS was equipped with two measurement units and IEDs able to detect the direction of the fault currents and communicate with DMS besides the adjacent IEDs. This scheme configured a DMS with decentralized architecture able to provide more flexibility and rapidity of intervention [36,37].

Fault management and the strategy in opening the breakers differs according to the network configuration, meshed or radial. Three-phase short circuit faults are the simplest to be identified and handled. If the network is managed in radial configuration and no distributed generation (DG) exists, the fault is fed only from the primary substation. In this case, the nodes that are located downstream the fault will not detect any fault current. This fault condition is unambiguous and enables fast fault localization. When a reclosing branch is installed in the SSS, the IED is alerted for reconfiguring the network in order to minimize the impact of the fault. In case of meshed networks, the operating characteristic of the directional relays for a three-phase short circuit fault can be depicted as shown in Figure 3. Depending on the phase of the current, it is easy to find the position of the faults analyzing the module and phase of the current.

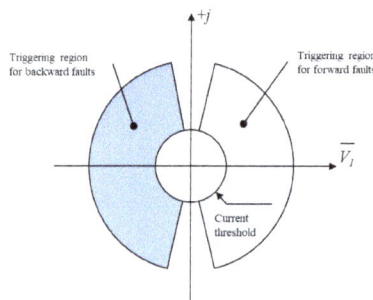

Figure 3. Operating characteristic of the directional criteria for three-phase short circuit faults.

In this proposed application, a smart selectivity scheme is assumed for three-phase inextinguishable faults. Those faults, even though, are the less frequent and the most critical in distribution networks. In fact, the resonance grounding (neutral grounded with arc suppression coils—Petersen coil), currently used in many European countries, permits choking the fault current below the level of self-extinction

(<35–50 A) by compensating the capacitive fault current of the network. By this action, all transient faults can be cleared without feeder tripping.

Considering an example of radial network such as the one represented in Figure 4, in case of three-phase short circuit located between nodes 4 and 5, the fault current will flow from the feeding high voltage (HV) substation A.

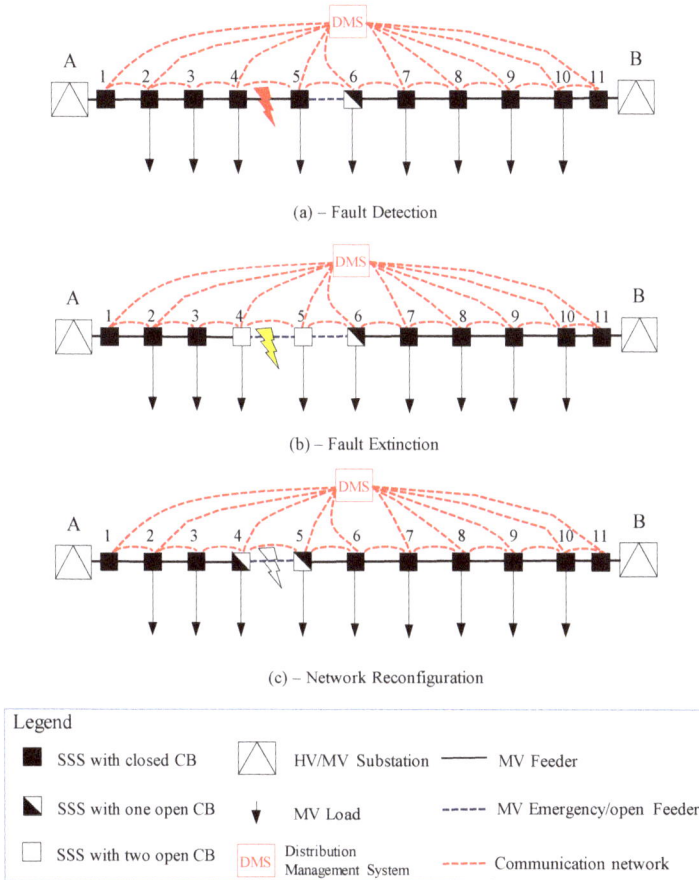

(a) – Fault Detection

(b) – Fault Extinction

(c) – Network Reconfiguration

Figure 4. Radial distribution network reconfiguration managed with emergency tie.

In case of fault on branch 4–5 in Figure 3a, the DMS, subsequently to processing information exchanged with the IED$_S$ deployed in SSS, has to:

(1) locate the fault (branch 4–5);

(2) open the first circuit breaker upstream the fault (SSS 4) in order to extinguish the fault, providing selectivity (Figure 3b); the load at node 5 is unserved until network reconfiguration is completed.

(3) Operate the emergency tie (5 and 6, in order to minimize the out of service area. Opening circuit breaker on node 5, on the side of the fault, guarantees that the second HV/MV substation does not feed the fault, and the fault is cleaned (Figure 3c).

(4) Operate closing of circuit breaker on node 6 permits the reconfiguration of the network and restoring the service to the load at node 5.

In case of fault, when the fault current exceeds the threshold, the IEDs of SSS are activated and send the measured current (module and phase) to the neighboring IEDs positioned in the adjacent SSS (Figure 5) in order to provide a fast localization of the fault.

Each IED that measures the outbound short circuit current will provide a message, e.g., a Generic Object Oriented Substation Event (GOOSE) message using IEC 61850 standard protocol, of the recognized fault to the DMS, and another message to adjacent IEDs. The IED receives a waiting signal from opening the corresponding circuit breaker (CB), the selectivity is obtained and the location of the fault is reached where the IED downstream the fault does not receive any waiting message. After that, the DMS has to communicate with peripheral units sequentially to perform the following actions:

a.   opening the CB (e.g., the outbound CB of the SSS4 in Figure 5) at the SSS upstream the fault;
b.   opening the CB (e.g., the inbound CB of the SS5 in Figure 5) at the SSS downstream the fault;
c.   closing of CB that permits the reconfiguration of the network.

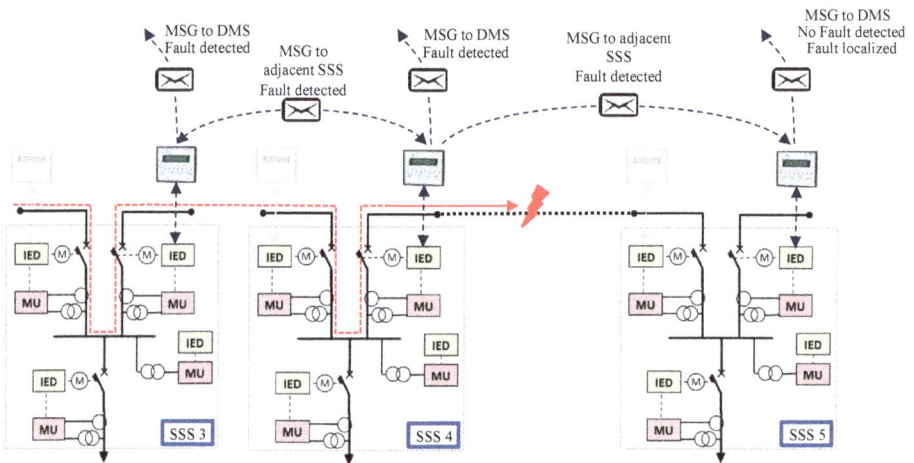

Figure 5. Short circuit current direction during three-phase fault in radial network.

In case of meshed (closed loop) network as the one represented in Figure 6, when a short circuit fault occurs between nodes 4 and 5, the fault current flows from both HV/MV substations A and B.

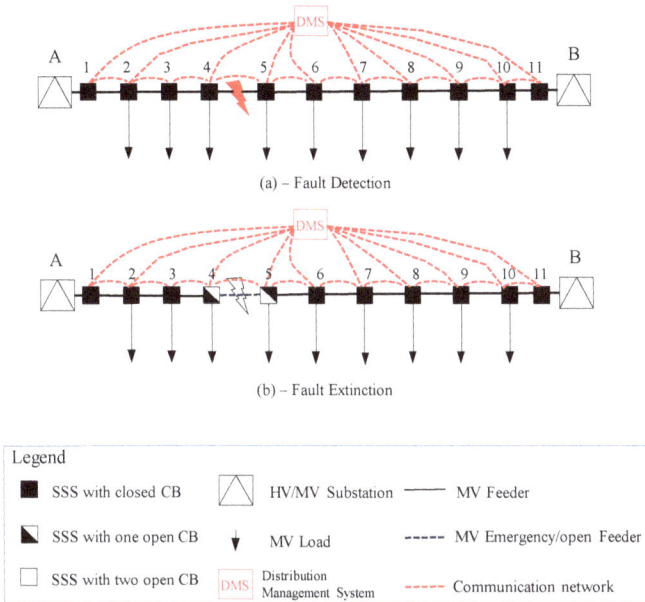

Figure 6. Fault detection and extinction in meshed distribution network.

In this case, each IED that measures the outbound short circuit current from the SSS provides a message of the recognized fault to the DMS, and another message to adjacent IEDs. Itself, it receives a wait signal from opening the corresponding CB, and the location of the fault is reached when two adjacent IEDs register currents with opposed phases. After that, the IEDs send the message to DMS to order the opening of the two CBs (e.g., the outbound CB of the SSS 4 and the inbound CB of SSS 5 in Figure 7) on both sides upstream the fault.

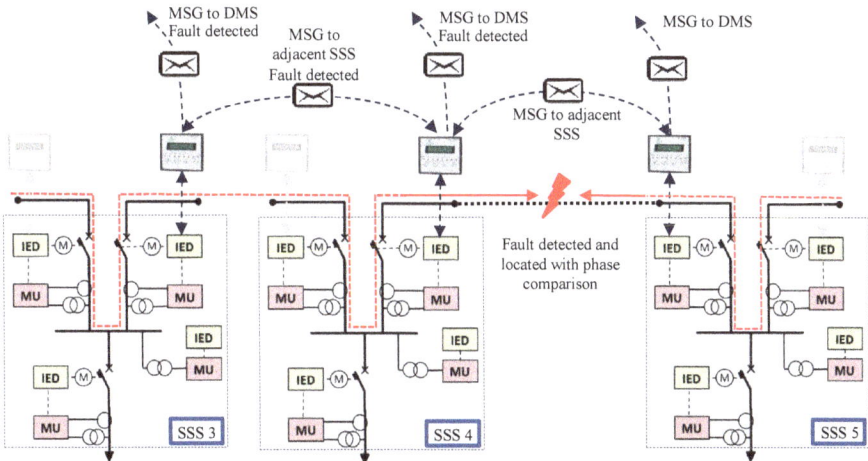

Figure 7. Short circuit current direction during three phase fault with meshed network.

2.3. LTE Communication Technology for Smart Grid Operation

The smart grid concept requires flexible communication architecture that allows power network devices such as sensors, smart meters, IEDs, and protection relays to exchange data in order to achieve an efficient operation of the electrical distribution system. A wide range of communication technologies, both wired and wireless, is nowadays available for building the communication infrastructure supporting smart grid data exchange. Peculiar features characterize each technology, for example data rate, coverage, installation and maintenance costs, reliability, exposure to cyber attacks, etc. Choosing appropriate technology is therefore crucial. Wireless technologies appear as key candidates in building the Smart Grid communications network due to their low installation costs and ease of deployment. Spreading of mobile telecommunication devices has stimulated a committed research over existing communication technologies. GSM, GPRS, EDGE, UMTS are part of a continuous evolution that has led to LTE, which represents one of the fourth-generation mobile technologies (4G). According to International Telecommunication Union (ITU), 4G technologies require to comply certain characteristics, among others [38]:

- ability to inter-work with other radio technologies;
- high quality of service;
- data rate of 100 Mbps in motion and 1 Gbps with fixed installations;
- sharing of network resources, allowing multiple users per cell;
- scalable bandwidth from 1.4 to 20 MHz;
- packet switching IP networks;
- connection spectral efficiency of 15 bps/Hz in downlink and 6.75 bps/Hz in uplink;
- operating modes: frequency division duplex (FDD) and time division duplex (TDD).

Currently, LTE is the technology that most efficiently meets all these requirements permitting broad coverage, high throughput, D2D capability, and the more recent LTE-Advanced release (LTE-A) provides the users with performances that are comparable with wired DSL technology [4,5]. Finally, the main features that enable LTE for supporting Smart Grid communication are [4]:

- Use of licensed bands: Even though the use of licensed bands alone does not grant or prevent cyber attacks, the communication network is robust against cyber attacks and possible stealing of confidential data and permits a better handing of interferences if compared with technologies that operate on license-free bands.
- Mature and ubiquitous coverage: The communication network span over vast areas, thus permits to integrate even remote endpoints to the main power grid.
- High performance: High data rate, low latency, and high system reliability enable critical automation tasks within the distribution grid that are often associated with demanding QoS requirements, such as severe time constraints.
- Third-party operation: It relieves DSOs from having to run and maintain a dedicated communication infrastructure.

For the abovementioned reasons, smart-grid operation considering the 4G LTE communication system, assuming to use the existing public mobile communication system has been chosen.

3. Case Study, Results and Discussion

The objective of this study is to analyze the performances of LTE as a communication carrier for supporting data communication required for FLISR in smart distribution networks, permitting fault location, isolation and service restoration in an acceptable time. In Italy, for example, the regulation of distribution systems includes output-based incentives to DSOs related to the quality of service, and, in particular, short interruptions from 1 s up to 3 min [39] can be subject to penalties or incentives. The DSO then, in order to avoid a worsening of its power quality indices, has to limit the maximum interruption time to under 1 s during faults; for this reason, it needs tools for assessment of communication technologies for smart grids applications like the one presented in this paper.

The proposed FLISR was tested on a real distribution network formed by five feeders, supplied by a HV/MV primary substation, and interconnected with emergency ties that can be used for changing the network reconfiguration. The network under study extended for about 10 km and supplied, through 46 secondary substations, a mix of residential and commercial loads in an urban scenario (Figure 8a). The area was assumed being served by LTE public mobile network, and the distribution of towers/antennas (e.g., eNB nodes in Figure 8b) followed realistic georeferenced data. In Table 1, the major simulation parameters used for the LTE network are reported. A three-phase permanent fault was assumed in branch 8–9, the fault was detected by the protection system of the network and then, the network could be reconfigured for permitting DSO crews to repair the fault.

Table 1. LTE communication network parameters.

LTE Related Parameter	Value
3GPP standard version	Release 10
Channel model (ITU scenario)	Urban Macrocell
Carrier Frequency	1800 MHz
Channel Bandwidth	20 MHz
Antenna Gain e-NodeB	18 dBm
Thermal Noise	−101 dBm
UE Noise Figure	2 dBm
e-NodeB Noise Figure	5 dBm
Packet Size	216 B
Protocol	UDP

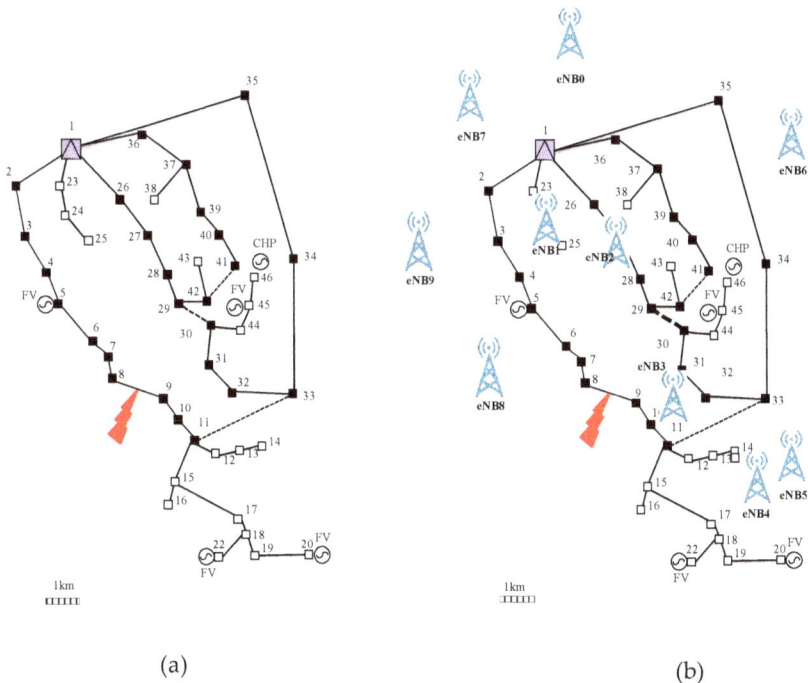

(a)   (b)

Figure 8. Case study: (**a**) Distribution network, (**b**) superimposed mobile LTE communication network.

For the sake of simplicity, but without loss of generality, the simulations shown in the following examples did not consider full implementation of IEC 61850 data model and parameters in IEDs and DMS simulated [28].

3.1. Radial Network Operation

In the radial operation of the network, the emergency tie between nodes 11 and 33 in Figure 8 was normally open. The co-simulation platform permitted simulating the detection and clearing of the fault condition, as well as the procedure for reconfiguring the network by closing the emergency for minimization of the network area out of service. This case study was of interest, for instance, for DSOs interested to know how much time was necessary to reconfigure the distribution network using LTE communication systems in a smart grid scenario.

The transient caused by a fault is shown in Figure 9. At node 8, after 164 ms from the fault, the IED triggered for opening the breaker, and extinguishing the fault current. The mechanical opening of the breaker was simulated by a time delay, which was randomly extracted from a Gaussian distribution (with mean of 0.2 s and standard deviation of 0.05 s) and the fault was extinguished after 330 ms. A message to the adjacent SSS (node 9) was sent for opening the switch and isolating the faulted network section. The node 11 waited for confirmation of the circuit breaker 9 opening that, due to the mechanical delay in the CB, arrived with a feedback packet at 464 ms. Afterwards, the DMS sent a message to IEDs at nodes 11 and 33 for closing the terminals of the emergency tie. The voltage profile at node 11 showed that the network was reconfigured after 748 ms.

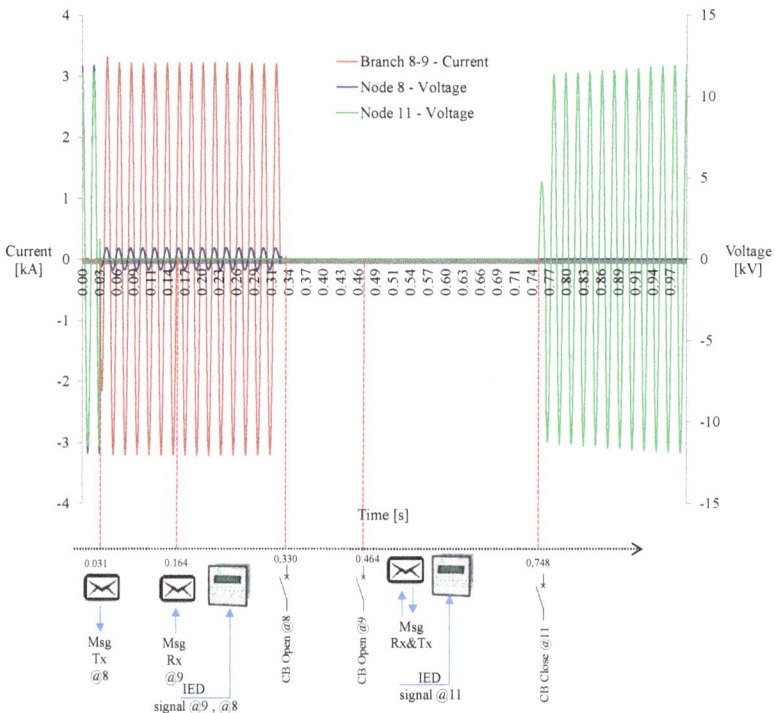

Figure 9. Voltage/current profiles: Voltage profile at node 8, and current profile in branch 8–9.

3.2. Meshed Network Operation

The second case analyzed considered a closed loop network operation. This means that the network was operated with switches at substations 8 and 9 normally closed, and the short circuit was fed by two sides.

Figure 10 shows the voltage at node 8, and the current that flowed in the branch between nodes 8 and 9. At 31 ms the overcurrent caused by the short circuit was detected by the IED in the SSS at node 8 that sent a message with current and phase measurement to the neighbors (see Tables 2 and 3).

Table 2. Node 8 current module/phase measurements.

	Current [kA]	Phase [rad]
Infeed	1.772	2.920 (cosφ = −0.976)
Outfeed	−1.786	0.220 (cosφ = 0.976)

At 56 ms, the substation 64 received the message from SSS at node 61, at 57 ms it received the message from SSS in node 67.

Table 3. Node 9 received current module/phase measurements.

	Time Rx [ms]	Current [kA]	Phase [rad]
Preceding SSS (node 8)	56	−1.057	0.077 (cosφ = 0.997)
Following SSS (node 10)	57	−1.433	0.259 (cosφ = 0.967)

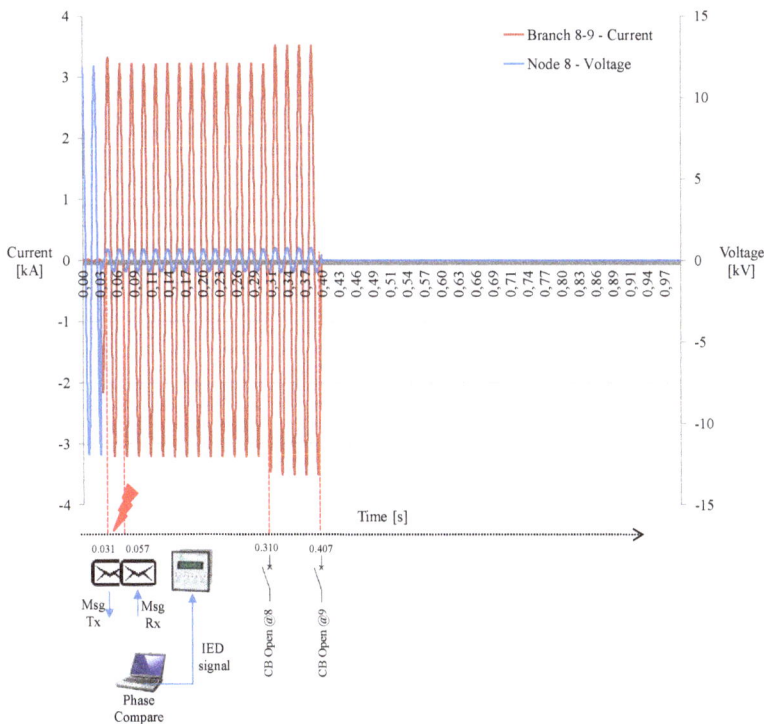

Figure 10. Voltage profile at node 8, and current profile in branch 8–9.

The DMS performed a phase comparison between the local measurements received allowing locating the fault. Opening of the two CB was completed in 377 ms. The communication took 27 ms, and 350 ms of mechanical delay was also considered. The network was reconfigured after 407 ms.

3.3. Backgroud Traffic Analysis

The simulations were executed in a more realistic situation considering the same network but different background conditions, in order to verify the LTE performances when a number of user are contemporarily connected to the same communication network. The first case considered a variable number of generic Mobile Broadband Users that were contemporarily active in the LTE network (MBBU). Several cases were considered with 0 (ideal case), to 50 MBBU per cell. Figure 11 shows the performances in terms of packet delivery ratios and coverage with variable active users per cell for typical LTE usage in a dense urban scenario with optimal exploitation of LTE network [40]. Overcharging the LTE network caused reduction in coverage that decreases to 90%, and a slight increase of the transmission delay of the packets. The coverage further decreased with more than 30 users per cell, what showed the congestion of the network and poor performances of the User Datagram Protocol (UDP) protocol in terms of packet loss. It has to be underlined that the low latency and packet drop could be obtained with an efficient planning of communication system cells [41].

Finally, in Figure 12, the performances variation of LTE in terms of delay and throughput are reported with a variable number of background traffic due to contemporary served MBBU. The delay increased on average from 28 ms up to 45 ms (with a maximum value observed during the repetitions of the scenario of 67 ms) in the scenario with 10 MBBU per cell. In the scenarios with more than 20 MBBU per cell the delay did not grow and was asymptotically held below 45 ms. This behavior was due to the congestion of LTE network over 20 MBBU per cell, that caused the rejection of new connections keeping the delays approximately unvaried. The saturation of the throughput for 20 up to 50 MBBU per cell demonstrated that the volume of data exchanged with the LTE network did not increase when charging the network over the number of 10 MBBU per cell.

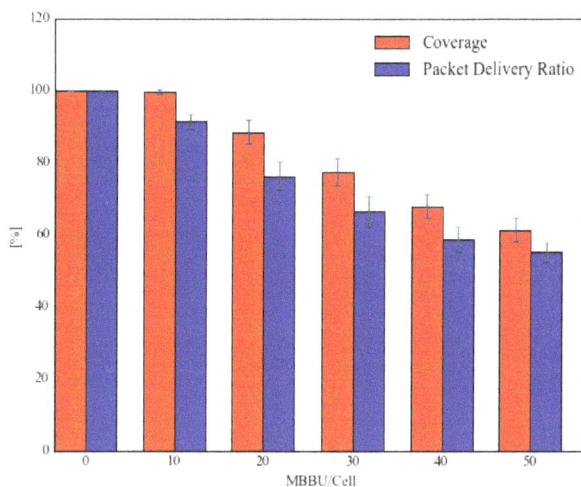

Figure 11. Coverage and packet delivery ratio in background traffic scenario.

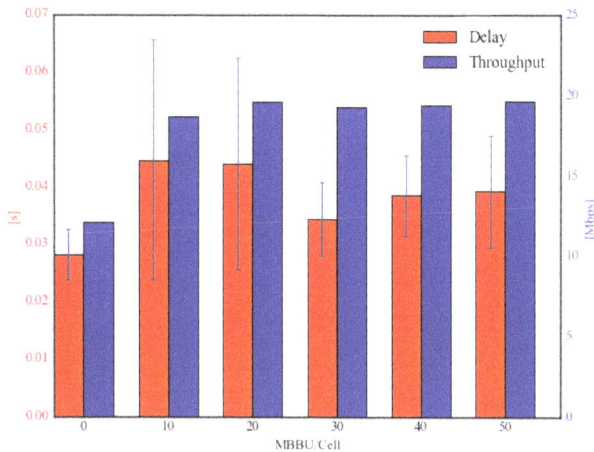

Figure 12. Communication delay and throughput in background traffic scenario.

4. Conclusions

The implementation of an automatic fast reconfiguration scheme of the electric distribution network strongly relies on the performance of the communication system. Compared to wired dedicated solutions, such as PLC or Fiber Optics, public wireless technologies offer an easier implementation of a communication link among IEDs at lower costs, but the doubt on their performance degradation due to the sharing with other users retards their exploitation. In this paper, the LTE communication technology has been tested with different traffic background conditions on a realistic case study using a co-simulation tool. According to the first results presented in this paper, LTE was technically able to start the intervention of protection devices in less than 100 ms, and showed to be adequate for FLISR applications related to Smart Grid implementation.

Author Contributions: F.P., G.C., E.G., and S.C. conceived the idea behind this research, designed the simulations and wrote the paper; M.G. developed the software, performed the simulations and wrote some parts of the manuscript; all the authors analyzed the data and proofread the paper.

Funding: This work has been partly financed by the Research Fund for the Italian Electrical System under the Contract Agreement between RSE S.p.A. and the Ministry of Economic Development—General Directorate for Nuclear Energy, Renewable Energy and Energy Efficiency in compliance with the Decree of 8 March 2006.

Conflicts of Interest: The authors declare no conflict of interest.

References

1. D'adamo, C.; Abbey, C.; Jupe, S.; Buchholz, B.; Khattabi, M.; Pilo, F. Development and operation of active distribution networks: Results of CIGRE C6.11 working group. In Proceedings of the 21st International Conference on Electricity Distribution, Frankfurt, Germany, 6–9 June 2011; Volume 6.
2. Control and Automation Systems for Electricity Distribution Networks (EDN) of the Future. Available online: https://e-cigre.org/publication/711-control-and-automation-systems-for-electricity-distribution-networks-edn-of-the-future (accessed on 17 April 2018).
3. Garau, M.; Celli, G.; Ghiani, E.; Soma, G.G.; Pilo, F.; Corti, S. ICT reliability modelling in co-simulation of smart distribution networks. In Proceedings of the 2015 IEEE 1st International Forum on Research and Technologies for Society and Industry Leveraging a Better Tomorrow (RTSI), Turin, Italy, 6–18 September 2015; pp. 365–370. [CrossRef]
4. Kalalas, C.; Thrybom, L.; Alonso-Zarate, J. Cellular Communications for Smart Grid Neighborhood Area Networks: A Survey. *IEEE Access* **2016**, *4*, 1469–1493. [CrossRef]

5. Garau, M.; Celli, G.; Ghiani, E.; Pilo, F.; Corti, S. Evaluation of Smart Grid Communication Technologies with a Co-Simulation Platform. *IEEE Wirel. Commun.* **2017**, *24*, 42–49. [CrossRef]

6. Patel, A.; Aparicio, J.; Tas, N.; Loiacono, M.; Rosca, J. Assessing communications technology options for smart grid applications. In Proceedings of the 2011 IEEE International Conference on Smart Grid Communications (SmartGridComm), Brussels, Belgium, 17–20 October 2011; pp. 126–131. [CrossRef]

7. Brown, J.; Khan, J.Y. Key performance aspects of an LTE FDD based Smart Grid communications network. *Comput. Commun.* **2013**, *36*, 551–561. [CrossRef]

8. Markkula, J.; Haapola, J. Impact of smart grid traffic peak loads on shared LTE network performance. In Proceedings of the 2013 IEEE International Conference on Communications (ICC), Budapest, Hungary, 9–13 June 2013; pp. 4046–4051. [CrossRef]

9. Cheng, P.; Wang, L.; Zhen, B.; Wang, S. Feasibility study of applying LTE to Smart Grid. In Proceedings of the 2011 IEEE First International Workshop on Smart Grid Modeling and Simulation (SGMS), Brussels, Belgium, 17 October 2011; pp. 108–113. [CrossRef]

10. Andrén, F.P.; Strasser, T.I.; Kastner, W. Engineering Smart Grids: Applying Model-Driven Development from Use Case Design to Deployment. *Energies* **2017**, *10*, 374. [CrossRef]

11. Stifter, M.; Kazmi, J.H.; Andrén, F.; Strasser, T. Co-simulation of power systems, communication and controls. In Proceedings of the 2014 Workshop on Modeling and Simulation of Cyber-Physical Energy Systems (MSCPES), Berlin, Germany, 14 April 2014; pp. 1–6. [CrossRef]

12. Li, W.; Ferdowsi, M.; Stevic, M.; Monti, A.; Ponci, F. Cosimulation for Smart Grid Communications. *IEEE Trans. Ind. Inform.* **2014**, *10*, 2374–2384. [CrossRef]

13. Hopkinson, K.; Wang, X.; Giovanini, R.; Thorp, J.; Birman, K.; Coury, D. EPOCHS: A platform for agent-based electric power and communication simulation built from commercial off-the-shelf components. *IEEE Trans. Power Syst.* **2006**, *21*, 548–558. [CrossRef]

14. Lin, H.; Veda, S.S.; Shukla, S.S.; Mili, L.; Thorp, J. GECO: Global Event-Driven Co-Simulation Framework for Interconnected Power System and Communication Network. *IEEE Trans. Smart Grid* **2012**, *3*, 1444–1456. [CrossRef]

15. Mets, K.; Verschueren, T.; Develder, C.; Vandoorn, T.L.; Vandevelde, L. Integrated simulation of power and communication networks for smart grid applications. In Proceedings of the 2011 IEEE 16th International Workshop on Computer Aided Modeling and Design of Communication Links and Networks (CAMAD), Kyoto, Japan, 10–11 June 2011; pp. 61–65. [CrossRef]

16. Cintuglu, M.H.; Ma, T.; Mohammed, O. Protection of Autonomous Microgrids using Agent-Based Distributed Communication. *IEEE Trans. Power Deliv.* **2016**, *32*, 351–360. [CrossRef]

17. Mueller, S.C.; Georg, H.; Nutaro, J.J.; Widl, E.; Deng, Y.; Palensky, P.; Awais, M.U.; Chenine, M.; Kuch, M.; Stifter, M.; et al. Interfacing Power System and ICT Simulators: Challenges, State-of-the-Art, and Case Studies. *IEEE Trans. Smart Grid* **2016**, *9*, 14–24. [CrossRef]

18. Palensky, P.; van der Meer, A.A.; Lopez, C.D.; Joseph, A.; Pan, K. Applied Cosimulation of intelligent power systems: Implementation, usage, and examples. *IEEE Ind. Electron. Mag.* **2017**, *11*, 6–21. [CrossRef]

19. Kazmi, J.H.; Latif, A.; Ahmad, I.; Palensky, P.; Gawlik, W. A flexible smart grid co-simulation environment for cyber-physical interdependence analysis. In Proceedings of the 2016 Workshop on Modeling and Simulation of Cyber-Physical Energy Systems (MSCPES), Vienna, Austria, 11 April 2016; pp. 1–6. [CrossRef]

20. Bottura, R.; Babazadeh, D.; Zhu, K.; Borghetti, A.; Nordström, L.; Nucci, C.A. SITL and HLA Co-simulation Platforms: Tools forAnalysis of the Integrated ICT and Electric Power System. In Proceedings of the 2013 IEEE EUROCON, Zagreb, Croatia, 1–4 July 2013.

21. Ahmad, I.; Kazmi, J.H.; Shahzad, M.; Palensky, P.; Gawlik, W. Co-simulation framework based on power system, AI and communication tools for evaluating smart grid applications. In Proceedings of the 2015 IEEE Innovative Smart Grid Technologies—Asia (ISGT ASIA), Bangkok, Thailand, 3–6 November 2015; pp. 1–6. [CrossRef]

22. Moulema, P.; Yu, W.; Griffith, D.; Golmie, N. On Effectiveness of Smart Grid Applications Using Co-Simulation. In Proceedings of the 2015 24th International Conference on Computer Communication and Networks (ICCCN), Las Vegas, NV, USA, 3–6 August 2015; pp. 1–8. [CrossRef]

23. Sadi, M.A.H.; Ali, M.H.; Dasgupta, D.; Abercrombie, R.K.; Kher, S. Co-Simulation Platform for Characterizing Cyber Attacks in Cyber Physical Systems. In Proceedings of the 2015 IEEE Symposium Series on Computational Intelligence, Cape Town, South Africa, 7–10 December 2015; pp. 1244–1251. [CrossRef]

24. Sadi, M.A.H.; Ali, M.H.; Dasgupta, D.; Abercrombie, R.K. OPNET/Simulink Based Testbed for Disturbance Detection in the Smart Grid. In Proceedings of the 10th Annual Cyber and Information Security Research Conference, CISR '15, Oak Ridge, TN, USA, 7–9 April 2015; ACM: New York, NY, USA, 2015; pp. 17:1–17:4. [CrossRef]

25. Dong, X.; Lin, H.; Tan, R.; Iyer, R.K.; Kalbarczyk, Z. Software-Defined Networking for Smart Grid Resilience: Opportunities and Challenges. In Proceedings of the 1st ACM Workshop on Cyber-Physical System Security, Denver, CO, USA, 12–16 October 2015; ACM: New York, NY, USA, 2015; pp. 61–68. [CrossRef]

26. Venkataramanan, V.; Srivastava, A.; Hahn, A. Real-time co-simulation testbed for microgrid cyber-physical analysis. In Proceedings of the 2016 Workshop on Modeling and Simulation of Cyber-Physical Energy Systems (MSCPES), Vienna, Austria, 11 April 2016; pp. 1–6. [CrossRef]

27. Virdis, A.; Stea, G.; Nardini, G. Simulating LTE/LTE-Advanced Networks with SimuLTE. In *Simulation and Modeling Methodologies, Technologies and Applications*; Springer: Cham, Switzerland, 2015; pp. 83–105.

28. Alvarez de Sotomayor, A.; Della Giustina, D.; Massa, G.; Dedè, A.; Ramos, F.; Barbato, A. IEC 61850-based adaptive protection system for the MV distribution smart grid. *Sustain. Energy Grids Netw.* **2017**. [CrossRef]

29. Dedè, A.; Giustina, D.D.; Massa, G.; Cremaschini, L. Toward a New Standard for Secondary Substations: The Viewpoint of a Distribution Utility. *IEEE Trans. Power Deliv.* **2017**, *32*, 1123–1132. [CrossRef]

30. Coppo, M.; Pelacchi, P.; Pilo, F.; Pisano, G.; Soma, G.G.; Turri, R. The Italian smart grid pilot projects: Selection and assessment of the test beds for the regulation of smart electricity distribution. *Electr. Power Syst. Res.* **2015**, *120*, 136–149. [CrossRef]

31. Botton, S.; Cavalletto, L.; Marmeggi, F. Schema project-innovative criteria for management and operation of a closed ring MV network. In Proceedings of the 22nd International Conference and Exhibition on Electricity Distribution (CIRED 2013), Stockholm, Sweden, 10–13 June 2013; pp. 1–4. [CrossRef]

32. Wolter, D.; Zdrallek, M.; Stötzel, M.; Schacherer, C.; Mladenovic, I.; Biller, M. Impact of meshed grid topologies on distribution grid planning and operation. *CIRED-Open Access Proc. J.* **2017**, *2017*, 2338–2341. [CrossRef]

33. Hadjsaid, N.; Alvarez-Hérault, M.C.; Caire, R.; Raison, B.; Descloux, J.; Bienia, W. Novel architectures and operation modes of distribution network to increase DG integration. In Proceedings of the IEEE PES General Meeting, Providence, RI, USA, 25–29 July 2010; pp. 1–6. [CrossRef]

34. Celli, G.; Pilo, F.; Pisano, G.; Allegranza, V.; Cicoria, R.; Iaria, A. Meshed vs. radial MV distribution network in presence of large amount of DG. In Proceedings of the IEEE PES Power Systems Conference and Exposition, New York, NY, USA, 10–13 October 2004; Volume 2, pp. 709–714. [CrossRef]

35. Jiang, J.; Qian, Y. Distributed Communication Architecture for Smart Grid Applications. *IEEE Commun. Mag.* **2016**, *54*, 60–67. [CrossRef]

36. Matos, M.A.; Seca, L.; Madureira, A.G.; Soares, F.J.; Bessa, R.J.; Pereira, J.; Peças Lopes, J. Control and Management Architectures. In *Smart Grid Handbook*; John Wiley & Sons, Ltd.: Hoboken, NJ, USA, 2016; ISBN 978-1-118-75547-1.

37. D'adamo, C.; Valtorta, G.; Consiglio, L.; Cerretti, A.; D'orazio, L.; Malerba, A.; Marmeggi, F. Smart fault selection: New operational criteria and challenges for the large-scale deployment in e-distribuzione's network. *CIRED-Open Access Proc. J.* **2017**, *2017*, 1475–1478. [CrossRef]

38. Korowajczuk, L. *LTE, WiMAX and WLAN Network Design, Optimization and Performance Analysis*, 1st ed.; Korowajczuk, L., Ed.; Wiley: Chichester, UK, 2011; ISBN 978-0-470-74149-8.

39. Aeegsi, A. per l'Energia E., il Gas ed il Sistema Idrico. Testo Integrato Della Regolazione Della Qualità Dei Servizi di Distribuzione e Misura Dell'energia Elettrica. 2014. Available online: https://www.autorita.energia.it/allegati/docs/11/198-11argtiqe_new.pdf (accessed on 18 May 2018).

40. Auer, G.; Giannini, V.; Desset, C.; Godor, I.; Skillermark, P.; Olsson, M.; Imran, M.A.; Sabella, D.; Gonzalez, M.J.; Blume, O.; et al. How much energy is needed to run a wireless network? *IEEE Wirel. Commun.* **2011**, *18*, 40–49. [CrossRef]

41. Saxena, N.; Roy, A.; Kim, H. Efficient 5G Small Cell Planning With eMBMS for Optimal Demand Response in Smart Grids. *IEEE Trans. Ind. Inform.* **2017**, *13*, 1471–1481. [CrossRef]

Article

Real-Time Simulation and Hardware-in-the-Loop Testbed for Distribution Synchrophasor Applications

Matthias Stifter [1,*,†], **Jose Cordova** [2,†], **Jawad Kazmi** [1] and **Reza Arghandeh** [2]

1   Center for Energy, Austrian Institute of Technology, Vienna 1210, Austria; jawad.kazmi@ait.ac.at
2   Electrical & Computer Engineering Department, Center for Advanced Power Systems,
     Florida State University, Tallahassee, FL 32306, USA; jdc13b@my.fsu.edu (J.C.); reza@caps.fsu.edu (R.A.)
*   Correspondence: matthias.stifter@ait.ac.at
†   These authors contributed equally to this work.

Received: 26 February 2018; Accepted: 5 April 2018; Published: 10 April 2018

Abstract: With the advent of Distribution Phasor Measurement Units (D-PMUs) and Micro-Synchrophasors (Micro-PMUs), the situational awareness in power distribution systems is going to the next level using time-synchronization. However, designing, analyzing, and testing of such accurate measurement devices are still challenging. Due to the lack of available knowledge and sufficient history for synchrophasors' applications at the power distribution level, the realistic simulation, and validation environments are essential for D-PMU development and deployment. This paper presents a vendor agnostic PMU real-time simulation and hardware-in-the-Loop (PMU-RTS-HIL) testbed, which helps in multiple PMUs validation and studies. The network of real and virtual PMUs was built in a full time-synchronized environment for PMU applications' validation. The proposed testbed also includes an emulated communication network (CNS) layer to replicate bandwidth, packet loss and collisions conditions inherent to the PMUs data streams' issues. Experimental results demonstrate the flexibility and scalability of the developed PMU-RTS-HIL testbed by producing large amounts of measurements under typical normal and abnormal distribution grid operation conditions.

Keywords: real-time simulation; hardware-in-the-Loop; synchrophasors; micro-synchrophasors; distribution phasor measurement units; distribution grid; time synchronization

---

1. Introduction

The emergence of new cyber and physical technologies in smart grids including distributed energy resources, transportation electrification, and modern communication networks in connected environments, add more complexity to power distribution networks. Using high resolution and accurate measurement devices such as Distribution Phasor Measurement Units (D-PMU) and the recently introduced Micro-Synchrophasors (Micro-PMU) [1] expand the situational awareness toward distribution levels and the grid edges. Moreover, an effort for developing an open source low-cost PMU device is presented in [2,3]. The actual data from D-PMUs enable visualization and observation of phenomena, which were not observable with past technologies. This leads to novel applications in power distribution networks [4]. Recent works show the advantages of using D-PMU data for distribution network topology detection [5–7], distribution state estimation [8,9], phase label identification [10], fault detection [11,12], and network modeling [13].

Therefore, the emerging need for high resolution and time-synchronized measurement data brings more attention to the need for realistic simulation environments to designing, testing, and validating D-PMUs. As opposed to classic system studies, the implementation of hardware-in-the-loop (HIL) testbeds enables proof of concept and experimental validation of different hardware and software solutions [14]. These HIL configurations combine digital real-time simulators for calculation while interacting with the actual physical devices [15,16].

Few HIL testbeds have been developed for the purpose of testing synchrophasor devices being mostly at the transmission level. For example, in [17,18], a real-time hardware-in-the-loop setup is proposed for compliance testing, where reference signal is generated by a Simulink model and amplified to be sensed via potential transformer (PT) and current transformer (CT) for the device-under-test. This setup lacks time synchronization between signal source and the PMUs, leaving the overall procedure for static compliance testing unsynchronized. In the case of dynamic testing, PMUs are synchronized via Inter-Range Instrumentation Group-Time Code Format B (IRIG-B) signals from a grandmaster clock as a one time-synchronization source. An additional calibrator device is producing the reference phasors with a GPS clock accuracy of ± 100 ns. Typical dynamic testing principles to obtain frequency error (FE), rate of change of frequency error (RFE), total vector error (TVE), as well as response related rates (e.g., delay, overshoot) are positive/negative frequency ramp and magnitude steps or unbalanced magnitude steps. The National SCADA testbed (NSTB) [19] focuses on potential cyber attacks to the communication infrastructure at the transmission level. The testbed has an HIL configuration with RTDS for simulating power systems along with established industry communication structures. The National Renewable Energy Laboratory (NREL) has developed a testbed that emulates the communication and power distribution networks able to interact with field equipment [20]. In [21], a wide-area-monitoring-system cyber-physical testbed was developed that has an HIL-based simulation with different communication protocols and PMU from different vendors. Tests were developed in a 9-bus transmission simulated system. A SCADA software/hardware testbed with RTDS and Opal-RT used as real-time simulators were proposed in [22]. The testbed has three real PMUs as part of their HIL configuration. It also has a variety of substations' communication protocols and a number of PDCs for the measurements streams. The testbed in [23] includes substation communication layers for distribution and transmission networks. Work in [24] presents an HIL testbed using RTDS with real PMUs from different vendors with real-time streams to virtual PDCs. Another initiative is presented in [25], with a GPS signals and PMU streams that are fully simulated capable of streaming with different established communication protocols. A cyber-physical system was introduced in [26] where the communication layer and the power systems were virtually simulated with OPNET and RT-LAB (Opal-RT), respectively, while the cyber-physical structure was simulated with MATLAB. In [13], a PMU framework with real-field data was used to validate the synthesized network modeling proposed in [27]. Comparisons between a simulated model in Opal-RT and real-field measurements were presented.

From the discussion above, it can be observed that related works are based on an HIL configuration to emulate most of the real-field scenarios inherent to power systems and in some cases with the communication infrastructure that collects and distributes data and/or control signals to the different intelligent electronic devices (IEDs) in the network. However, to the knowledge of the authors, none of the available works have been specifically focused on time synchronization issues and the integration of actual and virtual PMUs for power distribution level while considering the impact of latencies in the communication infrastructure and multi-vendors GPS, PMU hardware and firmware interoperability issues. The deployed PMUs in the field and their data sets show many issues regarding data quality even under normal daily operation (e.g., loss of GPS synchronization, GPS antenna malfunction, firmware bugs, etc.) [4]. Moreover, different vendors present their PMU solutions with various firmware, time reference, and phasor computation algorithms with confidential setting for firmware. Therefore, the proposed HIL configurations in this paper with actual and virtual PMU streams helps with exploring the impact of data accuracy and quality of D-PMU measurement data. Additionally, the effect of data traffic and time latencies in PMU streams becomes important when typical smart grid challenges such as mitigating transient stability issues and performing state estimation have latency requirements of 100 ms to 1 s [28]. Adhikari et al. [21] suggest that the lack of suited HIL testbeds is an impediment for creating industry-level standards for hardware, software, PMU components, and protocols.

The experimental setup presented in this paper also builds realistic validation scenarios using distribution network models with real and virtual PMU data streams. Our testbed (hereafter referred to as *PMU-RTS-HIL*) is a fully time-synchronized network of commercial PMUs different vendors acquiring real-time measurements using Opal-RT real-time simulator and hardware-in-the-loop setup. Our main contributions are listed as follows:

- First, we have built a time-synchronized and scalable environment that includes multiple PMUs from different vendors coupled with multiple virtual PMUs, in a hybrid hardware-in-the-loop-software-in-the-loop (HIL-SIL) configuration. This is a non-trivial task as synchronization is required between the model in the real-time simulator (RTS), the input/output interfacing (FPGA), and the GPS clock signals of virtual and real PMUs. This setup replicates the interoperability between a fleet of PMUs in an effort to compare ideal/reference and real-field cases for resilient and reliable distribution monitoring systems.
- Second, we have developed the setup of a simulated communication layer that resembles the traffic and latencies inherent to our hybrid network of virtual and physical PMUs' data streams. It is crucial to study the impact of time delays on synchrophasor data flows for operational applications by comparing ideal measurement devices (virtual PMUs) and the actual PMUs from different vendors. Moreover, additional latencies caused by different firmware, computing algorithms and time references may be observed.
- Third, the recent emphasis on PMU based applications for power distribution network brings more need for developing an environment to study different aspects of a cyber-physical network of PMU devices. The proposed PMU-RTS-HIL testbed will fill the gap for such developing environments and help researchers to create and test more PMU based algorithms.
- Fourth, the proposed PMU-RTS-HIL makes it possible to compare actual PMU from different vendors and validate their performance.

The rest of this paper is organized as follows: in Section 2, we give an overview of the components used and architecture for the development of the PMU-RTS-HIL testbed. Section 3 provides application results performed with the PMU-RTS-HIL testbed with detailed explanation and discussion of the testbed capabilities. We finalize the paper with our tests' conclusions in Section 4.

2. PMU Real-Time Simulation Hardware-in-the-Loop (PMU-RTS-HIL) Testbed Architecture

This section presents the technical overview of the PMU-RTS-HIL testbed, the components used for the setup and capabilities beyond our preliminary experimental results. In general, the fundamentals of the PMU-RTS-HIL testbed consist of the following specifications:

- Real-time power system model setup: the PMU-RTS-HIL testbed consists of an experimental setup providing realistic scenarios in a distribution test feeder model with a network of time-synchronized actual and virtual PMU data streams using hybrid real-time simulation capabilities.
- Communication network: the PMU-RTS-HIL testbed includes a communication layer resembling wide area monitoring systems (WAMS) such as PMU, and a Phasor Data Concentrator (PDC) to emulate real-field measurements in distribution networks. More importantly, a simulated communication layer infrastructure was developed to measure and analyze latencies and traffic congestions of the PMU streams.
- Streaming Data Analysis and Data Repository: The ultimate goal of the PMU-RTS-HIL testbed is building a data repository for data mining and analysis from the real-time platform implemented in HIL and SIL. This stage includes the database setup for the storage of the events monitored.

2.1. PMU-RTS-HIL Setup Overview

Figure 1 shows the configuration of the evaluation framework for the experimental setup of real-time PMU data streaming under fault conditions. A real-time simulated distribution grid (e.g., IEEE test feeders [29]) was modeled in the multi-core Opal-RT real-time simulator provided by the SmartEST facilities. The RTS target is connected to three PMUs from different industry established vendors. Additionally, there are a total number of six virtual PMUs (vPMUs) using a PMU model provided by Opal-RT company that complies with the C37.118 communication protocol.

The network of virtual and actual PMUs operates under normal conditions prior to possible occurrence of different fault types (balanced and unbalanced), changes in frequencies, normal and abnormal switching, etc. To obtain random scenarios for analysis, detection and classification testing purposes. Signals coming from the real-time environment can be obtained from the FPGA OP5142 console on the target. Virtual and real PMUs stream under a full time-synchronized environment for measurement comparison purposes. The communication setup complies with the IEC 61850 and the phasor magnitude, and angle measurements are then streamed under the IEEE standard C37.118 protocol. In order to understand the communication dependencies of fault detection and/or other abnormal distribution side events, a communication network layer is simulated resembling the different physical latencies experienced when sending real-field measurement data. The PMU-RTS-HIL testbed used the network layer under the CORE environment provided by SmartEST. An open-source phasor data concentrator (i.e., OpenPDC) is used to retrieve the synchrophasor readings and store them in the database with support for free alternatives such as PostgreSQL. Figure 1b shows an actual picture of some components of the testbed: the RTS target and its FPGA I/O console, the graphical user interface for the CNS, and an actual PMU used for the experiments.

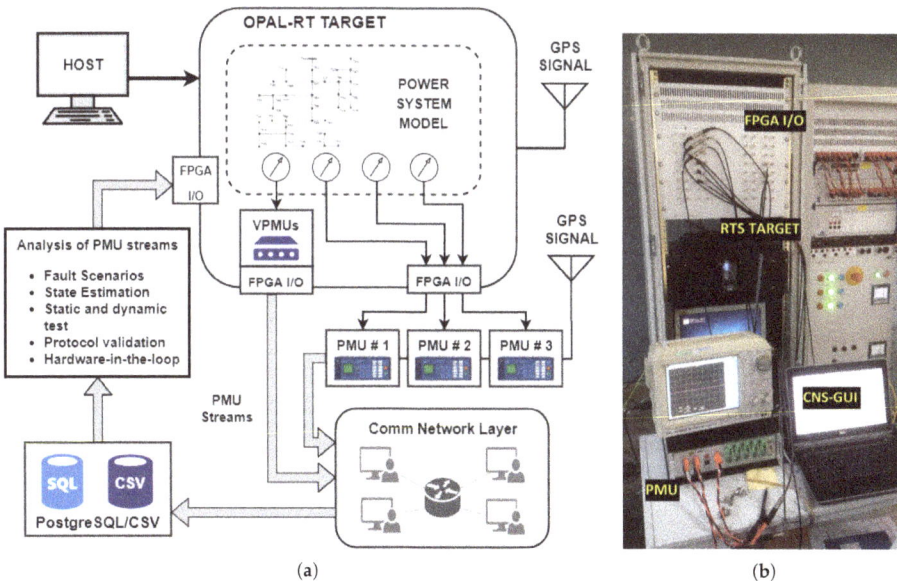

Figure 1. PMU - Real Time Simulation - Hardware-in-the-Loop Testbed (PMU-RTS-HIL): (**a**) schematic connection diagram; (**b**) physical setup.

2.2. Reference Clock and Synchronization

In order to develop a HIL-SIL testbed environment, GPS and computational environment clock synchronization becomes key for the integration of real and virtual PMUs. A first attempt to have synchronized environment was using a one PPS-based signal from a GPS antenna via SMA connector, signaling using NMEA data stream to the serial PC port. However, the RTS drivers did not support the additional NMEA stream tagging the UTC timestamp for the card. Therefore, this setup was able to provide time synchronization within nanoseconds (ns), but without knowing any UTC time reference. Given this limitation, this setup was discarded.

A Precision Time Protocol (PTP)—defined in the IEEE 1588 standard—is a network-based standard that provides nanoseconds of synchronization making it a perfect fit for PMU synchronization applications. Therefore, one of the most used approaches for different time clocks synchronization is to utilize a GPS-locked PTP Master for generating a clock standardized signals while a network interface card synchronizes the different local hardware clocks [30]. In the PMU-RTS-HIL testbed, a RSG2488 Ruggedcom [31] functions as an IEEE 1588 master clock. In order to synchronize the Opal-RT real-time simulator (RTS) and the virtual PMUs with real PMUs, it is necessary to use a high precision oscillator like the Oregano syn1588® PCIe NIC with OCXO Oszillator PCI-express card. This enables the real-time target to use the IEEE 1588 based time synchronization of the given time clock (e.g., GPS) with the needed precision. This configuration makes the system more accurate than utilizing regular Network Time Protocol (NTP). Since the real-time simulation also uses FPGA output for the real PMUs to provide voltages and currents, the FPGA hardware is connected via an adapter card to the oscillator. Finally, when the real-time simulation is configured to use the FPGA clock, it enables the simulation (virtual PMU-C37.118 slave) and the analog output to be synchronized to the given external time source. As shown in Figure 2, the GPS antenna is directly wired to the Siemens Ruggedcom RSG2488, generating the PTP signal. The Oregano card syn1588® PCIe NIC was installed in the RTS target in order to provide the correct timestamps to the internal clock adapter. The clock signal is fed to the internal clock adapter and then transmitted through the Real-Time System Integrator, which is used to share and exchange timing and control signal between the devices and the simulation. The same clock signal is used by the FPGA OP5142 to produce a full-synced environment for virtual and real PMUs to be tested. As an example, Figure 3 depicts the phasors measurement before the Oregano card time-synchronization was integrated in the testbed. It can be observed that they are not in synch if the RTS system and the virtual PMU are not locked to the same clock.

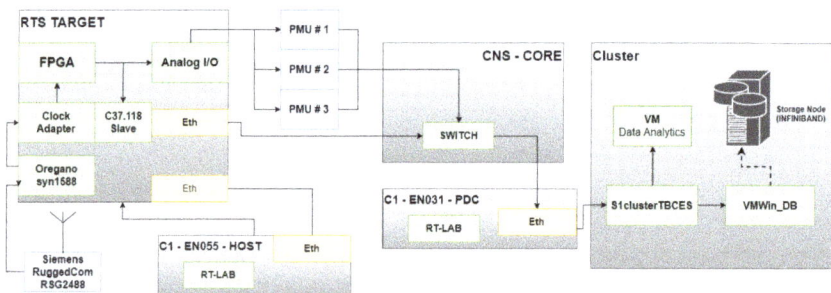

Figure 2. Synchronization and communication network setup for synchronized simulation and analog output to GPS clock.

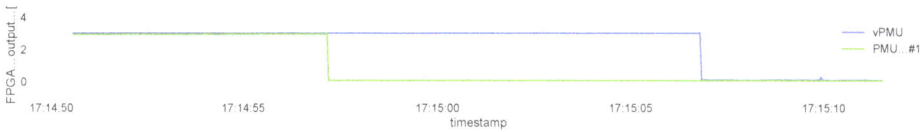

Figure 3. Non-synchronized step response of virtual and real PMUs.

2.3. Communication Network Model

The PMU-RTS-HIL testbed is capable of streaming real-time data through a communication infrastructure as shown in Figure 2. The RTS target and host take charge of the real-time response computation of the power system model and transmit measurements via the FPGA I/O interface to the physical PMU devices while an Opal-RT PMU virtual block assigns an IP address to each one of the virtual PMUs in the power system model. Each stream complies with the C37.118 communication protocol and then goes through the communication network simulation (CNS). To make the behavior of the communication layer more realistic, the communication network emulator CORE (Common Open Research Emulator) [32] is used to model the network topology shown in Figure 4. The CORE is a powerful and feature rich emulator that was first developed by the Boeing Research and Technology and now is being maintained and further developed by the US Naval Research Laboratory. The CORE runs in real time and further provides the capability to connect the emulated network with a physical network. Under the hood, CORE exploits the virtualization capabilities available in most Linux-based operating systems and each of the components in a network model being emulated with the CORE is rendered as a Linux container (LXC/LXD).

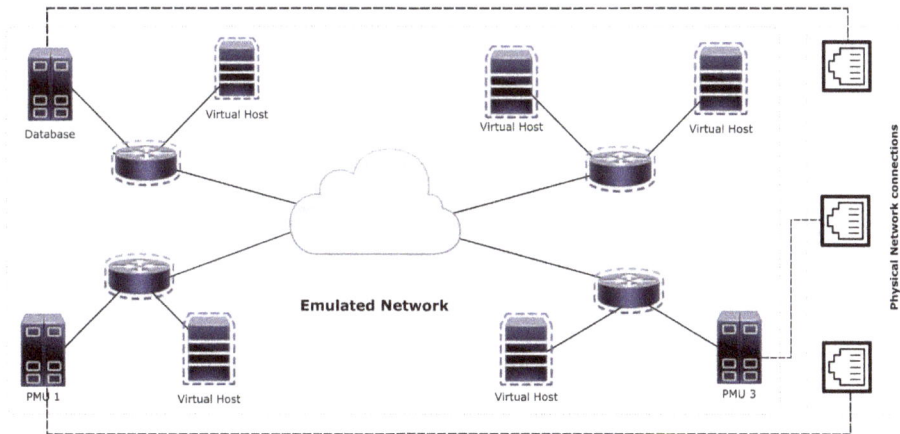

Figure 4. The communication network model consisting of different subnets with virtual (dotted outlined) and real hosts. These subnets are connected to each other through routers. The cloud here is used to represent a wide area network. The Ethernet port schematic shows the nodes that are connected to the physical nodes.

The model (see Figure 4) consists of different subnets with virtual (represented with dotted outline) and real nodes. These subnets are connected to each other through (virtual) routers. The cloud represents a wide area network. For this setup, the model was developed with full IPv4 protocol stack. Furthermore, the three real nodes (labeled PMU 1, PMU 3 and Database) represent the PMU #1 and #3 and the OpnePDC database host, and are physically connected to them through the USB Ethernet adapters on the host machine. All traffic including the C37.118 streams from these physical nodes passes through the emulated network and thus the communication parameters set for the emulated network affect the communication behavior in the physical network. For PMU-RTS-HIL, we have used CORE to evaluate different network scenarios. Moreover, the CNS setup was implemented to emulate the real distribution network latencies. Bandwidth, packet loss and collisions may be applied to the PMU data streams in order to evaluate the impact on delay and availability of the data. Delay time are measured as time difference between creation (measurement timestamps) and creation of the tuple in the database. This feature was developed in order to test various effects such as the output adapter batch processing parameter, which has a direct influence on the round-trip time of the data streams. All data is gathered in a Phasor Data Concentrator (PDC), which arranges it for their storage in the database. Finally, a data repository environment was built in a virtual machine in a Cluster with a database managed through PostgreSQL along with a local CSV historian. This virtual machine also serves as a working station for data analytics.

2.4. Streaming Data Repository and Data Analysis

The streaming data repository and analysis are shown in Figure 5. The AIT Energy Cluster provided the necessary processing capabilities to store and analyze PMU data streams from OpenPDC or direct measurements. The scalable network file system is based on GlusterFS, a large distributed storage solution for data analytics and other bandwidth intensive tasks. Interconnection is provided via fast high bandwidth networks, based on Infiniband technology.

The OpenPDC software was used for concentrating and streaming phasor data taking input streams from PMUs with various settings and protocol standards. It was also used to convert and stream PMU data to various connectors, namely PostgreSQL, local historian and CSV file. The graphical interface and visualization supports the setup and verification of the experiments performed. The cluster provides also a commercial distributed, analytical database for real-time analysis capabilities.

In order to analyze the data by performing statistical analysis and visualization, the user group used mainly two different open source software programs: Python and R. These tools provide live code, equations, visualization and comments on the codes used for the analysis. It is perfect for understanding and visualizing different programming languages in a fast and legible manner.

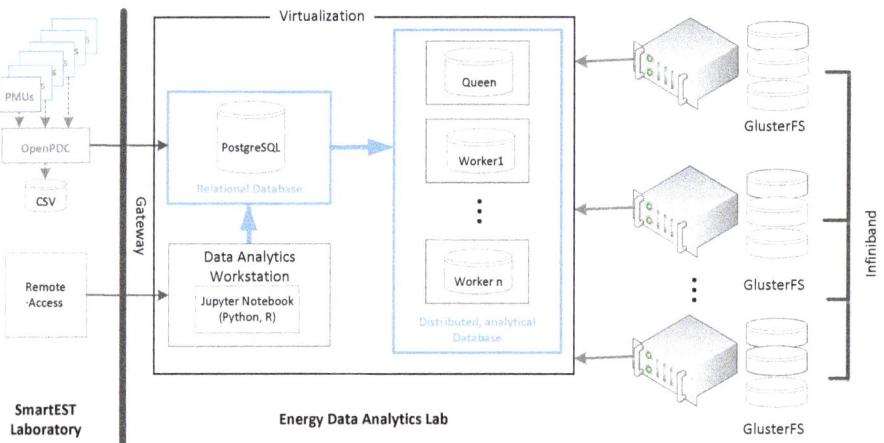

Figure 5. Data processing in the AIT Energy Data Analytics Cluster, with a conventional relational and the option for distributed, analytical database for real-time processing capabilities.

2.5. Scalability and Flexibility

In this section, we discuss the PMU-RTS-HIL has shown flexibility and scalability to model different power systems models and power quality events while supporting real PMU devices from diverse manufacturers.

Scalability of the testbed depends on the specific module as follows:

- Real time simulator: Relationship between the number of nodes vs. the number of cores is basically nonlinear as it depends on a lot of factors (e.g., simulation time step), especially with the usage of SSN (State Space Node) solver. Since this part of the testbed is a commercially available RTS, there are various options to scale up the necessary computational requirements for larger networks.
- Real PMUs: The connection to the RTS is realized via analog output channels of the system's FPGA. In the case of 2×16 channels, a total of five real PMUs (six phasors per PMU) can be fully connected (voltage and current) or 10 PMUs if only the three phase voltages are connected.
- Virtual PMUs: There needs to be a distinction between the processing power to simulate the C37.118 slaves in real time and the bandwidth needed to communicate the streams. In the first case, it can be roughly assumed that 10 vPMUs can be handled per CPU core. In the latter case, again, a rough estimation for a report rate of 50, a need for 100 kbps, 200 kbps respective for a reporting rate of 100 per seconds can be assumed. This would theoretically lead to five PMU streams for a 10 Mbit network adapter or 50 PMUs for a 100 Mbit adapter [28]. If latency is also considered with respect to requirements of application requirements, it would be advisable to distribute streams among available adapters and limit them to a maximum of 10 per network card.
- Communication network simulation: CORE is practically scalable in the sense that a simulated network can be partitioned and distributed among multiple nodes, splitting and connecting them via network links. CORE is able to handle the emulation of several 100k packets per host.
- PMU Data Concentrator and database acess: OpenPDC is able to handle a reasonable number of phasors and can also be run on parallel hosts, thus scale is no problem. Handling the streaming to the database, it is a best practice to setup multiple output adapters in parallel, since each would have a separate database connector process. This makes it possible to parallelize database access as well.

Flexibility is possible by modeling different communication and electrical network scenarios. Besides the various ways of supporting tests as defined in the standard, the versatility is in the combination of different communication and electrical network scenarios and their impact on various applications. In [28], the latency and data requirements for smart grid applications give an idea of how the interdependency between communication and application influences the correct operation and how this testbed supports their validation. Further examples for flexible usage of the testbed are described in detail in the following Section 3.

3. Application Examples for the PMU-RTS-HIL Testbed Validation

In this section, we present some of the experimental setups performed to demonstrate the PMU-RTS-HIL testbed capabilities. We begin by presenting the synchronized phasor measurements. Then, we have included the analysis of use cases such as abnormal events (electrical faults), PMU streams' latencies under the CNS, and the Rate of Change of Frequency (ROCOF). The use cases help in validating the PMU-RTS-HIL testbed by examining the D-PMUs performance in different applications. Table 1 summarizes the different experiments performed using the PMU-RTS-HIL testbed, which are detailed in the following sections.

Table 1. PMU-RTS-HIL use cases' capabilities.

Use Case	Description
Fault events simulation	Single-line-to-ground Line-to-line Three-line-to-ground
Communication network simulation	Data streams under different bandwidth packet loss and collisions
Static and dynamic tests Development support Machine Learning Application	Changes in magnitude, ROCOF Synchronization, calibration, protocol testing Fault detection in distribution networks

3.1. Network of Synchronized PMU Measurements

The main goal of the PMU-RTS-HIL testbed is providing a fully GPS clock synchronized network of PMUs for measurement analysis. Figure 6 shows the different measurements obtained from the same phase by a vPMU and the three different physical PMUs used. The PTP signal is utilized to provide the same clock base between the physical PMUs connected to FPGA I/O console (HIL configuration) and the internal RTS clock streaming data coming from the vPMU.

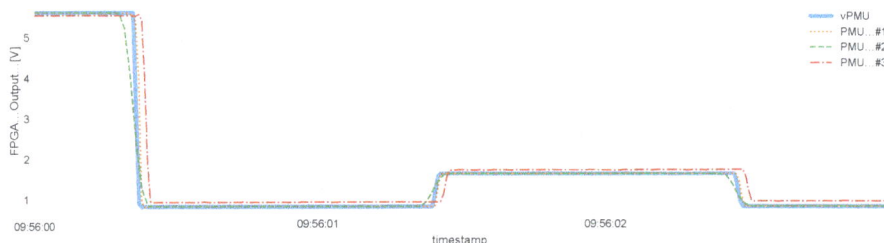

Figure 6. Time synchronized magnitudes of one phase from one virtual and three real PMUs during a fault connected to the same node. The vPMU acts as the reference signal. Note: PMU #2 has a reporting rate of 50 and the other two a reporting rate of 100.

One of the challenges of setting up this configuration was encountered in the form of a 'time gap' of exactly 36 s between the virtual PMU and the real PMU measurements. Although they were perfectly synchronized with nanoseconds accuracy, they were not on the second base. This was due to the PTP using TAI as its time base—which includes leap seconds taking the slowdown of Earth's rotation into account—whereas PMUs use UTC as their time base. Since the beginning of 2017, the UTC-TAI offset is -37 s [33], and older PTP driver stacks have still 36 s as in our setup. Even passing a parameter with an offset of 36 s to the Oregano driver will be ignored when running the oscillator card as a PTP slave. It was then discovered that the Grand Master Clock of the RSG2488 Ruggedcom becomes locked to the external GPS signal ignoring the UTCoffset configuration mentioned previously, and hence passing the TAI timestamp to the RTS. The problem was assessed in a post-processing step where the virtual PMUs were shifted backwards in time by 36 s in order to sync TAI and UTC based timestamps.

3.2. Sequence of Fault Events Simulation with Test Automation Script

The PMU-RTS-HIL testbed is able to simulate different power systems models and a number of fault events powered by an API Python environment. Test automation scripts control the number and type of faults created along with different parameters such as fault impedance, location and duration.

Presented first in [34], the IEEE 37-Nodes Test Feeder (see Figure 7) is part of a testbed composed of several real-life test feeders that provides the essential components and characteristics of a distribution system such as unbalanced load conditions and a considerable number of nodes and laterals. In order to show some of the capabilities of the PMU-RTS-HIL testbed, different experiments were performed using this test feeder model.

Figures 8 and 9 show the IEEE 37-nodes test feeder model and the PMU models in RT-Lab/Simulink environment, respectively. The model consists of the Simulink prototype of the test feeder that uses the state-space nodal solver (SSN). As this solver uses state-space equations, it can be used for delay-free parallelization with higher order discretization [15]. Consequently, the SSN solver splits the model into sections, each assigned to one core in the RTS target.

Figure 8 shows the virtual and the real PMU configuration blocks used in RT-LAB environment. With this setup, the node measurements become outputs in the FPGA OP5142. Each PMU has its own configuration such as IP address and port designation utilizing a C37.118 protocol. Figure 9 shows a look inside the subsystem formed by all virtual PMUs used in the experiments. The PMUs are setup with a 100 samples per second reporting rate and the measurements are stored by PostgreSQL on a virtual machine in Cluster and also in a local CSV historian. This interface is used to obtain the data set with different fault scenarios and results for identifying faults based on the PMU measurements.

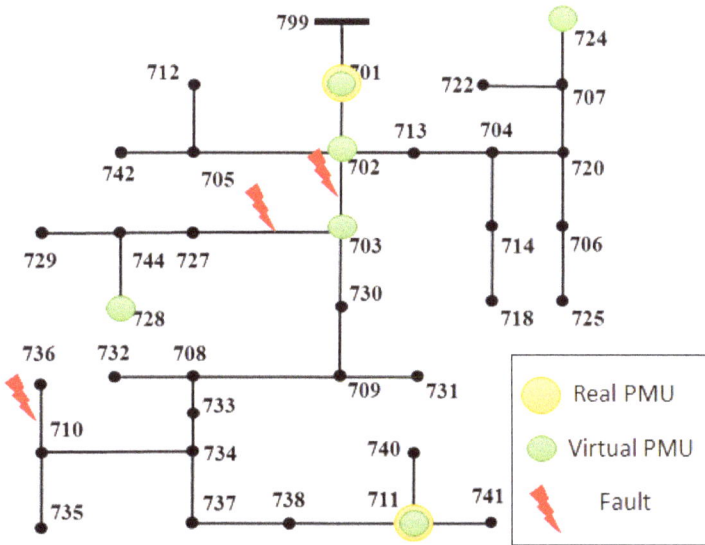

Figure 7. One-line diagram of the IEEE 37-Nodes Test Feeder [34].

IEEE 37 nodes test feeder

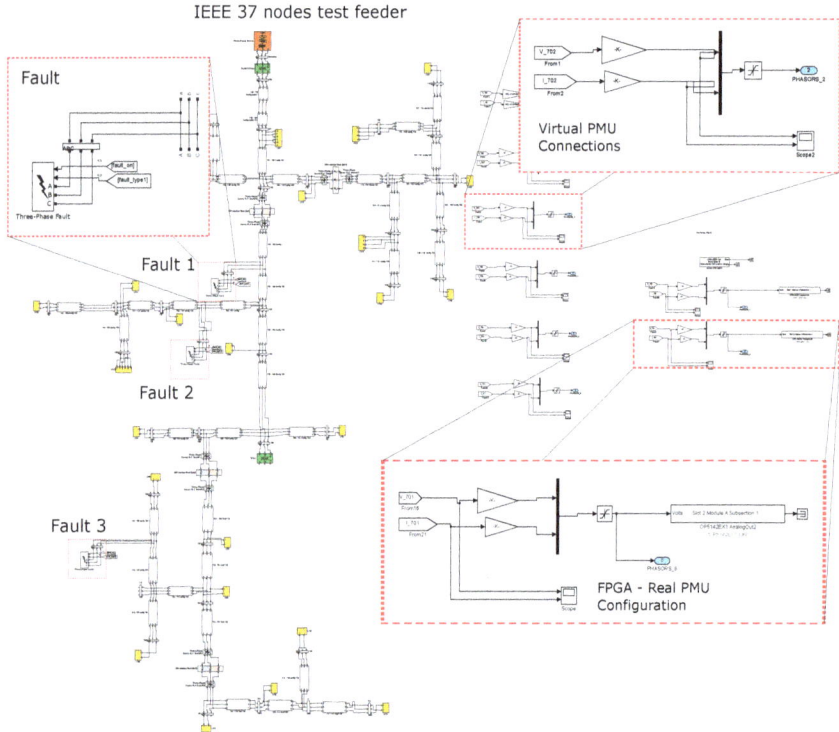

Figure 8. IEEE 37-nodes test feeder Opal-RT Model. The model is split into five subparts with the Opal-RT SSN solver configuration. Connectors for the virtual PMUs and the FPGA analog outputs are on the right side.

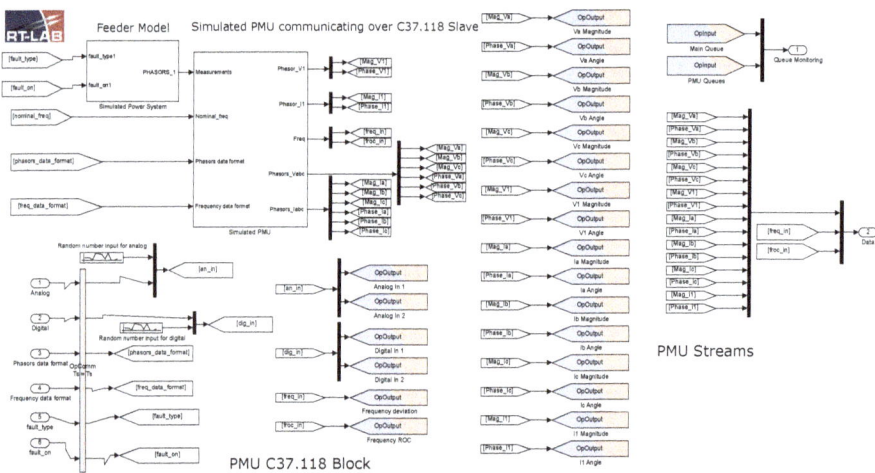

Figure 9. RT-lab/simulink model for the PMU C37.118 slave.

The PMU-RTS-HIL testbed uses the 37-Nodes test feeder model for simulating fault sequences of different types, locations and impedances to generate a dataset, which is used to analyze change frequency during the power disturbances. Additionally, the test feeder was used to create a data repository that may be used for machine learning algorithm validation. The RTS is capable of using an API in Python to perform automated sequences such as controlled switching, creating faults, changing loads, etc. As an example, a combination of different fault scenarios was created with the following parameters:

- Fault location: lines 702–703, 703–727 and 710–736.
- Fault types are line-to-ground, line-to-line and all three-lines-to-ground, A–G, A–B and ABC–G, respectively.
- Fault distance on the line: faults are placed the lines having an impact on the fault impedances.
- Fault impedance to ground: 0 Ω, 5 Ω, 10 Ω, 25 Ω and 50 Ω.

Figure 10 shows the time series of the fault sequences for different fault scenarios for the real and simulated measurements. In can be observed that a fault sequence has been introduced to the system from second 23:05:45 until second 23:05:47, where the system is running under normal operation conditions. Then, a single-line-to-ground takes place in phase A, showing the a voltage drop, which then follows by a line-to-line fault (phases B and C) and a three-phase-to-ground, consecutively. It is worth noting that all measurements are fully synchronized between different PMU vendors and the virtual PMU built in real-time environment.

It is worth noting that the upper time-series in Figure 10 shows the virtual PMU measurements depicting a steeper and cleaner transition between fault events and states. The three other graphs below show the real PMU measurements taken in an HIL setup. The difference in the transitions is due to different signal processing algorithms of the devices and their communication module.

Figure 11 shows two different sets of fault sequences, where it can be seen that vPMUs and PMUs #1 and #3 present a similar behavior following the sequence consistently. However, some problems with the correct configuration of PMU #2 have been encountered. It is not possible to use the high sensitive analog inputs together with the PMU streaming functionality of the device. Therefore, the measurements were taken from a low voltage range ($\hat{V} = \pm 15$ V) provided by the FPGA I/O, which translated into oscillations of ± 0.1 V. This is depicted in the bottom panel of Figure 11. This issue was confirmed by the PMU vendor and is currently working on a firmware that allows the PMU to stream data while using the analog inputs.

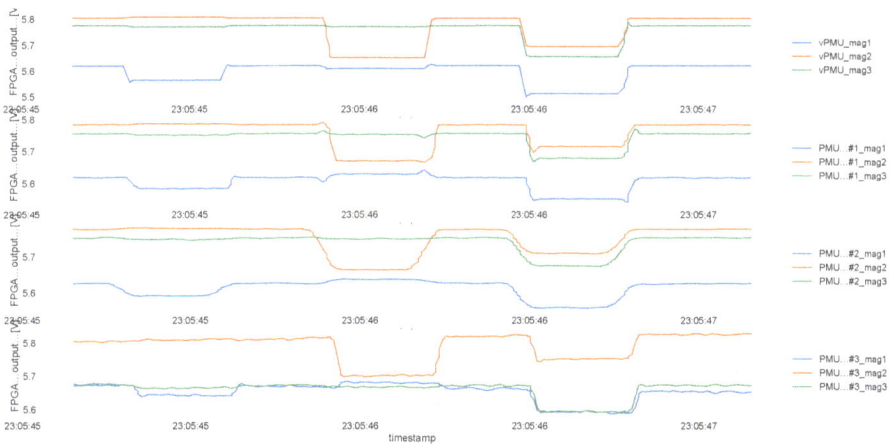

Figure 10. Fault scenario sequence: vPMU measurements; PMU #1 measurements; PMU #2 measurements; PMU #3 measurements.

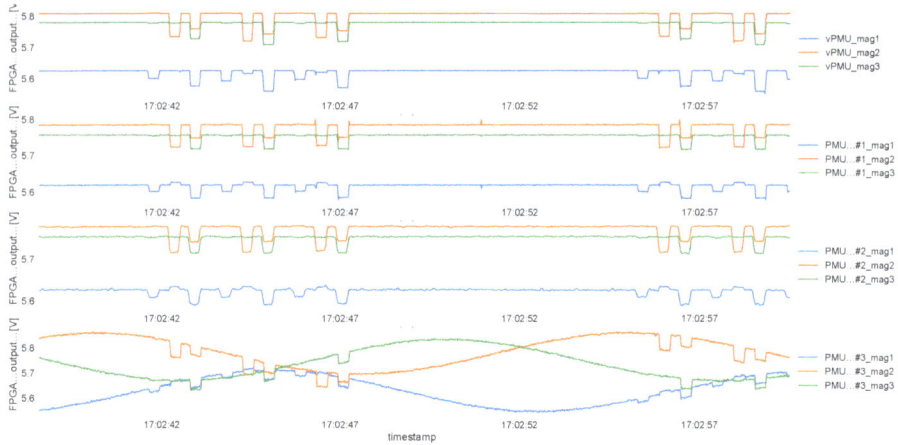

Figure 11. Comparison of measurements of three real PMUs and vPMU under a 50 Ω impedance fault.

Simulated measurement quantities need to be scaled to match the analog output voltage range capabilities of the FPGAs. Phasors are defined as sinusodial waveform: $x(t) = X_m \cos(\omega t + \Phi)$ and represented as Phasor:

$$\mathbf{X} = \frac{\hat{X}}{\sqrt{2}} e^{j\Phi} = \frac{\hat{X}}{\sqrt{2}} (\cos \Phi + j \sin \Phi) = X_r + jX_i, \tag{1}$$

where the magnitude is $\hat{X}/\sqrt{2}$ and X_r and X_i denote the complex values in rectangular form. Defining the magnitude of the simulated quantity as U_{RTS} and the desired range at the analog output as $\hat{U}_{FPGA} = \pm 15$ V, we need a scaling factor of

$$s = \hat{U}_{FPGA}/(\sqrt{2} \cdot U_{RTS}). \tag{2}$$

For the measurement range transformation, for example, the scale for the network model used for the fault use case was calculated as $s = 15$ V$/(\sqrt{2} \cdot 4.8)$ kV has been used.

3.3. Communication Network Simulation Analysis

The PMU-RTS-HIL testbed includes an emulated communication layer that is capable of introducing behaviors such as the packet loss, latencies, and collisions in addition to validating different topologies and protocols with a modeled communication infrastructure. As stated in Section 1, different operation events require different response latencies, making this layer an important part when observing PMU-grid interaction dynamics.

The experiments were carried out by configuring the communication emulation with varying parameter value for e.g., bandwidth and latencies and/or emulating the scenario when background traffic is generated between the hosts. The measurements were performed while changing the state of the power grid through the change in the power Opal-RT/FPGA Output sequence triggered in the console as shown in Table 2. The effects of altering the communication parameters are perhaps more visible in terms of communication delay. To evaluate this hypothesis, the potential time delay and availability of measurements to upstream processing are measured and recorded by calculating the difference between the timestamps of the measurement and the timestamps of the respective database tuple (when the record is written in the database). These recorded results are then used for further analysis. A subset of these results can be seen in Figure 12.

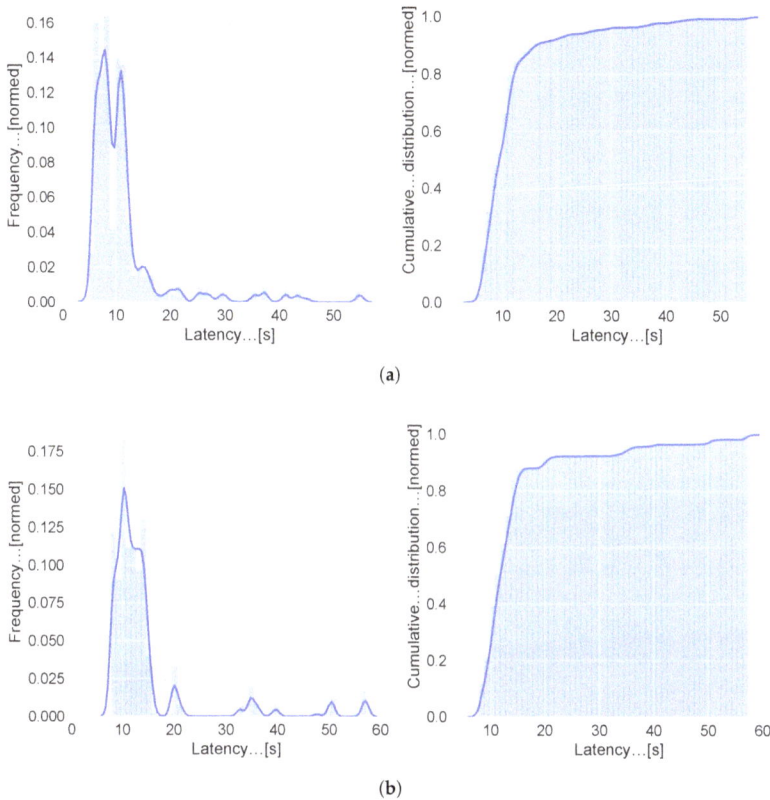

Figure 12. Example of analysis of measurement delays: Histogram and fitted distribution probability of delay between measurement creation and database storage (**left**) and cumulative distribution of measurement delays (**right**) for (**a**) PMU #1 and (**b**) PMU #3.

Table 2. Opal-RT/FPGA power output sequence for CNS.

Sequence Step	1	2	3	4	5	6	7	8	9	10
U (V)	230	230	230	230	230	110	110	110	110	110
P (W)	2300	4600	4600	4600	2300	2300	4600	4600	4600	2300
Q (VAR)	0	0	2300	4600	0	0	0	2300	4600	0

For this experiment, the communication model was emulated with full IPv4 protocol stack. There were no additional or artificial delays in the model. All the communication links were set as per IEEE 802.3-2008 Gigabit Ethernet (full-duplex having a bandwidth of 1 Gigabit). The virtual host remains idle for most of the emulation time and there was no noticeable background traffic. The physical network that was used as the part of the emulated network consisted of two PMUs, a control terminal and a switch. The host machine had multiple USB Ethernet adapters through which respective nodes were connected to the emulated network.

The database was configured to tag a new record with a 'created' timestamp. The time difference to the phasor timestamp enables the analysis of queuing times and gives a first estimate of average processing times and data availability. Effects like the PDC output adapter batch processing parameter (e.g., 1000 data points per batch insert) have direct influence on the round-trip time and need to be configured.

As an example, Figure 12a,b show the histogram and fitted probability distribution as well as the cumulative distribution of the time delay when PMUs #1 and #3 are streamed via the CNS. It could be seen that both PMUs have similar characteristics of the round-trip time. The important insight here is the random occurrences of higher delays, which have direct impact on the processing distribution system application. To evaluate the impact of the communication network on the delay, the experiment has been repeated with different communication channel properties (e.g., packet collision and loss).

3.4. Rate of Change of Frequency (ROCOF)

In addition to the proposed experimental setups, the PMU-RTS-HIL testbed is capable of performing static and dynamic compliance testing of PMUs according to "IEEE C37.242-2013 IEEE Guide for Synchronization, Calibration, Testing, and Installation of Phasor Measurement Units (PMUs) for Power System Protection and Control". The virtual PMU can act as a reference, since it provides exact values of the real-time simulated quantity.

One of the standardized tests is the "Rate of change of frequency" (ROCOF), which certifies that the PMU measurements comply with specific rates for reporting changes in the power grid's frequency. The PMU-RTS-HIL testbed is capable of performing the ROCOF along with communication network simulation, which plays an important role if time delays and latencies are present. When it comes to real setup, additional communication delays are introduced according to the communication network properties, e.g., a frequency control system (e.g, primary control reserve), which takes in the frequency changes from the PMU network. The total delay and quality of service of the network communication influences the dynamic and stability of the control loop.

A typical setup to test signal reporting characteristics (e.g., latency) of the "Rate of Change of Frequency" (ROCOF) measurement propagation as part of a frequency control system is depicted in Figure 13. It consists of a frequency reference signal generator that is connected to a power amplifier stage (e.g., Spitzenberger and Spieß PAS 1000 in order to have real-field network voltage levels, PMUs and the communication network simulation that models the network for data streams to the analysis, storage and processing platform. Additionally, a trigger is created to tag the exact start time of the frequency change for evaluation of the latency. As shown in Figure 13, the frequency signal is propagated by the linear operating amplifier, with neglectable signal latency run-times, and then sensed by the PMU devices. A phasor data concentrator (e.g., OpenPMU) is configured to store the frequency responses in a database (e.g., PostgreSQL).

For the test sequence, the frequency signal is programmed to start with 50 Hz to change frequency by 2 Hz increase for a one second wait for 10 s and reduce by 2 Hz within 1 s. Figure 14a shows the ramp stage of the frequency (in Hz) during the ROCOF test while Figure 14b shows the rate of change (df/dt, Hz/s). In both figures, various signal processing problems are shown (e.g., spikes, oscillation, magnitude), which has been discovered during the test as part of the development support and which have been reported back and fixed by the vendors accordingly.

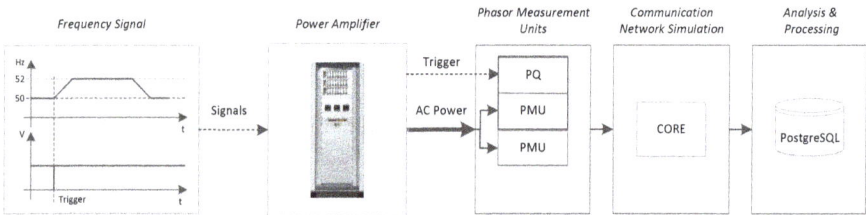

Figure 13. Setup for testing with mains voltage levels by connecting PMUs to AC Power amplifier for e.g., ROCOF test.

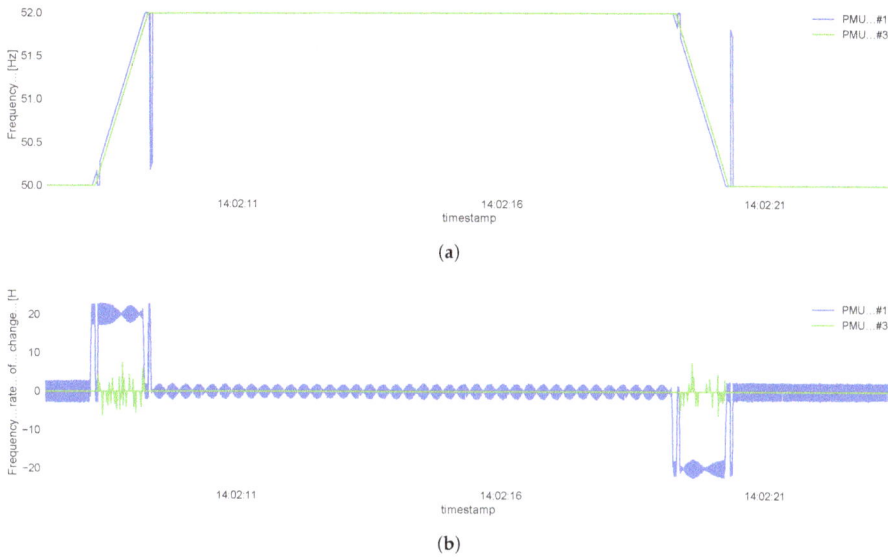

Figure 14. ROCOF Test: (**a**) resulting frequency measurements of two PMUs following a frequency ramp; (**b**) resulting frequency delta/deviations of two PMUs following a frequency ramp. Note: This example shows signal processing errors that have been reported and fixed as part of the validation test.

3.5. Machine Learning Application: PMU-Based Fault Detection

In addition to the previously mentioned applications, the testbed can be utilized to generate large datasets to implement novel PMU-based machine learning algorithms.

In order to validate the testbed capabilities, we present an experiment for comparing the PMU-RTS-HIL testbed with the work developed by the authors in [11]. The authors developed a hierarchical clustering novel algorithm based on simulated PMU measurements. The algorithm is based on a shape-preserving algorithm that obtains similarity distances under the Fisher–Rao metric that are used to detect and identify electrical faults in a distribution system. The details of the algorithm go beyond the scope of this paper and the authors would like to refer the reader to [11]. Therefore, we would present the differences in technical and physical setups of the experiment to validate our testbed's time-synchronized and HIL capabilities.

In [11], a non-Opal-RT state-of-the-art real-time simulator was used to model the IEEE 13-nodes test feeder. We have developed a setup that resembles the mentioned experiment for the IEEE 37-nodes test feeder case. For comparison purposes, Table 3 shows the different setup utilized for the fault detection experiment. Our proposed testbed (Case 2) includes 3 well-established industrial physical PMUs with GPS time synchronization. Additionally, it includes the communication layer simulated in CORE with IEEE C37.118 standard compliance. In contrast, case 1 does not have real PMUs connected and no communication layer. Furthermore, in case 1, fault detection was performed with voltage and current magnitudes only whereas the PMU-RTS-HIL provides the phasor measurement with both magnitude and angle values.

Table 3. IEEE 37-nodes test feeder modeling for fault detection under different platforms.

Feature	Case 1: Non-Opal-RT Simulator	Case 2: PMU-RTS-HIL
Real PMUs	-	3
Virtual PMUs	6	6
Synchronization	RTS system clock	GPS / PTP
Measurements	Voltage and Current magnitudes	Voltage and Current phasors
Communication simulation	-	C37.118 compliance in CORE

PMU measurements in both setups were used to feed the machine learning algorithm and classify their fault type. Approximately 300 fault events were created with three different fault types: single-line-to-ground, line-to-line and three-line-to-ground. In case 1, only voltage and current magnitudes were used when using the fault classification algorithm. Measurements with the proposed PMU-RTS-HIL consisted of both magnitudes and angles of the voltage and current signals. Table 4 shows the total prediction error for both experiment setups. The total prediction error (*TPE*) is defined as follows:

$$TPE = \frac{(ME + FAE)}{Total\ number\ of\ events} * 100\%, \tag{3}$$

where *ME* is the misdetection error and *FAE* is the false alarm error. It can be observed that the proposed PMU-RTS-HIL testbed gives a better overall classification of the fault type in the IEEE 37-nodes test feeder. Therefore, the PMU-RTS-HIL testbed can be utilized for machine learning applications while resembling real-field conditions that include communication and GPS synchronization.

Table 4. Fault detection total prediction error.

Case	Location 1 TPE (%)	Location 2 TPE (%)	Location 3 TPE (%)
Case 1	24.91	23.56	13.8
Case 2: PMU-RTS-HIL	14.47	26.93	10.77

4. Conclusions

In this paper, we presented a testbed that provides realistic scenarios of a distribution test feeder model with PMU data streams' simulations. The primary objective was providing a testbed for the integration of multiple PMUs from industry established vendors. Moreover, the PMU-RTS-HIL has the time-synchronization capabilities of supporting virtual PMUs in an Opal-RT environment with actual physical PMUs under a unique clock reference. Utilizing different PMU devices from multiple manufacturers inherently introduces dealing with different sampling rates, configurations, calculation algorithms, and different time synchronization references. In the first experiments, different synchronization issues between the virtual and real PMUs were mitigated. For power systems applications, time synchronization is crucial and developing a testbed of real field resemblance should include precise time stamps.

Experimental results show that the PMU-RTS-HIL is capable of creating an experimental setup providing realistic scenarios in a distribution test feeder model with simulating PMU data streams using an HIL setup. As a result, different sequences of a large number of electrical fault events were created in a data repository for further pattern recognition analysis.

Evaluating response and propagation time of measurements have been intended to be evaluated with this experiment setup. An example application to be utilized with PMU data can be the frequency control (e.g., primary control) of (virtual) rotating masses (e.g., generators, batteries). The experiment shows that the testbed is capable of performing this setup in an accurate manner.

Introducing communication simulation/emulation using CORE has enabled various aspects and additional dimensions of evaluating PMU applications for distribution systems. In the realized setup, we could investigate directly the impact on delays and packet drops. More specifically, scenarios related to cyber security can be analyzed and evaluated in detail. A consequent step would be to evaluate directly the impact on communication on the application layer (e.g., state estimation, fault identification).

The PMU-RTS-HIL has been shown to be flexible in its capabilities and may be scalable to different power systems models and power quality events, while it can also support PMU devices from diverse manufacturers. The proposed testbed has been extensively used for providing development support for three PMU vendors. Functions as well as new protocol implementations can be tested under various simulated conditions. This includes protocol formats, streaming behavior and measurement validation. Moreover, the testbed provided useful insight for the RTS and PMU vendors, and helped to improve their application suite.

Acknowledgments: The authors would like to thank the AIT Austrian Institute of Technology, Vienna for hosting the Transnational Access and providing laboratory infrastructure. Additionally, the authors would like to thank the following companies for their technical assistance and support in equipment: Artemes, Opal-RT, Oregano, Power Standards Lab, and Siemens. This research has been performed using the ERIGrid Research Infrastructure and is part of a project that has received funding from the European Union's Horizon 2020 Research and Innovation Programme under the Grant Agreement No. 654113. The support of the project 'European Research Infrastructure ERIGrid' and its partner AIT is very much appreciated.

Author Contributions: Reza Arghandeh, Jose Cordova and Matthias Stifter conceived and designed the experiments; Matthias Stifter was responsible for testbed synchronization, data handling and processing, Jose Cordova for modeling and real-time simulation and Reza Arghandeh was coordinating and designing the experiment scenarios, together experiments have been performed; Matthias Stifter and Jose Cordova analyzed the data; Jawad Kazmi contributed communication network simulation tools and modeling; Jose Cordova and Matthias Stifter wrote the paper with substantial contributions from all authors.

Conflicts of Interest: The authors declare no conflict of interest.

Abbreviations

The following abbreviations are used in this manuscript:

AIT	Austrian Institute of Technology
API	Application Programming Interfaces
CNS	Communication Network Simulation
CORE	Common Open Research Emulator
CSV	Comma Separated Values
CT	Current Transformer
DER	Distributed Energy Resource
D-PMU	Distribution Phasor Measurement Units
DRTS	Digital Real-Time Simulation
FE	Frequency Error
FPGA	Field Programmable Gate Array
FSU	Florida State University
GPS	Global Positioning System
HIL	Hardware-in-the-loop
HIL-SIL	Hardware-in-the-loop Software-in-the-loop
I/O	Input/Output
IAT	International Atomic Time
IED	Intelligent Electronic Devices
IRIG-B	Inter-Range Instrumentation Group - Time Code Format B
LXC/LXD	Linux Container
NIC	Network Interface Controller
NMEA	National Marine Electronics Association
NREL	National Renewable Energy Laboratory
NSTB	National SCADA Testbed
NTP	Network Time Protocol

Opal-RT	Opal-RT Technologies, Montreal, Canada
OPNET	OPNET Technologies, Inc, Maryland, USA
PCIe	Peripheral Component Interconnect Express
PDC	Phasor Data Concentrator
PHIL	Power Hardware-in-the-loop
PMU	Phasor Measurement Units
PMU-RTS-HIL	Phasor Measurement Unit Real-Time Simulation Hardware-in-the-loop
PMU-RTS-HIL	Phasor Measurement Units Real-Time-Simulation Hardware-in-the-loop
PPS	Pulse Per Second
PTP	Precision Time Protocol
RFE	Rate of Change of Frequency Error
ROCOF	Rate of Change of Frequency
RTDS	Real-time digital simulator
RTS	Real-Time Simulator
SCADA	Supervisory Control and Data Acquisition
SIL	Software-in-the-loop
SMA	Subminiature version A
SSN	State-Space Node
TAI	Temps Atomique International
TVE	Total Vector Error
micro-PMU	Micro Phasor Measurement Units
UTC	Coordinated Universal Time
vPMU	Virtual PMU
VT	Voltage Transformer
WAMS	Wide Area Monitori Ssystems

References

1. Meier, A.V.; Culler, D.; McEachern, A.; Arghandeh, R. Micro-synchrophasors for distribution systems. In Proceedings of the 2014 IEEE PES Innovative Smart Grid Technologies Conference (ISGT), Washington, DC, USA, 19–22 February 2014; pp. 1–5.
2. Laverty, D.M.; Best, R.J.; Brogan, P.; Khatib, I.A.; Vanfretti, L.; Morrow, D.J. The OpenPMU Platform for Open-Source Phasor Measurements. *IEEE Trans. Instrum. Meas.* **2013**, *62*, 701–709.
3. Laverty, D.M.; Vanfretti, L.; Khatib, I.A.; Applegreen, V.K.; Best, R.J.; Morrow, D.J. The OpenPMU Project: Challenges and perspectives. In Proceedings of the 2013 IEEE Power Energy Society General Meeting, Vancouver, BC, Canada, 21–25 July 2013; pp. 1–5.
4. Arghandeh, R.; Brady, K.; Brown, M.; Cotter, G.; Deka, D.; Hooshyar, H.; Jamei, M.; Kirkham, H.; McEachern, A.; Mehrmanesh, L.; et al. *Synchrophasor Monitoring for Distribution Systems: Technical Foundations and Applications*; North American SynchroPhasor Initiative: Albuquerque, NM, USA, 2018; doi:10.13140/RG.2.2.16825.26727.
5. Cavraro, G.; Arghandeh, R. Power Distribution Network Topology Detection with Time-Series Signature Verification Method. *IEEE Trans. Power Syst.* **2017**, doi:10.1109/TPWRS.2017.2779129.
6. Cavraro, G.; Arghandeh, R.; Poolla, K.; von Meier, A. Data-driven approach for distribution network topology detection. In Proceedings of the 2015 IEEE Power Energy Society General Meeting, Denver, CO, USA, 26–30 July 2015; pp. 1–5.
7. Arghandeh, R.; Gahr, M.; von Meier, A.; Cavraro, G.; Ruh, M.; Andersson, G. Topology detection in microgrids with micro-synchrophasors. In Proceedings of the 2015 IEEE Power Energy Society General Meeting, Denver, CO, USA, 26–30 July 2015; pp. 1–5.
8. Cordova, J.; Faruque, M.O. Fault location identification in smart distribution networks with Distributed Generation. In Proceedings of the 2015 North American Power Symposium (NAPS), Charlotte, NC, USA, 4–6 October 2015; pp. 1–7.
9. Li, W.; Vanfretti, L.; Chow, J.H. Pseudo-Dynamic Network Modeling for PMU-Based State Estimation of Hybrid AC/DC Grids. *IEEE Access* **2018**, *6*, 4006–4016.

10. Wen, M.H.F.; Arghandeh, R.; Meier, A.V.; Poolla, K.; Li, V.O.K. Phase identification in distribution networks with micro-synchrophasors. In Proceedings of the 2015 IEEE Power Energy Society General Meeting, Denver, CO, USA, 26–30 July 2015; pp. 1–5.

11. Cordova, J.; Arghandeh, R.; Zhou, Y.; Wesolowski, S.; Wu, W.; Matthias, S. Shape-based data analysis for event classification in power systems. In Proceedings of the 2017 IEEE Manchester PowerTech, Manchester, UK, 18–22 June 2017; pp. 1–6.

12. Zhou, Y.; Arghandeh, R.; Spanos, C.J. Online learning of Contextual Hidden Markov Models for temporal-spatial data analysis. In Proceedings of the 2016 IEEE 55th Conference on Decision and Control (CDC), Las Vegas, NV, USA, 12–14 December 2016; pp. 6335–6341.

13. Mahmood, F.; Vanfretti, L.; Pignati, M.; Hooshyar, H.; Sossan, F.; Paolone, M. Experimental Validation of a Steady State Model Synthesis Method for a Three-Phase Unbalanced Active Distribution Network Feeder. *IEEE Access* **2018**, *6*, 4042–4053.

14. Faruque, M.D.O.; Strasser, T.; Lauss, G.; Jalili-Marandi, V.; Forsyth, P.; Dufour, C.; Dinavahi, V.; Monti, A.; Kotsampopoulos, P.; Martinez, J.A.; et al. Real-Time Simulation Technologies for Power Systems Design, Testing, and Analysis. *IEEE Power Energy Technol. Syst. J.* **2015**, *2*, 63–73.

15. Lauss, G.F.; Faruque, M.O.; Schoder, K.; Dufour, C.; Viehweider, A.; Langston, J. Characteristics and Design of Power Hardware-in-the-Loop Simulations for Electrical Power Systems. *IEEE Trans. Ind. Electron.* **2016**, *63*, 406–417.

16. Ibarra, L.; Rosales, A.; Ponce, P.; Molina, A.; Ayyanar, R. Overview of Real-Time Simulation as a Supporting Effort to Smart-Grid Attainment. *Energies* **2017**, *10*, 817.

17. Almas, M.S.; Kilter, J.; Vanfretti, L. Experiences with Steady-State PMU Compliance Testing Using Standard Relay Testing Equipment. In Proceedings of the 2014 Electric Power Quality and Supply Reliability Conference (PQ), Rakvere, Estonia, 11–13 June 2014; pp. 103–110.

18. Kilter, J.; Palu, I.; Almas, M.S.; Vanfretti, L. Experiences with Dynamic PMU Compliance Testing Using Standard Relay Testing Equipment. In Proceedings of the 2015 IEEE Power Energy Society Innovative Smart Grid Technologies Conference (ISGT), Washington, DC, USA, 18–20 February 2015; pp. 1–5.

19. Idaho National Laboratory (INL). *INL Cyber Security Research—Idaho National Laboratory Research Fact Sheet*; INL: Idaho Falls, ID, USA, 2005.

20. National Renewable Energy Laboratory (NREL). Grid Modernization | Grid Simulation and Power Hardware-in-the-Loop. Available online: https://www.nrel.gov/grid/simulation-phil.html (accessed on 1 February 2018).

21. Adhikari, U.; Morris, T.; Pan, S. WAMS Cyber-Physical Test Bed for Power System, Cybersecurity Study, and Data Mining. *IEEE Trans. Smart Grid* **2017**, *8*, 2744–2753.

22. Ashok, A.; Krishnaswamy, S.; Govindarasu, M. PowerCyber: A remotely accessible testbed for Cyber Physical security of the Smart Grid. In Proceedings of the 2016 IEEE Power Energy Society Innovative Smart Grid Technologies Conference (ISGT), Minneapolis, MN, USA, 6–9 September 2016; pp. 1–5.

23. Sun, C.C.; Hong, J.; Liu, C.C. A co-simulation environment for integrated cyber and power systems. In Proceedings of the 2015 IEEE International Conference on Smart Grid Communications (SmartGridComm), Miami, FL, USA, 2–5 November 2015; pp. 133–138.

24. Soudbakhsh, D.; Chakrabortty, A.; Annaswamy, A.M. A delay-aware cyber-physical architecture for wide-area control of power systems. *Control Eng. Pract.* **2017**, *60*, 171–182.

25. Aghamolki, H.G.; Miao, Z.; Fan, L. A hardware-in-the-loop SCADA testbed. In Proceedings of the 2015 North American Power Symposium (NAPS), Charlotte, NC, USA, 4–6 October 2015; pp. 1–6.

26. Tang, Y.; Tai, W.; Liu, Z.; Li, M.; Wang, Q.; Liang, Y.; Huang, L. A Hardware-in-the-Loop Based Co-Simulation Platform of Cyber-Physical Power Systems for Wide Area Protection Applications. *Appl. Sci.* **2017**, *7*, 1279.

27. Mahmood, F.; Hooshyar, H.; Lavenius, J.; Bidadfar, A.; Lund, P.; Vanfretti, L. Real-Time Reduced Steady-State Model Synthesis of Active Distribution Networks Using PMU Measurements. *IEEE Trans. Power Deliv.* **2017**, *32*, 546–555.

28. Kansal, P.; Bose, A. Bandwidth and Latency Requirements for Smart Transmission Grid Applications. *IEEE Trans. Smart Grid* **2012**, *3*, 1344–1352.

29. Schneider, K.P.; Mather, B.; Pal, B.C.; Ten, C.W.; Shirek, G.; Zhu, H.; Fuller, J.; Pereira, J.L.R.; Ochoa, L.; Araujo, L.; et al. Analytic Considerations and Design Basis for the IEEE Distribution Test Feeders. *IEEE Trans. Power Syst.* **2017**, doi:10.1109/TPWRS.2017.2760011.

30. syn1588® PCIe NIC—Oregano Systems. Available online: http://www.oreganosystems.at/?page_id=71 (accessed on 1 February 2018).
31. RSG2488—Industrial Communication Siemens. Available online: http://w3.siemens.com/mcms/industrial-communication/en/rugged-communication/pages/ruggedcom.aspx (accessed on 1 February 2018).
32. Ahrenholz, J.; Danilov, C.; Henderson, T.R.; Kim, J.H. CORE: A real-time network emulator. In Proceedings of the MILCOM 2008 IEEE Military Communications Conference, San Diego, CA, USA, 16–19 November 2008; pp. 1–7.
33. Bulletin C 52 Paris: IERS. Available online: https://hpiers.obspm.fr/eoppc/bul/bulc/bulletinc.52 (accessed on 31 December 2018).
34. Kersting, W.H. Radial distribution test feeders. In Proceedings of the 2001 IEEE Power Engineering Society Winter Meeting Conference (Cat. No. 01CH37194), Columbus, OH, USA, 28 January–1 Feruary 2001; Volume 2, pp. 908–912.

![energies logo]

energies

MDPI

Article

Initialization and Synchronization of Power Hardware-In-The-Loop Simulations: A Great Britain Network Case Study

Efren Guillo-Sansano *, Mazheruddin H. Syed, Andrew J. Roscoe and Graeme M. Burt

Institute for Energy and Environment, Electronic and Electrical Engineering Department,
University of Strathclyde, Glasgow G1 1XW, UK; mazheruddin.syed@strath.ac.uk (M.H.S.);
andrew.j.roscoe@strath.ac.uk (A.J.R.); graeme.burt@strath.ac.uk (G.M.B.)
* Correspondence: efren.guillo-sansano@strath.ac.uk; Tel.: +44-(0)-141-444-7280

Received: 3 April 2018; Accepted: 25 April 2018; Published: 28 April 2018

Abstract: The hardware under test (HUT) in a power hardware in the loop (PHIL) implementation can have a significant effect on overall system stability. In some cases, the system under investigation will be unstable unless the HUT is already connected and operating. Accordingly, initialization of the real-time simulation can be difficult, and may lead to abnormal parameters of frequency and voltage. Therefore, a method to initialize the simulation appropriately without the HUT is proposed in this contribution. Once the initialization is accomplished a synchronization process is also proposed. The synchronization process depends on the selected method for initialization and therefore both methods need to be compatible. In this contribution, a recommended practice for the initialization of PHIL simulations for synchronous power systems is presented. Experimental validation of the proposed method for a Great Britain network case study demonstrates the effectiveness of the approach.

Keywords: PHIL (power hardware in the loop); simulation initialization; synchronization; time delay; synchronous power system; stability; accuracy

---

1. Introduction

Electrical power systems are under continuous development, accelerated by regulations enforced to mitigate climate change, the need to enhance efficiency and the substantial technology evolution. Power systems are evolving into a more variable and difficult to predict system with a mix of novel and complex components, such as renewable energy sources or power electronics components, and conventional components with well-known behavior. The interaction between such components is an important area of research to achieve a resilient and secure power system.

For the assessment of novel complex components, the interactions between modern and legacy power system components and the validation of novel control algorithms for future power systems, hardware-in-the-loop (HIL) techniques are proving to be a useful approach [1]. Depending upon the validation objectives and infrastructure available, HIL is broadly classified into two categories: (i) controller-HIL (CHIL), if the HUT is a controller or low power component (such as protection devices), and (ii) Power-HIL (PHIL) when the HUT is a high-power component requiring amplification of the simulated signal in order to be coupled together.

Specifically, PHIL is gaining attention internationally due to its good performance for testing power and energy systems at reduced cost and risk [2–5]. Typically, PHIL has been utilized in a range of applications including: (i) where a component (such as PV inverter) is physically available and it is computationally more efficient to utilize the component within a PHIL setup rather than developing a detailed and accurate model of the component, (ii) where novel power components need to be tested before their wide scale deployment and (iii) where the interactions of modern components with the grid need to be captured to understand the implication of its deployment. In all these applications, the HUT represents a relatively small portion of the network compared to the grid emulation i.e., the rest of the system being simulated on the Digital Real-Time Simulator (DRTS) [6–8]. However, this balance in PHIL between hardware and software is insufficient when it comes to validating wide area monitoring, protection and control (WAMPAC)—an area of increasing interest given the recent advancements in phasor measurement units (PMU) [9]. Such validation would increasingly require a rebalancing such that the HUT is composed of a larger portion of the test network.

An example representation of a PHIL implementation is presented in Figure 1. This shows a PHIL implementation comprising a virtually simulated network implemented within a DRTS, a hardware component (the HUT), and the power interface used for interconnecting both subsystems [10]. The HUT connected to the simulation can represent generation or load components, this may consist of many devices interconnected or just a simple significant device. The power interface allows for the interconnection of the two subsystems. The conventional approach of setting up a particular PHIL simulation involves the following steps:

- The power network within the DRTS is initialized, allowing for it to achieve steady state (referred to as initialization in this work).
- Interface signals from the initialized DRTS simulation are reproduced by the power interface.
- The HUT response to the reproduced signals is measured and fed back to the DRTS to complete the loop (referred to as synchronization in this work).

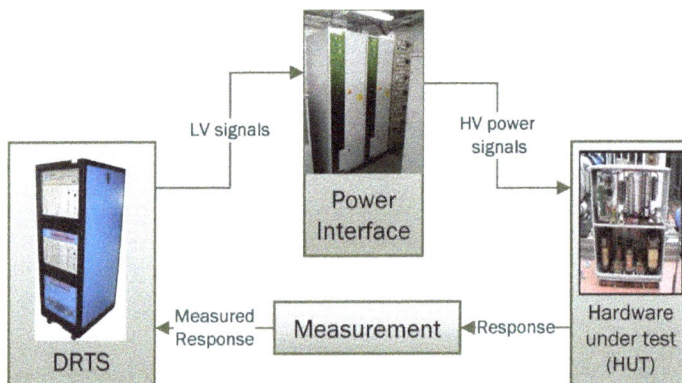

Figure 1. PHIL implementation.

For studies of synchronous power systems, the load and generation conditions along with the power transfer at points of interest are selected from known scenarios. This allows for testing under known stress conditions of the network or scenarios of interest. For example, a previously measured pre-fault condition of the network may be considered, where a novel control algorithm can be tested in order to analyze if the performance of such a controller could have improved the response to the event. Therefore, when a PHIL simulation is initialized and synchronized, it is important to ensure that the conditions at the different buses of the test network are comparable to that of a pure simulation.

In cases were the HUT is relatively small compared to the DRTS simulated power system [6–8], the DRTS simulation can be initialized without the HUT, hence the power network within the DRTS performs as a stiff grid whose voltage and frequency are not dependent upon the HUT to be interconnected. Then, the HUT is typically synchronized with the DRTS simulation by means of a simple switching action, closing the loop between the HUT and DRTS. Operation of the switch always introduces transient, however, in the cases of a stiff simulated grid or a modest HUT, this transient does not pose a significant risk for a stable operating point to be achieved at the start of the study. The processes of initialization and synchronization of PHIL are thus relatively straightforward.

However, for cases where the network is not a stiff grid and the HUT is significant for the grid (either to be able to initialize without it, or to remain stable if it is directly connected) an initialization procedure for the simulated part of the system as well as a reliable synchronization procedure is required.

In this paper, the initialization and synchronization of a PHIL simulation where the HUT represents a larger portion of the test network is investigated. The various possible options for initialization and synchronization of such PHIL setup are presented and their applicability, advantages and disadvantages discussed. A process of initialization and synchronization using a controlled current source is further evaluated by means of two case studies undertaken on a reduced dynamic model of the Great Britain (GB) power system.

2. PHIL Initialization and Synchronization

A number of studies have investigated the stability of PHIL simulations [11–15], where the main findings include the establishment of stability thresholds imposed by the interface algorithms used for the PHIL implementation. Improvements to alleviate the identified stability limitations have been proposed in [12,16,17]. However, these studies investigate the stability of an operational PHIL setup, assuming a successful initialization of simulation and hardware have already been established independently and straightforward synchronization has been achieved. In the following sub-sections, this assumption is shown to be limiting and options to address the resulting challenge are explored.

2.1. The Challenge

The issues associated with initialization and synchronization of a PHIL simulation arise when the HUT represents a more significant portion of the test network, and these conditions can broadly be classified into two: (i) where the test network to be represented by the HUT is critical for initialization of the DRTS simulation and (ii) where the test network to be represented by the HUT affects the voltage and frequency of the DRTS simulation. These are elaborated as follows:

- *HUT critical for initialization*: In such cases, the initialization and synchronization of the PHIL experiment present a paradoxical scenario where the DRTS simulation cannot be initialized without the hardware currents, while the hardware currents cannot be produced without the DRTS simulation being initialized. To elaborate, the DRTS simulation will fail to initialize due to a lack of generation or load leading to not enough synchronizing torque in the simulated network. Without the DRTS simulation initialized, the power interface will not be capable of reproducing the interface signals and therefore the HUT response cannot be synchronized. On the other hand, reproducing the interface signal during the initialization of DRTS is risky as the signal might not be suitable for reproduction or may be over the safety limits of the power amplifier and HUT.
- *HUT affects voltage and frequency*: Here, the HUT is not critical (the simulation can start without it connected) but still significant as to affect the frequency and voltage considerably triggering control actions from the components in the simulation, leading to a modified initial state of the system. This can also result in an impractical voltage and frequency levels for the initialization of the HUT.

2.2. Initialization of DRTS Simulation

A solution is therefore required to overcome the aforementioned problems, and allow PHIL simulations to be commenced even in these situations. For this purpose, the use of an auxiliary emulated HUT component is required. Four possible options to enable the initialization of difficult PHIL simulations by emulating the HUT component have been identified:

1. *Detailed simulation of HUT*: a detailed model of the HUT can be included as part of the simulation for establishing the initial conditions of the DRTS simulation. However, developing a detailed model of the HUT can be an arduous task, and considering that the expected power flows at the PCC can typically be estimated, simpler solutions can be utilized for the initialization process.

2. *AC voltage source*: readily available in every power system simulation tool, voltage source models can be utilized to initialize the simulated test network for PHIL simulations, emulating the HUT. However, as AC voltage sources act as infinite sources, the power flow of the network at the PCC cannot be controlled. This would lead to, an unsuccessful initialization, as the state of the network is no longer the intended for the test scenario. Additionally, with the change in power flows, new stability analyses would need to be undertaken as the system state under which the HUT was intended to be connected is no longer the same, unless an adjustment of the power setpoints is performed until power exchange with the infinite bus is brought to zero.

3. *Synchronous generator*: a synchronous generator model can control the active power at its output terminals for emulating the HUT required active power transfers at the PCC, this being controlled by means of a simple set-point. The reactive power of a synchronous generator is controlled by manipulating the excitation system. Either manual tuning of the voltage reference to the exciter or developing a simple PI control is required to attain the required reactive power flow at the PCC.

4. *AC Controlled Current Source*: for the emulation of the HUT power transfer at the PCC, a controlled current source allows for a straightforward implementation with high accuracy. This implementation will only require the measured voltage and the P and Q set points at the PCC for generating the current signals as shown:

$$I_d = \frac{P_{ref} V_d - Q_{ref} V_q}{V_d{}^2 + V_q{}^2} \tag{1}$$

$$I_q = \frac{P_{ref} V_q + Q_{ref} V_d}{V_d{}^2 + V_q{}^2} \tag{2}$$

where I_d is the direct axis current, I_q is the quadrature axis current, P_{ref} and Q_{ref} are the reference active and reactive powers to be injected at the PCC respectively, V_d is the direct axis voltage at the PCC and V_q is the quadrature axis voltage at PCC. The direct and quadrature axis voltages required can be obtained with Park's transformation as:

$$\begin{bmatrix} V_d \\ V_q \\ V_0 \end{bmatrix} = \frac{2}{3} \begin{bmatrix} \cos(\theta) & \cos(\theta - 2\pi/3) & \cos(\theta + 2\pi/3) \\ -\sin(\theta) & -\sin(\theta - 2\pi/3) & -\sin(\theta + 2\pi/3) \\ 1/2 & 1/2 & 1/2 \end{bmatrix} \begin{bmatrix} V_a \\ V_b \\ V_c \end{bmatrix} \tag{3}$$

The three phase currents required for the current controlled source can be obtained from the quadrature and direct currents using the inverse Park's transformation as:

$$\begin{bmatrix} I_a \\ I_b \\ I_c \end{bmatrix} = \begin{bmatrix} \cos(\theta) & -\sin(\theta) & 1 \\ \cos(\theta - 2\pi/3) & -\sin(\theta - 2\pi/3) & 1 \\ \cos(\theta + 2\pi/3) & -\sin(\theta + 2\pi/3) & 1 \end{bmatrix} \begin{bmatrix} I_d \\ I_q \\ I_0 \end{bmatrix} \tag{4}$$

In this manner, the initialization is straightforward and accurate when the power transfers at the PCC are known.

In this paper, the main focus is on situations where the HUT component contributes a significant active or reactive power, without which the simulated network cannot survive, due to under/over frequency/voltage. In this case, the simplest approach is to implement method (4), the controlled current source. This is because it is a conventional approach that can be implemented in simulation using well-known dq axis control techniques. These techniques are common to most conventional converter-connected generation, active front-end, and storage device technologies.

In some other scenarios, it may be that the HUT properties which are required to stabilise the power network are not so much absolute balances of fundamental active and reactive power, but other properties such as synchronising torque, grid stiffness, harmonic damping, etc. In these cases, which are out of the scope of the present paper, a simulated HUT using a current-source approach may not be appropriate or sufficient, and a voltage-source approach may be more suitable, along the lines of method (2) or (3). These types of solution could be explored in future work, and one potential solution has been referred to (within a simulation-only environment, without hardware) in [18].

Accordingly, the AC current controlled source is the ideal alternative for the initialization and has been implemented for its evaluation. A schematic of the current source configuration for initialization purposes of PHIL simulations is shown in Figure 2 for a six-area GB power system. This study presents four simulated areas, and two areas (Areas 1 and 2) represented in hardware. The initialization therefore sees the latter emulated by current sources.

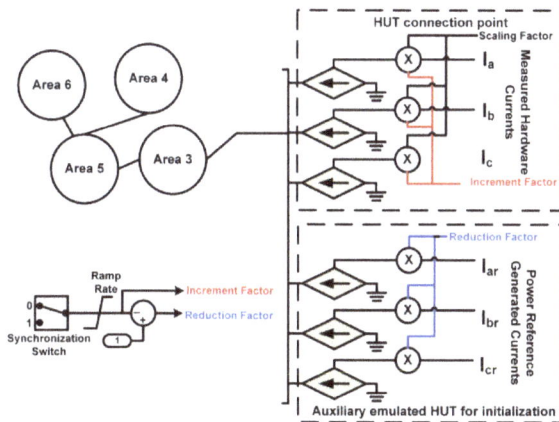

Figure 2. PHIL initialization and synchronization structure at DRTS side.

2.3. Synchronization

In typical PHIL simulations, the HUT is connected to the DRTS simulation (synchronized) through the action of closing a simple switch, thus closing the loop between the HUT and the DRTS. During the process of synchronization, the currents from the auxiliary emulated HUT utilized for initialization need to be replaced with the hardware currents (i.e., the measured response from the HUT). It is essential to ensure that during the process of synchronization the voltage and frequency of the network do not exceed the safety margins of the power interface and HUT. In addition, when any voltage or frequency control algorithms are implemented within the network, it is often desired that the synchronization of the HUT causes the least possible change in system frequency and voltage in order not to cause any undesired control actions. The synchronization method chosen will be dependent upon the initialization method selected. Therefore, in this section, the synchronization process for each of the initialization methods presented in previous section is discussed.

- *Detailed simulation of HUT*: while this could be the best option for the purpose of initialization of PHIL, assuming an accurate enough model of HUT is available, for the purpose of synchronization, a dispatching algorithm to reduce the generation and load of the emulated HUT would be required to avoid the frequency going to abnormal values when the HUT is first connected. It can therefore be said that, utilizing a detailed model of the HUT is very challenging for initialization and synchronization of PHIL setups due to the requirement of developing dedicated HUT models and dispatch algorithms.

- *AC voltage source*: Apart from the fact that the AC voltage source is not the ideal approach for initialization due to its response as an infinite source, similarly, the power output of the voltage source cannot be controlled and the process can lead to an erroneous synchronization.
- *Synchronous generator*: In order to attain a smooth transition from the auxiliary emulated HUT (the synchronous generator) and the HUT, a complex control would be required (for governor and excitation system) to ensure least deviation in frequency and voltage during the process. This controller would be a generic solution that can be reused, however, would be limited to scenarios where the HUT effectively emulates generation.
- *AC Controlled Current Source*: if a controlled current source is utilized, the synchronization can be achieved with a proposed simple logic as presented in Figure 2. The synchronization process is begun by means of a synchronization switch that inversely ramps up and down both controlled current sources. The ramp rate can be chosen such that it doesn't create any oscillations or transients on the system, once the currents from the auxiliary emulated HUT are reduced to zero and the currents from the HUT are fully connected to the simulation, the system is synchronized.

From the above discussion, it can be deduced that for the purpose of initialization and synchronization of PHIL setups, where the HUT represents a significant part of the test network, the most convenient option available is to utilize a controlled current source. It performs ideally under all scenarios, with accurate performance and a reliable, straightforward implementation. Furthermore, it is a generic approach that can be utilized when the HUT emulates net generation or load.

A proposed process for performing the initialization and synchronization is shown in Figure 3 with a flowchart. This process assumes that the PHIL simulation is stable for the interface algorithm chosen. This should be ensured before the PHIL simulation initialization and synchronization procedure is begun.

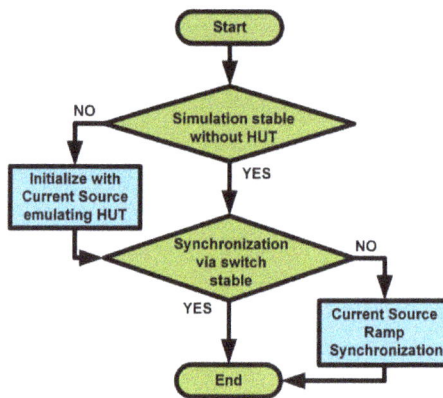

Figure 3. Initialization and synchronization flowchart.

3. Experimental Setup for PHIL

In this section, the different components of the PHIL setup used for the validation of the proposed initialization and synchronization process are described. First the real-time simulation model of the Great Britain (GB) reference power system developed within RSCAD (a power system simulation software from RTDS Technologies, Winnipeg, MB, Canada) is described, the characteristics of the power interface used for the interconnection of the simulated system with the hardware is presented, the HUT utilized is detailed and finally the existing time delay within the setup is considered for the synchronization.

3.1. GB Power System.

A reduced six-bus dynamic model of the GB power system has been chosen as the simulated network for this PHIL setup. A single line diagram of the GB power system is shown in Figure 4a. The choice of a six-area model is based on the GB National Electricity Transmission System (NETS) boundary map presented in the Electricity Ten Year Statement (ETYS) of National Grid, Transmission System Operator (TSO) of GB, where the GB transmission network is grouped into six regions [19]. These regions have been developed around major generation sources, power flow corridors and load centres [19]. The model is based on real power flow data of the six regions. Each area is built as the combination of a lumped generator and load. The area wise generation, area wise active and reactive power load, and inter-area power flows have been presented in Tables 1 and 2. The model has been developed in RSCAD and simulated in real-time using a Real Time Digital Simulator from RTDS Technologies.

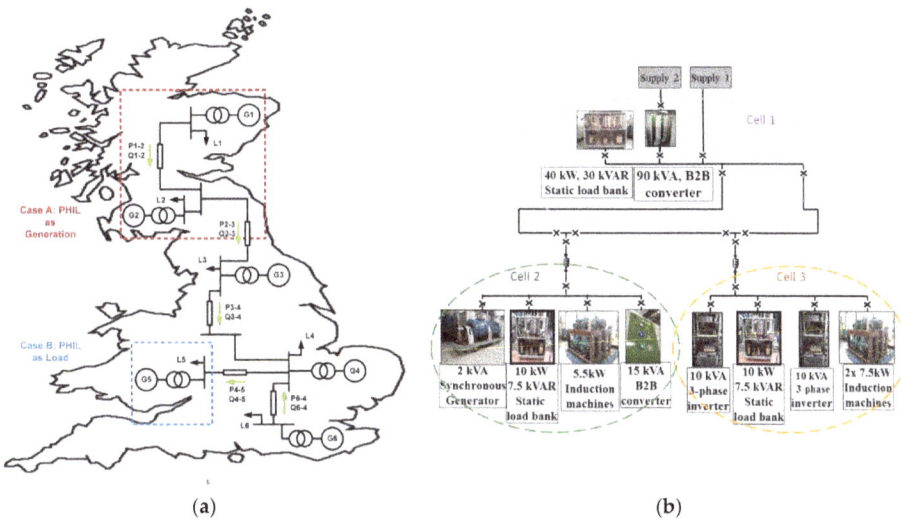

(a)   (b)

Figure 4. (a) Reduced six-bus dynamic model of the GB power system, (b) Dynamic power system laboratory (HUT).

Table 1. Area wise capacity and initial load condition.

Capacity and Loading Conditions	Area 1	Area 2	Area 3	Area 4	Area 5	Area 6
Area wise generation capacity (MVA)	11,000	20,000	9160	5500	15,500	2000
Area wise active power load (MW)	8468	12,548	8398	2150	26,852	100
Area wise reactive power load (MVAr)	4109	6077	4067	1041	13,005	500

Table 2. Inter-area power flows.

Inter-area active power flow (MW)	P1-2	P2-3	P3-4	P4-5	P6-4
	2097	8900	9105	13,080	970

Inter-area reactive power flow (MVAr)	Q1-2	Q2-3	Q3-4	Q4-5	Q6-4
	1328	4257	5025	7088	155

3.2. Power Interface

The power interface is composed of a 90 kVA back-to-back converter unit with a switching frequency of 8 kHz responsible for amplifying the signal received from the DRTS. Different interface algorithms have been described in the literature [11,17,20,21]. For the purpose of this paper, the voltage ideal transformer method (ITM) has been selected due to its straightforward implementation and good stability performance. Accordingly, the power interface amplifies and reproduces the voltages received from the simulation and therefore is controlled as a voltage source. At the same time, in the configuration used for this experiment, the power interface is also responsible for measuring the response of the HUT and sending it back to the DRTS for closing the loop with the simulation. Having analog-to-digital and digital-to-analog conversions at both ends, analog interface signals are exchanged between the power interface and the DRTS.

3.3. HUT

The Dynamic Power Systems Laboratory (DPSL) will be utilized as the HUT within this paper for the example case study. DPSL comprises a reconfigurable 125 kVA, 400 V three-phase AC power network with multiple controllable supplies and loads with flexible control systems and interfaces. The one-line diagram of DPSL is presented in Figure 4b. The laboratory network is designed such that it can be split into three separate power islands (represented as cells in Figure 4b) under independent control, or as a centralized interconnected system.

For the example case study, only cell 2 and 3 of Figure 4b will form the HUT, while the 90 kVA unit from cell 1 will represent the power interface. Since the laboratory equipment is relatively small compared to the required power transfers at the PCC, the response of the HUT (the measured currents at the hardware PCC) will be scaled up proportionally to match the test scenario.

The scaling of the response does not impact the simulation results; however, the scaling of the response does make the system more sensitive to oscillations. Small oscillations within the HUT can result in large power oscillations within the simulated system. This can lead the system to instability or cause the implemented controllers to malfunction. This further demonstrates the need for appropriate measures to be undertaken for initialization and synchronization of large synchronous power systems that become sensitive to oscillations under scaled PHIL setups.

3.4. Time Delay Compensation

The synchronization process is completed when the measured response of the HUT is injected into the simulation. However, typically the hardware response (measured currents at the PCC) is delayed compared to the reference currents generated during initialization. The entire PHIL process: the processing time of the DRTS, the communication between DRTS and power interface, the power interface processing time for amplifying the signal and the feedback loop measurement and communication, contributes to this delay. For this work, a phase compensation method, as reported in [22], is utilized for compensating the time delay. If the delay is not compensated, it is not possible to ensure the equality of the reference power and the injected power and therefore the fidelity of the simulation. Once the delay has been accounted for, the hardware and simulation can be accurately synchronized.

4. Experimental Assessment and Validation

In this section, an assessment of the proposed initialization and synchronization procedures for PHIL simulations is performed. Two case studies have been developed for this purpose, first a case with a significant HUT that affects the PHIL initialization and synchronization if no measure is in place, and afterwards a case in which the simulation could be initialized without HUT, although with erroneous voltage and frequency parameters.

Case Study A: HUT critical for the stability of the real-time simulation.

For the first case study, the HUT is to represent Area 1 and 2 of the GB power system (as shown in Figure 4a) while the remaining areas will be part of the simulated network on the DRTS. As can be observed from Table 2, there is an active and reactive power flow from Area 2 to Area 3. This power flow needs to be matched by the HUT, effectively emulating generation of active and reactive power. Cell 2 and Cell 3 of the DPSL represent Area 2 and Area 1 respectively, while the 90kVA back-to-back converter is used as the power interface.

Initialization

The active and reactive power flow from Area 2 to Area 3 is 8900 MW and 4257 MVAR. Without the HUT representing Area 1 and 2, the remainder of the simulated GB power system (Area 3–6) within RSCAD, fails to initialize due to a lack of generation supposed to be produced by Area 1 and 2. The HUT emulates a large generation portion and the DRTS simulation model is unstable without the HUT. Hence, the process shown in Figure 3 is followed, were initialization of the RT system with an initialization technique is required. The AC current source mode for initialization is implemented due to its simplicity and its suitability to perform the synchronization process.

The schematic of the used methodology for initialization and synchronization of this PHIL implementation is shown in Figure 2. As can be observed from the figure, the auxiliary emulated HUT (the controlled current source in this case) is utilized to reproduce currents generated from the power reference i.e., 8900 MW and 4257 MVAR. Once the RT simulation is initialized and attained steady state, the PCC voltages (scaled down to 400 V) are reproduced in the laboratory using the power interface. Trying to reproduce the simulation voltages when the simulation is not yet stabilized or initialized properly, can damage the hardware components as large transients or/and oscillations can be present. With stable voltage being reproduced at the laboratory, the HUT components are then connected and initialized.

Synchronization

The main objective of the synchronization is to replace the auxiliary currents generated by the controlled current sources with the measured HUT currents. This is intended to be performed so that there is least change in frequency and voltage during the process of synchronization and least impact on the stability of the simulated network. After the DRTS is successfully initialized, the voltage from the simulation is reproduced by the power interface, allowing for the set reference power to be injected into the PCC. In this case, the net generation is produced by the 15 kVA back-to-back converter, the synchronous generator in cell 1 and the 10 kVA inverter in cell 2. A load bank has been added in each of the cells to represent the local loads within the two areas. By using these three power sources, the maximum power injection is limited to 27 kVA. The parameters used for the different hardware components are listed in Table 3. In order to represent the 8900 MW and 4257 MVAR from Area 2, the hardware currents are scaled by means of a scaling factor. For this work the scaling factor is chosen as $k = 9 \times 10^5$.

Table 3. Power setpoints for hardware components for case study A.

Area	Component	P (W)	Q (Var)
	15 kVA B2B Inverter	7000	3600
Cell 2	2 kVA Synchronous Generator	1500	0
	Load Bank 2	−1500	0
Cell 3	10 kVA Inverter	6600	1100
	Load Bank 3	−3300	0

Consequently, the measured HUT currents, when injected into the simulation, should be equal to the auxiliary emulated HUT currents. In this case, the currents are being generated by small scale power converters which produce a considerable number of harmonics. The amplitude of the harmonic components when scaled up is not typical of transmission levels. To mitigate this issue, the currents received at the DRTS are filtered by means of a low pass filter. The low pass filter cut off frequency is selected as 200 Hz in order to reduce the impact of the harmonics, thereby alleviating the sensitivity of the simulation to oscillations. However, the low pass filter increases the time delay of the received signal. Therefore, time delay compensation is required not only to compensate for the delay characteristics of PHIL structures, but more importantly in this case, the delay introduced by the filter. This is because the selected cut-off frequency attains a much larger delay than the typical delay of PHIL setups. The results for the compensation of the time delay in this implementation are presented in Figure 5a.

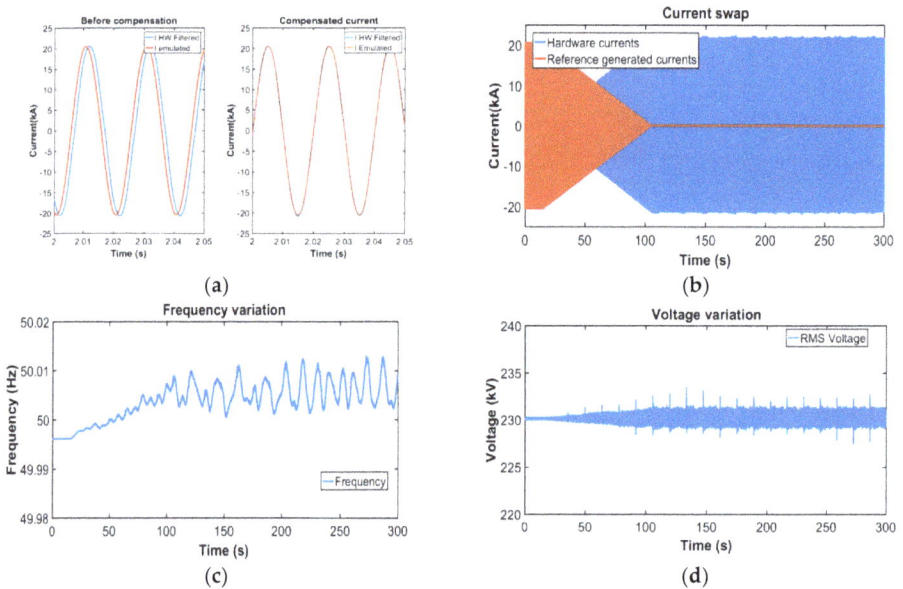

Figure 5. Results for study case A: HUT critical for stability of RT simulation, (**a**) time delay compensation, (**b**) currents swapping during synchronization, (**c**) frequency at PCC during synchronization and (**d**) voltage at PCC during synchronization.

Once the measured hardware currents (after scaling) and the auxiliary signals used for initialization are of the same magnitude and phase, the replacement process is initiated by means of the synchronization switch shown in Figure 2. The currents are then ramped up and down simultaneously over a period of 90 s as shown in Figure 5b. The frequency and voltage at PCC, during the process of synchronization, are presented in Figure 5c,d respectively. As can be observed from the two figures, although the HUT currents have been compensated for the time delay and filter delay and are approximately similar to the emulated currents (as shown in Figure 5a), there is an obvious change in both voltage and frequency during the process of synchronization. The change in frequency is less than 0.02% and the change in voltage is less than 0.05%. This is mainly due to small inaccuracies such as losses in small impedances, measurements inaccuracies and control inaccuracies of the hardware assets, although in this case the difference is not affecting to the test implementation.

The ramp rate utilized plays an important practical role in minimizing such impacts during the process of synchronization. Synchronizing without a ramp rate, risks transients being introduced. The ramp rate is dependent upon the acceptable variation on voltage and frequency during the process of synchronization. With the PHIL simulation fully synchronized and the power transfers set as required by the scenario, the testing of, for example, new control algorithms in a more realistic environment is made possible.

Case Study B: HUT Affecting Voltage and Frequency During Initialization.

For the second case study, the HUT represents Area 5 of the GB power system (as shown in Figure 4a). As can be observed from Table 2, there is an active and reactive power flow into Area 5. Therefore, in this case, the HUT effectively emulates a consumer area at the PCC. Cell 2, Cell 3, and the 40 kVA load bank of Cell 1 combined represent Area 4, while the 90 kVA power converter from Cell 1 is used as the power interface. The specific setpoints used for each of the components in each cell are presented in Table 4.

Table 4. Power setpoints for hardware components for case study B.

Area	Component	P (W)	Q (Var)
Cell 1	Load Bank 1	−14,000	−7900
Cell 2	15 kVA B2B Inverter	4500	−3000
	2 kVA Synchronous Generator	1000	0
	Load Bank 2	−9000	−3500
Cell 3	10 kVA Inverter	5200	0
	Load Bank 3	−9000	−3500
	Induction motor	−4700	−3000

Initialization

The schematic of the PHIL interconnection within RSCAD is presented in Figure 6a. The HUT emulates net consumption, and so the measured current will have opposite direction to case A and will be drawing power from the simulated power system (through the power interface). In order to represent the 13,080 MW and 7088 MVAR absorbed by Area 5, the hardware currents are scaled by means of a scaling factor. The scaling factor in this case is set to k = 5 × 10^5.

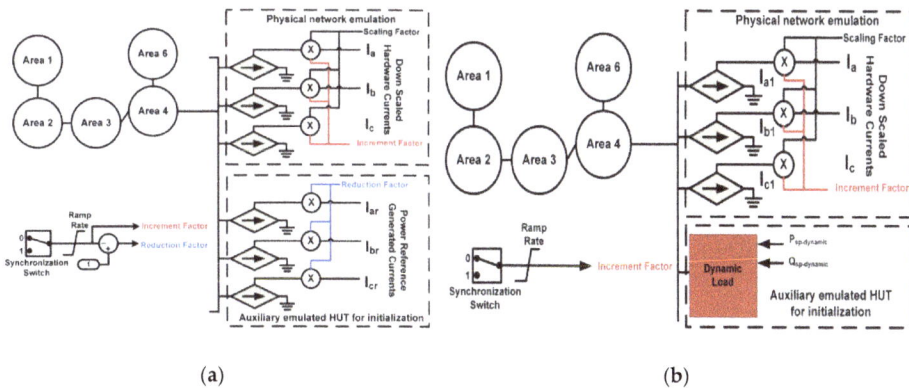

(a)   (b)

Figure 6. PHIL configuration for case study B: affecting voltage and frequency during initialization. (a) with current source, (b) with dynamic load.

In this case, the remainder of the GB power system (Areas 1–3, 4 and 6) simulated within RSCAD would initialize without Area 5, as there is enough generation to support the network, unlike case A, but the frequency of the network would be above nominal (as no dispatching algorithm is implemented). The value of the frequency deviation would depend upon the droop settings.

This is again undesirable as this would activate any frequency control algorithms implemented within the network. There is, of course, an option to de-activate the control during the process of synchronization. However, there might be hardware limitations on the value of frequency that can be sustained/emulated within the laboratory. To avoid such risks, an auxiliary component should be utilized to initialize the test network for PHIL simulations. A dynamic load or a controlled current source can be used for this purpose.

The schematic for initialization and synchronization of PHIL at the RTDS with the current source as the initialization and synchronization component is shown in Figure 6a. The reference currents for the auxiliary current source can be generated as presented in Case A Initialization. A dynamic load could be also used for the initialization, however depending on the simulation software used this can have different forms. For example, within Simulink simulation software the dynamic load is equivalent to a current source, hence the method would be equivalent to the current source method presented before (and therefore can be utilized for case A and B), while within RSCAD (RTDS simulation software tool) the dynamic load model does not allow for negative power set points (rendering it unusable for case A).

Synchronization

If the current sources are used for the initialization, the same process of synchronization as presented in Case A Synchronization is used. In this case, the dynamic load model could also be utilized indistinctly. However, if a dynamic load is being utilized, instead of using current set points, active and reactive power set points are required (as presented in Figure 5b), and these can be calculated as:

$$P_{SP\_dynamic} = P_{ref} - P_{HW} \tag{5}$$

$$Q_{SP\_dynamic} = Q_{ref} - Q_{HW} \tag{6}$$

where P_{HW} and Q_{HW} are the active and reactive power drawn by the hardware respectively, and can be calculated as:

$$P_{HW} = V_a I_{a1} + V_b I_{b1} + V_c I_{c1} \tag{7}$$

$$Q_{HW} = \frac{1}{\sqrt{3}} (V_a(I_{b1} - I_{c1}) + V_b(I_{c1} - I_{a1}) + V_c(I_{a1} - I_{b1})) \tag{8}$$

where V_{abc} are the three phase voltages at PCC and I_{abc1} are the currents measured at the PCC after the increment factor.

For this scenario, the current source model has been implemented. It can be observed from Figure 7 that by initializing and synchronizing the PHIL implementation with the current sources, the frequency and voltage at the PCC remain at the same steady state levels as before the connection with the HUT. Also, due to the nature of the hardware used and the real measurement devices, which can introduce some noise into the signals, the frequency and voltage waveforms show more realistic dynamics in comparison with a pure simulation.

(a) (b)

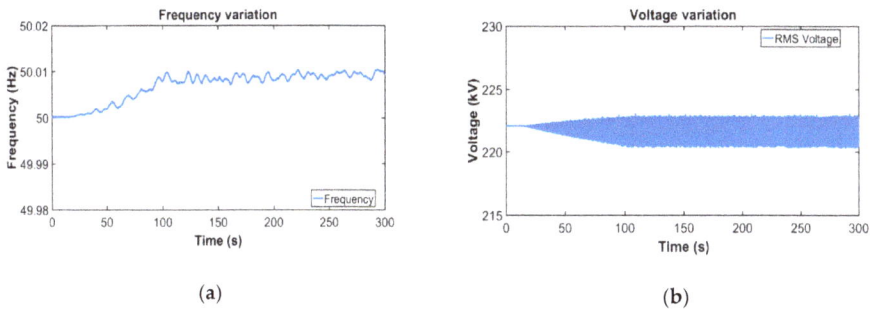

Figure 7. Results for case B: affecting voltage and frequency during initialization (**a**) Frequency variation, (**b**) voltage variation.

5. Conclusions

PHIL simulations can be complex to initialize and synchronize depending on the HUT and its significance to the dynamic behaviour and stability of the overall study system. This then represents a limiting factor on the range of scenarios for which PHIL can be applied. This paper presents a range of possible methodologies for enabling the initialization and synchronization of such scenarios. The investigation of these alternatives has led to the identification of a recommended approach that uses current sources for initialization and symmetrical ramping rates for synchronization of PHIL simulations. The resulting improved performance and extended range of feasible scenarios that can be studied have been validated by experimentation of two different and realistic scenarios for the GB power system. This will allow for safer and more stable PHIL simulations, and permit validation of a wider range of realistic simulations through PHIL to the betterment of new controllers and power components.

Author Contributions: E.G.-S. and M.H.S. conceived the research idea, performed the experiments and wrote the paper. A.J.R. and G.M.B. oversaw the work and proofread the paper.

Acknowledgments: This work was partly supported by the European Community's Horizon 2020 Program (H2020/2014–2020) under the project "ERIGrid: European Research Infrastructure supporting Smart Grid Systems Technology Development, Validation, and Roll Out" (Grant No. 654113) and from the European Union Seventh Framework Programme (FP7) under the project ELECTRA IRP (Grant No. 609687).

Conflicts of Interest: The authors declare no conflict of interest.

References

1. Strasser, T.; Andrén, F.P.; Lauss, G.; Bründlinger, R.; Brunner, H.; Moyo, C.; Seitl, C.; Rohjans, S.; Lehnhoff, S.; Palensky, P.; et al. *e & i Elektrotechnik und Informationstechnik*; Springer: Vienna, Austria, 2017; Volume 134, pp. 71–77.
2. Kotsampopoulos, P.C.; Lehfuss, F.; Lauss, G.F.; Bletterie, B.; Hatziargyriou, N.D. The Limitations of Digital Simulation and the Advantages of PHIL Testing in Studying Distributed Generation Provision of Ancillary Services. *IEEE Trans. Ind. Electron.* **2015**, *62*, 5502–5515. [CrossRef]
3. Edrington, C.S.; Steurer, M.; Langston, J.; El-Mezyani, T.; Schoder, K. Role of Power Hardware in the Loop in Modeling and Simulation for Experimentation in Power and Energy Systems. *Proc. IEEE* **2015**, *103*, 2401–2409. [CrossRef]
4. Kotsampopoulos, P.; Kleftakis, V.; Messinis, G.; Hatziargyriou, N. Design, development and operation of a PHIL environment for Distributed Energy Resources. In Proceedings of the IECON 2012—38th Annual Conference on IEEE Industrial Electronics Society, Montreal, QC, Canada, 25–28 October 2012; pp. 4765–4770.
5. Ren, W.; Steurer, M.; Woodruff, S. Applying Controller and Power Hardware-in-the-Loop Simulation in Designing and Prototyping Apparatuses for Future All Electric Ship. In Proceedings of the 2007 IEEE Electric Ship Technologies Symposium, Arlington, VA, USA, 21–23 May 2007; pp. 443–448.

6. Langston, J.; Schoder, K.; Steurer, M.; Faruque, O.; Hauer, J.; Bogdan, F.; Bravo, R.; Mather, B.; Katiraei, F. Power hardware-in-the-loop testing of a 500 kW photovoltaic array inverter. In Proceedings of the IECON 2012—38th Annual Conference on IEEE Industrial Electronics Society, Montreal, QC, Canada, 25–28 October 2012; pp. 4797–4802.

7. Naeckel, O.; Langston, J.; Steurer, M.; Fleming, F.; Paran, S.; Edrington, C.; Noe, M. Power Hardware-in-the-Loop Testing of an Air Coil Superconducting Fault Current Limiter Demonstrator. *IEEE Trans. Appl. Superconduct.* **2015**, *25*, 1–7. [CrossRef]

8. Kotsampopoulos, P.; Hatziargyriou, N.; Bletterie, B.; Lauss, G.; Strasser, T. Introduction of advanced testing procedures including PHIL for DG providing ancillary services. In Proceedings of the IECON 2013—39th Annual Conference of the IEEE Industrial Electronics Society, Vienna, Austria, 10–13 November 2013; pp. 5398–5404.

9. Zhu, K.; Chenine, M.; Nordstrom, L. ICT Architecture Impact on Wide Area Monitoring and Control Systems' Reliability. *IEEE Trans. Power Deliv.* **2011**, *26*, 2801–2808. [CrossRef]

10. Lauss, G.F.; Faruque, M.O.; Schoder, K.; Dufour, C.; Viehweider, A.; Langston, J. Characteristics and Design of Power Hardware-in-the-Loop Simulations for Electrical Power Systems. *IEEE Trans. Ind. Electron.* **2016**, *63*, 406–417. [CrossRef]

11. Ren, W.; Steurer, M.; Baldwin, T.L. Improve the Stability and the Accuracy of Power Hardware-in-the-Loop Simulation by Selecting Appropriate Interface Algorithms. *IEEE Trans. Ind. Appl.* **2008**, *44*, 1286–1294. [CrossRef]

12. Viehweider, A.; Lauss, G.; Felix, L. Stabilization of Power Hardware-in-the-Loop Simulations of Electric Energy Systems. *Simul. Model. Pract. Theory* **2011**, *19*, 1699–1708. [CrossRef]

13. Hatakeyama, T.; Riccobono, A.; Monti, A. Stability and accuracy analysis of power hardware in the loop system with different interface algorithms. In Proceedings of the 2016 IEEE 17th Workshop on Control and Modeling for Power Electronics (COMPEL), Trondheim, Norway, 27–30 June 2016; pp. 1–8.

14. Dargahi, M.; Ghosh, A.; Ledwich, G. Stability synthesis of power hardware-in-the-loop (PHIL) simulation. In Proceedings of the 2014 IEEE PES General Meeting | Conference & Exposition, National Harbor, MD, USA, 27–31 July 2014; pp. 1–5.

15. Brandl, R. Operational Range of Several Interface Algorithms for Different Power Hardware-In-The-Loop Setups. *Energies* **2017**, *10*, 1946. [CrossRef]

16. Liegmann, E.; Riccobono, A.; Monti, A. Wideband identification of impedance to improve accuracy and stability of power-hardware-in-the-loop simulations. In Proceedings of the 2016 IEEE International Workshop on Applied Measurements for Power Systems (AMPS), Aachen, Germany, 28–30 September 2016; pp. 1–6.

17. Lehfuss, F.; Lauss, G.; Strasser, T. Implementation of a multi-rating interface for Power-Hardware-in-the-Loop simulations. In Proceedings of the IECON 2012—38th Annual Conference on IEEE Industrial Electronics Society, Montreal, QC, Canada, 25–28 October 2012; pp. 4777–4782.

18. Yu, M.; Roscoe, A.J.; Dyśko, A.; Booth, C.D.; Ierna, R.; Zhu, J.; Urdal, H. Instantaneous Penetration Level Limits of Non-Synchronous Devices in the British Power System. *IET Renew. Power Gener.* **2016**, *11*, 1211–1217. [CrossRef]

19. National Grid. *Electricity Ten Year Statement 2016*; National Grid: Warwick, UK, 2016.

20. Lentijo, S.; D'Arco, S.; Monti, A. Comparing the Dynamic Performances of Power Hardware-in-the-Loop Interfaces. *IEEE Trans. Ind. Electron.* **2010**, *57*, 1195–1207. [CrossRef]

21. De Jong, E.; de Graff, R.; Vassen, P.; Crolla, P.; Roscoe, A.; Lefuss, F.; Lauss, G.; Kotsampopoulos, P.; Gafaro, F. *European White Book on Real-Time Power Hardware in the Loop Testing*; DERlab: Kassel, Germany, 2012.

22. Guillo-Sansano, E.; Roscoe, A.J.; Burt, G.M. Harmonic-by-harmonic time delay compensation method for PHIL simulation of low impedance power systems. In Proceedings of the 2015 International Symposium on Smart Electric Distribution Systems and Technologies (EDST), Vienna, Austria, 8–11 September 2015; pp. 560–565.

energies

MDPI

Article

Multi-Agent System with Plug and Play Feature for Distributed Secondary Control in Microgrid—Controller and Power Hardware-in-the-Loop Implementation

Tung-Lam Nguyen [1,*], Efren Guillo-Sansano [2], Mazheruddin H. Syed [2], Van-Hoa Nguyen [3], Steven M. Blair [2], Luis Reguera [2], Quoc-Tuan Tran [3], Raphael Caire [1], Graeme M. Burt [2], Catalin Gavriluta [4] and Ngoc-An Luu [5]

[1]   Univ. Grenoble Alpes, CNRS, Grenoble INP, G2Elab, F-38000 Grenoble, France; raphael.caire@g2elab.grenoble-inp.fr

[2]   Institute for Energy and Environment, Electronic and Electrical Engineering Department, University of Strathclyde, Glasgow G1 1XW, UK; efren.guillo-sansano@strath.ac.uk (E.G.-S.); mazheruddin.syed@strath.ac.uk (M.H.S.); steven.m.blair@strath.ac.uk (S.M.B.); luis.reguera@strath.ac.uk (L.R.); graeme.burt@strath.ac.uk (G.M.B.)

[3]   Alternative Energies and Atomic Energy Commission (CEA), National Institute for Solar Energy (INES), F-73375 Le Bourget-du-Lac, France; Vanhoa.NGUYEN@cea.fr (V.-H.N.); QuocTuan.TRAN@cea.fr (Q.-T.T.)

[4]   Austrian Institute of Technology, 1210 Vienna, Austria; Catalin.Gavriluta@ait.ac.at

[5]   Department of Electrical Engineering, University of Science and Technology—The University of Danang, Danang 550000, Vietnam; lnadhbk@gmail.com

*   Correspondence: tung-lam.nguyen@g2elab.grenoble-inp.fr; Tel.: +33-7-6878-5357

Received: 26 October 2018; Accepted: 16 November 2018; Published: 22 November 2018

Abstract: Distributed control and optimization strategies are a promising alternative approach to centralized control within microgrids. In this paper, a multi-agent system is developed to deal with the distributed secondary control of islanded microgrids. Two main challenges are identified in the coordination of a microgrid: (i) interoperability among equipment from different vendors; and (ii) online re-configuration of the network in the case of alteration of topology. To cope with these challenges, the agents are designed to communicate with physical devices via the industrial standard IEC 61850 and incorporate a plug and play feature. This allows interoperability within a microgrid at agent layer as well as allows for online re-configuration upon topology alteration. A test case of distributed frequency control of islanded microgrid with various scenarios was conducted to validate the operation of proposed approach under controller and power hardware-in-the-loop environment, comprising prototypical hardware agent systems and realistic communications network.

Keywords: distributed control; microgrid; hardware-in-the-loop; average consensus; multi-agent system

---

1. Introduction

Microgrids (MGs) are considered as the major component of the future power system to deal with the proliferation of distributed generators (DGs) in low and medium voltage grids. In general, a MG is a system integrated by DGs, controllable and non-controllable loads, energy storage systems (ESS) and control and communication infrastructure. The MG can operate in islanded mode or grid-connected mode [1]. By moving the generation closer to loads, MGs aid in reducing power transmission losses. MGs also improve the reliability of the system with the ability to switch to islanded mode during system disturbances and faults.

MGs introduce many advantages to both utility and consumers. However, control and management of MGs induce significant challenges in term of coordination and aggregation.

Centralized schemes, which are common in conventional power systems, may no longer be suitable for significantly larger numbers of DG units due to many reasons [2], e.g., excessive computation in the central unit due to numerous controllable loads and generators, reliability and security of the central controller, frequent mutation of grid due to installation of new DGs and loads, unwillingness to share data of participant actors, etc. Decentralized strategies are highly scalable and robust because controllers only need local information and ignore coordination with others. However, the system controlled in decentralized way can hardly reach network-wide optimum operation. Distributed approach is considered as the best alternative for the control and management of the next generation of power systems. In this approach, the central unit is eliminated and local controllers coordinate with nearby units to reach global optima. The main advantage of the distributed approach [2–4] is that the MGs can avoid system failure because the single central unit for controlling whole system is neglected. Moreover, it presents enhanced cyber-security and reduced communication distances. Furthermore, with the ability to perform in parallel, the computational load can be shared and condensed significantly. Finally, the privacy of sensitive information of loads of DERs could be inherited in the global operation.

Multi-agent system (MAS) is an advanced technology that has been recently applied in various areas of science and engineering, including smart grids. In the power systems domain, agent-based approaches have been applied in a wide range of applications such as load shedding, secondary control or optimal power flow [5–7]. Agent with properties as autonomous, social, reactive and proactive [7] are ideally suitable for distributed control implementation. The focus of this paper is the distributed secondary frequency control strategy for islanded MGs to restore grid network to normal state under various disturbances using MAS. Distributed control algorithms with an upper layer of agents have been presented recently in refs. [7,8]. In refs. [9–12], novel distributed algorithms where proposed and validated using pure power systems simulation tools. Some other works have utilized MAS runtime environments (such as Java Agent Development Framework (JADE)) to implement the distributed control system [13,14].

The interactions between entities (controllers, agents, devices, etc.) in MGs can lead to unexpected behaviors, thus advanced platforms for testing are required to evaluate the performance of MGs before the real deployment. The Hardware-in-the-loop (HIL) is an effective methodology to investigate MGs [15–17]. HIL enhances the validation of components in power systems by conducting controller Hardware-in-the-loop (CHIL) and power Hardware-in-the-loop (PHIL) [18]. CHIL is performed to validate protection and controller devices. In the meanwhile, PHIL is used to validate the operation of power devices as well as the dynamic interactions between them. The HIL implementation has been used in refs. [6,19,20] for the investigation in distributed control in MGs. However, the combination of CHIL and PHIL or the realistic communication environment is rarely reported.

In distributed control, communication plays an important role as system performance (e.g., local optimization and global convergence time) depends heavily on the information exchange among agents [21]. In order to ensure seamless communications in MAS, it is required that the system possesses and maintains a high level of inter-agent interoperability. Interoperability allows the network to seamlessly and autonomously integrate all components of power, distribution, management, and communication while minimizing human intervention. It has a direct impact on the cost of installation and integration and also introduces the ability to easily connect and integrate new components and systems. It allows the substitution/improvement of a component in the network without any problem to the overall operation of the integrated system [22]. It is however not an easy task due to the existence of a variety of vendors and communication interfaces in the framework of micro-grid. Standards or regulations can be used to bridge the gap but are not necessarily sufficient to ensure interoperability. In some cases, systems implementing the same standard may fail to interoperate because of the variability in the practical implementations.

Interoperability can be considered in several evaluation models and in terms of different technical and conceptual levels (e.g., semantic, syntactic, dynamic and physical) [23]. As in the Smart Grid

Interoperability Maturity Model (SGIMM) [24], ultimate goal of interoperability is the concept of "plug-and-play": the system is able to configure and integrate a component into the system by simply plugging it in. An automatic process determines the nature of the connected component to properly configure and operate. Achieving plug-and-play is not easy, and in the particular context of distributed control in micro-grid with MAS, several important challenges are highlighted:

- Firstly, in MGs, infrastructure may be supplied by different vendors and may be compliant to different protocols. Agents are required to be able to transfer data with local controllers and measurement system through various standardized or commercialized industrial protocols, while on the other hand, has to comply with the inter-agent communication protocols.
- Secondly, in distribution network of MG, the structure of grid and the total capacity of ESSs may change/be upgraded progressively along with the increase of loads and renewable energy sources. Furthermore, ESS is an element which requires regular maintenance and replacement. The corresponding agent has to be activated or deactivated accordingly to the state of the ESS. The local control algorithm (intra-agent) needs to be flexible enough to adapt to this frequent alteration of structure and capacity without major re-configuration.
- Not only at local level, the alteration of topology is also a critical obstacle that needs to be solved to achieve "Plug and Play" capacity at system level. The micro-grid operation is based on the consensus processes of the agents which tries to find a global solution based on limited information acquired from the neighbourhood. Consensus algorithms are introduced mathematically and often adapted to a certain network topology. Therefore, the integration or removal of an agent in the network (or alteration of topology) requires a throughout re-configuration or adaptation of the entire network.
- Last but not least, the asynchronous interaction (inter-agent) under influence of various type of uncertainties in a real communications network is much more complex and is not yet covered in the mathematical model. The performance of the real system may be derived from the theoretical one if this aspect is not considered during the design and validation process. However, in aforementioned research, the communication network is typically ignored. In ref. [25], the data transfer latency is considered, as deterministic time delays which does not accurately reflect realistic communications networks. Furthermore, the design of agents and the interactions among the agents as well as with controllers and devices were ambiguous and unspecific.

The above challenges are tackled in this paper. Particularly, we propose a method to implement interoperability within a MG with plug and play feature at the agent layer of distributed control scheme. The main contributions of this paper are twofold:

- We develop a multi-agent system with "plug and play" capacity for distributed secondary control of frequency in islanded MGs. Firstly, a multi-layer structure is proposed to describe thoroughly the MG system operating with agents. The structure consists of three layers: Device layer, Control layer and Agent layer. The agent, which is an autonomous program with server/client structure, is designed to process an average consensus algorithm and send proper signal to inverter controller in a distributed scheme. The agent is also equipped with the ability of collecting and broadcasting messages via the industrial protocol IEC 61850. The "Plug and Play" capacity is realized at the agent layer, as the system will automatically adapt to the alteration of topology (integration of new agent or removal of an agent) and react accordingly to maintain seamless operation.
- The proposed distributed secondary control is implemented in a laboratory platform based on the propose in [18] with controller and power-hardware-in-the-loop (C/PHIL) setup, incorporating realistic communications network with the impact of uncertainties considered. The performance of system under realistic condition shows that the agents are able to resist to disturbances and to self-configure under alteration of grid topology.

The paper is organized as following: Section 2 presents layer structure of a MG and average consensus algorithm. Section 3 describes the design of agents with plug-and-play feature operating as highest layer in the structure. Section 4 provides a laboratory platform with controller and power HIL setup to simulate a test case autonomous MG. A testing procedure and experimental results is also presented to validate the operation of agent system in a physical communication network. Section 5 concludes the paper while also highlighting some aspects worthy of consideration in future.

2. MAS Based Multi-Layer Architecture for Distributed Secondary Control in MG

The hierarchical structure, which is comprised of primary, secondary and tertiary level, is commonly used to control in MG [26–28]. The primary control level is used to stabilize frequency and voltage when disturbances occur by using only local measurements. The system in this control level has a fast response to reach the steady-state. However, there are deviations of frequency and voltage compared with the nominal values. The secondary control with global information is implemented to restore the frequency and voltage. At the top layer, the power flow to the main grid and optimized operation within grids is managed in the tertiary control level. This paper deals with the problem of secondary control in MG. In particular, we propose a three layer structure based on MAS for distributed secondary control in MG.

The architecture consists of three layers: Device layer containing the physical components and electrical connections, Control layer corresponding to alterations in the system operation, and providing control signals to the Device layer and finally Agent layer which receives measurements from corresponding devices, communicates, calculates and then returns proper signals to controllers. This architecture shows distinctly the relationship between agents and power system components while emphasizes on the communications network, which introduces increasingly important impact to the modern grid. Figure 1 illustrates the three-layer structure, where devices, controllers and agents are shown.

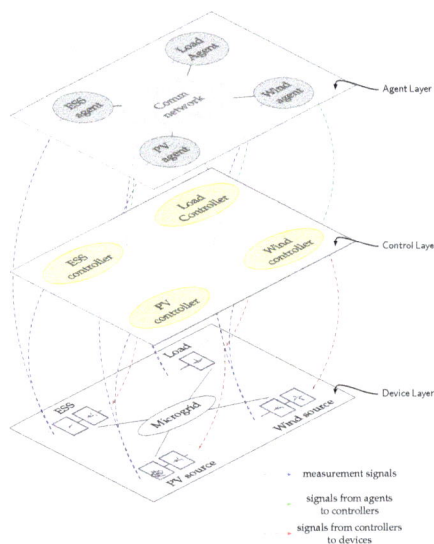

Figure 1. Layer structure.

Depending upon the objective and requirement of the device under control, the controller in Control layer either needs information from its agent and device or only its device. For instance, MPPT controller for PV source or primary control for inverter require local measurements from Device layer. However, with the secondary and tertiary controls of inverters, the additional signals from Agent layer are mandatory. In this paper, Control layer comprises controllers of inverters operating in parallel as grid-forming sources. The parallel configuration of voltage source inverters (VSIs) allows connection of multiple VSIs to form a MG thus facilitating the scalability and improving the redundancy and reliability of system. The control of each inverter is responsible for sharing the total load demand according to the rated power and resisting the instability of frequency and voltage in MGs. When a disturbance in the MG happens, the primary control of DGs will active immediately and the system is stabilized. Then, agents in Agent layer will send signals to compensate any errors with respect to the nominal condition through the secondary control. The local controller of DG in Control layer includes primary and secondary control level in the hierarchical control structure. The primary control adjusts the references of frequency and voltage to provide to the inner control loop of the inverter. The secondary control is conducted to fix the frequency and voltage to their nominal values after any changing of the system. This paper will focus only on controlling frequency, the proportional integral (PI) controller is used to deal with the steady-state error.

Classical approaches employ the MG's central controller which receives Δf from the measurement of a single point of the grid. The setpoint of secondary control unit is then distributed to local controllers of primary control. However in the distributed control manner, this centralized unit is eliminated. In our proposal, each local controller corresponds to an agent. The measurement devices are required to provide local frequency deviation to the connected agent which communicates with others to return the global Δf. An agent-based consensus algorithm is applied in the Agent layer and the processes in all agents converge at a same consensus value after a number of iterations. This consensus value is also the average frequency deviation transferred to PI controller.

In the proposed distributed frequency control system, each agent needs only local information but could return global results by using the average consensus algorithm. The algorithm also ensures that the signals are sent to the local controllers concurrently and those signals have the identical values as in the case of the centralized strategy.

A consensus algorithm is an interaction rule for a specific objective. The rule describes the information exchange between a entity and its neighbors in the communication network. It is assumed that each agent receives an input value at initial. The average consensus problem is the distributed computational problem of finding the average of the set of initial values by using only local and adjacent information. Consider a network with N nodes, the initial value at node i is $x^i(0) \in \mathbb{R}$. Node i only communicate with node $j \in \mathcal{N}_i$ in a constraint network. The goal of the algorithm is: firstly, each node compute the average of initial values, $\frac{1}{N}\sum_{i=1}^{N} x^i(0)$ and secondly, all nodes reach consensus on this value at the same time.

$$\lim_{t \to \infty} x^i(t) = \frac{1}{N} \sum_{i=1}^{N} x^i(0) \qquad \forall i \in \mathcal{V} \tag{1}$$

Equation (2) introduces a standard algorithm to solve the average consensus problem following the iteration update.

$$x^i(t+1) = \sum_{j=1}^{N} W_{i,j} x^j(t) \qquad i = 1, ..., N \tag{2}$$

where $t \in \mathbb{N}$ are the iteration steps and $W \in \mathbb{R}^{N \times N}$ is the weight matrix. Each node uses only local and neighborhood information, hence, $W_{i,j} = 0$ if $j \notin \mathcal{N}_i$ and $j \neq i$. To simplify the expression of the algorithm, let us define the column vector of $x^i(t)$

$$X(t) = \begin{bmatrix} x^1(t) & x^2(t) & \cdots & x^N(t) \end{bmatrix}$$

Then Equation (2) can be rewritten as:

$$X(t+1) = WX(t) \tag{3}$$

Assuming that consensus state is achieved at iteration t_0, from Equation (3) we can imply that $X(t_0) = W^{t_0}x(0)$. The necessary and sufficient condition for the convergence is:

$$\lim_{t \to \infty} W^t = \frac{\mathbf{1}\mathbf{1}^T}{N} \tag{4}$$

where $\mathbf{1}$ is the vector consisting of only ones. There exists various ways to determine the weight matrix. In this work, we choose the Metropolis rule [29] because of its stability, adaptability to topology changes and near-optimal performance. The element of weight matrix is found as Equation (5)

$$w_{ij} = \begin{cases} \frac{1}{max(n_i+1, n_j+1)}, & \text{if } i \in \mathcal{N}_j \\ 0, & \text{if } i \notin \mathcal{N}_j \\ 1 - \sum_{i \in \mathcal{N}_j} a_{ij}, & \text{if } i = j \end{cases} \tag{5}$$

where $n_i = |\mathcal{N}_i|$, $w_{ij} \in [0, 1]$.

3. Design of Agent with the Plug and Play Feature

In this section, we introduce the design of agents with plug and play feature for distributed secondary control in an islanded MG with multiple grid-forming inverters. The MG includes a number of ESSs with power electric inverter interface operated in parallel. All of the ESSs participate in regulating frequency and voltage to keep the grid in the steady state. In this work, we focus on frequency control. Due to multi-master strategy, the coordination between inverters in the grid is mandatory. The operation of an ESS, which is connected to grid through an inverter based interface, is separated into three parts in the proposed layer structure, as described in Figure 2. The PI controllers of inverters requires setpoints from agents to recover the frequency to normal once disturbances occur in the MG. The agent in this work is designed to implement the average consensus algorithm presented in previous section. The process of the algorithm is iterative. The state in initial iteration of an agent is the input of the agent, which is frequency deviation sensed locally from the device layer. Agent output, serving as feedback to the controller, is the average of inputs of all agents in the system. The output is collected after a specific number of iterations.

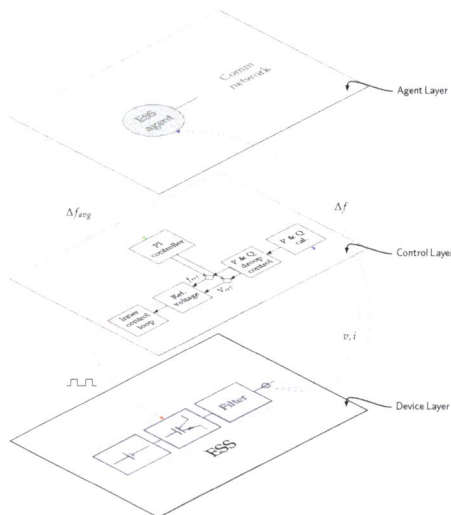

Figure 2. An inverter based interface ESS in the layer structure.

The iterative process in an agent is described in Algorithm 1. The agent conducts consecutive consensus loop. A loop is begun from Iteration 0 when the agent receives the measurement from devices and is finished at Iteration t_0. Upon reaching the consensus state at Iteration t_0, the agent sends its final state to the corresponding controller and immediately jumps to a new loop at Iteration 0 again.

Algorithm 1 The average consensus process in Agent i.

1: $t = 0$ ▷ begin a loop at initial iteration
2: $\mathcal{N}^i \leftarrow \mathcal{N}_0^i$ ▷ list of neighborhood agents
3: $n_i = |\mathcal{N}_0^i|$ ▷ number of neighbors
4: $x^i \leftarrow x_0^i$ ▷ obtain initial state from Device layer, this state is value of frequency deviation measured
     locally at node i
5: distribute the initial value and number of neighbors to all neighbors
6: collect the initial value and number of neighbors from neighbor agents x_j, n_j $j \in \mathcal{N}^i$
7: $w_{ij} = \begin{cases} \frac{1}{max(n_i+1,n_j+1)}, & \text{if } i \in \mathcal{N}^j \\ 1 - \sum_{i \in \mathcal{N}^j} a_{ij}, & \text{if } i = j \end{cases}$ ▷ calculate elements of weight matrix involved in Agent i and
     its neighbors using Metropolis rule
8: collect the initial value of neighbors
9: $t = t + 1$ ▷ move to Iteration 1
10: **while** $t < t_0$ **do** ▷ t_0 is the number of iteration needed to reach the consensus state
11:      $x^i(t) = \sum_{j \in \mathcal{N}_i} w_{ij} x^j(t-1)$ ▷ update the state at Iteration t
12:      distribute the updated state at Iteration t to all neighbors
13:      collect the state of all neighbors at Iteration t
14:      $t = t + 1$ ▷ move to next iteration
15: **return** $x^i(t_0)$ ▷ consensus value which is the average of measured frequency deviation
16: send the consensus value to Control layer (to PI controller) ▷ finish the current loop
17: **redo** from step 1 ▷ start a new loop

Intuitively, what happens in agents is separated into three phases:

1. Initialization phase: Each agent receives initial state which is its local frequency deviation. Data are transferred from Device layer to Agent layer.
2. Updating state phase: States of next iterations in each agent are updated using the agent current state and neighbors' states following Metropolis rule. An agent will move from Iteration t to Iteration $t + 1$ if and only if it collects information from all neighbors at Iteration t. Data are then transferred internally within the Agent layer.
3. Returning value phase: At a specific iteration, all agents finish consensus process loop and send the same average value of frequency deviation to controllers. Data are transferred from Agent layer to Control layer.

The calculation for each iteration relies on information being received from neighbors. The consensus processes in agents are therefore almost at the same iteration (not always at the same iteration due to minor differences introduced by time taken to exchange data amongst the agents). It can be imagined that all agents are on a line and all elements in this line march ahead from an iteration to the next iteration together in a "lock-step" manner. If an issue occurs with any element, this line will stop moving on until the issue is fixed.

In MGs, the topology and the total capacity may change subject to the increase in load and fluctuations in renewable energy sources. The global consensus based operation of the MG has to be capable of adapting to this frequent alteration of structure and capacity without major re-configuration. In our research, we design the agent system with the capability of plug and play operation, i.e., the network and the algorithm need to automatically detect and adapt to addition and/or removal of agents.

Figure 3 describes the logic implemented within agents when Agent *i* is shut down owing to its corresponding ESS *i* being out of service. We also consider agents who are connected with Agent *i*. Agent *j* is one of the neighbors of Agent *i*. When obtaining signal from Device layer and knowing that the ESS it handles was tripped out, Agent *i* triggers its process of shutting down. It sends signals to all neighbors to inform its status before stopping. In term of Agent *j* (as well as other neighbors of Agent *i*), when receiving the alert from Agent *i* at Iteration *t*, it will pause the process of updating state and start the reconfiguration process. Because Agent *j* lost one neighbor, the neighbors of Agent *j* also have to recompute the weight matrix elements. The agent system are paused at Iteration *t* until all involved agents finish modifying and return to the updating process.

Figure 4 presents the mechanism of an Agent *i* and its neighbors when the Agent *i* is added into the operating multi-agent system (an ESS is installed to MG). The unknown integration of Agent *i* in the agent network may cause the disturbance to the involved agents. Agent *j* is one neighbor of Agent *i*. The task of all involved agents in this case is more complicated because Agent *i* has no information about current iteration of agent system that may break the synchronization and accuracy in computation of agents. We propose a way to overcome this challenge as follows: Once Agent *i* is notified that its corresponding ESS (ESS *i*) is connected to the MG, it will inform its neighbors about its appearance in the agent system and require the current Iteration *t* of system in reverse. Simultaneously, Agent *i* takes parameters from its neighbors to compute weight matrix elements. Neighbor *j* deals with this scenario in the similar way when Agent *i* is removed. An additional step in this case is only that Agent *j* broadcasts the current iteration to Agent *i*. The multi-agent system after that moving to next iterations and operating normally.

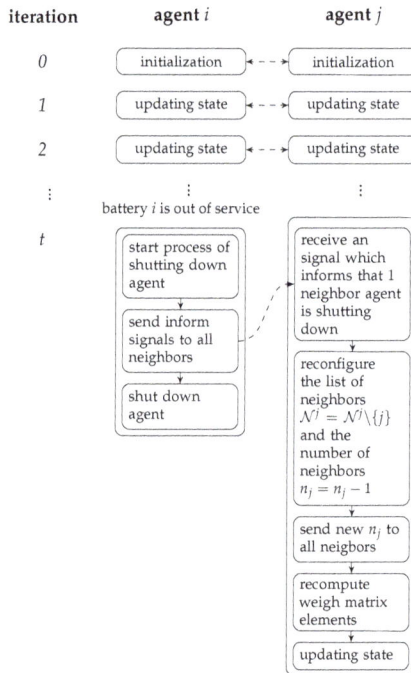

Figure 3. Algorithm of agent system when an ESS is out of MG.

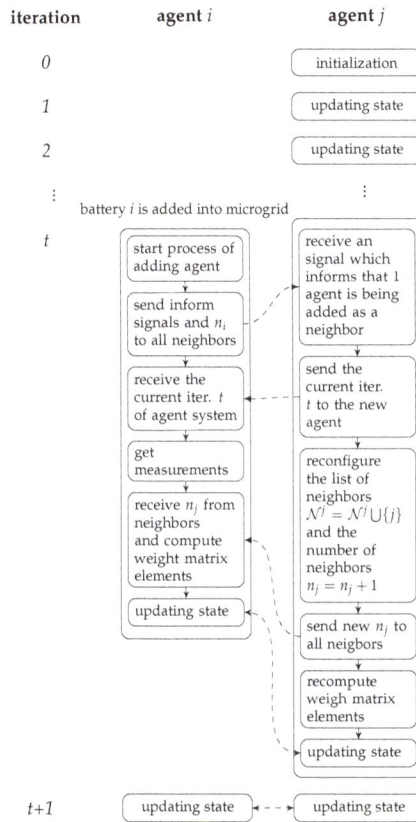

Figure 4. Algorithm of agent system when an ESS is added to MG.

This proposal minimizes human intervention in network operation upon alterations of topology due to addition or removal of agents. While infrastructure of MG can be supplied by various vendors and uses various multiple protocols, the proposed system can ensure interoperability at Agent layer and therefore facilitates the integration and coordination of assets in MG. To demonstrate the proposed architecture and its plug and play feature, in the following section, a case study of distributed frequency control in MG is presented. The case-study is implemented on a laboratory platform using controller and power HIL environment incorporating real communications network.

4. Validation

To validate the distributed control algorithm with the proposed architecture of agents, we consider a MG as depicted in Figure 5.

4.1. Platform Design for Validation of Distributed Control in MG

The experimental platform for this case-study (as shown in Figure 6) consists of two main groups of components: firstly a PHIL capability with a power inverter as the power component and secondly the CHIL setup with a MAS performed in a realistic communications network.

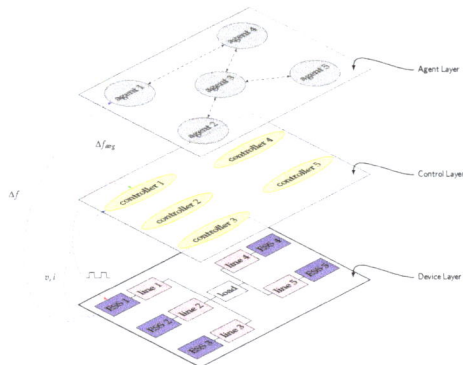

Figure 5. The MG case study in the layer structure.

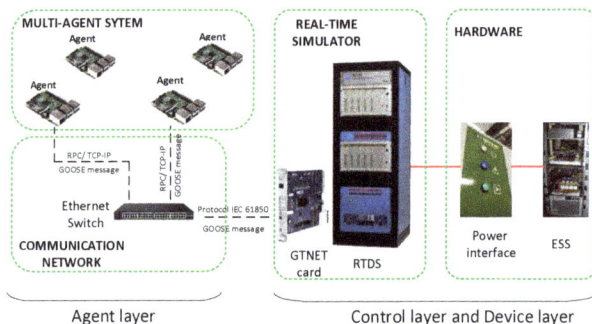

Figure 6. The designed laboratory platform.

4.1.1. PHIL with Power Inverter

This part of the platform covers the Device layer and the Control layer in the layer structure. The MG (Device layer) and local controllers (Control layer) of the inverters were implemented in Real Time Digital Power System Simulator (RTDS). The testing of the distributed control can be significantly improved by adding one more layer of reality to the experiments by driving real hardware components with the controller under test. For the introduction of real dynamics into the test case, the system includes two sections: one section is composed of one ESS which is the hardware battery, emulated by the inverter, with its corresponding local controller, and one section is composed of the remaining elements of the MG simulated within the real-time simulator from RTDS Technologies.

4.1.2. CHIL with MAS and Realistic Communications Network

Agent layer in the layer structure is represented by this part of the platform. A cluster of Raspberry PIs (RPI) with custom distributed control embedded forms the multi-agent system. The RPIs connect to the laboratory communications network through an Ethernet switch. The communications among agents is in a client/server manner and can be configured to correspond under any network topology. Each agent is a server that waits for incoming messages and dispatches them to the corresponding method calls but is also a client of neighboring servers. To implement the consensus algorithm in the MAS, we use the *aiomas* open source framework (https://aiomas.readthedocs.io). Each agent waits for input, then compute outputs based on the input and internal states. Specifically, in our case, input of an agent is sensed from local devices and adjacent information is received from nearby agents. Agents run the average consensus process in parallel and return the convergent values to the controllers. Moreover, the inter-agent data are transmitted through physical local area network.

Therefore, the convergence of distributed implementation and its impact to the power system operation is evaluated in a more realistic manner. Moreover, the disturbance of communication (i.e., latency, packet loss, cyber-attack, etc.) could be integrated directly to analyze the performance and the stability of system. Figure 7 explicitly illustrates the structure of agents that we designed. Beside of RPC server/client, asynchronous processes run in agents for implementing the algorithms as well as transferring data with outside devices.

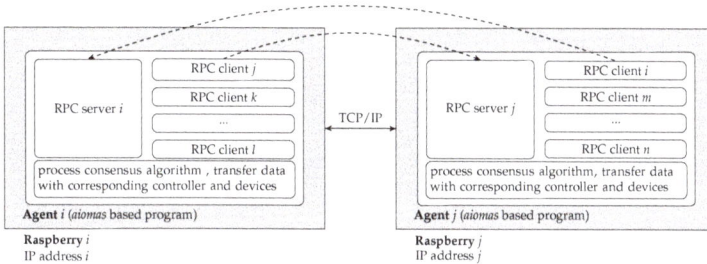

Figure 7. Structure of agents.

4.1.3. Interfaces between Agents and RTDS

In the context of this paper, RPIs transfer information to RTDS through GTNETcard using IEC 61850 GOOSE (Generic Object Oriented Substation Event) protocol. IEC 61850 is an industrial protocol which improve interoperability, reduce the time required for sending real-time data and its approach is closer to industrial applications [30]. The objective is to provide utilities with a common advanced protocol for transferring information of inverter-based DERs from different vendors. The signals are distributed to make sure that RPIs can assess only local data.

4.2. Testing Procedure

A case study of secondary control in an autonomous MG was implemented to verify the operation of the designed MAS by using the laboratory platform. The test case MG includes one load which is supplied by five ESSs. Each ESS is interfaced with the MG through a power inverter and a filter. ESS 1 is the hardware battery controlled by an emulating controller. The inverters operate in parallel, and are controlled as grid-forming converters, controlling grid frequency and voltage. The operation of an ESS and its inverter, filter as well as its local controller with agent is explicitly presented in Section 2. The parameters of inverter droop controllers were chosen to distinguish clearly the transient behavior of local frequency. The selection is also appropriate with respect to real world deployment. Many ESSs with various power capacities can be installed into the system. The rated active and droop coefficient of the controllers are presented in Table 1. The proportional gains and the integral time constants of secondary controllers of all inverters are identical.

The proposed layer structure is used to describe the test system as Figure 5. By separating the system into distinct layers, we can have a thorough overview of the system and see how data are transferred between devices, controllers and agents. For simplicity, Figure 5 only shows data flows of ESS 1 and its controller and agent. Data flows for other ESSs are identical. The controllers of inverters are decentralized as illustrated in Control layer because they only contact with local units. The system information can be obtained via agents. In the distributed manner, the information agents receive is not global but only from adjacent agents.

Table 1. Parameters of ESS inverter controllers.

	Parameter	Value	Unit
Inverter 1	P_0^1	3	kW
	k_P^1	100	Hz/kW
Inverter 2	P_0^2	8	kW
	k_P^2	200	Hz/kW
Inverter 3	P_0^3	11	kW
	k_P^3	50	Hz/kW
Inverter 4	P_0^4	10	kW
	k_P^4	100	Hz/kW
Inverter 5	P_0^5	9	kW
	k_P^5	250	Hz/kW
Secondary controllers	K_p	0.01	
	K_i	0.12	

The performance of MAS is proven by showing that the system is stable and frequency is controlled under various changes of the MG. Eight different scenarios with alteration of network topology are emulated as illustrated in Figure 8. Initially, the MG operates in a steady state, i.e., all ESSs are connected to the system. The connection among agents is presented in Figure 8a. In Scenarios 1 and 2, we increase and decrease load power respectively. In Scenario 3, ESS 5 is disconnected leading to the removal of Agent 5 out of the MAS as in Figure 8b. Then, the load is changed in Scenarios 4 and 5 to verify the operation of the agent system after removing one unit. In Scenario 6, we trip ESS 1 to test the adaption ability of MAS with hardware device. In this scenario, the MAS operates with only three agents, as shown in Figure 8c. Finally, in Scenario 7, ESS 1 is reconnected to the system and in Scenario 8, the load is changed again to justify the operation of MAS upon addition of an agent. The experiment procedure was carried out continuously and throughout from Scenario 1 to Scenario 8 to prove ability of on-line self configuration of agents. A scenario commences when the previous scenario is completed and the system has reached steady state (i.e., frequency has returned to its nominal value).

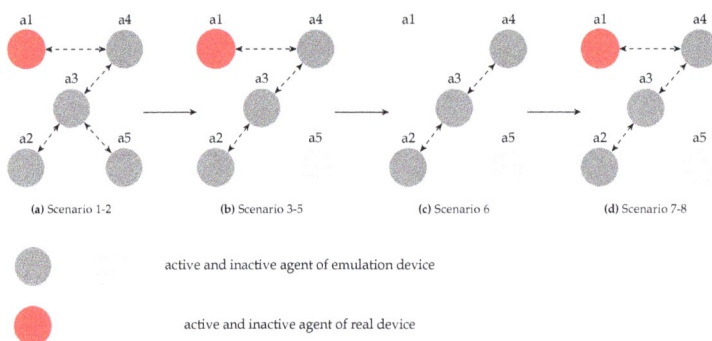

(a) Scenario 1-2   (b) Scenario 3-5   (c) Scenario 6   (d) Scenario 7-8

active and inactive agent of emulation device

active and inactive agent of real device

Figure 8. Topology of MAS in ech scenario.

4.3. Experimental Results

Efficiency of the proposed distributed control is verified by observing the following: (1) iterative state values in each agent to show convergence performance of consensus process; (2) values of signals controllers receives from agents; (3) local frequency measurements; (4) ESS active power outputs; and (5) consensus computation performance.

4.3.1. Step Change of Load Active Power

In the first two scenarios, when the MG consists of five ESSs, changes in active power of the load are simulated to examine the operation of system under fixed (normal) condition of network topology. Firstly, in Scenario 1, the load power was increased from 27 kW to 36 kW. Then, Scenario 2 was implemented by reducing the load power to 25 kW. The results of the two scenarios are presented in Figures 9 and 10 respectively.

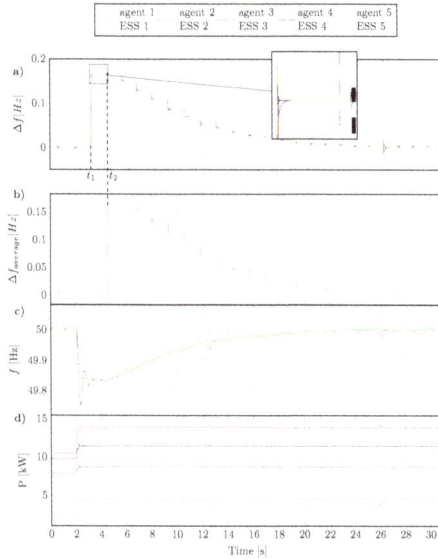

Figure 9. Scenario 1. (**a**) The state values of agents; (**b**) Consensus values sent to controllers from agents; (**c**) Frequency values measured at output of ESSs; and (**d**) Active power output of ESSs.

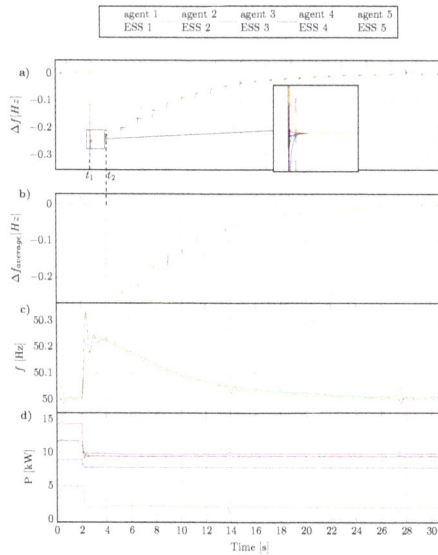

Figure 10. Scenario 2. (**a**) The state values of agents; (**b**) Consensus values sent to controllers from agents; (**c**) Frequency values measured at output of ESSs; and (**d**) Active power output of ESSs.

As in Figures 9d and 10d, at the time the load was altered, power outputs of all ESSs are changed accordingly to compensate the unbalanced power following the droop rule to stabilize system. Specifically, during about first 2 s, where the inverters were under only primary control, a steady-state frequency deviation from the nominal value exists as observed in Figures 9c and 10c.

To express thoroughly the computation of a consensus loop process in agents, we consider a duration from t_1 to t_2 as illustrated in Figures 9a and 10a. Agents process the calculation as presented in Algorithm 1. At t_1, corresponding to Step 1 in the algorithm, all agents receive new initial states and the local frequency deviations. The agents then exchange information, conduct the calculation, and obtain the convergence at t_2, corresponding to Step 17 in the algorithm. The considered process was finished when the results (state values at t_2) were sent to the controllers. The new consensus loop was begun upon receiving new initial state by means of updating the measurements.

The statistics of the calculation time for a consensus process are shown in Figure 11. The time is collected based on logging operation of agents. The values is not immutable but fluctuates in the range mainly from 1.17 s to 1.26 s. This is because the agents communicate in a real physical network environment in the laboratory. Even though the delay for transporting data may be nonsensical due to short distances between agents (raspberry PIs in the designed platform), the performance of transferring data has closely approached to practical network implementation.

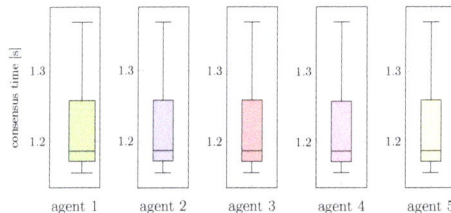

Figure 11. Time of one loop consensus process Scenarios 1–2.

It can be observed that, although communications delays are uncertain, consensus time is still nearly analogous in all agents. Moreover, five traces of controller inputs (Figures 9b and 10b) overlap demonstrating synchronous operation of MAS to send the average values of frequency errors to secondary controllers concurrently. The MAS takes the role similar to a central controller in the centralized control regime to send global information to local controllers. The controller of an ESS inverter is a closed loop control system with feedback signal is the frequency error. Instead of receiving continuously from the central controller, the PI controller in secondary control level of this system updates the feedback from the agent. However, as above analysis, the agent need time for completing a consensus loop process and reach the final state. The sample rate of feedback signal depends on the calculation time in agent and the time for sending data. The network quality and computation performance of agents therefore can significantly affect to the control system which in turn may cause instabilities in the grid. Hence, the selection of parameters of PI controllers plays an important role in controlling the system. Frequency of the grid system under proposed distributed control was restored accurately to its nominal value within approximately 30 s of the occurrence of the disturbance.

4.3.2. Disconnecting an ESS

In Scenario 3, ESS 5 is tripped and in turn Agent 5 will be out of service. At this time, the topology of MAS network was transformed by excluding one node and one connection line as in Figure 8b. Agent 3 loses a neighbor, thus values of elements of weight matrix are no longer correct. The modification for Agent 3 is mandatory for proper consensus process in MAS. The reconfiguration process in Agent 3 is triggered when receiving inform message from Agent 5 as described in Section 3. Figure 12 presents the results of system in Scenario 3. When ESS 5 is disconnected, the remaining ESSs have to increase of power output to share the power previously supplied by ESS 5. The frequency is reduced as the result of droop controllers. Figure 12a shows the convergence of consensus processes in agents which proves the capability of on-line adaptability of Agent 3 when its neighbor—Agent 5—is removed. The frequency of system is controlled to return to reference value after the trip event occurs.

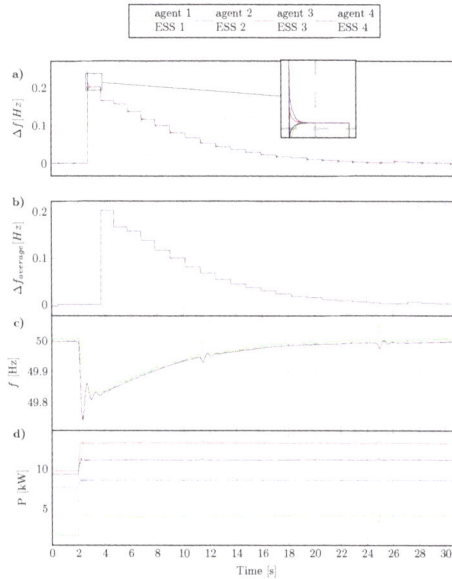

Figure 12. Scenario 3. (**a**) The state values of agents; (**b**) Consensus values sent to controllers from agents; (**c**) Frequency values measured at output of ESSs; and (**d**) Active power output of ESSs.

To inevitably verify the operation of agents against disturbances, two different events of changing load power were conducted in two successive scenarios. In Scenario 4, the load power is increased to 30 kW and, in Scenario 5, it is reduced to 20 kW. The results are presented respectively in Figures 13 and 14 to ascertain the agent-based distributed algorithm. Similar to the results collected in previous scenarios, the system frequency is gradually restored and kept steady at 50 Hz as initial state. The time for a consensus loop process of the MAS with 4 agents depicted in Figure 15. Each process needs approximately 1.1 s to be accomplished. Compared with the first scenarios, it is slightly faster due to the reduction of MAS complexity and the decreasing quantity of agents.

Scenario 6 was implemented to check the resilience of system when the physical hardware ESS was disconnected from the grid. Agent 1 was shut down along with ESS 1 and eliminated from neighbor list of Agent 4. Agent 4, as aforementioned, also reconfigured itself to adapt to the new condition. The topology of MAS is switched as Figure 8c with only three agents and two communication lines. To ensure firmly the safety of devices, the real ESS was not tripped out abruptly. Alternatively, we declined gradually the load power to zero. Therefore, as can be seen from the results depicted in Figure 16, the remaining ESSs did not change immediately but increased slowly and reached to stable values after about 20 s. Although there are significant differences in implementation, the system was still robust to disturbances under distributed control with the MAS. The convergence of computation in agents was assured to send precise signals to secondary controllers. Figure 17 shows the performance of consensus processes in the agents. Agents computed faster (mainly about 0.86 s) yet ensured to reach the consensus state. The system with consistently chosen PI parameters was proved to be stable under various changes of feedback signals.

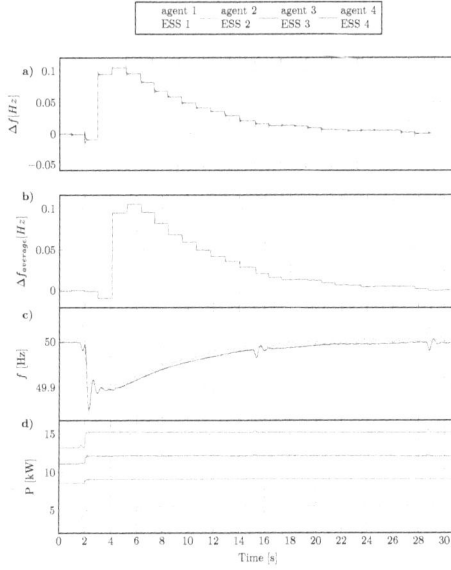

Figure 13. Scenario 4. (**a**) The state values of agents; (**b**) Consensus values sent to controllers from agents; (**c**) Frequency values measured at output of ESSs; and (**d**) Active power output of ESSs.

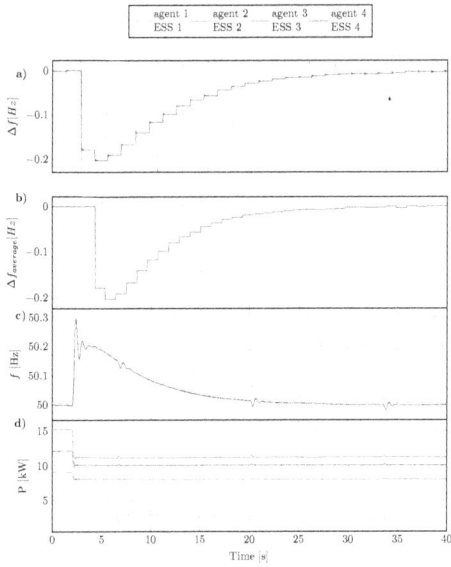

Figure 14. Scenario 5. (**a**) The state values of agents; (**b**) Consensus values sent to controllers from agents; (**c**) Frequency values measured at output of ESSs; and (**d**) Active power output of ESSs.

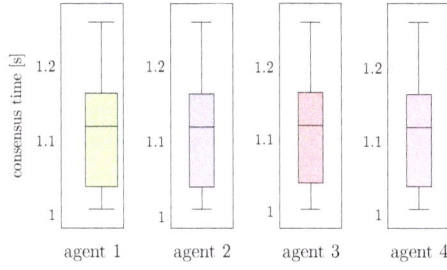

Figure 15. Time of one loop consensus process Scenarios 3-5.

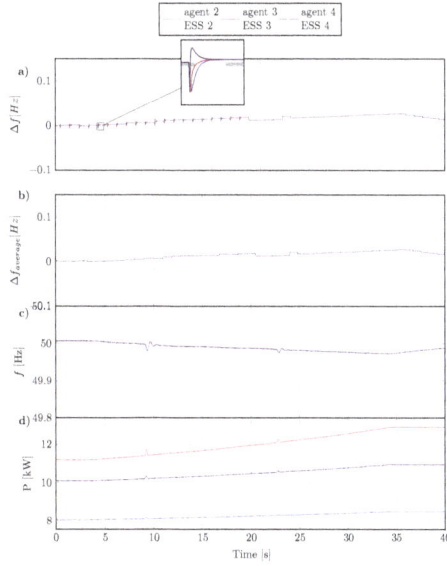

Figure 16. Scenario 6. (**a**) The state values of agents; (**b**) Consensus values sent to controllers from agents; (**c**) Frequency values measured at output of ESSs; and (**d**) Active power output of ESSs.

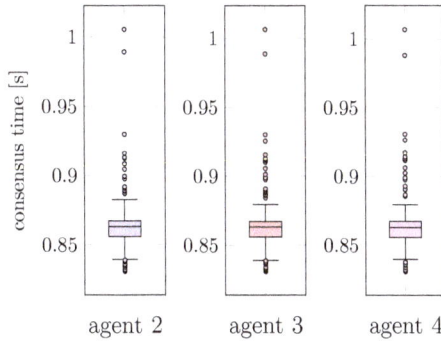

Figure 17. Time of one loop consensus process Scenario 6.

4.3.3. Connecting an ESS to MG

In Scenario 7, we check the implementation of the MAS as well as the operation of the test case when one more agent is added. In this case, we reconnected the hardware ESS which was tripped out in previous scenarios and the system comes back to the state as in Scenarios 3–5. Agent 1 was activated again and the connection link between Agent 1 and Agent 4 was also established as Figure 8d. As described in Section 3, the agents in this case have to handle more complex tasks to embed a new agent into an operating MAS. Agent 1 was started together with ESS 1 and it instantly informed its appearance in the network to neighbor (Agent 4). It is noted that in order to avoid a high transient current and cause adverse effects to devices when connecting ESS 1 to the grid, instead of closing a breaker, we adjusted the reference power P_0 in controller of Inverter 1 to increase gradually the power output of ESS 1 to desired value. The results of this scenario are shown in Figure 18.

Figure 18a illustrates the convergence of state values in MAS after adding new agent. Before proceeding the consensus computation, at initial phase, Agent 1 waited for feedback from neighbors (Agent 4) to seek current iteration of MAS. Agent 4 also included Agent 1 to be one of its neighbors. Although the secondary process for regularizing frequency was prolonged due to the connection procedure of physical ESS, the results expressed that the system was still under robust control to be stabilized in the nominal state. An power increase of load was then conducted to prove the proper operation of the system in the new state as shown in Figure 19.

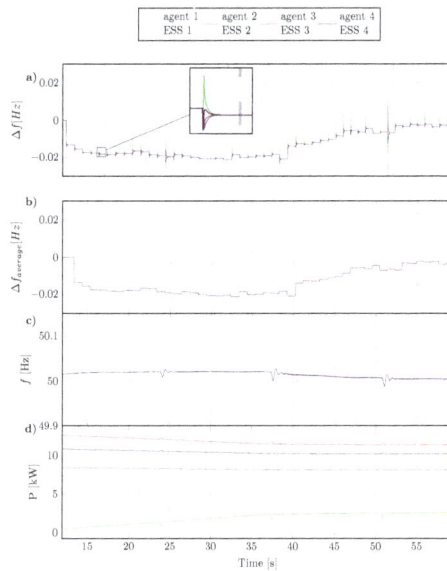

Figure 18. Scenario 7. (**a**) The state values of agents; (**b**) Consensus values sent to controllers from agents; (**c**) Frequency values measured at output of ESSs; and (**d**) Active power output of ESSs.

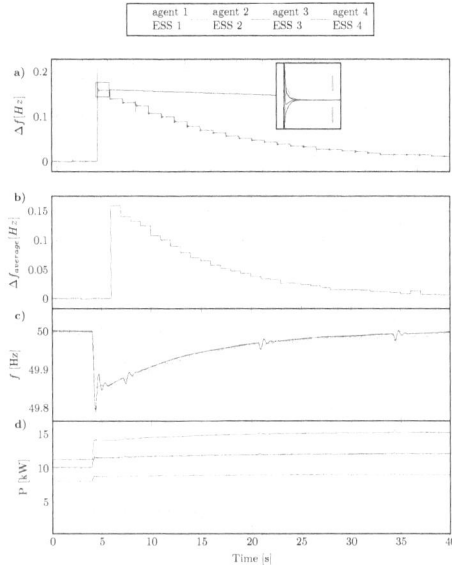

Figure 19. Scenario 8. (**a**) The state values of agents; (**b**) Consensus values sent to controllers from agents; (**c**) Frequency values measured at output of ESSs; and (**d**) Active power output of ESSs.

5. Conclusions

This paper proposed a multi-agent system with plug and play feature for operation in distributed secondary frequency control of islanded MG. The agents are designed to interface devices from different vendors via the industrial protocol IEC 61850. The agents implement an average consensus algorithm by using local measurements and interacting with neighbors to return the identical average signals of frequency deviations to local controllers. The proposed approach is resilient to various disturbances in MG and provides plug and play feature that enable online re-configuration of the network in the case of alteration of topology, hence, interoperability at agent layer. A test case autonomous MG with various scenarios was conducted to validate the operating of the proposed agent under PHIL setup, hardware agent system and realistic communication network. The results have shown that under alterations of network topology, the recovery of grid frequency was always ensured with the MAS control. Communications between agents and consensus time were also investigated with real implementation.

In the future work, the ability to work with large scale system will be investigated and the experimental platform will be improved to deal with more cyber-security issues.

Author Contributions: T.-L.N. conceived the research idea with suggestions from C.G. and N.-A.L.; T.-L.N., E.G.-S., M.H.S. performed the experiments and analyzed the data; S.M.B. and L.R. contributed to the HIL setup; T.-L.N., E.G.-S., M.H.S. and V.-H.N. wrote the paper; and Q.-T.T., R.C and G.M.B. oversaw the work and proofread the paper.

Funding: This work was partly supported by: (1) the European Community's Horizon 2020 Program (H2020/2014–2020) under the project "ERIGrid: European Research Infrastructure supporting Smart Grid Systems Technology Development, Validation, and Roll Out" (Grant No. 654113); and (2) the European Union Seventh Framework Programme (FP7) under the project ELECTRA IRP (Grant No. 609687).

Conflicts of Interest: The authors declare no conflict of interest.

References

1. Hatziargyriou, N. *Microgrids: Architectures and Control*; John Wiley & Sons: Hoboken, NJ, USA, 2014; pp. 1–344. [CrossRef]
2. Yazdanian, M.; Mehrizi-Sani, A. Distributed control techniques in microgrids. *IEEE Trans. Smart Grid* **2014**, *5*, 2901–2909. [CrossRef]
3. Olivares, D.E.; Mehrizi-Sani, A.; Etemadi, A.H.; Cañizares, C.A.; Iravani, R.; Kazerani, M.; Hajimiragha, A.H.; Gomis-Bellmunt, O.; Saeedifard, M.; Palma-Behnke, R.; et al. Trends in microgrid control. *IEEE Trans. Smart Grid* **2014**, *5*, 1905–1919. [CrossRef]
4. Guerrero, J.M.; Chandorkar, M.; Lee, T.; Loh, P.C. Advanced Control Architectures for Intelligent Microgrids; Part I: Decentralized and Hierarchical Control. *IEEE Trans. Ind. Electron.* **2013**, *60*, 1254–1262. [CrossRef]
5. Gomez-Sanz, J.J.; Garcia-Rodriguez, S.; Cuartero-Soler, N.; Hernandez-Callejo, L. Reviewing microgrids from a multi-agent systems perspective. *Energies* **2014**, *7*, 3355–3382. [CrossRef]
6. Colson, C.M.; Nehrir, M.H. Comprehensive real-time microgrid power management and control with distributed agents. *IEEE Trans. Smart Grid* **2013**, *4*, 617–627. [CrossRef]
7. Kantamneni, A.; Brown, L.E.; Parker, G.; Weaver, W.W. Survey of multi-agent systems for microgrid control. *Eng. Appl. Artif. Intell.* **2015**. [CrossRef]
8. Dragicevic, T.; Wu, D.; Shafiee, Q.; Meng, L. Distributed and Decentralized Control Architectures for Converter-Interfaced Microgrids. *Chin. J. Electr. Eng.* **2017**, *3*, 41–52.
9. Shafiee, Q.; Stefanovic, C.; Dragicevic, T.; Popovski, P.; Vasquez, J.C.; Guerrero, J.M. Robust Networked Control Scheme for Distributed Secondary Control of Islanded Microgrids. *IEEE Trans. Ind. Electron.* **2014**, *61*, 5363–5374. [CrossRef]
10. Dehkordi, N.M.; Sadati, N.; Hamzeh, M. Fully Distributed Cooperative Secondary Frequency and Voltage Control of Islanded Microgrids. *IEEE Trans. Energy Convers.* **2017**, *32*, 675–685. [CrossRef]
11. Zhang, G.; Li, C.; Qi, D.; Xin, H. Distributed Estimation and Secondary Control of Autonomous Microgrid. *IEEE Trans. Power Syst.* **2017**, *32*, 989–998. [CrossRef]
12. Dehkordi, N.M.; Sadati, N.; Hamzeh, M. Distributed Robust Finite-Time Secondary Voltage and Frequency Control of Islanded Microgrids. *IEEE Trans. Power Syst.* **2017**, *32*, 3648–3659. [CrossRef]
13. Raju, L.; Milton, R.S.; Mahadevan, S. Multi agent systems based distributed control and automation of micro-grid using MACSimJX. In Proceedings of the 2016 10th International Conference on Intelligent Systems and Control (ISCO), Coimbatore, India, 7–8 January 2016; pp. 1–6. [CrossRef]
14. Harmouch, F.Z.; Krami, N.; Benhaddou, D.; Hmina, N.; Zayer, E.; Margoum, E.H. Survey of multiagents systems application in Microgrids. In Proceedings of the 2016 International Conference on Electrical and Information Technologies (ICEIT), Tangiers, Morocco, 4–7 May 2016; pp. 270–275. [CrossRef]
15. Nelson, A.; Chakraborty, S.; Wang, D.; Singh, P.; Cui, Q.; Yang, L.; Suryanarayanan, S. Cyber-physical test platform for microgrids: Combining hardware, hardware-in-the-loop, and network-simulator-in-the-loop. In Proceedings of the IEEE Power and Energy Society General Meeting, Boston, MA, USA, 17–21 July 2016. [CrossRef]
16. Kotsampopoulos, P.C.; Lehfuss, F.; Lauss, G.F.; Bletterie, B.; Hatziargyriou, N.D. The limitations of digital simulation and the advantages of PHIL testing in studying distributed generation provision of ancillary services. *IEEE Trans. Ind. Electron.* **2015**. [CrossRef]
17. Guillo-Sansano, E.; Syed, M.H.; Roscoe, A.J.; Burt, G.M. Initialization and Synchronization of Power Hardware-In-The-Loop Simulations: A Great Britain Network Case Study. *Energies* **2018**, *11*. [CrossRef]
18. Nguyen, T.; Guillo-Sansano, E.; Syed, M.H.; Blair, S.M.; Reguera, L.; Tran, Q.; Caire, R.; Burt, G.M.; Gavriluta, C.; Nguyen, V. Systems Level Validation of a Distributed Frequency Control Algorithm. In Proceedings of the 2018 IEEE International Conference on Environment and Electrical Engineering and 2018 IEEE Industrial and Commercial Power Systems Europe (EEEIC/I CPS Europe), Palermo, Italy, 12–15 June 2018; pp. 1–6. [CrossRef]
19. Shafiee, Q.; Guerrero, J.M.; Vasquez, J.C. Distributed secondary control for islanded microgrids-a novel approach. *IEEE Trans. Power Electron.* **2014**. [CrossRef]
20. Chen, M.; Syed, M.H.; Sansano, E.G.; McArthur, S.D.; Burt, G.M.; Kockar, I. Distributed negotiation in future power networks: Rapid prototyping using multi-agent system. In Proceedings of the IEEE PES Innovative Smart Grid Technologies Conference Europe, Ljubljana, Slovenia, 9–12 October 2017. [CrossRef]

21. Guo, J.; Hug, G.; Tonguz, O.K. On the Role of Communications Plane in Distributed Optimization of Power Systems. *IEEE Trans. Ind. Inform.* **2017**. [CrossRef]

22. Nguyen, V.H.; Tran, Q.T.; Besanger, Y. SCADA as a service approach for interoperability of micro-grid platforms. *Sustain. Energy Grids Netw.* **2016**, *8*, 26–36. [CrossRef]

23. Nguyen, V.H.; Besanger, Y.; Tran, Q.T.; Nguyen, T.L. On Conceptual Structuration and Coupling Methods of Co-Simulation Frameworks in Cyber-Physical Energy System Validation. *Energies* **2017**, *10*, 1977. [CrossRef]

24. Tolk, A.; Muguira, J. The levels of conceptual interoperability model. In Proceedings of the 2003 Fall Simulation Interoperability Workshop, Orlando, FL, USA, 14–19 September 2003.

25. Lu, X.; Yu, X.; Lai, J.; Wang, Y.; Guerrero, J.M. A Novel Distributed Secondary Coordination Control Approach for Islanded Microgrids. *IEEE Trans. Smart Grid* **2017**. [CrossRef]

26. Han, Y.; Li, H.; Shen, P.; Coelho, E.; Guerrero, J. Review of Active and Reactive Power Sharing Strategies in Hierarchical Controlled Microgrids. *IEEE Trans. Power Electron.* **2016**, *8993*. [CrossRef]

27. Bidram, A.; Davoudi, A. Hierarchical structure of microgrids control system. *IEEE Trans. Smart Grid* **2012**, *3*, 1963–1976. [CrossRef]

28. Guerrero, J.M.; Vasquez, J.C.; Matas, J.; de Vicuna, L.G.; Castilla, M. Hierarchical Control of Droop-Controlled AC and DC Microgrids—A General Approach Toward Standardization. *IEEE Trans. Ind. Electron.* **2011**, *58*, 158–172. [CrossRef]

29. Sayed, A.H. Adaptive Networks. *Proc. IEEE* **2014**, *102*, 460–497. [CrossRef]

30. Blair, S.M.; Coffele, F.; Booth, C.D.; Burt, G.M. An Open Platform for Rapid-Prototyping Protection and Control Schemes With IEC 61850. *IEEE Trans. Power Deliv.* **2013**, *28*, 1103–1110. [CrossRef]

Article

Comparison of Power Hardware-in-the-Loop Approaches for the Testing of Smart Grid Controls

Falko Ebe [1,*], Basem Idlbi [1], David E. Stakic [1], Shuo Chen [1], Christoph Kondzialka [1], Matthias Casel [1], Gerd Heilscher [1], Christian Seitl [2], Roland Bründlinger [2] and Thomas I. Strasser [2]

[1] Smart Grids Research Group, Ulm University of Applied Sciences, 89075 Ulm, Germany; idlbi@hs-ulm.de (B.I.); stakic@hs-ulm.de (D.E.S.); chen@hs-ulm.de (S.C.); kondzialka@hs-ulm.de (C.K.); casel@hs-ulm.de (M.C.); heilscher@hs-ulm.de (G.H.)

[2] AIT Austrian Institute of Technology, Electric Energy Systems—Center for Energy, 1210 Vienna, Austria; christian.seitl@ait.ac.at (C.S.); roland.bruendlinger@ait.ac.at (R.B.); thomas.strasser@ait.ac.at (T.I.S.)

* Correspondence: ebe@hs-ulm.de; Tel.: +49-731-50-28350

Received: 25 October 2018; Accepted: 26 November 2018; Published: 3 December 2018

check for
updates

Abstract: The fundamental changes in the energy sector, due to the rise of renewable energy resources and the possibilities of the digitalisation process, result in the demand for new methodologies for testing Smart Grid concepts and control strategies. Using the Power Hardware-in-the-Loop (PHIL) methodology is one of the key elements for such evaluations. PHIL and other in-the-loop concepts cannot be considered as plug'n'play and, for a wider adoption, the obstacles have to be reduced. This paper presents the comparison of two different setups for the evaluation of components and systems focused on undisturbed operational conditions. The first setup is a conventional PHIL setup and the second is a simplified setup based on a quasi-dynamic PHIL (QDPHIL) approach which involves fast and continuously steady state load flow calculations. A case study which analyses a simple superimposed voltage control algorithm gives an example for the actual usage of the quasi-dynamic setup. Furthermore, this article also provides a comparison and discussion of the achieved results with the two setups and it concludes with an outlook about further research.

Keywords: Hardware-in-the-Loop; Software-in-the-Loop; Power-Hardware-in-the-Loop; Quasi-Dynamic Power-Hardware-in-the-Loop; smart grids; real-time simulation; validation and testing; decentralised energy system; smart grids control strategies

1. Introduction

The energy system has become a fundamental pillar of our society and has to be changed dramatically. The Three Ds—Decarbonisation, Decentralisation and Digitalisation—headline the major requirements of Western society these days [1]. The ongoing process of climate change speeds up under ominous conditions. New research results of the Intergovernmental Panel on Climate Change strengthen assumptions of the negative effects on human society [2,3].These effects were provoked by the humanity itself by emitting carbon dioxide in large amounts from burning fossil resources (i.e., oil, gas, and coal). On 4 November 2016 in Paris, the world community committed to an agreement that will be keeping global temperature well below two degrees Celsius above pre-industrial level [4]. To achieve this goal, the entirety of human society will have to change their energy supply infrastructure to a renewable energy-based system.

The transition towards renewable energy and decentralised power supply leads to several challenges for the electrical grid operation, particularly on the distribution level [5,6]. Handling the large number of different characteristics of a fluctuating renewable generation and orchestrating the flexibility of decentralised loads and generation units in context of the distribution grids are complex challenges. To avoid potential problems in the distribution grid, such as voltage violations and line or power transformer overloading, Distribution System Operators (DSO) need to undertake smart operation and planning strategies, based on Information and Communication Technology (ICT) [7].

Examples and background to these ideas are given in [8–11] which supposedly will result in more resilient infrastructure [12,13] and are the context of various research projects [14–17]. This resulting change from capital expenditure- to operational expenditure-dominated grid operation should supposedly result in lower overall costs. This was also suggested by ref [18].

Digitalisation in a decentralised energy grid produces many data. This is what a complex ICT system is supposed to take care of. Taking into account that the energy grid is a sensitive system with massive consequences for society in case of a collapse, the ICT has to be tested in a save environment of a laboratory before deployment into the real world. Different test approaches have to be evaluated for specific use cases and boundaries [19]. The "Three Ds" will headline not only the major requirements for the transformation of the energy system, but also the motivation to develop the approach of this paper. To summarise, the question where the answers were sought for was: How can one test and validate systems which are performing a smart grid control strategy?

For the laboratory testing, one seeks for an environment with a variety of parameters to control. Such a setup can be described with the term of real-time Hardware-in-the-Loop (HIL) concepts [20–22]. The basic idea behind it is to place a system in an environment in which all inputs and outputs can be controlled. In general terms, this means that a test candidate (system, component or algorithm) is placed in an environment, where the adjacent systems are simulated, and the behaviour at the points of interface is emulated. Figure 1 depicts this approach in a generic way. Based on the reaction of the System-under-Test (SUT), the simulated environment adopts accordingly.

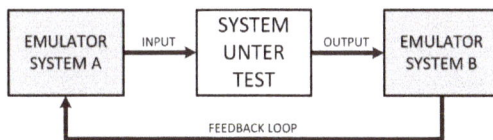

Figure 1. Generic setup of a X-in-the-loop as a concept for system testing.

For the testing of power system components, an additional interface is necessary, i.e. the Power Interface (PI), as it can provide voltages (or currents) in the typical range of a power hardware device. Linear amplifier, switched-mode amplifier and coupled motor-generator assemblies are typically used [23,24]. In this article, two representatives from the switched-mode and linear amplifier category are utilised for the experiments. Beside the physical properties of the PI, the used Interface Algorithms (IA) are relevant for the success, as these algorithms deal with the occurring stability problems. This is discussed in detail in [25–27]. Combined with the issues regarding the stability of a PHIL experiment, as discussed in [26,27], such an experiment cannot be considered as *plug'n'play*. A simplification to enable a boarder range of research institutes to carry out PHIL experiments is the transition from PHIL to Quasi-Dynamic PHIL (QDPHIL). The main idea is to increase the time step which will enable the possibility to use steady state load flow simulation tools for the process. The typical temporal resolution of a PHIL experiment is in the range of 10–100 µs [24], where, as with the implemented QDPHIL setup, a time step below one second is intended and realised. Through this simplification, the accuracy is supposedly reduced and the system setup needs to be analysed and the functionality validated.

This analysis is the scope of this paper. Details on the actual PHIL and QDPHIL setup used in this paper are given in Section 3. The transition from real-time calculation to steady state is discussed in more detail in Section 3.4. The QDPHIL concept will presumably enable the testing of superimposed control regimes as well as local control regimes such as a energy management systems. On the other hand, examples for topics which are in the scope examined with PHIL but not with QDPHIL experiments are given in [28–30]. All of the referenced experiments involve operation states which can be summarised as "disturbed operation", whereas QDPHIL focuses on undisturbed operation. A detailed definition of requirements for the QDPHIL setup is given in Section 3.1. In [31], QDPHIL and quasi-static PHIL are mentioned as topics of current investigation.

The remaining part of this contribution is organised according to the following schema: In Section 2, a more detailed introduction to PHIL setups and smart grid testing in general is given. This includes the discussion of multi-domain Smart Grid testing and the application of the HIL concept to other domains such as ICT. The used PHIL and QDPHIL setups are introduced in Section 3. Section 4 describes the chosen validation methodology for comparing them. Section 5 presents the results of the carried out experiments and Section 6 describes the application of the proposed test methodology at the example of a case study. The results of the comparison are discussed in Section 7. Finally, this article is concluded by Section 8 with an outlook for further steps. The results of this paper are based on the work of the "Smart beats Copper" project within the ERIGrid Trans-national Access funding framework [32].

2. Overview of Validation Approaches of Smart Grid Systems

2.1. Testing Methodologies and Concepts in General

Testing such a complex system as described in the Introduction is often a challenging and complex task as it spans a diverse field of areas of investigation. These areas range from a single component test up to the test of a complete system [19]. As a second variation, the domain of test can vary as well and range from small scale systems in the premise of the costumer to medium or to large scale power plants. They can be communication tests, function testing for complex systems, or long-term stability evaluation. To structure these aspects, one can refer to the Smart Grid Architecture Model (SGAM) [33], which is structured in its layers (i.e., component, communication, data model, function and business), its domains (generation, transmission, distribution, DER (Distributed Energy Resources) and costumer premise) and its zones (i.e., process, field, station, operation, enterprise and market). For each layer, a variety of examinations can be undertaken. Therefore, the object under investigation can range from components to complex systems.

For a complex system, such as smart grids, a multiple stage approach is advisable. These stages can be generalised in the following five categories which consecutively build upon each step. These processes are based on the suggestions in [19,34]:

- *Simulation-based Test:* Small-scale to large-scale tests of concepts in a simulated environment during an early phase of the development.
- *Controller Target Test:* Small-scale test with the control functions implemented on the actual target.
- *Laboratory Test:* Small-scale test with a limited number of components in a controlled environment. Complexity reaches from basic component tests up to complex system tests.
- *Pilot Test:* Small-scale test with a limited amount of components in an uncontrolled environment to evaluate the difficulty of a real-world application. Typically, up to medium complexity functions are tested.
- *Field Test:* Medium-scale test with a few hundred up to a few thousand components in a real environment for preparing a large scale implementation of the tested system.

2.2. Introduction to PHIL

A typical PHIL setup consists of three main parts, that are depicted in Figure 2 and described hereafter: the SUT, the PI and the Simulation Module [20–22,25,26]:

- *System-under-Test (may also be referred to as Hardware-under-Test (HUT), Device-under-Test (DUT) or Equipment-under-Test (EUT)):* In the domain of smart grids, it is typically some kind of generation or consumption unit whose properties or functions are the subject of the examination. Around this element the PHIL systems are set up.

- *Power Interface:* Provides the ability to set the operating points for the real electrical system (e.g., voltage). There are basically three main types of PI: linear power amplifier, switched-mode power amplifier and generator based PI [23,24]. The dependent parameters need to be measured and are fed back as analogue signals or digital values into the simulation process. Furthermore, a system setup can be distinguished by the set output variable of the PI, which can be either voltage or current [23]. The feedback of the measured values forms a closed-loop to synchronise the HIL simulations and SIL simulations, whereas in open-loop tests a feedback of the measurements is not part of the system setups.
- *Simulation Module:* The simulation system—usually a Digital Real-Time Simulator (DRTS)—receives the measured values and calculates new set points, which are sent to the power interface as analogue signals or digital values based on the type of the power interface. The simulation tools are distinguished by their calculation method. Typical methods are transient calculation, phasor calculation or steady state calculation. Besides the different methodologies of the simulation, the used IA is of particular importance. An overview of the different concepts and corresponding algorithms is given in [23,26,27].

Details on the possibilities for the calculation methodologies are given in Table 1.

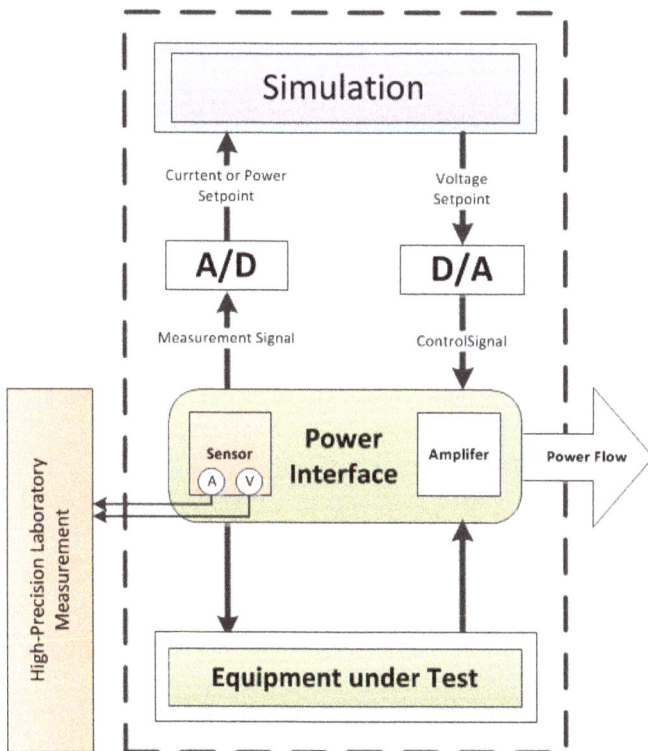

Figure 2. Generalised overview of a PHIL setup. The power interface applies a voltage value received from the simulation. The power interface exchanges electrical power with the EUT according to the applied voltage and the properties of the EUT.

Table 1. Comparison of different methods for the calculation of energy transfer and control processes.

Name	Description	References
Transient Calculation	Besides the fundamental frequency, all transient components of the input signals are used for the calculation of the voltage signals. This is also considered as electromagnetic transient (EMT) simulation.	([35], Ch. 13) [36,37]
Phasor Calculation Fundamental Frequency	In contrast to the transient calculation, one special case is considering only the fundamental fractions of the analogue input signals. In the context of PHIL, this can be used to ensure that SUT is operated under conditions with good properties in respect to power quality issues.	[37]
Effective Value Calculation	The effective value calculation method only considers the effective value of the input signals, which is typically used for steady state simulation. This is also illustrated as root mean square.	[37,38]

2.3. Smart Grid Testing and Usage of Additional X-in-the-Loop Concepts

Besides the PHIL concepts, there are further domains which can be simulated and controlled relevant for a holistic system testing of a smart grid strategy. They are listed and explained in Table 2. As the discussion of the trends in the previous section has shown, digitalisation is one of the key elements for future system design. When setting up a test environment for smart grid control strategies, communication components and communication protocols are as important as the actual electrical grid properties. Considering these aspects, the concept depicted in Figure 3 was developed and used in the case study discussed in Section 6. The experiment combined the implemented QDPHIL setup as well as SIL and ICT Hardware-in-the-Loop (ICTHIL) aspects for the test of a superimposed voltage control. This experiments gives a practical example for the use of QDPHIL system setup.

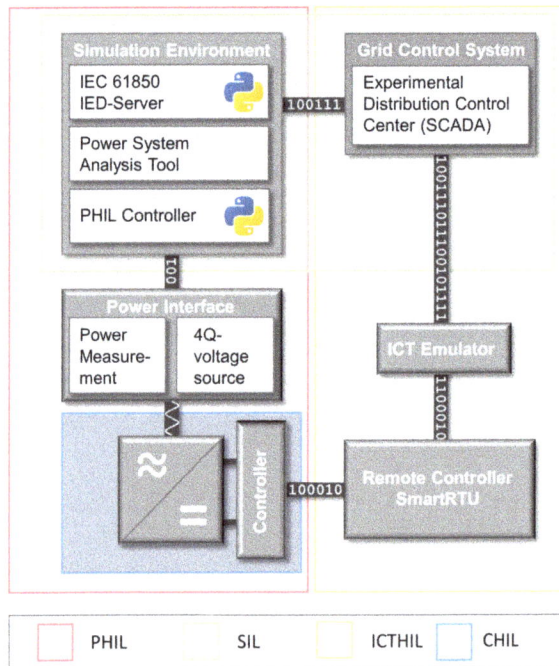

Figure 3. Schematic overview of generic setup for the test and validation of Smart Grid control concepts. Different aspects of the X-in-the-Loop concept are combined or can be used to examine a single topic.

Table 2. Comparison of different domains using the X-in-the-loop concept for system testing with focus on the energy sector.

Name	Description	References
Hardware-in-the-Loop (HIL)	The subject of the system testing can be any physical component that can provide small scale input and output signals. HIL can also be seen as general term and can be specified by the following test categories	[21,22]
Software-in-the-Loop (SIL)	Software components are the subject of a SIL test setup. The input and output signals are the calls and responses of the adjacent software components that are emulated.	[39]
Controller Hardware-in-the-Loop (CHIL)	Subjects are the controller boards of a power hardware device. A CHIL setup will therefore simulate the mechanical, electrical and power electronic parts of the power hardware device, which are controlled by the tested controller board. In contrast to a PHIL-setup, the controller board is introduced to the small signals which are typically provided by the power hardware. The reaction of the controller board to a disturbance can be examined and resulting voltages and currents can be evaluated regarding the relevant specification. CHIL tests are typically undertaken before lab testing of the complete device.	[22]
Power Hardware-in-the-Loop (PHIL)	The subject of this system is a component from the power system domain. In contrast to HIL, the input and output signals can be typical power systems levels, e.g., a common low-voltage connection with 400 V_{AC} and up to a nominal power of 1 MVA.	[20–22]
ICT Hardware-in-the-Loop (ICTHIL)	A complex smart grid system is typically based on communication and relies on package-based communication connection like IP/TCP or UDP, which often suffers from impairments like latency, package drop, package reorder or corruption of data in real-world applications. This means that, when testing a system, these effects have to be reproduced as well. Therefore, it is, e.g., necessary to precisely weaken the transmission path. This effect can be achieved by using a physical network emulator. If the communication network is simulated as well, the term Network-Simulation-in-the-Loop (NSIL) can be used.	[39–41]

3. Proposed System Setup

This section starts with the description of the requirements for the QDPHIL setup and the two implemented system setups which are compared in this contribution. This section is concluded with a discussion of the difference of phasor calculation use in an PHIL setup and the steady state calculation of the QDPHIL Setup.

3.1. Requirements for the Proposed Use

The planned use of a switched-mode PI as key component is for system testings of complex smart grid systems. The PI is available in both institutions involved in this contribution. The primary use of these PI are system function testings with a given test profile of the voltage which involves voltage swell, Q(U)-control, etc. It is suggested that the PI can also be used in a QDPHIL setup as described in the previous section, when the main goal is to analyse the system behaviour and not to focus on the dynamic behaviour of a single component involved. Regarding the set point of the PI of the PHIL and GDPHIL experiments, the setups were only designed for voltage set points, consequently only current and power, can be used as feedback signal. Systems that are in the scope of the test environment feature at least one of the characteristics stated in Table 3. Based on the characterisation of the tested systems, the requirements for QDPHIL setup are stated in the lower part of the table.

Table 3. Overview of the requirements for the proposed use.

Category	Domain	Requirement
Systems Properties	Involved Domains	SGAM Domain *Distribution, Decentralised Energy Resource* or *Customer Premise*
	Communication	communication over WAN (e.g., BPL or Mobil 4G)
	Communication	multiple communication protocols are used and converter
	Control	central and decentralised control
	Control	autonomous or partially autonomous control of multiple systems
	Timing	update and control cycles > 30 s
Requirements	Timing	PHIL cycle time $t_{PHIL} \leq 1$ s
	Accuracy	only sinusoidal waveform
	Accuracy	low voltage deviation $\Delta U \leq 1 V_{RMS}$
	Operation Mode	only voltage type interface algorithm

3.2. PHIL System Setup A: Opal-RT RT-LAB with Spitzenberger Spies PAS

The system analysed first was based on a linear signal amplifier (i.e., Spitzenberger Spies PAS [42]) as power interface which simulates a grid connection point from the simulated grid. The PI receives an analogue voltage signal from the DRTS (i.e., Opal-RT [43]), which generates the signal according to the parameters of the simulated grid connection point (e.g., voltage U_{pcc}). These parameters are calculated within a real-time dynamic simulation with a high-time resolution of typically 10–50 μs, based on a code-generated model converted from a MATLAB/Simulink model. To close the simulation loop, the DRTS is fed back with analogue current measurements, which are converted internally to digital values to be entered into the simulation model. These transient measurements correspond to the feed-in power of the PV inverter (i.e., EUT: Fronius PV inverter, [44]), which is connected physically to the power interface representing the Point of Common Coupling (PCC) of the grid. As the IA for the setup, a voltage type Interface utilising an Ideal Transformer Model (ITM) [23] was selected. For the purpose of recording the simulation results, an external measurement system was connected to the PCC (cf. [45]). The described system is illustrated schematically in Figure 4 along with the second setup, which is discussed in the following section. In the course of the following sections, this setup is referred to as *System Setup A*.

3.3. QDPHIL System Setup B: Digsilent PowerFactory 2018 with Regatron TC.ACS

This setup proposes the utilisation of the steady-state load flow calculation of the power system analysis software (i.e., DIgSILENT PowerFactory [46]). The calculated voltage value at a predefined bus bar in PowerFactory is passed as a new voltage set point to a PI, which in this case is a switched-mode voltage source from Regatron (i.e., Type: TC.ACS [47]). For sending the set values, an interface was programmed in C# language that utilises the API provided by the manufacturer to communicate with PI. A PV inverter (cf. [44]) is connected physically to the voltage source as EUT and feeds in power to the grid. The active and reactive power of the PV inverter is captured by a measurement device (cf. Janitza UMG96 [48]). The RMS measurements are fed back to the simulation in PowerFactory as digital values to form a closed loop. For this measurement feedback, an interface was programmed in Python utilising Modbus/TCP functionality of the measurement device. The system setup is illustrated schematically in Figure 4. To control the voltage in the simulation of PowerFactory, an external function was programmed in Python. To control the voltage in the steady state load flow calculations of PowerFactory, another external Python function was developed for the control and synchronisation of these calculations. In the course of the following sections, this setup is referred to as *System Setup B*.

3.4. Discussion of the Mathematical Difference of PHIL System A and QDPHIL System B

The used steady state simulation approach in System Setup B differs significantly from the phasor calculation used in the Opal-RT system of System Setup A, especially when comparing the cycle times. Regarding the calculation time, a transient PHIL simulation setup should achieve around a couple of microseconds. For System Setup B, an approximately 400 ms calculation time was realised in the laboratory experiments. This difference in calculation method is described in the expression below, where the left part represents the instantaneous value of the voltage and the right part represents the steady state value of voltage:

$$u_{grid}\,(dt) \rightarrow U_{grid,RMS}\,(\Delta t) \tag{1}$$

where dt is typically 10–40 μs, and Δt is approximately 400 ms. The second proposed method could be named as the Quasi-Static Power-Hardware-in-the-Loop [31] in contrast to the classical PHIL setup which can be referred to as Real-Time Power-Hardware-in-the-Loop. Depending on the purpose of the examinations, which could correspond to the described requirements (cf. Section 3.1), the use of the Quasi-Static PHIL could be advisable.

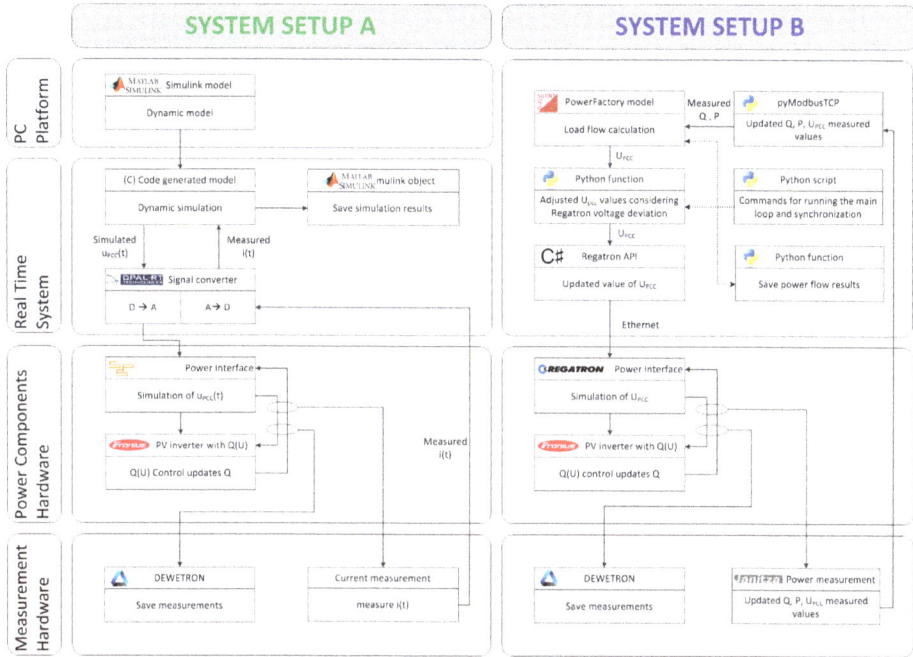

Figure 4. Comparison of the architecture of the used system setups: *System Setup A* is an Opal-RT RT-LAB based system with Spitzenberger Spies PAS and *System Setup B* is a Digsilent PowerFactory based system with Regatron TC.ACS.

3.4.1. Phasor Calculation

The mathematical equations to describe the used voltage step for evaluation of the system setups are presented in this section. For a modelled grid with a slack node and a single impedance between the slack node and the point of common coupling, the voltage at the PCC can be expressed in a simplified manner as addition of the voltage at the slack node and the voltage rise due to the power feed-in of the inverter over the impedance. To comply with the test procedure, as described in the following section (please refer to Section 4.1), a step function in the slack element was assumed. For the phaser calculation used in System Setup A, this relates to the following equations for each of the three phases:

$$
\begin{pmatrix} u_{AN,PCC}(dt) \\ u_{BN,PCC}(dt) \\ u_{CN,PCC}(dt) \end{pmatrix} = \begin{pmatrix} u_{AN,Slack}(dt) \\ u_{BN,Slack}(dt) \\ u_{CN,Slack}(dt) \end{pmatrix} + \begin{pmatrix} \triangle u_{AN}(dt) \\ \triangle u_{BN}(dt) \\ \triangle u_{CN}(dt) \end{pmatrix} \tag{2}
$$

with a sinusoidal voltage source at a fixed frequency and a fixed phase angle deference θ of $\pm120°$

$$
\begin{aligned}
u_{AN,PCC}(t) &= U_{nom} \times k_{step} \times \sin(2 \times \pi \times f_{nom} \times t + \theta_{AN}) \\
u_{BN,PCC}(t) &= U_{nom} \times k_{step} \times \sin(2 \times \pi \times f_{nom} \times t + \theta_{BN}) \\
u_{CN,PCC}(t) &= U_{nom} \times k_{step} \times \sin(2 \times \pi \times f_{nom} \times t + \theta_{CN})
\end{aligned} \tag{3}
$$

The assumed step can be expressed with the help of a multiplier k_{step} which changes its value depending on the calculation time. In this case, a change at the relative time $t = 10$ ms is used.

$$
k_{step} \begin{cases} 1.04 & t \le 10\,\text{ms} \\ 1.08 & t > 10\,\text{ms} \end{cases} \tag{4}
$$

The voltage raise due to impedance can be determined by the following equation. In the phasor calculation, the actual value of the current is used for the calculation.

$$
\begin{aligned}
\triangle u_{AN}(t) &= Z_{grid} \times i_{AN}(t) = (R_{grid} + j \times X_{grid}) \times \sin(2 \times \pi \times f_{nom} \times t + \theta_{AN} + \cos(\phi)_{load}) \\
\triangle u_{BN}(t) &= Z_{grid} \times i_{BN}(t) = (R_{grid} + j \times X_{grid}) \times \sin(2 \times \pi \times f_{nom} \times t + \theta_{AN} + \cos(\phi)_{load}) \\
\triangle u_{CN}(t) &= Z_{grid} \times i_{CN}(t) = (R_{grid} + j \times X_{grid}) \times \sin(2 \times \pi \times f_{nom} \times t + \theta_{AN} + \cos(\phi)_{load})
\end{aligned}
\tag{5}
$$

3.4.2. Steady State Calculation

The steady state calculation can be carried out symmetrically. The used PI, the measurement device and the simulation program can also calculate the voltage in an unsymmetrical manner. Therefore, the equation changes to the examination of RMS values. The calculation methodology was adapted from [49].

$$
U_{RMS,PCC}(t) = U_{RMS,Slack}(t) + \triangle U_{RMS,GRID}
\tag{6}
$$

Containing the slack voltage defined as step function with the same limits as the multiplier k_{step} for the phaser calculation:

$$
U_{RMS,Slack}(t)
\begin{cases}
1.04 \; p.u. \times U_{nom} & t \le 10 \, \text{ms} \\
1.08 \; p.u. \times U_{nom} & t > 10 \, \text{ms}
\end{cases}
\tag{7}
$$

The additional voltage needed to transfer power over the line segment is represented with the following equation. The variables change from a current to an active/reactive power.

$$
\triangle U_{RMS,GRID}(P_{PCC}(t), Q_{PCC}(t)) = \frac{P_{PCC}(t) \times R_{grid} + Q_{PCC}(t) \times X_{grid}}{U_{nom}}
\tag{8}
$$

3.4.3. Comparison of Phasor and Steady State calculation

A more detailed look into that context is provided in Figure 5 which depicts a constant voltage rise from 180 V_{RMS} to 240 V_{RMS}. The ideal result is drawn as the blue dashed line. The behaviour of System Setup A is depicted as a stem diagram, where each stem represents a new set-point for the amplifier part of the power interface. The actual limit for the minimum step size is significantly lower than it is depicted in the diagram. The used step size is chosen for representation purpose. Each of these set points shows the result of a new calculation. In contrast to the small time steps of System Setup A, System Setup B sends RMS-set points to the PI with a bigger cycling time. This is depicted as the red line. The change from one set point to the next is one step, as illustrated. This is discussed with measured signals in the following section. These properties are only observed for the used system. Other systems have not been tested regarding this behaviour.

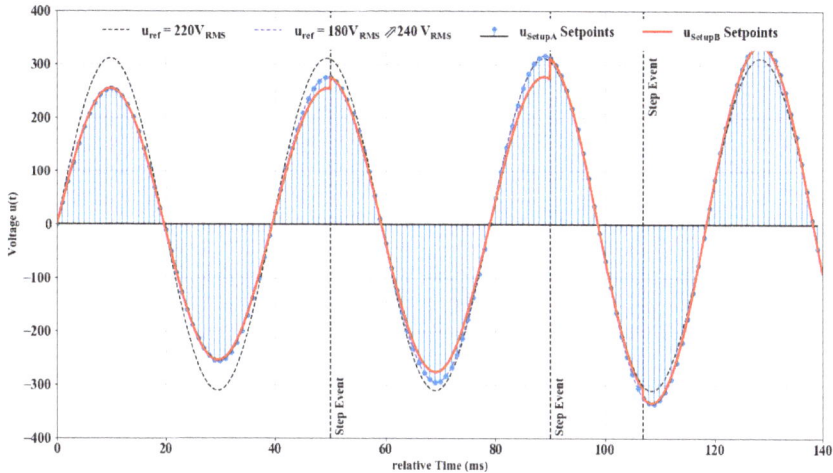

Figure 5. Visualisation of voltage rise event from 180 V_{RMS} to 240 V_{RMS} in a time frame of 140 ms. The update principles for the different power interface and the corresponding simulation step resolution are illustrated. For System Setup A, an update rate of 1 ms of the $u(dt)$ is illustrated. In comparison to System Setup B, discrete update steps are illustrated. The used resolution does not correspond to the actual used setups and is for illustration purpose only.

4. Validation Methodology

This article focuses on the comparison of the two system setups as introduced in Section 3. It is proposed that both systems can be used to evaluate the performance of smart grid components, solutions, and technologies as well as new approaches with high renewable energy feed-in if the requirements in Section 3.1 are met. For the validation of the QDPHIL setup, a two-stage process was chosen. In the first phase, the proposed system setup was tested and compared against the PHIL setup. Therefore, three test cases were used to characterise the differences in the response of the system to the test case scenarios. In the second phase, the QDPHIL system setup was used to conduct an experiment to gather results from a actual application of the QDPHIL concept. The results are presented in Section 6. A complex scenario was chosen which involves SIL components as well.

4.1. Test Methodology

The examination of both used system setups was carried out by comparing the response of the systems within three different scenarios. The scenarios reflect the most extreme possible variation as well as variations typically observed in real world test cases of the QDPHIL setup. The test cases are:

- *Step Function:* Voltage deviation at the slack bus bar from 1.04 p.u. to 1.08 p.u. in a step. This represents the most extreme change of a parameter in the modelled system. It was chosen to evaluate the difference in the dynamic behaviour of the systems in the time domain of seconds.
- *Transient Behaviour of the Step Function:* The voltage change of the two setups was evaluated regarding the transient behaviour of the system.
- *Ramp Function:* Voltage deviation at the slack bus bar from 1.04 p.u. to 1.08 p.u. as a voltage ramp over the course of 1–10 min. These are typical changes which were deployed to examine the stability of systems in varying conditions.

The individually used experiments are depicted in Figure 6 which also shows the system setups and the used grid model. The depicted experiments are grouped into three parts. These are the previous described tests for the comparison which are called Phase 1. Phase 1 is accompanied with a preliminary test to characterise the system setups, which is referred to as Phase 0. The subsequent tests are subject to a case study, which is called Phase 2 and is discussed in Section 6. These tests were also executed within the ERIGrid TA project "Smart beats Copper" at the host facility AIT in Vienna, Austria. Additional tests with System Setup B were executed at the Smart Grid Lab of Ulm University of Applied Sciences, Ulm, Germany.

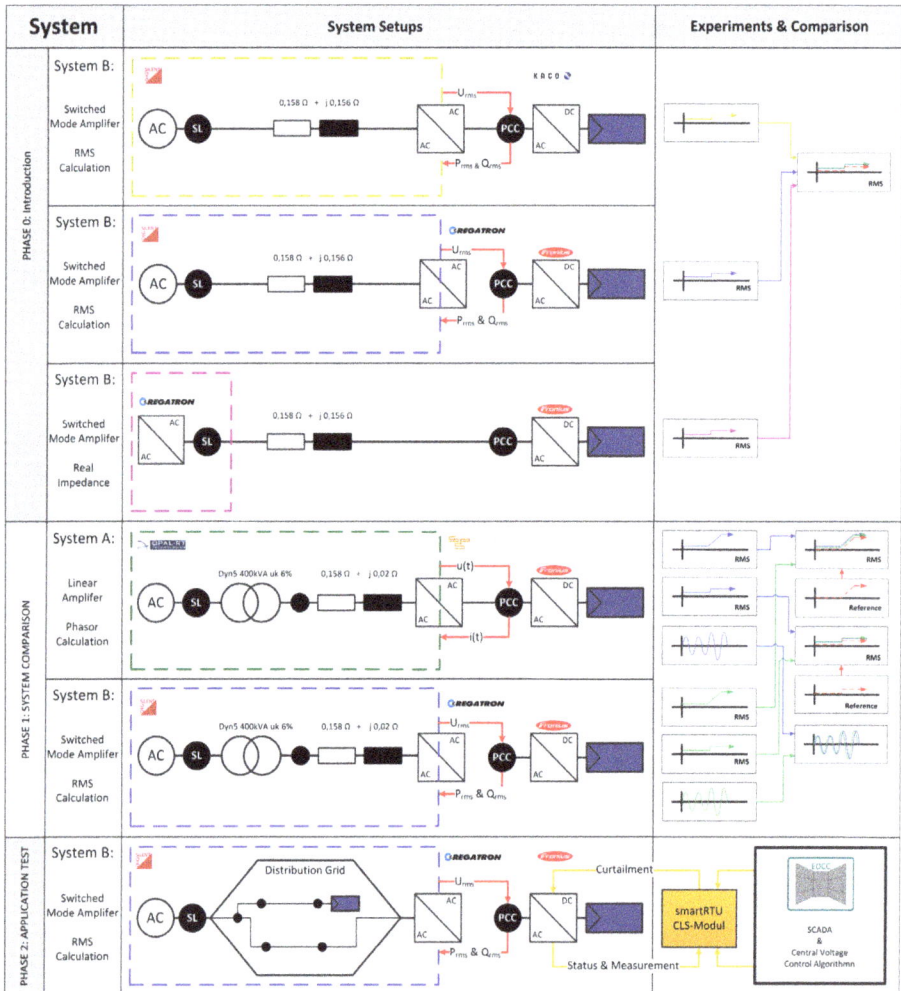

Figure 6. Schematic overview of the test scenarios for the comparison of the system response, which is split into three phases. The system setups and the implemented grid models as well as the individual experiments and the carried comparisons are shown. The colour of dashed box corresponds with with the graphs in the following section.

4.2. Used Grid Model

The proposed system setups were investigated in Phase 1 when simulating a simplified grid model which consists of a regional Low-Voltage (LV) transformer, a power line represented as an impedance and a connection point (i.e., grid node), where a PV inverter is connected as depicted in Figure 6. It should be taken into consideration that the developed system can also be used for simulating large-scale grids, however, a simple grid was adopted for this study to avoid complexity in the simulation results.

4.3. Used EUT for the Test

The proposed system setups were tested considering a PV inverter as the EUT. Two inverters were used for the experiments, which was due to their availability at the different involved laboratories. The models Symo 20.0-3-M with nominal apparent power of 20 kVA and Symo 15.0-3-M with a nominal apparent power of 15 kVA from the manufacturer Fronius were used. Both systems originate from the same series and have the properties described in [44]. In addition, the function Q(U) for voltage control was configured in adaption to the suggestions by Elbs and Pardatscher [50] which again results in the parametrised Q(U) curve. Due to the reactive power priority of the implemented algorithm, the limit was lowered by 5% to prevent an oscillation of the controller. The selected set points are extreme scenarios as most of the relevant grid codes (e.g., [49,51]) or related documents rarely expect capabilities exceeding the limit of $cos\varphi = 0.9$ which corresponds to a limit of ± 0.42 Q/S_{nomm}. The Q(U) controller has the option to set a time constant T_C. Different time constants were chosen. All chosen time constants represent challenging scenarios as the value suggested by the manufacturer is 5 s. An overview of the chosen parameters for the different tests is provided in Table 4.

Table 4. Overview of the implemented step points parameterised for the system.

Parameter	Fronius Symo 20.0-3-M		Fronius Symo 15.0-3-M	
	U (p.u.)	Q/S_{nomm} (-)	U (p.u.)	Q/S_{nomm} (-)
Setpoint 1	0.98	+1.00	0.98	+0.95
Setpoint 2	1.02	0.0	1.02	0.0
Setpoint 3	1.05	0.0	1.05	0.0
Setpoint 4	1.08	−1.00	1.08	−0.95
Time Constant T_C	10 ms		10 ms, 50 ms, 1 s	

4.4. Used Test Scenarios for Phase 1

To examine the performance of the developed systems, two simulation scenarios were applied. In the first scenario, a voltage increase in single steps was implemented by means of increasing the set voltage in the slack bus bar. Such a scenario of voltage increase can be observed in practice, if a regional transformer changes the position of its On-Load Tap Changer (OLTC). In the second scenario, a gradual voltage increase in a ramp was implemented by means of increasing the set voltage in the slack bus bar. Such a scenario can be observed in practice if voltages in the medium-voltage grid increase according to natural increase of PV feed-in during a clear sky day.

5. Phases 0 and 1: Results of the Comparison of the System Properties of System Setup A and B

The comparison of the two system setups is the subject of this section. The preliminary tests of Phase 0 introduced the system properties. Sections 5.2 and 5.3 discuss the transient and dynamic behaviour of the different systems setups when handling the voltage step. The discussion of the system behaviour for the modelled voltage ramp is given in the following section. The analysis of the cycles, especially of System Setup B, is outlined in Section 5.5. Section 7 concludes the Phases 1 and 2 discussion.

5.1. Introduction to the Compared Systems

As a first step, a basic analysis of the behaviour of System Setup B is presented. The system was tested first with a non-PHIL configuration and a physical impedance between the PI and the EUT. In a next step, this was compared with the corresponding PHIL configuration. The impedance is typically used for flicker tests according to IEC 61000-3-3 [52], and is part of the SmartEST facility at AIT. As shown in Figure 7, an oscillation behaviour was present for both configurations. It was assumed that the oscillation was caused by the searching algorithm of Maximum Power Point Tracking (MPPT) of the Fronius PV inverter. For additional comparison, the pre-step part of a comparable setup with another type of inverter [53] is depicted. The voltage was more stable with this specific product in comparison to the results of the PV inverter. As this oscillation was caused by varying power in-feed, it was present in both configurations.

The difference between the two configurations could be seen most explicitly at the small voltage step of 2.5 s of the PHIL-configuration. Besides, a voltage deviation was present for both configurations. This offset ranged from approximately 1 V_{RMS} to 3 V_{RMS}. Active and reactive power were fairly constant before and after the step which caused a change in reactive power due to the implemented Q(U)-control.

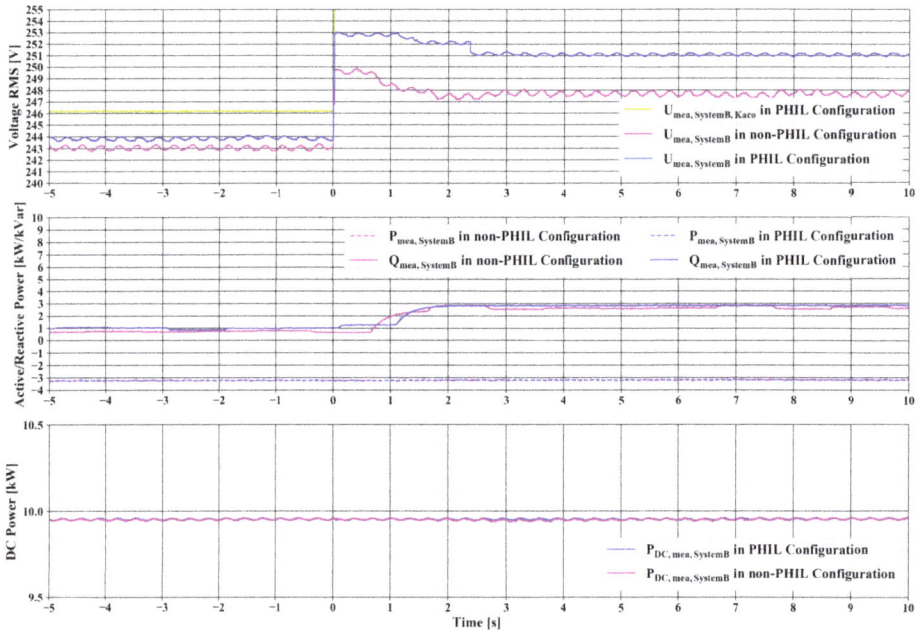

Figure 7. Comparison of the different system responses to the deviation voltage step. The est compared a non-PHIL setup with a real impedance to the PHIL setup with emulation of this impedance. The depicted values are RMS-values calculated on a one-period time frame. The DC-power is also shown.

5.2. Comparison of the Transient System Response for Voltage Step

This section focuses on the comparison of the two setups described in Section 3.4. At first, the comparison of the used systems was made by analysing their transient behaviour. The trend of the voltage and the current of the two three-phase systems are depicted in Figures 8 and 9, respectively. Both systems changed from one given waveform, which corresponded to a value of 240.6 V_{RMS}, to another waveform with a corresponding value of 249.7 V_{RMS} and from 232.9 V_{RMS} to 241.3 V_{RMS} within roughly 100 μs, respectively. For both systems, no lasting disturbance in waveform of the voltage could be observed. The given ideal sinusoidal waveform for the voltage was generated by both systems with neglectible fractions of no fundamental parts. The capabilities of both systems to generate a varying amount of non-fundamental disturbance have not been examined further in this contribution.

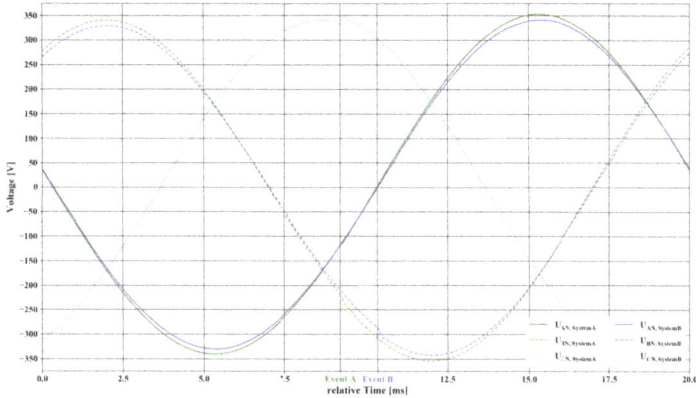

Figure 8. Comparison of the different transient systems response to the step of the voltage. The voltage of both three-phase systems are overlayed and synchronised. The occurrence of every step event is indicated for each setup separately. Neither system shows a significant amount of disturbance.

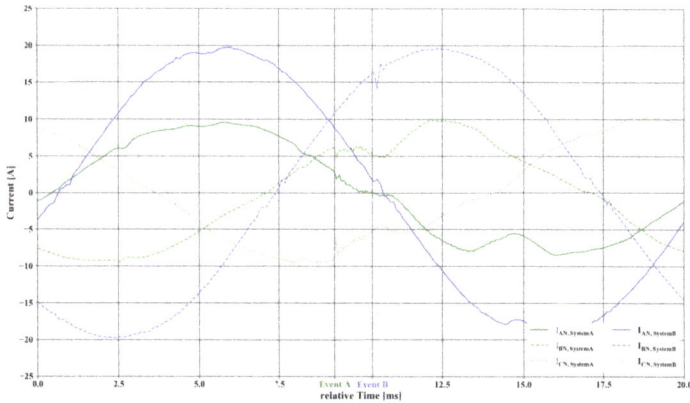

Figure 9. Comparison of the different transient system responses to the step for the current feed-in by the PV inverter. The currents of both three-phase systems are overlayed and synchronised. The occurrence of the step event is indicated separately for each setup.

5.3. Comparison of the Dynamic System Response for Voltage Step

The evaluation depicted in Figure 10 focuses on the dynamic behaviour of the systems for the test scenario. This change of perspective can also be understood as change from the evaluation of the individual power interface to the evaluation of the complete PHIL setup. Statistical key parameters are listed in Table 5 to quantify the observations made in the figure. The one-period RMS values provided by the independent measurement device were used for this evaluation.

As a reference, the steady state values for the used grid model and implemented Q(U)-control are the parameters for the constructed curve which is plotted as a red dashed line. The experiments were carried out in individual runs. This means that the measured curves are aligned with the ideal step function. Therefore, an offset between the parameterised step and the actual step cannot be observed in the graphs. This offset was limited to a maximum of one simulation step. For System Setup B, this was limited to a maximum of 420 ms. Details regarding the simulation time are given in Section 5.5.

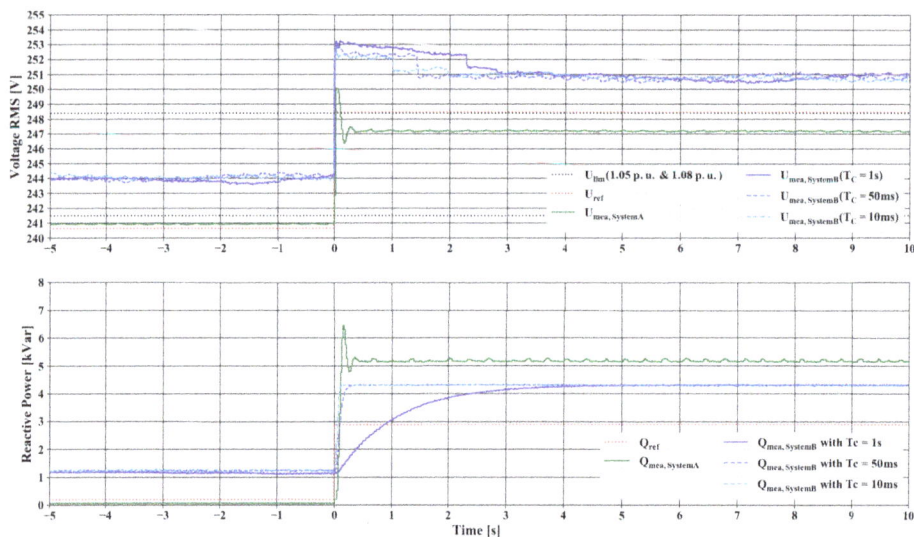

Figure 10. Comparison of the different system responses to the deviation voltage step. The reference curve is constructed by using the steady state results for given Q(U)-control. For System Setup B, varying time constants T_C for the Q(U) controller of PV inverter are presented. The depicted values are RMS-values calculated on a one-period time frame. The dashed line represents the values for 1.05 p.u. as well as 1.08 p.u. which in turn are the setpoints for the break of the slope of the Q(U) controller.

Table 5. Statistical key parameter for the characterisation of system response for the step change of Phase A of the system.

Key Parameter	Time Frame	Dimension	System Setup A	System Setup B [†]
ME(U)	PS	(V)	0.287	3.439
ME(U)	PT	(V)	−1.266	2.311
CC(U)	CT	(-)	0.991	0.994
DELTA(U)	PS	(V)	0.086	0.519
DELTA(U)	PT	(V)	0.153	0.660
STD(U)	PS	(V)	0.022	0.093
STD(U)	PT	(V)	0.029	0.138
$T_{Stabilisation}$	CT	(s)	0.620	3.732
MEAN(P) [STD(P)]	PS	(kW)	−1.613 [0.007]	−1.631 [0.023]
MEAN(P) [STD(P)]	PT	(kW)	−1.535 [0.012]	−1.547 [0.011]
MEAN(Q) [STD(Q)]	PS	(kvar)	+0.064 [0.003]	+1.238 [0.032]
MEAN(Q) [STD(Q)]	PT	(kvar)	+5.171 [0.033]	+4.310 [0.014]
$f_{Oscillation}$	CT	(Hz)	3.570	3.570

NA: not applicable ; [†] T_C = 10 ms is used; PS: pre-step; PT: post-stabilisation; CT: complete time frame.

Three interesting aspects could be observed regarding the step functions: (i) the voltage deviation for the steady state situations before and after the step; (ii) the constancy of selected setpoints and the overall stabilisation time; and (iii) the stabilisation time, which is highly affected by the implemented control algorithm (e.g., Q(U) control) and the selected parameters.

The varying time constants of the Q(U) control could be observed when looking at the different responses of the PV inverter to the voltage step. For System Setups A and B, the inverter reacted immediately with a small time constant with a reactive power in-feed when the voltage changed. Smaller time constants resulted in faster reach of a steady state. For both setups, the reaction of the inverter was immediate when small time constants were used. In contrast to that was the calculated reaction to the changed reactive power in-feed by the system setups. For System Setup A, the reaction of the voltage happened immediately due to small time steps of the PHIL-setup, whereas, for System Setup B, the reaction was delayed in more discrete steps, which corresponded to the cycle time. As described in Section 4.3, the time constant suggested by the manufacturer is $T_C = 5$ s, which will result in a highly damped system response. This seems to be a preferred strategy for DSO to prevent interference with multiple deployed systems.

Looking at the results of this test scenario, both systems showed a fairly linear offset to the reference value. This effect was reflected in the *Mean Error (ME)*. However, the overall result was quite pleasurable showing a high *Correlation Coefficient (CC)* for both systems with a negligible difference between them. This occurrence could be expected as the calculated time for the time periods before and after the step function was included. The presented values were for Phase A to neutral of the three-phase system. Therefore, the readings were only a third of the in-feed of the three phase inverter. The oscillation already observed in Section 5.1 was quantified by the maximum spread *(DELTA)* and the resulting *Standard Deviation (STD)* of the voltage readings. As presented in the previous section, this related to the MPPT algorithm of the used PV inverters. As a result of the comparison of these values, one could see that System Setup A was more constant than System Setup B. The further examination also showed that these oscillations have a relatively constant subfundamental peak of around 3.570 Hz. This behaviour was also present for System Setup A but with a significantly lower amplitude. This also diminished when changing the voltage setpoint, and the oscillation was overlayed with other effects. Together with further analyses of the PI, it seemed that the immanent impedance of System Setup B was significantly higher than of System Setup A.

The upper dashed line represents the 1.08 p.u. value and, as one can see, the voltage of System Setup B never dropped below this parameterised limit of the Q(U) control. According to the given droop curve of the Q(U) controller, this resulted in a constant reactive power.

5.4. Comparison of the System Response for Voltage Ramp

In comparison to the previous section where the focus lay on the comparison of the two systems in regard to the transient and dynamic behaviour when changing the voltage in an extreme manner, this section presents the comparison of both systems in the context of a slowly but constantly rising voltage level and the interaction of the Q(U) control. The results are shown in Figure 11. The corresponding statistical key figures are given in Table 6 (Statistical Key Figures for the Evaluation are shown in Appendix A). In contrast to the short term evaluation of the step function, this test scenario was executed with a variety of durations with up to 10 min of rise time. The scenario with a rise time of 10 min is outlined in the current section. To evaluate this rather long-term scenario, two different active power levels were chosen. This is due to the system properties of System Setup A, as the complete feed-in energy was dissipated by the PI. In the course of the experiment, issues regarding the overheating and thermal shutdown had to be tackled.

The reference curves for the ramp illustrate what is expected in theory. The red dashed line represents the constantly raising voltage level of the upstream grid with the inverter working against the voltage raise with feeding-in reactive power. When comparing the measured values, both systems showed a voltage deviation from the calculated value. For System Setup B, this deviation lay within the same order as it already could have been observed in the step experiment. With System Setup A, the voltage changed from over-voltage in respect to the reference curve to under-voltage due to a higher reactive power feed-in. The individual oscillation appeared in the same order of magnitude as in the previous scenario. The reactive power response of System Setup B became in the bigger time frame of this analysis quite pleasing. For System Setup A, the reactive in-feed was too high due to a different parameterisation.

The active in-feed power remained constant for both setups. The vertical line which is annotated with "Event A" represents the point where the power interface of System Setup A shuts down due to over-temperature.

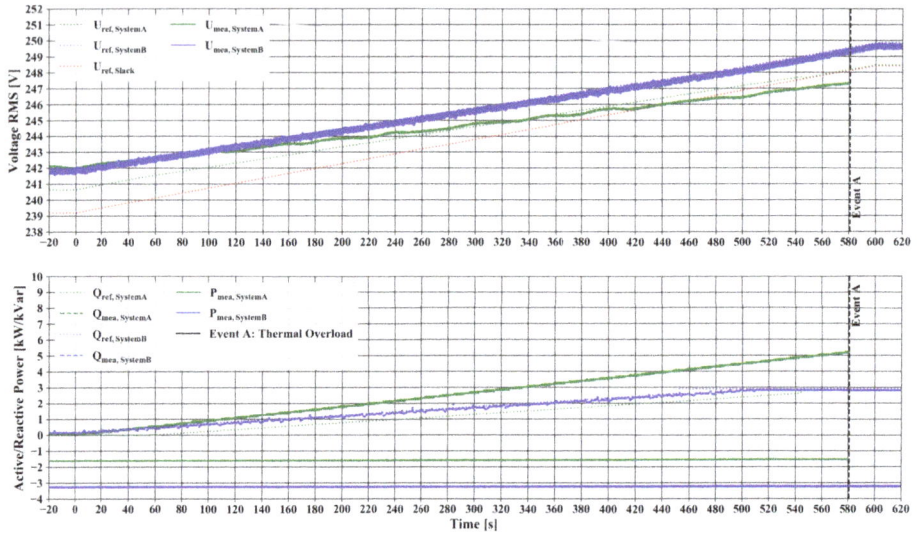

Figure 11. Comparison of the different system responses to the deviation voltage ramp with the raise from 1.04 to 1.08 p.u. at the slack node over the course of 10 min. The thermal overload of the PI of System Setup A is indicated.

Table 6. Statistical key parameter for the characterisation of system response for the ramp change. Comparison of ideal simulation and system output.

Key Parameter	Time Frame	Dimension	System Setup A	System Setup B
ME(U)	CT	(V)	0.210	−0.200
CC(U)	CT	(-)	0.999	0.999
DELTA(U)	CT	(V)	2.417	0.659
STD(U)	CT	(V)	0.660	0.113

CT: complete time frame.

5.5. Analysis of the Cycle Time for the Examined Setups

System Setup A was able to meet the time requirements due to its real-time control loop with a cycle time well below 50 μs. For the QDPHIL system setup, the cycle time was significantly higher and required a more detailed analysis. As shown in Figure 12, the total cycle time remained under one second. The cycle time t_{PHIL} for System Setup B consisted of the control loop in the main loop, the calculation of the new setpoints in PowerFactory, the control of the switched-mode amplifier Section 3.3 and the feedback of the measured values. The main control algorithm was implemented as a python script, which took about 20–30 ms runtime. Although the underlying interpreter required a bit more time to execute than a compiled program, new power interfaces can be quickly and flexibly integrated into the process using Python. This also applies to PowerFactory, which is controlled by a dedicated Python interface. Besides the python interface, there are several other ways to solve this task. Other possibilities such as *OPC UA* are discussed in detail in [54]. The solution was chosen because the whole functionality of PowerFactory's internal scripting language (DPL) can also be addressed using the Python interface. Since Python has established itself as the standard in the given research infrastructure, it offers the possibility to use this interface also for the investigation of other research questions. Due to the simple grid model (see Section 4.2), the load flow calculation including feedback took 50 ms (±5), which can increase with larger modelled systems. The feedback contained the new set point for the switched mode amplifier. Its manufacturer provides a C# library, with which the new set point value can be communicated to the device via Ethernet.

The time needed to execute the command could not be further optimised and lies at 250 ms (±50). One possibility to actually achieve faster response times is to use the option of analog control of the network simulator in "amplifier mode". To access the C# interface, a helper application was used, which is coupled to the python main loop via a named pipe. Subsequently, a new actual value was queried by Modbus/TCP from the measuring device and thus the control loop started again. This last step required 100 ms (±20). Overall, a t_{PHIL} with a median of 420 ms could be achieved with System Setup B. The cycle time shown for different experiments is depicted as violin plot in Figure 12. The density representation shows two and three clusters for the cycle time with the highest density around the median. The second cluster is in range of 0.2–0.3 s. Due to the added IEC61850 simulator function used in the case study, this cluster is moved towards the median.

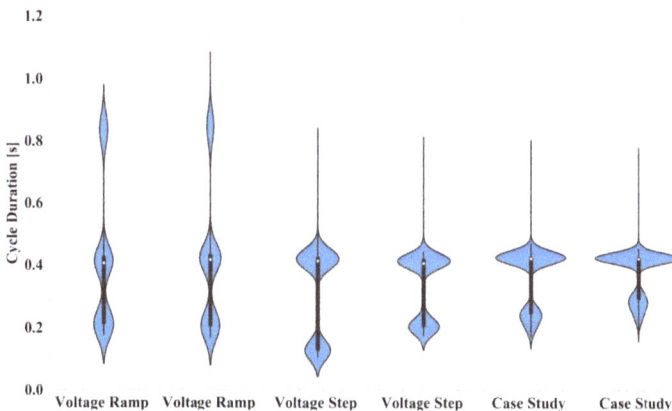

Figure 12. Illustration of the cycle time t_{PHIL} for System Setup B for use in the evaluations of Phase 1 as well as for use in Phase 2. The white circle represents the median of the violin plot, the bold line represent the interquartile range and the blue areas represent the density distribution.

6. Phase 2: Test of A Complex Smart Grid Control Regime

In the following section, the experiment is introduced briefly which complies with the requirements given in Section 3.1. This experiment was the main example use case when designing the comparison of the previous section. Therefore, this section describes the actual process of setting up and carrying out this experiment. The scope of the contribution is therefore widened and enriched with real results. Moreover, finding a flawless control algorithm was not the aim of the carried out experiment. The chosen algorithm is not suitable for this application but shows the necessity of the complete testing procedure. Preliminary results are presented in [55] and are extended within this section.

6.1. Description of the Case Study Experiment

6.1.1. System under Test

The purpose of this case study experiment was to observe the performance and reaction of the EUT responding to smart grid control strategies in a controlled environment. In this case, the EUT was a Fronius PV inverter with its active power feed-in regulated by a coordinated voltage control strategy. The inverter has a nominal active power of 20 kW; however, due to active power restriction, the nominal value was limited to 10 kW during the experiment. The Q(U) setting applied to the EUT was the same as in the previous Phase 1. To distinguish the EUT from the SUT, the SGAM Layer representation in Figure 13 can be used. The SUT can be identified as the complete construction shown in the diagram which consists of a Supervisory Control and Data Acquisition (SCADA) system, a fleet management for the smartRTUs (also referred to as Controllable Local Systems (CLS) Management), traditional RTUs and smartRTUs and the depicted primary equipments.

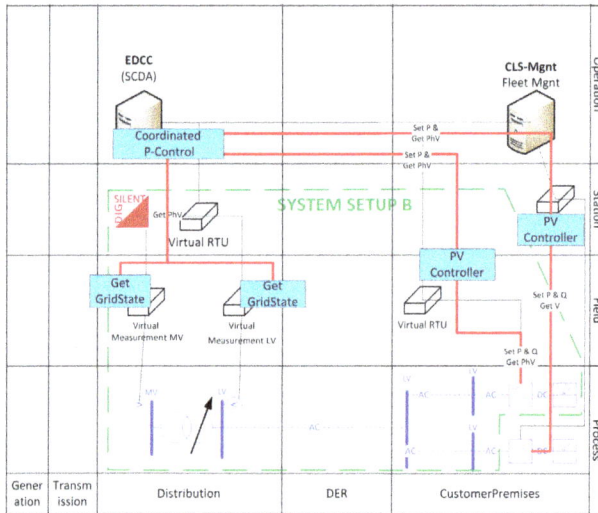

Figure 13. Combined SGAM-layer-diagram of the implemented case study. The base layer depicts the real and simulated components. The boundary of the PHIL setup is illustrated as a green dashed line. The utilised functions are depicted in blue boxes. This diagram uses the template provided by [56].

6.1.2. Communication and Control Infrastructure

The control infrastructure consists of a grid control system, a Remote Terminal Unit (RTU) that acts as a protocol gateway, a fleet management system for the RTU and the controlled PV inverter. As grid control system, the experimental Distribution Control Center (EDCC) at Ulm University of applied Science is used. The operational measurements from the PV inverter are gathered by the RTU via the *SunSpec* protocol [57] and transmitted to the control system via the IEC 61850-7-4 [58] and IEC 61850-7-420 [59] standard. This concept complies with German regulations regarding the use of the SmartMeter infrastructure. Further information is given in [55,60].

6.1.3. Control Function

The virtual Intelligent Electronic Device (IED) Server transmits the grid information via Virtual Private Network (VPN) to the EDCC, where a voltage control algorithm was implemented and configured as a visual basic script. As a result of interacted hardware measurement and software simulation, a power curtailment setpoint, represented as a percentage value with respect to nominal active power, was sent back to the RTU and was subsequently forwarded to the PV inverter. The voltage regulation would be triggered when the voltage was bigger than 1.05 p.u. The active power curtailment was transmitted to the inverter in gradual steps by 10% an exceeded 0.005 p.u. The used control algorithm can be described by the following equation:

$$P_{Limit} = P_{nominal} * f_{curtailment}(U_{PCC}) \begin{cases} f = 0.0 & U_{PCC} \geq 1.080 \,\text{p.u.} \\ f = 0.1 & U_{PCC} \geq 1.075 \,\text{p.u.} \\ \quad \vdots \\ f = 0.9 & U_{PCC} \geq 1.055 \,\text{p.u.} \\ f = 1.0 & U_{PCC} \leq 1.050 \,\text{p.u.} \end{cases} \tag{9}$$

It might appear that the algorithm is not suitable for the task assigned as it would lead to an oscillation behaviour. Nonetheless, the algorithm was chosen fully aware of this circumstance. A clearly defined oscillation is easily observable and the effect is hard to be superimposed by other effects.

6.2. Results of the Case Study Experiment

The described methodology was used to test and evaluate the setup described in the previous section. The simulated OLTC of the grid was actuated to step through the voltage range and to trigger the two (Q(U) and P(U)) control regimes. Figure 14 shows the results of the executed experiment as a plot of active andreactive power over the voltage at the PCC of the inverter. The described steps can be seen in clusters of data points in the diagram. The characteristics for the local Q(U) and the superior P(U) control are represented as lines as well. The green coloured area represents the anticipated deviation for the Q(U) control due to output filters and error of the measurements from the inverter as well as from the independent measurement. Nevertheless, the measured values for Q(U) control remained close to the desired characteristics. Looking at the P(U) control governed by the SCADA system, it can be noticed that the deactivated P(U) control resulted in stable operating points at different set points of the simulated voltage regulator of the transformer. The in-feed power of the inverter was not curtailed. With activated P(U) control, the purple coloured area represents the undesired voltage range. As the control-algorithm is a very basic lookup-table, the voltage has to be within this area for at least one transmitted period.

In the active power and voltage plot in Figure 14, a few control actions of the control algorithm are depicted. The value pairs form four main clusters, which are inherent to the used setup. The direct causes of the change from one cluster to another are described in the following list:

- *(A → B)* The voltage is within the undesired range. In addition, curtailment command is transmitted after the executed cycle of the control algorithm. There is a decrease in feed-in power due to the curtailment command, and slightly reduced voltage level due to the inherent properties of the voltage source (please refer to the previous section).
- *(B → C)* Due to a new voltage set point for voltage source, the voltage readings decrease after a load flow calculation is carried out.
- *(C → D)* With the next performance of the tested control algorithm, the voltage is in an acceptable range, and the curtailment is retracted. Due to the inherent properties of the voltage source, the voltage increases slightly.
- *(D → A)* With the next carried out load flow calculation, the system state returns to A.

The analysis showed that the proposed setup with a simple test algorithm is working. This is visualised in the time-series plot of Figure 15, where the measurements of the independent measurement device and the transmitted readings of the SCADA system are compared. Due to the automatically sent commands of the SCADA system, the field gateway passed the curtailment to the inverter. The inverter limited the feed-in to the desired value. This led to a change in voltage level due to the PHIL setup. As result of the experiment, the postulated oscillation occurred; nonetheless, the control algorithm would not be suitable for a real grid operation. When comparing the measurement and the transmitted values of the SCADA system, the discretisation evoked by the technical constraints was clearly observable.

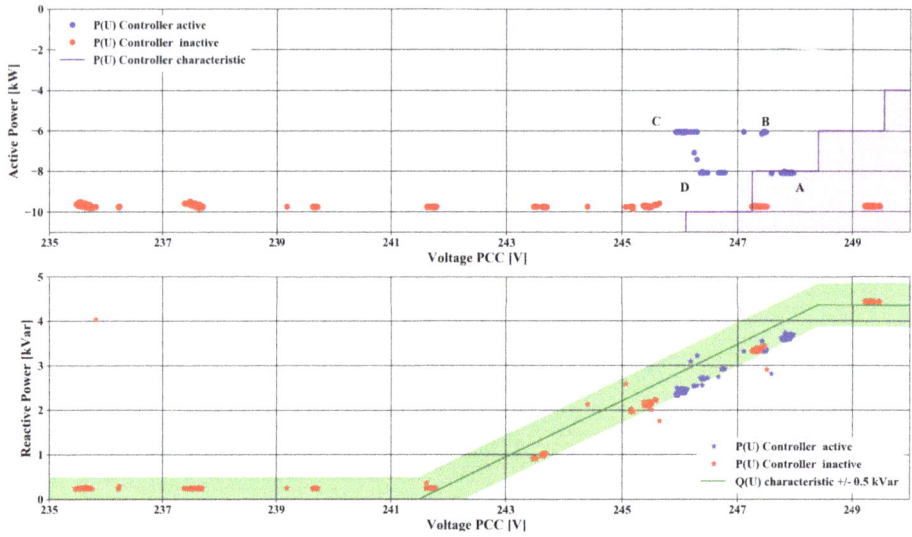

Figure 14. Comparison of the coordinated P(U) control governed by the EDCC and the Q(U) control of the inverter. Control characteristics are represented by the purple and green lines, respectively.

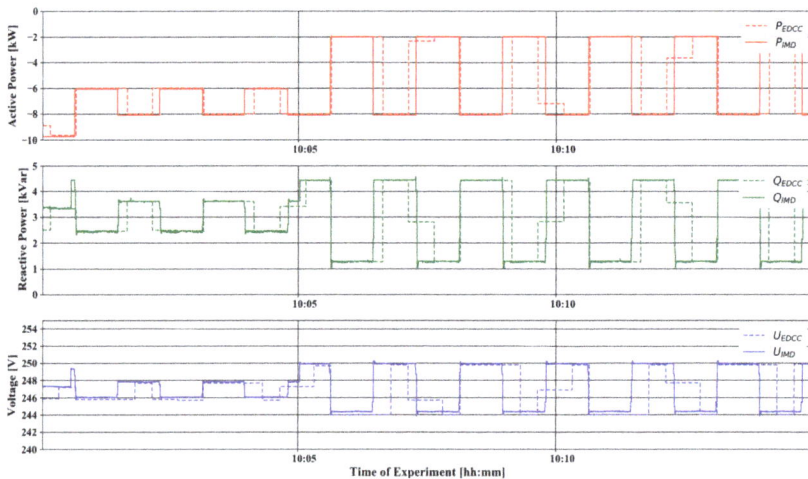

Figure 15. Transmitted Readings (EDCC) and independent measurements (IMD) for a time period with activated P(U) control algorithm.

7. Discussion of the Results

The experiments of Phase 1 showed that the System Setups A and B have been able to provide three-phase voltage systems, which are suitable to synchronise the given PV inverters, and started to operate properly. The modelled grid was used for a load flow calculation, and the extracted voltage signal was forwarded to the power interface. The Q(U)-function of the inverter was also tested. The biggest draw back observed in the carried out experiments was the significant voltage deviation for System Setup B which ranged from 2.3 to 3.4 V_{RMS} for the step function experiment. Based on further examinations, we assume that the offset was caused by an immanent impedance. In these examinations, a proportional offset from the idle state voltage setpoint to the idle state voltage present at the terminals of the power interface occurred. When changing the consumed or feeding power independent of voltage set point, the voltage offset changed as well. Therefore, it is expected that the offset could be compensated with simple correction equations, which do not even introduce another controller to the system. Alternatively, the PI of System Setup B could be replaced with the superior PI of System Setup A, which featured a lower impedance and would therefore have reduced the undesired effects. Nonetheless, the dynamic properties would stay the same. Both experiments showed the correlation coefficient is >0.99 for both of the system setups. Since all of experiments presented in this contribution have in common that they utilised voltage setpoints, and a current or power, as feedback loops, the results presumably are not applicable for current type experiments. Regarding the cycle time of the given QDPHIL setup, a median of 0.42 s could be achieved. The communication between the simulation and the PI has a big share of approximately 250 ms of this process and offers the possibility for further improvements.

The occurred automatic power shut-off to prevent overheating of the PI of System Setup A could be tackled by using an additional load to prevent the feed-in energy being dissipated within the PI. This would also alter the Thevenin-equivalent of the setup, which might be undesirable. Regarding the quality of the provided voltage signals, both systems generated negligible amounts of harmonics. To represent real grid situations, introducing a realistic amount of harmonics must be considered.

Regarding the soft facts, involving setup and training, System Setup B seemed less challenging as the options were more limited. In addition, the software PowerFactory is one of the standard tools used at Ulm University of Applied Science not only for real-time simulation but also for scenario and time series analysis of distribution grids. Using those models and calculation algorithms in pure simulation experiments and in PHIL testing is beneficial. The models had to be setup up only once and results from the simulation were helpful in the PHIL experiment design. In addition, the results from the PHIL experiment could improve the models implemented in the simulation environment.

Another difference between the PI of the two setups was the maximal frequency at which the systems could reproduce harmonics. For System Setup A, this was limited to 30 kHz, whereas, for System Setup B, this was limited to 5 kHz. This aspect was not a prominent requirement for the contemplated usage, however. The described aspects are given in a short overview in Table 7.

Table 7. Findings of experiments of Phase 0 and Phase 1.

	System Setup A	System Setup B
Pros (+)	small cycle time ($\leq 50\,\mu s$) multiple calculation methods (phasor simulation, transient simulation) low immanent impedance	regenerates energy during the test
Neutral (#)	specific usage of the simulation tool	sufficient cycle time ($\leq 1\,s$) variation in cycle time only voltage type experiments
Cons (-)	over heating occurs dissipates energy during test run	one calculation method (steady state) high immanent impedance

8. Conclusions and Outlook

The conclusions of this work are manifold and can be divided into the following parts. The first part summarises the findings of the voltage step and ramp experiments carried out in Phase 1, while the second part discusses the findings of the case study (Phase 2). In the third part, the conclusions are provided and finally the planned improvements and the utilisation of the entire setup is summed up.

The contribution presents results for a comparison of two PHIL setups—these were called System Setup A for the combination of an Opal-RT system and Spitzenberger Spies PAS PI and System Setup B for the combination of a Digsilent PowerFactory instance and a Regatron TC.ACS PI. The use case—testing of coordinated smart grid control function in an undisturbed operation—which was examined with both setups is only a subset of the capabilities of classical PHIL setups. The expected outcome of the experiments was that both systems are capable of solving this task. The fundamental difference between both systems were the calculation principles which are phasor calculation for System Setup A and the use of steady-state simulation for System Setup B. For this comparison, two scenarios were considered: (i) change of voltage level with a step; and (ii) a ramp function in the modelled grid at slack element. As a result, different levels of accuracy and continuity of the output signal were observed. The difference of the systems are apparent for the reaction of the Q(U) controller of the PV inverter within the test environment. If the time constant of the Q(U) controller is significantly lower than the cycle time, System Setup B shows discrete steps of the voltage and the stabilisation time is prolonged. As an overall statement, System Setup B is suitable for the use case described in Section 3.1. One issue which occurred during the test was an offset between the set-point sent to the PI and the actual voltage measured at the terminals. It is suspected that this problem is caused by the high immanent impedance of System Setup B. This problem needs to be tackled.

In the second part of this contribution, an actual utilisation of System Setup B is presented, which shows the intended use for a combined PHIL and SIL testing of complex smart grid control functions. The use of the SGAM presentation provided a basis for a common understanding of the presented test. This is in line with the holistic testing description introduced by the ERIgrid Consortium [61]. Due to the description of the planned scenario, the different facilities were able to implement sub-functions in an efficient manner and were able to test the relevant sub-functions before the actual experiment. As the outcome of this case study, the tested central control algorithm is not useful for actual grid operation, as it was anticipated and deliberately chosen at the stage of test design. However, the gathered data provided a good basis for the validation of the process. In addition, the carried out experiment is a successful pilot test for the application of the RTU. It showed clearly that gathering of relevant measured values at the grid connection point and the generation unit is possible with the implemented SunSpec to the IEC 61850 converter. The other way round was tested as well and the curtailment of the PV inverter was successfully carried out.

As a conclusion, it can be stated that there was a significant difference in the working principle of the two system setups. Both systems could provide a function essential for the evaluation of EUT and SUT with respect to the requirements stated in Section 3.1. The PHIL System Setup A was more accurate and showed more capabilities than the QDPHIL System Setup B but this came at the price of a more complex system regarding modelling, setup and operation than it was required in System Setup B. In addition, the expenditures for purchase and maintenance have been higher for System Setup A compared to System Setup B. Regarding the use of electric energy, the PI of System Setup B was able to feed back into the grid, whereas the linear PI of System Setup A dissipated the energy.

As further steps, System Setup B will be improved to solve the issues regarding the occurring offset without the use of an additional controller. As first evaluations suggest the effects are fairly linear and are likely caused by a high immanent impedance of the PI, further analyses and engineering will be necessary. These will also include considerations regarding the implementation of the immanent impedance in the grid model. The implemented System Setup B is and will be used at the Ulm University of Applied Science in the course of different research projects. The setup will be mainly used for pilot testing of applications involving small decentralised control units in combination with the German Smart Meter Infrastructure, which consists of the combination of Smart Meter, Smart Meter Gateway and CLS-Module to enable information gathering and control of small generation units such as PV inverters, as tested in the case study of Section 6. These measurements are necessary in preparation of a broader deployment in the demonstration project C/sells [62]. As specific suggestions for future work, the examination of effect of different interface algorithms and a detailed study of the influence communication interface between simulation and power interface for the quasi-static PHIL setups can be given. In this context, the implementation of the IA transmission line model is considered. This would enable the utilisation of a available and standardised line impedance network which is typically used for tests according to IEC 61000 series standard.

Author Contributions: F.E., T.I.S., R.B., and G.H. conceived and designed the experiments; F.E., C.S., M.C., S.C., B.I., and C.K. performed the experiments; F.E., S.C. and M.C. analysed the data; C.K. implemented and analysed the interface for the switch mode amplifier; S.C. implemented and operated the CLS-Modules and contributed to the case study; F.E., B.I., and T.I.S. wrote the paper; and C.S., D.E.S., M.C., S.C., C.K., R.B., and G.H. also contributed by editing this paper.

Funding: This contribution presents the results of the Trans-national Access (TA) activity "Smart beats Cooper", which is supported by the European Community's Horizon 2020 Program (H2020/2014–2020) under project "ERIGrid" (Grant Agreement No. 654113) . In this context, the methodologies and results of the following research projects haven been used and tested within the following projects:

- The project "C/sells" (Grant Agreement No.: BMWi 03SIN136) within the "SINTEG" funding program by the Federal Ministry for Economic Affairs and Energy which addresses the construction of the intelligent energy future. C/sells is one of five showcase regions which are jointly funded with an overall project volume of 200 million Euro to foster this purpose. Three features of future energy grids form C/sells guiding principle: cellularity, participation and diversity. As fundamental principles, they constitute the guide rails for the different activities and single segments by means of which C/sells brings energy transition to reality.
- The project CLS-AppBW (Grant Agreement No: BWSGD 16014-16) within the Program "Smart Grids und Speicher" of the federal state Baden-Württemberg.
- The project "ESOSEG" within the program "Future-proof power grids" of the German Federal Ministry of Economics and Energy (BMWi 0325811C).

Conflicts of Interest: The authors declare no conflict of interest.

Abbreviations

The following abbreviations are used in this manuscript:

CC	Correlation Coefficient
CHIL	Controller Hardware-in-the-Loop
CLS	Controllable Local System
DER	Distributed Energy Resources
DUT	Device-under-Test
DRTS	Digital Real-Time Simulator
DSO	Distribution System Operator
EDCC	Experimental Distribution Control Center
EMT	Electromagnetic Transient (Simulation)
EUT	Equipment-under-Test
HIL	Hardware-in-the-Loop
HUT	Hardware-under-Test
IA	Interface Algorithm
ICT	Information and Communication Technology
ICTHIL	Information-and-Communication-Technology-in-the-Loop
IED	Intelligent Electronic Device
IMD	Independent Measurement Device
IPCC	Intergovernmental Panel on Climate Change
ITM	Ideal Transformer Model
ME	Mean Error
MEA	Measurement
MPPT	Maximum Power Point Tracking
OLTC	On-Load-Tap-Changer
PCC	Point of Common Coupling
PHIL	Power Hardware-in-the-Loop
PI	Power Interface
PV	Photovoltaic
RMS	Root Mean Square
RTU	Remote Terminal Unit
SCADA	Supervisory Control and Data Acquisition
SGAM	Smart Grid Architecture Model
SIL	Software-in-the-Loop
STD	Standard Deviation
SUT	System-under-Test
QDPHIL	Quasi-Dynamic Power Hardware-in-the-Loop

Appendix A. Statistical Key Figures for the Evaluation

Besides the visualisation of the short- and medium-term changes of the system properties, these properties are described by the following statistical key figures. Starting off with the *Mean Error (ME)*:

$$ME = \frac{\sum_t^n u(t)_{ref} - u(t)_{measure}}{n} \tag{A1}$$

The Pearson *Correlation Coefficient (CC)*:

$$CC = \frac{\frac{1}{n} * \sum_t^n (U_{ref}(t) - \bar{U}_{ref}) * (U_{measure}(t) - \bar{U}_{measure})}{\sqrt{\frac{1}{n} \sum_t^n (U_{ref}(t) - \bar{U}_{ref})^2} * \sqrt{\frac{1}{n} * (U_{measure}(t) - \bar{U}_{measure})^2}} \tag{A2}$$

Standard functions are used for the characterisation of time series with f as an arbitrary/any measured parameter:

$$MEAN = \frac{\sum_t^n f(t)_{measure}}{n} \tag{A3}$$

$$STD = \sqrt{\frac{\sum_t^n f(t)_{Mea} - MEAN(f_{Mea})}{n}} \tag{A4}$$

For the voltage ramp scenario, this equation is changed to the following function, which uses the STD function onto the difference of the reference values and the measured value

$$STD_{Ramp} = \sqrt{\frac{\sum_t^n (f(t)_{Mea} - f(t)_{Ref}) - MEAN(f_{Mea} - f_{Ref})}{n}} \tag{A5}$$

Variance of the parameter for steady state scenario (*DELTA* and *STD*): DELTA expresses the median value of the range for one of the oscillation periods that was observed.

$$DELTA = MAX(U_{RMS}) - MIN(U_{RMS}) \tag{A6}$$

For the the examined voltage ramp:

$$DELTA_{Ramp} = MAX(U_{RMS,mea} - U_{RMS,Ref}) - MIN(U_{RMS} - U_{RMS,Ref}) \tag{A7}$$

For the properties of the EUT, the following parameters have additionally been calculated:
The oscillation with a frequency below the fundamental frequency of the grid is calculated by determination of the median cycle time of one half wave.

$$f_{Osci} = \frac{1}{2 * MEDIAN(\Delta t_i)} \tag{A8}$$

The time Δt_i is determined by extracting the oscillation of the RMS values by subtracting the mean value of the voltage from the individual reading of the voltage. By using the signum function, this can be used in a general manner:

$$K_{sgn}(t) = sgn(U_{Mea}(t) - U_{Mea,Mean}) \tag{A9}$$

The $K_{sgn}(t)$ series is split into individual groups by the following schema:

$$G_{Ksgn}(t) = \begin{cases} 0 & t \equiv 0 \\ G_{Ksgn}(t-1) & K_{sgn}(t) \equiv K_{sgn}(t-1) \\ G_{Ksgn}(t-1) + 1 & K_{sgn}(t) \neq K_{sgn}(t-1) \end{cases} \tag{A10}$$

The difference between the first and last time stamp in each group is used to calculate the Δt_i:

$$\Delta t_i(g) = MAX(t()) - MIN(t()) \tag{A11}$$

References

1. Di Silvestre, M.L.; Favuzza, S.; Riva Sanseverino, E.; Zizzo, G. How Decarbonization, Digitalization and Decentralization are changing key power infrastructures. *Renew. Sustain. Energy Rev.* **2018**, *93*, 483–498. [CrossRef]
2. Field, C.B. *Climate Change 2014: Impacts, Adaptation, and Vulnerability: Working Group II Contribution to the Fifth Assessment Report of the Intergovernmental Panel on Climate Change*; Cambridge University Press: Cambridge, UK, 2014.
3. Stocker, T. *Climate Change 2013: The Physical Science Basis: Working Group I Contribution to the Fifth Assessment Report of the Intergovernmental Panel on Climate Change*; Cambridge University Press: Cambridge, UK, 2014.
4. United Nations. *Paris Agreement*; United Nations: New York, NY, USA, 2015.
5. Farhangi, H. The path of the smart grid. *IEEE Power Energy Mag.* **2010**, *8*, 18–28. [CrossRef]
6. Liserre, M.; Sauter, T.; Hung, J. Future Energy Systems: Integrating Renewable Energy Sources into the Smart Power Grid Through Industrial Electronics. *IEEE Ind. Electron. Mag.* **2010**, *4*, 18–37. [CrossRef]
7. Gungor, V.C.; Sahin, D.; Kocak, T.; Ergut, S.; Buccella, C.; Cecati, C.; Hancke, G.P. Smart Grid Technologies: Communication Technologies and Standards. *IEEE Trans. Ind. Inform.* **2011**, *7*, 529–539. [CrossRef]
8. Benz, T. *The Cellular Approach: The Basis of Successful, Cross-Regional Energy Transition*; VDE-Study; VDE Verband der Elektrotechnik, Elektronik, Informationstechnik e. V: Frankfurt, Germany, 2015.
9. Martini, L.; Brunner, H.; Rodriguez, E.; Caerts, C.; Strasser, T.I.; Burt, G.M. Grid of the future and the need for a decentralised control architecture: The web-of-cells concept. *CIRED-Open Access Proc. J.* **2017**, *2017*, 1162–1166. [CrossRef]
10. Kariniotakis, G.; Martini, L.; Caerts, C.; Brunner, H.; Retiere, N. Challenges, innovative architectures and control strategies for future networks: The Web-of-Cells, fractal grids and other concepts. *CIRED-Open Access Proc. J.* **2017**, *2017*, 2149–2152. [CrossRef]
11. Lund, P. The Danish Cell Project—Part 1: Background and General Approach. In Proceedings of the 2007 IEEE Power Engineering Society General Meeting, Tampa, FL, USA, 24–28 June 2007; pp. 1–6.
12. Bitkom Bundesverband Informationswirtschaft, Telekommunikation und neue Medien e.V. *Ausfallsicherheit des Energieversorgungssystems—Von der Robustheit zur Resilienz*; Bitkom: Berlin, Germany, 2018.
13. Scharte, B.; Thoma, K. Resilienz—Ingenieurwissenschaftliche Perspektive. In *Multidisziplinäre Perspektiven der Resilienzforschung*; Wink, R., Ed.; Springer Fachmedien Wiesbaden: Wiesbaden, Germany, 2016; pp. 123–150.
14. Ebe, F.; Idlbi, B.; Stakic, D.; Heilscher, G.; Birkelbach, J.; Gutekunst, F.; Enzenhöfer, R. Abstimmungskaskade als Werkzeug für eine zellbasierte Infrastruktur—Allgemeingültiger Prozess für die Koordinierung von Netzzellen. In Proceedings of the Tagungsunterlagen zukünftige Stromnetze für erneuerbare Energien, Berlin, Germany, 30–31 January 2018.
15. Wang, H.; Kraiczy, M.; Schmidt, S.; Wirtz, F.; Toebermann, J.C.; Ernst, B.; Kaempf, E.; Braun, M. Reactive Power Management at the Network Interface of EHV- and HV Level: Assessment of Technical and Economic Potential Based on a Case Study for Bayernwerk Netz GmbH. In Proceedings of the International ETG Congress 2017, Bonn, Germany, 28–29 November 2017.
16. Vogt, M.; Marten, F.; Lower, L.; Horst, D.; Brauns, K.; Fetzer, D.; Menke, J.H.; Troncia, M.; Hegemann, J.; Tobermann, C.; Braun, M. Evaluation of interactions between multiple grid operators based on sparse grid knowledge in context of a smart grid co-simulation environment. In Proceedings of the 2015 IEEE Eindhoven PowerTech, Eindhoven, The Netherlands, 29 June–2 July 2015; pp. 1–6.
17. Drayer, E.; Hegemann, J.; Gehler, S.; Braun, M. Resilient distribution grids—Cyber threat scenarios and test environment. In Proceedings of the 2016 IEEE PES Innovative Smart Grid Technologies Conference Europe (ISGT-Europe), Ljubljana, Slovenia, 9–12 October 2016; pp. 1–6.
18. *Dena-Verteilnetzstudie: Ausbau- und Innovationsbedarf der Strom Veteilnetze in Deutschland bis 2030*; DeNA: Berlin, Germany, 2018.

19. Strasser, T.; Pröstl Andrén, F.; Lauss, G.; Bründlinger, R.; Brunner, H.; Moyo, C.; Seitl, C.; Rohjans, S.; Lehnhoff, S.; Palensky, P.; et al. Towards holistic power distribution system validation and testing—An overview and discussion of different possibilities. *e & i Elektrotechnik und Informationstechnik* **2017**, *134*, 71–77.

20. De Jong, E.; Graaff, R.; Vaessen, P.; Crolla, P.; Roscoe, A.; Lehfuss, F.; Lauss, G.; Kotsampopoulos, P.; Gafaro, F. *European White Book on Real-Time Power Hardware-in-the-Loop Testing*; DERlab e.V.: Kassel, Germany, 2012.

21. Omar Faruque, M.D.; Strasser, T.; Lauss, G.; Jalili-Marandi, V.; Forsyth, P.; Dufour, C.; Dinavahi, V.; Monti, A.; Kotsampopoulos, P.; Martinez, J.A.; et al. Real-Time Simulation Technologies for Power Systems Design, Testing, and Analysis. *IEEE Power Energy Technol. Syst. J.* **2015**, *2*, 63–73. [CrossRef]

22. Guillaud, X.; Faruque, M.O.; Teninge, A.; Hariri, A.H.; Vanfretti, L.; Paolone, M.; Dinavahi, V.; Mitra, P.; Lauss, G.; Dufour, C.; et al. Applications of Real-Time Simulation Technologies in Power and Energy Systems. *IEEE Power Energy Technol. Syst. J.* **2015**, *2*, 103–115. [CrossRef]

23. Lauss, G.; Faruque, M.O.; Schoder, K.; Dufour, C.; Viehweider, A.; Langston, J. Characteristics and Design of Power Hardware-in-the-Loop Simulations for Electrical Power Systems. *IEEE Trans. Ind. Electron.* **2016**, *63*, 406–417. [CrossRef]

24. Roscoe, A.J.; Mackay, A.; Burt, G.M.; McDonald, J.R. Architecture of a Network-in-the-Loop Environment for Characterizing AC Power-System Behavior. *IEEE Trans. Ind. Electron.* **2010**, *57*, 1245–1253. [CrossRef]

25. Lehfuss, F.; Lauss, G.; Kotsampopoulos, P.; Hatziargyriou, N.; Crolla, P.; Roscoe, A. Comparison of multiple power amplification types for power Hardware-in-the-Loop applications. In Proceedings of the 2012 IEEE Complexity in Engineering (COMPENG), Aachen, Germany, 11–13 June 2012; pp. 1–6.

26. Brandl, R. Operational Range of Several Interface Algorithms for Different Power Hardware-In-The-Loop Setups. *Energies* **2017**, *10*, 1946. [CrossRef]

27. Ren, W.; Steurer, M.; Baldwin, T.L. Improve the Stability and the Accuracy of Power Hardware-in-the-Loop Simulation by Selecting Appropriate Interface Algorithms. *IEEE Trans. Ind. Appl.* **2008**, *44*, 1286–1294. [CrossRef]

28. Ou, T.C.; Lu, K.H.; Huang, C.J. Improvement of Transient Stability in a Hybrid Power Multi-System Using a Designed NIDC (Novel Intelligent Damping Controller). *Energies* **2017**, *10*, 488. [CrossRef]

29. Helmedag, A.; Isermann, T.; Monti, A. Fault ride through certification of wind turbines based on a power hardware in the loop setup. In Proceedings of the 2013 IEEE International Workshop on Applied Measurements for Power Systems (AMPS), Aachen, Germany, 25–27 September 2013; pp. 150–155.

30. Kleftakis, V.; Rigas, A.; Vassilakis, A.; Kotsampopoulos, P.; Hatziargyriou, N. Power-Hardware-in-the-loop simulation of a D-STATCOM equipped MV network interfaced to an actual PV inverter. In Proceedings of the IEEE PES ISGT Europe 2013, Copenhagen, Denmark, 6–9 October 2013; pp. 1–5.

31. Brandl, R. Improved Methods for Real-time Simulation and Hardware-in-the-loop. In Proceedings of the International Conference on Integration of Renewable and Distributed Energy Resources (IRED), Vienna, Austria, 16–19 October 2018.

32. Strasser, T.I.; Moyo, C.; Bründlinger, R.; Lehnhoff, S.; Blank, M.; Palensky, P.; van der Meer, A.A.; Heussen, K.; Gehrke, O.; Rodriguez, J.E.; et al. An Integrated Research Infrastructure for Validating Cyber-Physical Energy Systems. In *Industrial Applications of Holonic and Multi-Agent Systems*; Mařík, V., Wahlster, W., Strasser, T., Kadera, P., Eds.; Springer International Publishing: Cham, Switzerland, 2017; pp. 157–170.

33. CEN-CENELEC-ETSI Smart Grid Coordination Group. *Smart Grid Reference Architecture*; CEN-CENELEC-ETSI Smart Grid Coordination Group: Brussels, Belgium, 2012.

34. Maniatopoulos, M.; Lagos, D.; Kotsampopoulos, P.; Hatziargyriou, N. Combined control and power hardware in-the-loop simulation for testing smart grid control algorithms. *IET Gener. Trans. Distrib.* **2017**, *11*, 3009–3018. [CrossRef]

35. Kundur, P.S.; Balu, N.J. (Eds.) *Power System Stability and Control*; The EPRI Power System Engineering Series; McGraw-Hill: New York, NY, USA, 1994.

36. Mahseredjian, J.; Dinavahi, V.; Martinez, J.A. Simulation Tools for Electromagnetic Transients in Power Systems: Overview and Challenges. *IEEE Trans. Power Deliv.* **2009**, *24*, 1657–1669. [CrossRef]

37. Andrén, F.; Lehfuss, F.; Jonke, P.; Strasser, T.; Rikos, E.; Kotsampopoulos, P.; Moutis, P.; Belloni, F.; Sandroni, C.; Tornelli, C.; et al. DERri Common Reference Model for Distributed Energy Resources—Modeling scheme, reference implementations and validation of results. *e & i Elektrotechnik und Informationstechnik* **2014**, *131*, 378–385.

38. Martinez, J.A.; Mahseredjian, J. Load flow calculations in distribution systems with distributed resources. A review. In Proceedings of the 2011 IEEE Power and Energy Society General Meeting, Detroit, MI, USA, 24–29 July 2011; pp. 1–8.

39. Lundstrom, B.; Chakraborty, S.; Lauss, G.; Brundlinger, R.; Conklin, R. Evaluation of system-integrated smart grid devices using software- and hardware-in-the-loop. In Proceedings of the 2016 IEEE Power & Energy Society Innovative Smart Grid Technologies Conference (ISGT), Minneapolis, MN, USA, 6–9 September 2016; pp. 1–5.

40. Holt, C.; Kong, A.; St. Leger, A.; Bennett, D. Communications network emulation for smart grid test-bed. In Proceedings of the 2016 IEEE Power and Energy Society General Meeting (PESGM), Boston, MA, USA, 17–21 July 2016; pp. 1–5.

41. Babazadeh, D.; Chenine, M.; Zhu, K.; Nordstrom, L.; Al-Hammouri, A. A platform for wide area monitoring and control system ICT analysis and development. In Proceedings of the 2013 IEEE PowerTech Grenoble Conference, Grenoble, France, 16–20 June 2013; pp. 1–7.

42. Spitzenberger & Spies GmbH & Co. KG, Data Sheet: PAS Series of 4-Quadrant Amplifiers. Available online: https://www.spitzenberger.de/weblink/1002 (accessed on 24 August 2018).

43. Opal-RT Technologies Inc. Product Sheet:Real-Time Simulation Solutions for Power Grids and Power Electronics. Available online: https://www.opal-rt.com/wp-content/themes/enfold-opal/pdf/L00161_0260.pdf (accessed on 24 August 2018).

44. Fronius International GmbH, Data Sheet: FRONIUS Symo Series. Available online: http://www.fronius.com/~/downloads/Solar%20Energy/Datasheets/SE_DS_Fronius_Symo_DE.pdf (accessed on 24 August 2018).

45. Dewetron GmbH, Data Sheet: DEWE -800. Available online: https://www.dewetron.com/wp-content/uploads/2016/05/dewetron_dewe-800_e-1.pdf (accessed on 24 August 2018).

46. DIgSILENT GmbH, Data Sheet: PowerFactory 2018. Available online: https://www.digsilent.de/en/downloads.html (accessed on 24 August 2018).

47. Regatron AG, Data Sheet: TC.ACS Programmable AC Source-Sink Series. Available online: https://www.regatron.com/service/download/brochures/tc.acs-series-brochure.pdf (accessed on 24 August 2018).

48. Janitza Electronics GmbH, Data Sheet: Universal Measuring Device UMG 96—Operating Instructions. Available online: https://www.janitza.de/betriebsanleitungen.html?file=files/download/manuals/current/UMG96/Janitza-Manual-UMG96-all-versions-en.pdf (accessed on 24 August 2018).

49. Network Technology/Network Operation Forum at VDE (VDE⏐FNN). *VDE-AR-N 4105:2011-08 Power Generation Systems Connected to the Low-Voltage Distribution Network*; VDE-VERLAG GMBH: Berlin, Germany, 2010.

50. Elbs, C.; Pardatscher, R. *Einsatz der Q(U)-Regelung bei der Vorarlberger Energienetze GmbH*, Vorarlberger Energienetze: Bregenz, Austria, 2014.

51. Energie-Control Austria. *TOR D4: Parallelbetrieb von Erzeugungsanlagen mit Verteilernetzen*; Energie-Control: Vienna, Austria, 2016.

52. International Electrotechnical Commission. *IEC 61000-3-3:2013 Electromagnetic Compatibility (EMC)—Part 3-3: Limits-Limitation of Voltage Changes, Voltage Fluctuations and Flicker in Public Low-Voltage Supply Systems, for Equipment with Rated Current < 16 A Per Phase and Not Subject to Conditional Connection*; International Electrotechnical Commission: Geneva, Switzerland, 2013.

53. KACO New Energy GmbH, Data Sheet: Powador 12-20 TL3. Available online: http://kaco-newenergy.com/fileadmin/data/downloads/products/Powador_12.0-20.0_TL3/Data%20Sheets/DTS_PW_12-20_TL3_en.pdf (accessed on 24 August 2018).

54. Stifter, M.; Andren, F.; Schwalbe, R.; Tremmel, W. Interfacing PowerFactory: Co-simulation, Real-Time Simulation and Controller Hardware-in-the-Loop Applications. In *PowerFactory Applications for Power System Analysis*; Power Systems; Gonzalez-Longatt, F.M., Rueda, J.L., Eds.; Springer: Cham, Switzerland, 2014; pp. 343–366.

55. Ebe, F.; Idlbi, B.; Casel, M.; Kondzialka, C.; Chen, S.; Morris, J.; Heilscher, G.; Seitl, C.; Bründlinger, R.; Strasser, T. An approach for validating and testing micro grid and cell-based control concepts. In Proceedings of the CIRED Workshop 2018, Ljubljana, Slovenia, 7–8 June 2018.

56. Santodomingo, R.; Uslar, M.; Gottschlak, M.; Goering, A.; Nordstrom, L.; Valdenmaiier, G. The discern tool support for knowledge sharing in large smart grid projects. In Proceedings of the CIRFD Workshop 2016, Helsinki, Finland, 14–15 June 2016.

57. SunSpec Alliance. *SunSpec Technology Overview*; SunSpec Alliance: San Jose, CA, USA, 2015.

58. International Electrotechnical Commission. *IEC 61850 Part 7-4: Basic Communication Structure—Compatible Logical Node Classes and Data Object Classes*; International Electrotechnical Commission: Geneva, Switzerland, 2010.

59. International Electrotechnical Commission. *IEC 61850 Part 7-420: Basic Communication Structure—Distributed Energy Resources Logical Nodes*; International Electrotechnical Commission: Geneva, Switzerland, 2009.

60. Heilscher, G.; Chen, S.; Lorenz, H.; Ebe, F.; Kondzialka, C.; Kaufman, T.; Hess, S.; Wening, J. Secure Energy Information Network in Germany—Demonstration of Solar-, Storage- and E-Mobility Applications. In Proceedings of the 8th International Workshop on Integration of Solar Power into Power Systems, Stockholm, Sweden, 16–17 October 2018.

61. Heussen, K.; Morales Bondy, D.E.; Nguyen, V.H.; Blank, M.; Klingenberg, T.; Kulmala, A.; Abdulhadi, I.; Pala, D.; Rossi, M.; Carlini, C.; et al. *D-NA5.1 Smart Grid Configuration Validation Scenario Description Method*; ERIGrid: Vienna, Austria, 2017.

62. Smart Grids-Plattform Baden-Württemberg e.V. C/Sells-Guiding principle. Available online: https://www.csells.net/en/about-c-sells/guiding-principle.html (accessed on 24 August 2018).

![energies logo] *energies*

MDPI

Article

Experimental Validation of Peer-to-Peer Distributed Voltage Control System

Hamada Almasalma [1,2,*], Sander Claeys [1,2], Konstantin Mikhaylov [3], Jussi Haapola [3], Ari Pouttu [3] and Geert Deconinck [1,2]

[1]   Departement Elektrotechniek, KU Leuven, Kasteelpark Arenberg 10, 3001 Leuven, Belgium; sander.claeys@kuleuven.be (S.C.); geert.deconinck@kuleuven.be (G.D.)
[2]   EnergyVille Research Center, Thor Park 8310, 3600 Genk, Belgium
[3]   Centre for Wireless Communications (CWC), University of Oulu, 90014 Oulu, Finland; konstantin.mikhaylov@oulu.fi (K.M.); jussi.haapola@oulu.fi (J.H.); ari.pouttu@oulu.fi (A.P.)
*   Correspondence: hamada.almasalma@kuleuven.be

Received: 18 April 2018; Accepted: 16 May 2018; Published: 20 May 2018

Abstract: This paper presents experimental validation of a distributed optimization-based voltage control system. The dual-decomposition method is used in this paper to solve the voltage optimization problem in a fully distributed way. Device-to-device communication is implemented to enable peer-to-peer data exchange between agents of the proposed voltage control system. The paper presents the design, development and hardware setup of a laboratory-based testbed used to validate the performance of the proposed dual-decomposition-based peer-to-peer voltage control. The architecture of the setup consists of four layers: microgrid, control, communication, and monitoring. The key question motivating this research was whether distributed voltage control systems are a technically effective alternative to centralized ones. The results discussed in this paper show that distributed voltage control systems can indeed provide satisfactory regulation of the voltage profiles.

Keywords: peer-to-peer; distributed control; device-to-device communication; voltage control; experimentation

1. Introduction

Over the last decade, there has been a clear focus in the European Union (EU) on promoting low-carbon generation technologies and renewables. To ensure the EU meets its climate and energy goals, the 20-20-20 targets aim to cut the emission of greenhouse gasses by 20% compared to the 1990 level, achieve a 20% share of renewables in the total energy consumption, and improve the energy efficiency by 20%. In many countries, feed-in tariffs for eligible technologies have guaranteed returns for investors, and this along with other forms of market support have contributed to a reduction in technology costs and an increasing penetration of renewable energy resources (RESs) into distribution networks across Europe. This growth of renewable energy is expected to maintain since by 2030 the EU aims for 27% of the final energy consumption to come from renewable sources. The progress of the EU and its member states towards 2020 climate and energy targets are summarized in [1].

These trends are impacting the operation of distribution networks, making the Distribution System Operators' (DSOs) mission of providing secure electricity supply and high quality of service increasingly challenging. The historical "fit-and-forget" strategy of distribution networks was consistent with the unidirectional power flows from substations to end consumers and their predictable load profiles. When connecting significant amounts of RESs to the network, the assumption of unidirectional power flows is not always valid anymore. The generated power of RESs can reverse the power flows in the grid, what could lead to a rise of the voltage profiles beyond the allowed limits. Moreover, intermittent and unpredictable nature of renewables increases the complexity of controlling the distribution networks. A comprehensive overview of the impacts of the renewable energy and information and communications technology (ICT) driven energy transition on distribution networks is presented in [2]. To maintain a high security of supply and quality of service, DSOs have to find new strategies to control their networks.

A transition towards active management strategies would be capable of maintaining the voltage profiles of distribution networks within acceptable limits to comply with the European standard EN 50160 [3] while minimizing, deferring, or even avoiding any capacity upgrades. Additionally, valuable flexibility of prosumers can be embedded in the operational management of the networks, to allow the prosumers to participate in supporting the grid as kind of ancillary service. Details of the most effective and efficient ways for managing the future active distribution networks, to address the 21st century challenges of transitioning to low-carbon electricity, are discussed in [4].

The need for managing distribution networks actively by employing smart grid solutions and creating innovative investments and business models are the reasons for launching the EU funded Peer-to-Peer Smart Energy Distribution Networks (P2P-SmarTest) project. The project was launched in 2015 and continued until the end of 2017. The idea of the project consists in developing intelligent control, trading, and communication algorithms through a "Peer-to-Peer" concept; to facilitate the integration of demand side flexibility and to ensure optimal operation of RESs within the network while maintaining quality and security of supply. The deliverables of the project can be found on the website of the project [5]. In [6], the view to Peer-to-Peer (P2P) approach for smart grid operation adopted in P2P-SmarTest project is presented. The P2P control paradigm used in the project is presented in [7].

The approach adopted in P2P-SmarTest project to regulate voltage profiles of active distribution networks is based on distributed optimization techniques and P2P communication. Distributed optimization, as an alternative approach to solve challenges of the centralized optimization mechanism, has attracted increasing attention recently [8]. A Distributed optimization-based control system is characterized by the complete absence of a central controller. Every RES is considered to be an autonomous control agent where all agents are equally important. To overcome the absence of the central decision making controller, the agents communicate with each other in a P2P fashion. With communication, they are able to make the correct control decisions in every particular situation. Failure of one controller in distributed control system does not lead to an inability to control the system. The work in [9] describes fundamental concepts and approaches within the field of distributed control systems that are appropriate to power engineering applications.

Centralized control systems often suffer from serious computation, robustness, and communication issues for power networks with many controllable devices. Distributed control is perhaps the only viable strategy for such networks. Nevertheless, these centralized systems can achieve high performance. In a centralized control system, there is only one controller, which receives all necessary data, and based on all available information the multi-objective controller can achieve a globally optimal performance. An interesting question is whether P2P distributed control systems can achieve a comparable good performance to the centralized one. Most research studies appearing in the literature attempt to answer this question by means of simulators, as reviewed in [10–12]. For instance, in [13] a gradient descent method has been used to distribute a centralized optimization problem over agents participating in the voltage control, a push-sum gossip algorithm is implemented to enable P2P communication between the agents. Simulink (MATLAB, version R2016a, The MathWorks, Inc, Natick, MA, USA) has been used to model a 5-bus microgrid and to validate the performance of the proposed algorithm. In [14], a dual decomposition technique is used to design a P2P-based voltage control system. A backward/forward sweep power flow calculation algorithm, coded in MATLAB, has been used to model a low voltage, 62-bus, semiurban feeder and to test the ability of the algorithm to control the voltage effectively within limits. In [15], openDSS simulator (version 2017, EPRI, Palo Alto, CA) has been used to validate the effectiveness and robustness of a fully distributed voltage control algorithm that has been developed based on the Alternating Direction Method of Multipliers and consensus protocol (consensus ADMM). The same method has been used in [16] and validated using CVX software (version 2014, CVX Research, Inc., Stanford, CA, USA) (convex programming). Distributed Energy Storage Systems (DESSs) are used in [17] to control the voltage profiles of active distribution networks in a distributed way. The proposed methodology is based on network partitioning strategy. Linear programming and voltage sensitivities are used to define the areas for which each DESS maximizes its influence. To study the performance of the proposed algorithms, MATLAB has been used to code the algorithms and to model an IEEE 123 nodes test system. The concept of network partitioning is also used in [18] to implement a decentralized voltage control system that regulates reactive power of photovoltaic (PV) inverters. The proposed methodology of [18] is based on Lyapunov theory and has been validated via Matlab/Simulink environment.

The concepts of transactive energy (TE), home microgrids (H-MGs) and coalition formation are used in [19] to design an algorithm for optimal use of electrical/thermal energy distribution resources, while maximizing profit of H-MGs. The algorithm is based on an optimization problem in which an objective function is based on economic strategies, distribution limitations and the overall demand in the market structure. MATLAB was used to solve the optimization problems of the proposed algorithm. The same concepts have been used in [20] to design an optimal, autonomous, and distributed bidding-based energy optimization scheduling algorithm to maximize profit and energy balancing efficiency of H-MGs under residential loads. A comprehensive simulation study was carried out to reveal the effectiveness of the proposed method in lowering the market clearing price, increasing H-MG responsive load consumption, and promoting local generation. Optimal management system of battery energy storage is proposed in [21] to enhance the resilience of a PV-based commercial building while maintaining its operational cost at a minimum level. The methodology is based on linear optimization programming problem with Conditional Value at Risk (CVaR) incorporated in the objective function. CVaR is used to account for the uncertainty in the intermittent PV system generated power and that in the electricity price. MATLAB simulation studies were carried out to evaluate the performance of the proposed method.

There are few studies in existing literature addressing the experimental validation of distributed control algorithms. Experimental evaluations of real deployments are thus lacking. In [22], a gossip-based P2P voltage control has been tested in a pilot site, the work is part of the European Commission FP7 DREAM project. Six households were equipped with smart control agents, which measure the households' consumption and control the households' flexible loads. Each agent is connected to a local Wi-Fi router (internet gateway) and a virtual private network is then used to enable P2P communication between the neighboring agents. In [23], a multi-agent platform has been implemented and used to test a dual-decomposition-based optimization method for controlling the prosumers' flexibility. The distributed agents are implemented in Raspberry Pi computers. The agent-based control algorithm of each agent is implemented in Python and executed via Matlab calls. The setup is part of Local Intelligent Networks and Energy Active Region (LINEAR) project [24]. In [25], a gossiping P2P semantic overlay network is implemented by a toolbox, Agora+, enabling P2P communication between agents. The toolbox has been used to implement a distributed tertiary control algorithm, which allows groups of generators to operate at an economical optimum. In [26], distributed reactive power control has been implemented and tested using real power inverters. Each inverter is considered to be an agent where coordination between the agents is obtained by exchanging information via an IP-based communication network.

This paper discusses the results of the experimental validation of a dual-decomposition-based P2P voltage control algorithm developed within the P2P-SmarTest project. A simulation already demonstrated the effectiveness of this algorithm [14] and this paper demonstrates it experimentally. The voltage control problem is formulated as an optimization problem. The proposed method calculates the minimum change in reactive power and active power needed to maintain the voltages within the limits. The dual-decomposition method decomposes an optimization problem (with separable cost functions and coupled constraints) into sub-problems, suitable for distributed control. Dual-decomposition applies the theory of Lagrangian multipliers and duality to convert a centralized constraint optimization problem into a fully distributed constraint optimization problem. The proposed dual-decomposition method differs from the classical dual-decomposition. In classical dual-decomposition [27], there is a need for a central agent to calculate the Lagrangian multipliers (control signals), whereas in the proposed dual-decomposition method, the Lagrangian multipliers are calculated locally and each agent communicates its Lagrangian multipliers to the other agents in a P2P fashion.

Our main contributions can be summarized as follows. (1) We present the design, development and hardware setup of a laboratory-based P2P voltage control testbed; (2) Secondly, we propose the use of a fully distributed dual-decomposition method to design a P2P voltage control system; (3) Thirdly, we propose the use of Long Range Wide-area network (LoRaWAN) technology to design a device-to-device communication system. The device-to-device communication is used to enable P2P data interchange between agents of the proposed voltage control system; (4) Finally, we validate experimentally that the proposed P2P voltage control system can indeed provide satisfactory regulation of the voltage profiles.

The testbed presented in this paper provides realistic and pragmatic solution for evaluating P2P smart grid applications. The testbed is used to evaluate the performance of the proposed dual-decomposition-based voltage control system. It can also be used to evaluate other distributed applications for grid management. The testbed allows for re-using of the existing simulator code, while still facilitating accurate integration of power and communication effects on a real hardware platform.

The rest of this paper is organized as follows. The laboratory-based P2P voltage control testbed is described in Section 2. Section 3 presents the P2P-based voltage control algorithm. Drive of the inverters is presented in Section 4. The Device-to-Device (D2D) communication modules used to enable P2P communication between the agents are described in Section 5. Section 6 presents the experimental results and the key performance indicators. Finally, the paper is concluded in Section 7 with future work.

2. Testbed Architecture

The architecture of the P2P voltage control testbed is depicted in Figure 1. The testbed consists of four different layers which interact with each other: (1) microgrid layer; (2) control layer; (3) communication layer; and (4) monitoring layer. The microgrid layer consists of programmable inverters (label 1 in Figure 1); connected to DC power supplies (label 2). The inverters emulate prosumers with photovoltaic (PV) installations, they are connected to the grid by resistors in series with inductors (label 3). The resistors and inductors are used to emulate a low voltage feeder. The control layer consists of inner control systems (label 4) that drive the power inverters, and grid voltage support functions (GVSFs) that control the voltage profiles of the micogrid (label 6). The communication layer consists of D2D communication modules (label 7) that are used to disseminate the status of the voltage profiles in a P2P fashion. The monitoring layer consists of voltmeters (label 9) and data acquisition platform (label 11).

The P2P voltage control testbed consists of three types of agents: (1) actuators; (2) observers; and (3) a monitor. Each GVSF is connected to a D2D communication module and together they form an actuator agent (label 5). The actuator agents are connected to the programmable inverters through the inner control systems (control loops) and participate actively in voltage control by calculating the change in reactive power and active power that each inverter should follow to maintain the voltage profiles within specified limits. The set-points of the change in reactive and active power of the inverters are determined based on an optimization problem solved in a fully distributed way.

The observer agent (label 8) consists of a voltmeter connected to a D2D communication module through a Raspberry Pi (R.Pi) computer (label 10). The voltmeter periodically measures the voltage of its bus, and the R.Pi fetches the latest reading. The R.Pi of each voltmeter hosts a software that was developed for interfacing with both the communication module and the voltmeter. The R.Pi calculates the control signals (further referred to as Lagrangian multipliers) based on the latest voltage measurement, according to a procedure described later. These control signals are broadcasted through the D2D modules. The actuators communicate with the observers in a P2P fashion to receive the control signals. The actuators then determine how to react based on these control signals and based on their impact on the observed voltages (the impact on the voltages is expressed by voltage sensitivities). They also take into account the cost of dispatching a change in active and reactive power.

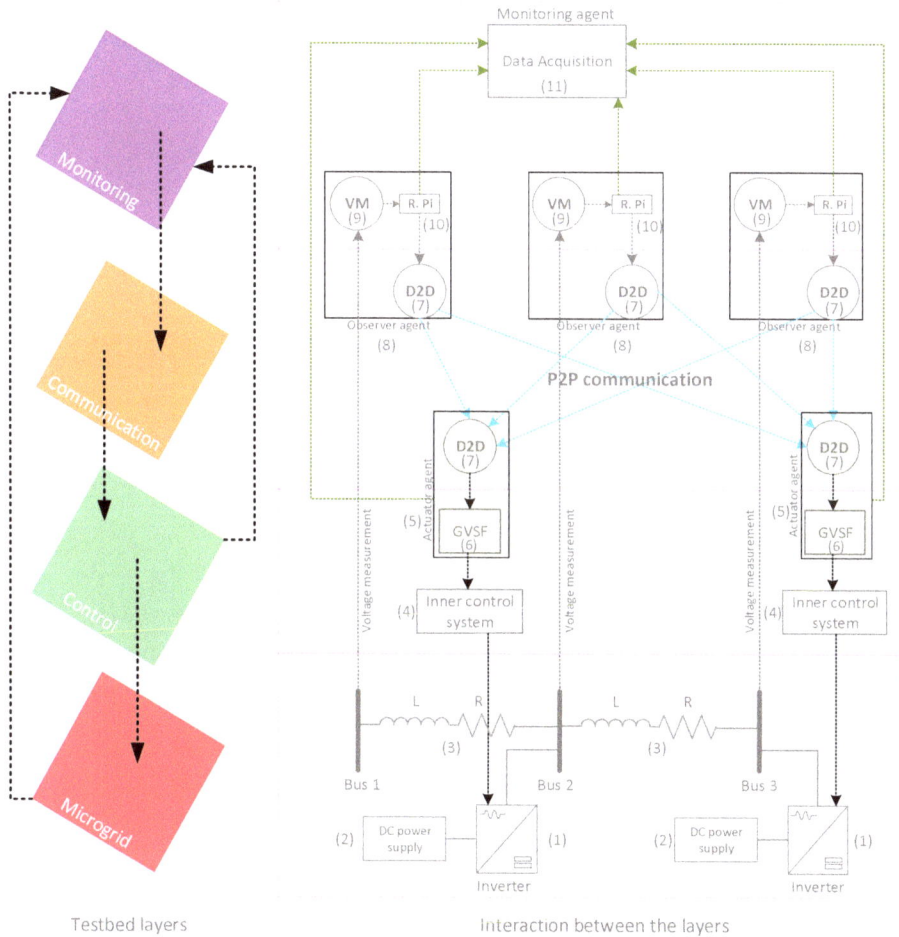

Figure 1. Multi-layer multi-agent architecture of the Peer-to-Peer (P2P) voltage control testbed (VM stands for voltmeter, Device-to-Device (D2D): device-to-device communication module, grid voltage support function (GVSF): grid voltage support function, Raspberry Pi (R.Pi): raspberry pi computer, **Right**: resistor, **Left**: inductor, labels 1 to 11 indicate the different parts of the testbed).

The third type of agent, the monitoring agent (label 11), represents a data acquisition platform. This additional agent is not required for the operation of the P2P voltage control algorithm. The observers and actuators record several variables from the algorithm that they execute, together with timestamps. These recordings are cached locally. Periodically, the observers and actuators transfer the cached recordings to the monitoring agent in a robust way. Therefore, even if the data acquisition network is temporarily offline, no data will be lost. The monitoring agent hosts a web service, through which all recorded data is visualized in several dashboards.

The overall schematic of the testbed is depicted in Figure 2. The microgrid is connected to the main grid through 400 V (line-to-line voltage (L-L)), 64 Amps (A) busbar.

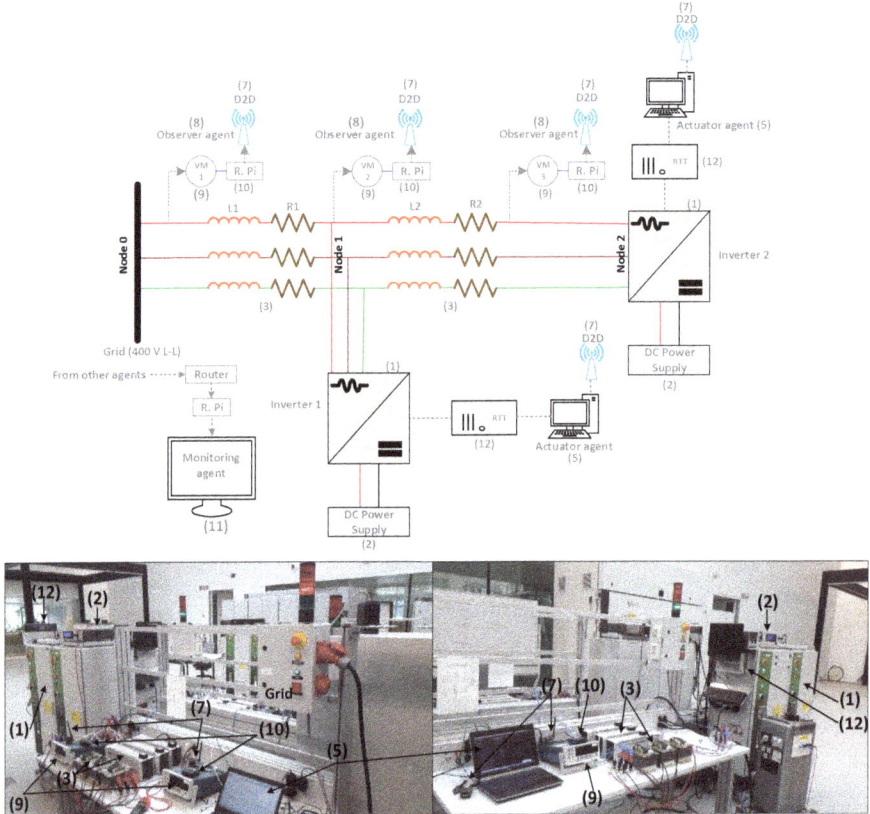

Figure 2. Schematic of the P2P voltage control testbed (VM stands for voltmeter, D2D: device-to-device communication module, **Right**: resistor, **Left**: inductor, Real-Time Target (RTT): real time target computer, R.Pi: raspberry pi computer, V: volt, L-L: line to line, labels (1)–(12) indicate the different parts of the testbed, labels **(1)**–**(11)** same as in Figure 1).

3. Dual-Decomposition-Based P2P Voltage Control Algorithm

The proposed P2P voltage control algorithm regulates the voltage within allowed limits based on an optimization problem. The algorithm uses a minimum change in reactive and active power consumption or injection of some participating inverters installed in the microgrid to control the voltage. The derivation of the algorithm is presented in [14] and here we present the algorithm in a more practical way.

Without compensation, each inverter injects a certain amount of active power into the system. In reality, this active power originates from the solar energy received by the photovoltaic cell. The inverter can additionally inject reactive power, as long as the total apparent power does not exceed the inverter rating. The inverter has an additional degree of freedom; it can curtail a fixed percentage of the active power. Therefore, the actuator agent can take two actions: reducing the active power (by an amount ΔP) and injecting or absorbing reactive power (by an amount ΔQ).

Each actuator agent solves the following optimization problem to find ΔP_d and ΔQ_d :

$$\underset{\Delta P, \Delta Q}{\text{argmin}} \quad c^P(\Delta P_d^{(t)})^2 + c^Q(\Delta Q_d^{(t)})^2 + \sum_{i=1}^{N}\left((\lambda_i^{\max})^{(t-1)} - (\lambda_i^{\min})^{(t-1)}\right)\left(v_{d,i}^P\Delta P_d^{(t)} + v_{d,i}^Q\Delta Q_d^{(t)}\right)$$

Subject to: $(-c_r)(P_d^{\text{profile}})^{(t)} \leq \Delta P_d^{(t)} \leq 0$ (1)

$$-\sqrt{(S_d)^2 - \left((P_d^{\text{profile}})^{(t)} + \Delta P_d^{(t)}\right)^2} \leq \Delta Q_d^{(t)} \leq \sqrt{(S_d)^2 - \left((P_d^{\text{profile}})^{(t)} + \Delta P_d^{(t)}\right)^2}$$

where $d \in \mathcal{D}$ is the number of the actuator agent ($\mathcal{D}$ is the set of actuators participating in the voltage control). $i \in \mathcal{N}$ is the number of the observer agent ($\mathcal{N}$ is the set of observers participating in the voltage control). $c^P(\Delta P_d^{(t)})^2$ represents the quadratic cost of a change in active power of inverter d with an amount ΔP_d at time step t, while $c^Q(\Delta Q_d^{(t)})^2$ represents the quadratic cost of a change in reactive power of inverter d with an amount ΔQ_d at time step t. c^P and c^Q are constant factors used to penalize the control variables ΔP_d and ΔQ_d. These factors define the priorities for the control actions. It is supposed that reactive power control of the inverter is cheaper than cutting its active power. Therefore, c^P should be greater than c^Q in a sense that gives priority of the control action to the reactive power. When the reactive power of the inverter is not sufficient, active power curtailment of the inverter will be used to regulate the system voltages. In our control system, we set $c^P = 200$ and $c^Q = 1$. Active power curtailment can be penalized more to minimize its use, but having higher c^P would decrease the speed of convergence when the curtailment is used to return the voltages back to the limits. It is worth mentioning that the factor c^Q can be calculated to incorporate losses on the network (related to reactive power compensation) and other costs. In reality, reactive power provision can lead to some additional losses in the network. An approximate cost factor can include the additional losses in the inverter [28]. Incorporating the grid losses however would require a more complete network model.

$v_{d,i}^P$ and $v_{d,i}^Q$ are the sensitivity of the voltage at bus i (observer i) to the change in the active power and reactive power (respectively) of inverter d. c_r is the curtailment factor. In this paper, c_r is set to 30%. In reality, c_r can be set based on how much the prosumer would like to curtail the active power. $(P_d^{\text{profile}})^{(t)}$ is the active power generated by inverter d at time step t. S_d is the rated apparent power of inverter d.

$(\lambda_i^{\max})^{(t-1)}$ and $(\lambda_i^{\min})^{(t-1)}$ are the control signals of violating the maximum and minimum (respectively) allowed voltage at bus i. They are calculated at the previous time step $t-1$ and considered in the optimization of time step t. Mathematically speaking, they represent the Lagrangian multipliers. Each observer measures the voltage at its bus and updates these control signals based on the following equations:

$$(\lambda_i^{\max})^{(t)} = \max\left(0, (\lambda_i^{\max})^{(t-1)} + \alpha\left((V_i^{\text{meas}})^{(t)} - V^{\max}\right)\right)$$
$$(\lambda_i^{\min})^{(t)} = \max\left(0, (\lambda_i^{\min})^{(t-1)} - \alpha\left((V_i^{\text{meas}})^{(t)} - V^{\min}\right)\right)$$ (2)

where $(\lambda_i^{\max})^{(t)}$ and $(\lambda_i^{\min})^{(t)}$ are the updated control signals calculated at time step t and considered in the optimization of time step $t+1$. $(V_i^{\text{meas}})^{(t)}$ is the measured voltage at bus i after applying the decisions $\Delta P_d^{(t)}$ and $\Delta Q_d^{(t)}$. $V^{\max}$ and $V^{\min}$ are the maximum and minimum allowed voltage, respectively. We set $V^{\max} = 1.1$ p.u. (per unit) and $V^{\min} = 0.9$ p.u. according to the European standard EN50160. The parameter α is the step size of the dual decomposition method. Because of the Karush-Kuhn–Tucker conditions (KKT), the Lagrangian multipliers cannot be smaller than zero. This explains the use of maximum operator in (2).

The control algorithm goes through the following steps:

1. Each observer agent measures the voltage. If the voltage exceeds the upper voltage limit, it will increase λ_i^{max}. If the voltage is lower than the upper limit, it will decrease λ_i^{max}, at most until it reaches zero. A similar procedure applies to λ_i^{min}. The parameter α determines how large the updates to the control signals will be.
2. The actuator agents receive updates of λ_i^{max} and λ_i^{min} periodically. They will adjust their compensation to take the new values of the control signals into account.
3. The voltage changes due to the actions of the actuator agents. The observer agents update again their λ_i^{max} and λ_i^{min}, and the whole process repeats. The communication from observer to actuator takes place through the D2D communication modules, while the feedback path goes through the electrical network.

From this explanation, it is clear that this process is based on feedback. As long as the voltage problem persists, the observer agents will increase the control signals to get more compensation from the actuator agents. The effect of α is similar to a gain in control theory. The trade-off in its selection is similar: a low value can lead to slow convergence, while a too large value can lead to instability.

4. Drive of the Rapid Prototyping Inverter with Voltage Support Function

4.1. PM15FM30C Triphase Module

DC/AC PM15FM30C Triphase rapid prototyping inverter modules are used in the testbed to emulate prosumers with PV installations. A schematic diagram of the PM15FM30C circuit is depicted in Figure 3. The PM15FM30C module mainly consists of:

1. A 15 kVA three phase inverter, consisting of three half-bridges with insulated-gate bipolar transistors (IGBTs).
2. A rectifier that can be connected directly to the AC voltage of the microgrid; it can be used to charge the DC bus in case one does not want to use a DC source.
3. An inductor-capacitor-inductor filter (LCL filter).
4. Three bypass resistors to limit the inrush current at the beginning of operation; these resistors are bypassed with a relay when the rapid prototyping module is running.
5. Current sensors to measure the current before and after the LCL filter.
6. Voltage sensors to measure the DC bus voltage and the AC voltage after the LCL filter at the grid side.
7. Control board to drive the IGBTs, control the switches K1-K6, and the fan of the module.

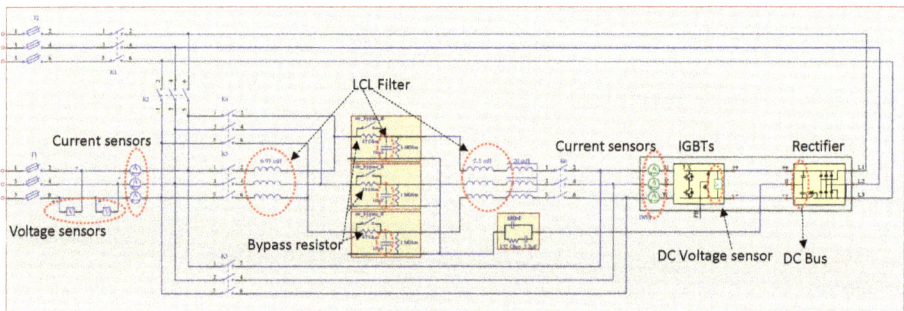

Figure 3. Circuit diagram of Triphase PM15FM30C rapid prototyping inverter.

The PM15FM30C inverter is programmed and operated through MATLAB/Simulink running on a computer. The computer communicates over Ethernet with an on-board PC-based Real-Time Target (RTT), which controls the Triphase power electronics as shown in Figure 4. Python (version 3.3, Python Software Foundation, Wilmington, DE, USA) has been used to code a software that manages the interface with the D2D communication module, fetches the Lagrangian multipliers from the D2D modules, stores the PV profiles, and solves a quadratic optimization problem with respect to local constraints as in (1). The software also manages the interface with MATLAB.

MATLAB exposes an interface to the Python software, which allows the Python software to directly execute scripts in MATLAB. The Python software uses a MATLAB script to push updates on the PV profiles and ΔP and ΔQ set-points to Simulink. MATLAB also manages the interface with RTT to control the switches and fan of the PM15FM30C.

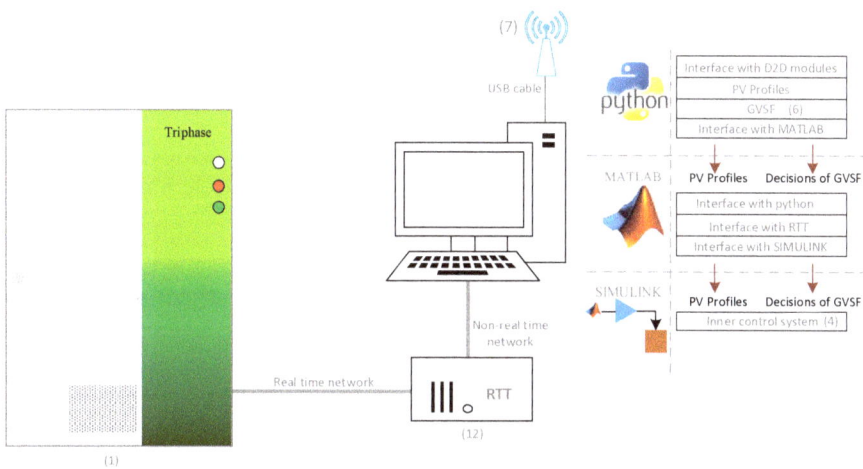

Figure 4. Actuator agent setup (labels same as in Figures 1 and 2).

4.2. Drive of PM15FM30C Triphase Module

The P2P voltage control algorithm represents a high level control system to coordinate the inverters in a distributed way aiming at regulating the voltages within the accepted limits. For the inverter to be able to follow the regulation of the P2P control algorithm, an internal control system has to be implemented and integrated with the GVSF. We have implemented a state-of-the-art current control loop, Phase-Locked-Loop (PLL), and Kalman filter to drive the inverter. The inner control system of the inverter is depicted in Figure 5.

The inverter uses current-mode control to control the active and reactive power. The line current is tightly regulated by the current control loop, through the inverter AC-side terminal voltage. Then, the active and reactive power are controlled by the phase angle and the amplitude of the inverter current with respect to a rotating frame that is synchronized with the point of common coupling (PCC) voltage using PLL. A Kalman filter is placed in front of the PLL in order to ensure that the PLL input at all times matches an ideal sinusoidal waveform as closely as possible, even when the voltage is highly distorted by the presence of harmonics. This ensures fast and low distortion operation of the PLL. Kalman filter is used in this work because it efficiently deals with the uncertainty of tuning its parameters.

The reference set-points of active and reactive power are calculated based on the PV profiles and the decisions of the GVSF as shown in Figure 5; then the reference set-points of active and reactive power are converted into d-q (direct-quadrature) reference set-points of the three phase current and these d-q set-points are used by the current control loop as a reference to control the d-component and q-component of the three phase current in order to follow the reference set-points of active and reactive power.

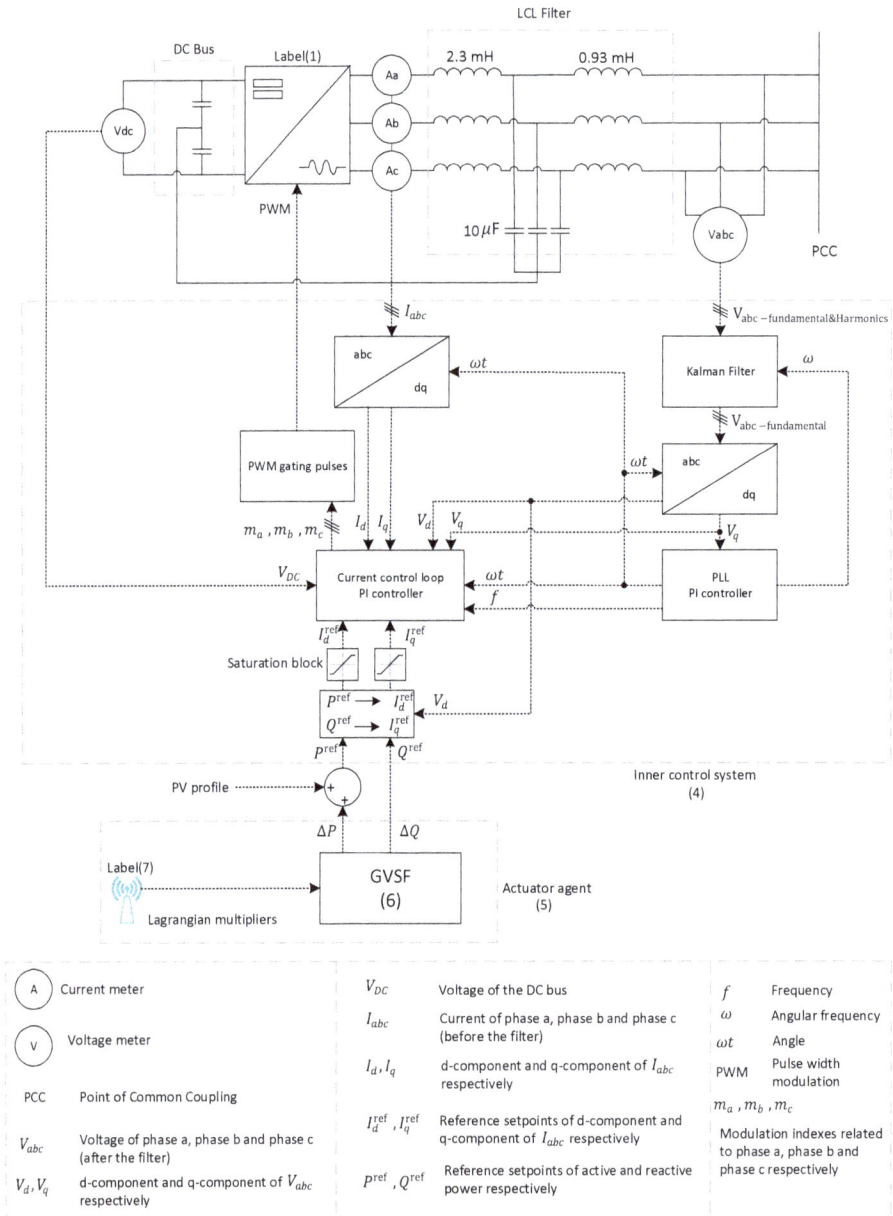

Figure 5. Inner control system of the inverter integrated with the GVSF (labels same as in Figures 1 and 2).

The details of the design and tuning of Kalman filter are presented in [29]. Chapter 8 in [30] presents the details of the design and tuning of the proportional-integral(PI) controllers used by the current control loop and PLL. The d-q transformation (of the current, voltage and power) can be also found in the same chapter.

It is worth mentioning that we have not used a voltage control loop, because the voltage of the DC bus is fixed by the DC source. For the inverter to be able to inject power from the DC side to the AC side, the DC bus has to be charged to a DC voltage higher than 650.6 V, which is the peak-to-peak voltage (phase to neutral) of the grid connection ($V_{pp} = 2 \times 230 \times \sqrt{2} = 650.6$).

5. Device-to-Device Communication

5.1. Background

Device-to-device communications typically refer to cellular communications technologies enabling direct transmission between proximate devices, without relaying information through the cellular base station [31]. However, D2D communications is not the exclusive domain of cellular networks and generally relates to the ability of peer devices to directly communicate with one another without having to relay the actual data through a central coordinator device, as e.g., used in [32,33]. The paper [31] presents a survey of the current state of the art for cellular D2D communications and points out that cellular D2D communications are much more efficient than communications on unlicensed spectrum as the communication interference is controllable at the licensed spectrum. The paper categorizes cellular D2D communications into four categories based on the level of control the base station has on them. The first category is device relaying with base station assisted controlled link. Here the base station allocates the channel resources for user equipment communications so that user equipment in poor coverage can maintain connectivity with the network. Direct communication between devices with base station assisted controlled link is the second category, where user equipment exchange data directly and some of these features have already been standardized by 3rd Generation Partnership Project (3GPP) release 13 proximity services and in release 14 for cellular vehicle-to-everything (C-V2X) communications (mode 3). The C-V2X uses outband communications as the actual data communications occur at the intelligent transportation system licensed radio frequency band and not on the cellular bands. The third category is relaying device with device assisted controlled link, where the user equipment communicate with one another using relays and without base station control. The fourth category is direct D2D with device assisted controlled link, where the user equipment communicate directly with one another without base station provision of control links. The paper [34] proposes a solution combining categories three and four for smart grid demand response scenarios for increased resiliency of smart grid operations.

The D2D communications required for P2P voltage control need not be based on cellular technologies and currently, no commercial-of-the-shelf cellular D2D chipsets are available. The key criteria for the selection of appropriate communication technology to adopt arise from the distances and placement of the observer and actuator agents. Common unlicensed band communication technologies like the IEEE 802.11 family (WiFi) or the IEEE 802.15.4 family (low rate personal area network) can be utilized if the distances between agents are within a few hundred meters and they have been installed outdoors or inside buildings near the exterior walls. Even then mesh type network where devices communicate in ad hoc fashion are required to ensure reliable connectivity. In other cases low power wide area (LPWA) communication technologies need to be utilized. Raza et al., [35] provide a survey on LPWA networks and claim that they represent a novel communication paradigm, which will complement traditional cellular and short range wireless technologies in addressing diverse requirements of Internet of Things (IoT) applications. This applies to smart grids in particular as LPWA technologies offer unique sets of features including wide-area connectivity for low power and low data rate devices, not provided by legacy wireless technologies. As an example, [33] proposes a gateway assisted D2D communications solution utilizing Long Range Wide-area network (LoRaWAN) technology and the work is a basis for the D2D communications scheme used in this paper. The work in [33] and the communications solution of this paper are not the same though; the work in [33] is similar to the second category whereas the solution applied in this paper is similar to the fourth category of cellular D2D communications, both using LoRaWAN technology.

5.2. D2D Communication Modules

The D2D communication modules are implemented based on a modular WSAN/IoT platform (wireless sensor and actuator network/Internet of Things) [36]. Each module is composed of three submodules stacked on top of each other, as shown in Figure 6. The radio submodule (the top submodule) hosts the RN2483 LoRaWAN radio transceiver (Microchip Technology, Chandler, AZ, USA). The main submodule (the middle submodule) includes the microcontroller (ST32F217, STMicroelectronics, Geneva, Switzerland), the power circuitry, and other peripherals. The USB submodule (the lower submodule) hosts an FTDI USB-UART chip (FT8U232AM, FTDI, Glasgow, UK) (FTDI: future technology devices international (semiconductor device company), USB: universal serial bus, UART: universal asynchronous receiver-transmitter). Additionally, each D2D module needs to have an 868 MHz SMA (SubMiniature version A) antenna. Also a mini or micro USB cable (any of these two, but only one at a time) should be connected to the USB submodule to interface with the agents (actuators and observers). The power required for the module's operation is also provided via the USB interface (maximum consumption is in the order of 200–300 mW).

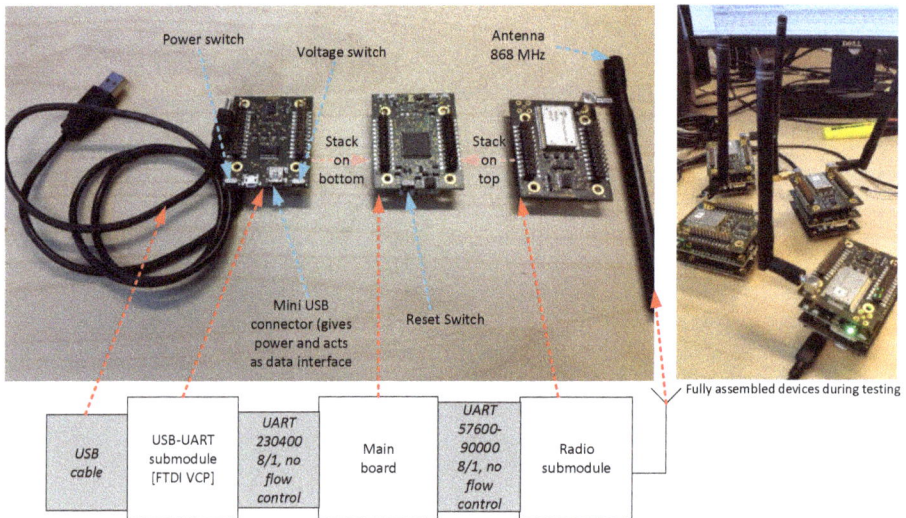

Figure 6. Structure of the D2D communication module (USB: universal serial bus, UART: universal asynchronous receiver-transmitter, FTDI: future technology devices international (semiconductor device company), VCP: virtual communication port).

5.2.1. Implemented Embedded Firmware

The application software is written in C (Dev-C++, Cambridge, MA, USA) and operates on top of the FreeRTOS embedded operation system (10.0.1, Real Time Engineers Ltd., Bristol, UK). The software has been developed using Eclipse (Kepler Service Release 2, Eclipse Foundation, Inc., Ottawa, ON, Canada) and compiled with GNU Compiler Collection (GCC) (7.1, Free Software Foundation, Boston, MA, USA) for Advanced RISC Machine (ARM) processors (ARMv8.3-A, Acorn, Cambridge, England, UK). The high level structural diagram of the embedded firmware is depicted in Figure 7 and it is composed of the three threads: main thread, radio thread, and UART thread. The main thread is initialized after the basic initialization procedures (setting clock, checking the module and configuring the peripherals, blinking LEDs). The main thread initializes the UART thread for communicating with the physically connected agent, the radio thread for controlling the radio transceiver, and the server data structure for storing the data from the agents. The UART drivers are implemented based on direct memory access (DMA) and use a timer to detect end of a packet. Due to this reason, agents should enable for at least a 5 ms idle time between the sequential UART packets.

The server data structure is implemented as a table listing the identifiers of the agents and the most recent data from them. The server structure is accessed and can be modified by either the radio thread or the UART thread. The D2D module can be configured to periodically report the complete table (i.e., the data from all other agents) to its agent.

The developed firmware implements a multi-stage error detection and correction system. In case of noncritical errors (e.g., wrong format of UART commands from the physically connected agent) the module recovers automatically. In case of severe mistakes (detected by the software or if the software hangs), the module reset procedure is initiated. After reset, the most recent state of the module is recovered. The restored data does not include the calculations made by the connected agent.

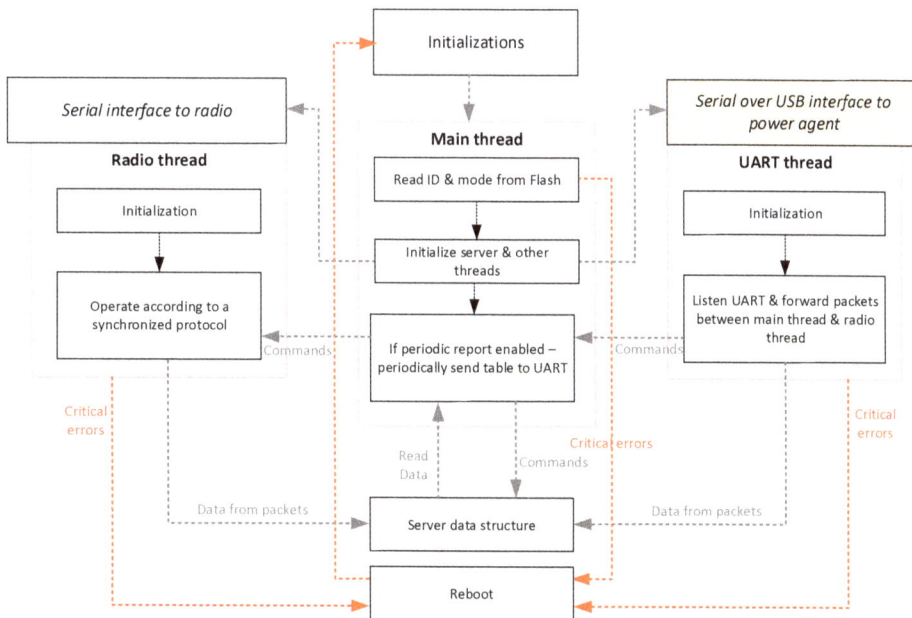

Figure 7. Structure of the embedded software.

5.2.2. Synchronized Protocol

The radio thread handles control over the radio transceiver and implements a synchronized radio protocol. The synchronized protocol is a simple slotted protocol, where each of the D2D modules is assigned a periodic time slot for transmission of its data and receiving the transmissions from the other modules in their respective slots, as shown in Figure 8.

The parameters of the protocol, namely the number of slots (M) in the superframe and the duration of each slot (T-slot) are hardcoded in the firmware and cannot be changed without reprogramming the module. Each module uses for its transmission the slot with the number equaling to its programmed identifier (i.e., a module with ID 1 will send in slot 1, etc.). Empirically it was found out that the need of using low-speed UART interface between the main module and the radio transceiver chipset and the slow operation of the chipset itself introduces substantial overheads (e.g., packet transmission, switching between transmit and receive, etc.). Due to this reason, the duration of one slot cannot be set below 150 ms.

When enabled the first time, the D2D module based on this protocol first scans the radio channel for several superframe periods. If it does no-t find any transmissions and it has data to send, it will start the transmission right away. If during scanning a module finds some transmissions ongoing, it will use this transmission as a reference for defining its designated slot. After each superframe, a module adjusts its synchronization. As a reference point for adjusting the synchronization, each module uses the timestamp of the packet with minimum identifier not exceeding the identifier of the module. If such a reference is not available, no compensation is applied. As a practical example, module 1 transmitting in slot 1 never adjusts its synchronization. If modules 2, 3 and 4 hear transmission of module 1, they will adjust their synchronization based on it. If module 5 does not hear module 1 but hears modules 2 and 3, it will adjust its synchronization based on the transmission of module 2.

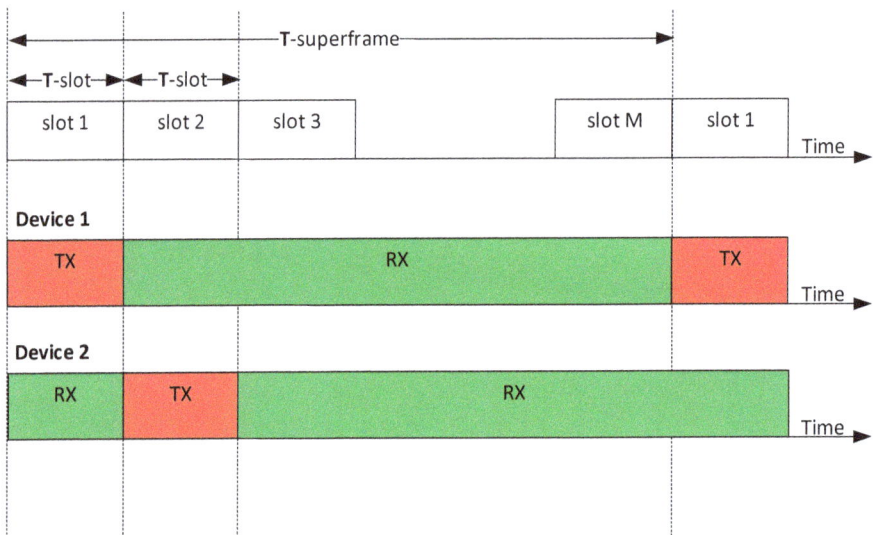

Figure 8. Illustration of the implemented synchronized protocol operation (TX: transmission, RX: reception).

6. Results of the Experiment

To test the performance of the P2P distributed voltage control system, one needs to create a voltage rise (or drop) problem and solve it in a P2P fashion. To create a voltage rise problem in a laboratory-based microgrid, a high-power injection from the inverters back to the grid can be used. Alternatively, the impedance of the feeder depicted in Figure 2 can be oversized to create such a problem with low-power injection. In the following experiments, R1 and R2 are set to 8 Ω, L1 and L2 are set to 5 mH. Figure 9 shows the generation profile applied at both inverters. The active power generation starts at zero, and increases to a maximum of 1200 W. At the higher generation, the voltage is expected to rise above the maximum voltage limit. To comply with the European standard EN 50160, the voltage limits V^{max} and V^{min} are enforced to be $\pm10\%$ of the nominal phase voltage.

Two experiments are carried out to compare the voltage profiles with and without voltage control. The comparison helps in quantifying the performance of the P2P voltage control.

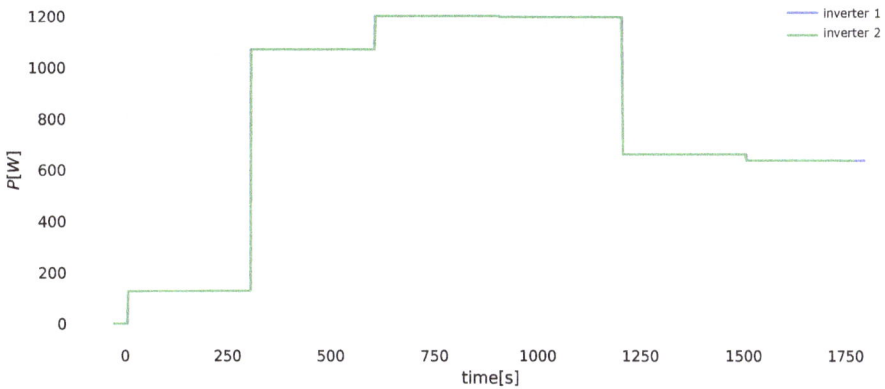

Figure 9. 30 min generation profiles of inverter 1 and inverter 2 (both inverters apply the same generation profile).

6.1. First Experiment: Without P2P Voltage Control

The generation profile described by Figure 9 is applied at both inverters of the setup. Figure 10 shows that this leads to voltages exceeding the upper limit of 1.1 p.u. at both the first and second node. The agents remained idle during this experiment.

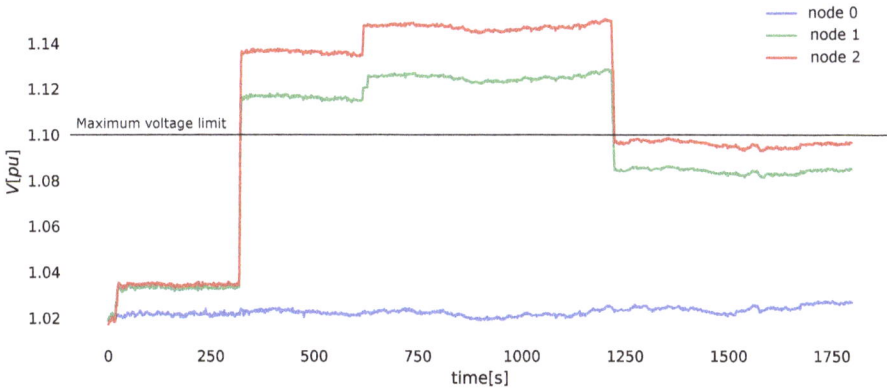

Figure 10. Voltage profiles without voltage control for the 3 nodes in the microgrid (the generation profile causes over-voltages up to 1.145 p.u.)

6.2. Second Experiment: With P2P Voltage Control

The inverters apply the same generation profile, but now the agents execute the distributed voltage control algorithm. This leads to the voltage profile shown by Figure 11. When an increase in generation causes an over-voltage issue, the agents bring the voltages back to the defined limits (±10%) within 3 min.

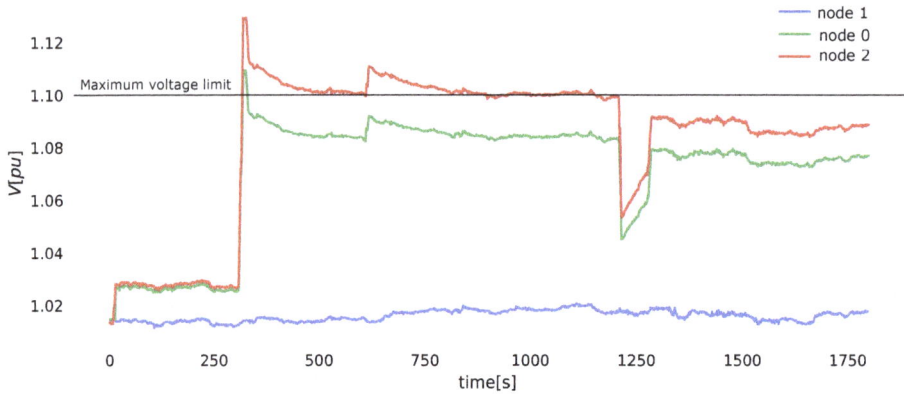

Figure 11. Voltage profiles with voltage control for the 3 nodes in the microgrid.

The actions of the observer and actuator agents are reflected in Figure 12. The evolution of the control signals over time are presented in Figure 12c,d. The control signals for under-voltages ($\lambda^{\min}$) are zero, because no under-voltages beyond the limits occur during this experiment. The control signals for over-voltages ($\lambda^{\max}$) however, increase sharply after an increase in the voltages above $V^{\max}$. One can notice that the control signals $\lambda^{\max}$ return back to zero when the voltages return back to normal values without compensation, due to a decrease in the generation profiles.

Figure 12a,b shows the actions taken by the actuator agents. As soon as an over-voltage occurs, nearly all reactive power is dispatched. This behaviour depends on the values of α, c^P and c^Q. The step size α controls mainly how fast the control signals will increase, and hence how fast compensation is dispatched. Since the cost of active power is set to be a lot higher than the cost of reactive power, the algorithm will dispatch first the available reactive power. α is set high enough to get a fast response in the active power dispatch. However, this causes the dispatch of reactive power to be nearly instantaneous. ΔP and ΔQ return back to zero when the voltages return back to normal values without compensation.

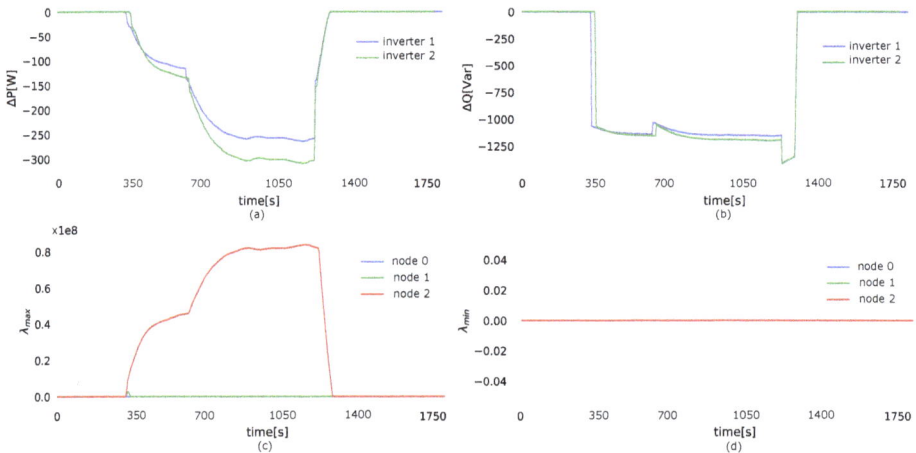

Figure 12. (**a**) Active power curtailment of inverter 1 and inverter 2; (**b**) Reactive power compensation of inverter 1 and inverter 2; (**c**) The control signals for over-voltages; (**d**) The control signals for under-voltages.

When the measured voltage V_i^{meas} is less than V^{max}, λ_i^{max} starts to decrease till it reaches zero. One can explain this based on Equation (2). The Lagrangian multipliers drop back to zero because the underlying profile of the inverters change. The active power injection drops, and the voltage drops with it. The Lagrangian multipliers adapt to the new situation. When both λ_i^{max} and λ_i^{min} are zero at each observer, problem (1) can be written as:

$$\underset{\Delta P, \Delta Q}{\text{argmin}} \quad c^{\text{P}}(\Delta P_d^{(t)})^2 + c^{\text{Q}}(\Delta Q_d^{(t)})^2 + \text{zero}$$

$$\text{Subject to the local constraints of problem (1)} \tag{3}$$

One can notice that the solution of the above optimization problem is: $\Delta P_d^{(t)} = 0$ and $\Delta Q_d^{(t)} = 0$. Hence, a stop mechanism can be designed to stop the solver of the optimization problem whenever the Lagrangian multipliers are zero at each observer. This should decrease the computational burden of the algorithm.

6.3. Key Performance Indicators

There are three key performance indicators (KPIs) considered in this work: (1) Convergence time; (2) Voltage quality; and (3) Communication delays.

The first KPI, convergence time, is a measure for how long it takes the algorithm to solve the voltage problem. Voltage quality reflects how well the control algorithm can mitigate the voltage rise (or drop) problems. Finally, the communication delays depend on the communication infrastructure. Below follows an explanation of how each of these KPIs is quantified in practice.

6.3.1. Convergence Time

The voltage control algorithm is online and adjusts itself continuously. When a change in the generation profile occurs, there are two possibilities: either there is a voltage problem or not. If there is no voltage problem, the control algorithm stays idle. However, if there is a voltage problem, then the agents start to undertake action. The observer agents change the control signals until the voltage problems are resolved. If they succeed, then the control signals converge to a stable value, and the voltages converge to a value within the limits. In this paper, we define the convergence time as the time it takes from a moment when the voltage exceeds the limits until the moment when the voltage is restored within the limits. As demonstrated in Figure 11, it takes the algorithm around 3 min to regulate the voltages within the defined limits, which is an acceptable time for voltage problems.

It is worth mentioning that the intervention time of the interface protection relay of Triphase inverter is much less than the convergence time. The intervention time of the interface protection relay of Triphase inverter is less than 1 ms. The Triphase inverter is configured to trip at 280 V. This means that there is 27 V as voltage margin, since the algorithm starts regulating the voltage when the PCC voltage is higher than 253 V. Hence, the inverter in our setup is able to correct the voltages before reaching 280 V. If an inverter trips at 253 V (maximum voltage defined by the standard EN 50160), then V^{max} of the proposed algorithm should be set to a value lower than 253 V (i.e., 240 V), in a way to make sure that the convergence time is sufficient to correct the PCC voltages before reaching the maximum voltage at which the inverter trips.

6.3.2. Voltage Quality

The voltage quality is quantified by the metric $E \geq 0$ defined by Equation (4). The metric E integrates the over and under voltages as shown in Figure 13. This means that both the duration of a voltage problem and its severity will increase the metric E. A value of zero is the best possible value and indicates that there are no over or under voltage issues: the higher the E, the worse the voltage problem.

$$E = \sum_{i=1}^{N} \int_{t_{start}}^{t_{end}} \left(\max\left(V_i - V^{max}, 0\right) + \max\left(V^{min} - V_i, 0\right) \right) dt \tag{4}$$

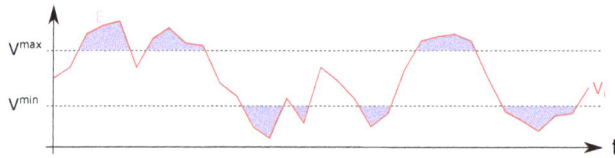

Figure 13. Voltage quality metric: sum of the surfaces above and below the voltage limits.

Table 1 shows a comparison between the regulated and the unregulated voltage profiles based on the voltage quality metric. E is the sum of the metrics E_0, E_1 and E_2 of the nodes 0, 1, and 2, respectively. The P2P voltage control reduced the metric E from 58.724 to 2.633. E of the regulated voltage profiles is slightly higher than zero, because it takes the algorithm some time until it has resolved the voltage issues.

Table 1. Voltage quality metrics of the regulated and unregulated voltage profiles.

V_i	E_i			E
	Node 0	Node 1	Node 2	
without control	0	19.719	39.005	58.724
with control	0	0.081	2.552	2.633

6.3.3. Communication Delays

For the observer agents, the delay is defined as the time between consecutive updates of their control signals, which they broadcast periodically to the actuator agents. For the actuator agents, the delay is defined as the time between consecutive updates of the set-points which are sent to the Triphase power hardware.

Figure 14 shows the delays between the iterations of the control algorithm, for each agent individually. The observer agents are implemented by dedicated single-board computers with few other processes running in the background. They manage to update the control signals every 1.5 s, with little deviation. The actuator agents however experience longer control delays, with large differences between both actuators. There are two main causes for these additional delays. Firstly, the actuator agents solve an optimization problem at each iteration. Secondly, the actuator agents are implemented by laptops. These laptops run additionally control software for the Triphase Rapid Prototyping Inverter System, which requires rather heavy processing. The laptop running actuator 2 is older, which shows in the performance. Adapting the implementation of the algorithm for the actuator agents can lower the delays. The lower limit for the delays is 1.5 s, which is the period with which the observer agents send updates of the control signal.

Overall, the delays are as expected. Only the delays for actuator 2 could be shorter to be in line with the other devices. Upgrading actuator 2 to hardware similar to actuator 1, should resolve these additional delays.

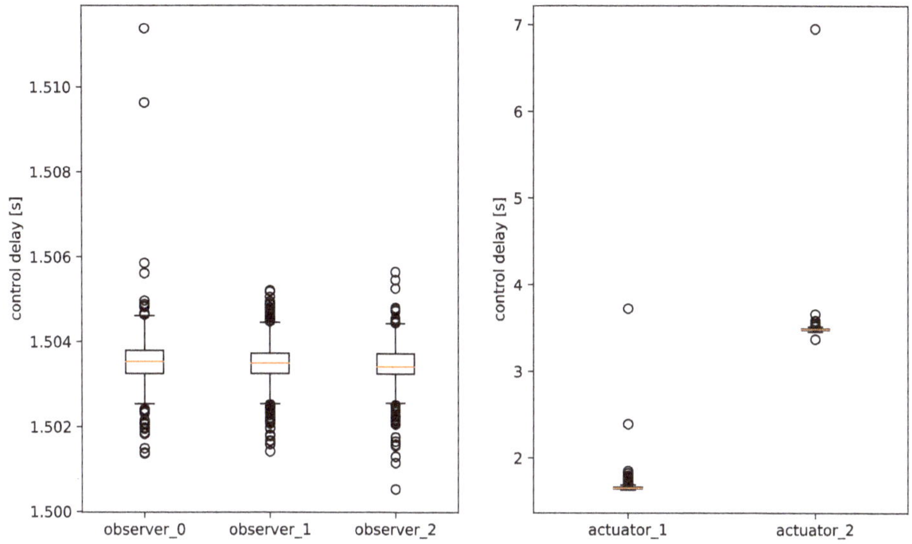

Figure 14. Communication delays: The observer agents update the control signals every 1.5 s, with very little deviation. The actuator agents issue their updates more slowly, with a significant difference between both actuator agents.

6.4. Discussion

The proposed P2P voltage control system managed to increase the voltage quality of the voltage profiles. Some over-voltage issues remain, because the control algorithm needs around 3 min to bring back the voltages within the limits. However, it is in line with the European standard EN50160 as all 10 min mean rms values of the voltages are within the range $[V^n - 10\%, V^n + 10\%]$, where $V^n = 1$ p.u.

The key question motivating this research, was whether fully distributed voltage control systems are a technically effective alternative to centralized ones. The results discussed in this paper show that fully distributed P2P voltage control systems can indeed provide satisfactory regulation of the voltage in distribution networks.

Technically, the P2P approach has shown good characteristics to be considered by DSOs to deliver high quality power to customers. The proposed P2P system could help in delivering easier access to prosumers' flexible supply and demand by making their active participation in the grid possible. This can be used to alleviate grid stress and defer or avoid grid upgrades, and consequently will help the DSOs to host more RESs.

7. Conclusions and Future Work

The dual-decomposition method and LoRaWAN D2D communication modules are used in this paper to design a P2P optimization-based voltage control system. A multi-agent, multi-layer microgrid testbed has been constructed at the EnergyVille premises to validate the performance of the proposed P2P system. Experimental results show the ability of the proposed system to solve the voltage rise problem within 3 min.

Future work includes expanding the P2P setup, by connecting the microgrid to a virtual network. The virtual network can be modelled using the Real-Time Digital Simulator (RTDS) [37]. Power Hardware-in-the-Loop (PHIL) can be used to connect the real inverters to virtual ones, and Control Hardware-in-the-Loop (CHIL) can be used to connect the real agents to virtual agents. The setup will be used to test distributed control algorithms that have faster convergence.

Energies **2018**, *11*, 1304

As demonstrated in [16], the Alternating Direction Method of Multipliers (ADMM) has a much faster convergence than the dual decomposition.

Author Contributions: H.A. conceived and designed the dual-decomposition-based peer-to-peer voltage control system. H.A. and S.A. designed the laboratory testbed, performed the experiments, and analyzed the data. K.M., J.H. and A.P. designed and implemented the device-to-device communication modules. H.A. wrote the paper. G.D. oversaw the work and proofread the paper.

Acknowledgments: This work is partially supported by: (1) P2P-SmarTest project (European Commission-Grant number 646469); and (2) an SB PhD fellowship from FWO-Vlaanderen.

Conflicts of Interest: The authors declare no conflict of interest.

References

1. European Environment Agency (EEA). *Trends and Projections in Europe 2017: Tracking Progress towards Europe's Climate and Energy Targets for 2020*; Publications Office of the European Union: Luxembourg, 2017.
2. Nijhuis, M.; Gibescu, M.; Cobben, J. Assessment of the impacts of the renewable energy and ICT driven energy transition on distribution networks. *Renew. Sustain. Energy Rev.* **2015**, *52*, 1003–1014. [CrossRef]
3. Carlo, M. Revision of European Standard EN 50160 on power quality: Reasons and solutions. In Proceedings of the 2010 14th International Conference on Harmonics and Quality of Power (ICHQP), Bergamo, Italy, 26–29 September 2010. [CrossRef]
4. Markets, R. *Market Design and Regulation During the Transition to Low-Carbon Power Systems*; International Energy Agency: Paris, France, 2016.
5. Website of P2P-SmarTest Project. Available online: www.p2psmartest-h2020.eu (accessed on 1 April 2018).
6. Pouttu, A.; Haapola, J.; Ahokangas, P.; Xu, Y.; Kopsakangas-Savolainen, M.; Porras, E.; Matamoros, J.; Kalalas, C.; Alonso-Zarate, J.; Gallego, F.D.; et al. P2P model for distributed energy trading, grid control and ICT for local smart grids. In Proceedings of the 2017 European Conference on Networks and Communications (EuCNC), Oulu, Finland, 12–15 June 2017; IEEE: Piscataway, NJ, USA, 2017; pp. 1–6.
7. Almasalma, H.; Engels, J.; Deconinck, G. Peer-to-peer control of microgrids. In Proceedings of the 8th IEEE Benelux PELS/PES/IAS Young Researchers Symposium, Eindhoven, The Netherlands, 12–13 May 2016; pp. 1–6.
8. Molzahn, D.K.; Dörfler, F.; Sandberg, H.; Low, S.H.; Chakrabarti, S.; Baldick, R.; Lavaei, J. A survey of distributed optimization and control algorithms for electric power systems. *IEEE Trans. Smart Grid* **2017**, *8*, 2941–2962. [CrossRef]
9. McArthur, S.D.; Davidson, E.M.; Catterson, V.M.; Dimeas, A.L.; Hatziargyriou, N.D.; Ponci, F.; Funabashi, T. Multi-agent systems for power engineering applications—Part I: Concepts, approaches, and technical challenges. *IEEE Trans. Power Syst.* **2007**, *22*, 1743–1752. [CrossRef]
10. Antoniadou-Plytaria, K.E.; Kouveliotis-Lysikatos, I.N.; Georgilakis, P.S.; Hatziargyriou, N.D. Distributed and decentralized voltage control of smart distribution networks: models, methods, and future research. *IEEE Trans. Smart Grid* **2017**, *8*, 2999–3008. [CrossRef]
11. Wang, Y.; Wang, S.; Wu, L. Distributed optimization approaches for emerging power systems operation: A review. *Electr. Power Syst. Res.* **2017**, *144*, 127–135. [CrossRef]
12. Yazdanian, M.; Mehrizi-Sani, A. Distributed control techniques in microgrids. *IEEE Trans. Smart Grid* **2014**, *5*, 2901–2909. [CrossRef]
13. Engels, J.; Almasalma, H.; Deconinck, G. A distributed gossip-based voltage control algorithm for peer-to-peer microgrids. In Proceedings of the 2016 IEEE International Conference on Smart Grid Communications (SmartGridComm), Sydney, NSW, Australia, 6–9 November 2016; IEEE: Piscataway, NJ, USA, 2016; pp. 370–375.
14. Almasalma, H.; Engels, J.; Deconinck, G. Dual-decomposition-based peer-to-peer voltage control for distribution networks. *CIRED-Open Access Proc. J.* **2017**, *2017*, 1718–1721. [CrossRef]
15. Liu, H.J.; Shi, W.; Zhu, H. Distributed Voltage Control in Distribution Networks: Online and Robust Implementations. *IEEE Trans. Smart Grid* **2017**. [CrossRef]
16. Šulc, P.; Backhaus, S.; Chertkov, M. Optimal distributed control of reactive power via the alternating direction method of multipliers. *IEEE Trans. Energy Convers.* **2014**, *29*, 968–977. [CrossRef]

17. Bahramipanah, M.; Cherkaoui, R.; Paolone, M. Decentralized voltage control of clustered active distribution network by means of energy storage systems. *Electr. Power Syst. Res.* **2016**, *136*, 370–382. [CrossRef]

18. Cagnano, A.; De Tuglie, E. A decentralized voltage controller involving PV generators based on Lyapunov theory. *Renew. Energy* **2016**, *86*, 664–674. [CrossRef]

19. Marzband, M.; Fouladfar, M.H.; Akorede, M.F.; Lightbody, G.; Pouresmaeil, E. Framework for smart transactive energy in home-microgrids considering coalition formation and demand side management. *Sustain. Cities Soc.* **2018**, *40*, 136–154. [CrossRef]

20. Marzband, M.; Azarinejadian, F.; Savaghebi, M.; Pouresmaeil, E.; Guerrero, J.M.; Lightbody, G. Smart transactive energy framework in grid-connected multiple home microgrids under independent and coalition operations. *Renew. Energy* **2018**, *126*, 95–106. [CrossRef]

21. Tavakoli, M.; Shokridehaki, F.; Akorede, M.F.; Marzband, M.; Vechiu, I.; Pouresmaeil, E. CVaR-based energy management scheme for optimal resilience and operational cost in commercial building microgrids. *Int. J. Electr. Power Energy Syst.* **2018**, *100*, 1–9. [CrossRef]

22. Kouveliotis-Lysikatos, I.; Koukoula, D.; Vlachos, I.; Dimeas, A.; Hatziargyriou, N.; Makrynikas, S. Decentralised distribution system operation techniques: Results from the Meltemi community smart grids pilot site. *CIRED-Open Access Proc. J.* **2017**, *2017*, 1673–1677. [CrossRef]

23. Iacovella, S.; Vingerhoets, P.; Deconinck, G.; Honeth, N.; Nordstrom, L. Multi-Agent platform for Grid and communication impact analysis of rapidly deployed demand response algorithms. In Proceedings of the 2016 IEEE International Energy Conference (ENERGYCON), Leuven, Belgium, 4–8 April 2016; IEEE: Piscataway, NJ, USA, 2016; pp. 1–6.

24. Dupont, B.; Vingerhoets, P.; Tant, P.; Vanthournout, K.; Cardinaels, W.; De Rybel, T.; Peeters, E.; Belmans, R. LINEAR breakthrough project: Large-scale implementation of smart grid technologies in distribution grids. In Proceedings of the 2012 3rd IEEE PES International Conference and Exhibition on Innovative Smart Grid Technologies (ISGT Europe), Berlin, Germany, 14–17 October 2012; IEEE: Piscataway, NJ, USA, 2012; pp. 1–8.

25. Vanthournout, K.; De Brabandere, K.; Haesen, E.; Van den Keybus, J.; Deconinck, G.; Belmans, R. Agora: Distributed tertiary control of distributed resources. In Proceedings of the 15th Power Systems Computation Conference (PSCC'05), Liège, Belgium, 22–26 August 2005; pp. 1–7.

26. Macken, K.J.; Vanthournout, K.; Van den Keybus, J.; Deconinck, G.; Belmans, R.J. Distributed control of renewable generation units with integrated active filter. *IEEE Trans. Power Electron.* **2004**, *19*, 1353–1360. [CrossRef]

27. Palomar, D.P.; Chiang, M. A tutorial on decomposition methods for network utility maximization. *IEEE J. Sel. Areas Commun.* **2006**, *24*, 1439–1451. [CrossRef]

28. Efkarpidis, N.; De Rybel, T.; Driesen, J. Optimization control scheme utilizing small-scale distributed generators and OLTC distribution transformers. *Sustain. Energy Grids Netw.* **2016**, *8*, 74–84. [CrossRef]

29. De Brabandere, K.; Loix, T.; Engelen, K.; Bolsens, B.; Van den Keybus, J.; Driesen, J.; Belmans, R. Design and operation of a phase-locked loop with Kalman estimator-based filter for single-phase applications. In Proceedings of the IECON 2006—32nd Annual Conference on IEEE Industrial Electronics, Paris, France, 6–10 November 2006; IEEE: Piscataway, NJ, USA, 2006; pp. 525–530.

30. Yazdani, A.; Iravani, R. *Voltage-Sourced Converters in Power Systems: Modeling, Control, and Applications*; John Wiley & Sons: Hoboken, NJ, USA, 2010.

31. Gandotra, P.; Jha, R.K.; Jain, S. A survey on device-to-device (D2D) communication: Architecture and security issues. *J. Netw. Comput. Appl.* **2017**, *78*, 9–29. [CrossRef]

32. Huang, P.K.; Stacey, R.J.; Li, Q.; Yang, R. *Wireless Device, Method, and Computer Readable Media for Spatial Reuse for Device-to-Device Links*; Technical Report US9794790B2; United States Patent and Trademark Office: Alexandria, VA, USA, 2017.

33. Mikhaylov, K.; Petäjäjärvi, J.; Haapola, J.; Pouttu, A. D2D Communications in LoRaWAN Low Power Wide Area Network: From Idea to Empirical Validation. In Proceedings of the 2017 IEEE International Conference on Communications Workshops (ICC Workshops), Paris, France, 21–25 May 2017.

34. Markkula, J.; Haapola, J. Ad Hoc LTE Method for Resilient Smart Grid Communications. *Wirel. Personal Commun.* **2018**, *98*, 3355–3375. [CrossRef]

35. Raza, U.; Kulkarni, P.; Sooriyabandara, M. Low Power Wide Area Networks: An Overview. *IEEE Commun. Surv. Tutor.* **2017**, *19*, 855–873. [CrossRef]

36. Mikhaylov, K.; Petäjäjärvi, J. Design and implementation of the plug&play enabled flexible modular wireless sensor and actuator network platform. *Asian J. Control* **2017**, *19*, 1392–1412.

37. Kotsampopoulos, P.C.; Kleftakis, V.A.; Hatziargyriou, N.D. Laboratory education of modern power systems using PHIL simulation. *IEEE Trans. Power Syst.* **2017**, *32*, 3992–4001. [CrossRef]

energies

MDPI

Article

Experimental Assessment of a Centralised Controller for High-RES Active Distribution Networks

Francisco de Paula García-López, Manuel Barragán-Villarejo *, Alejandro Marano-Marcolini, José María Maza-Ortega and José Luis Martínez-Ramos

Department of Electrical Engineering, Universidad de Sevilla, Camino de los Descubrimientos s/n, 41092 Seville, Spain; fdpgarcia@us.es (F.d.P.G.-L.); alejandromm@us.es (A.M.-M.); jmmaza@us.es (J.M.M.-O.); camel@us.es (J.L.M.-R.)

* Correspondence: manuelbarragan@us.es; Tel.: +34-954-481-281

Received: 31 October 2018; Accepted: 28 November 2018; Published: 1 December 2018

Abstract: This paper assesses the behaviour of active distribution networks with high penetration of renewable energy sources when the control is performed in a centralised manner. The control assets are the on-load tap changers of transformers at the primary substation, the reactive power injections of the renewable energy sources, and the active and reactive power exchanged between adjacent feeders when they are interconnected through a DC link. A scaled-down distribution network is used as the testbed to emulate the behaviour of an active distribution system with massive penetration of renewable energy resources. The laboratory testbed involves hardware devices, real-time control, and communication infrastructure. Several key performance indices are adopted to assess the effects of the different control actions on the system's operation. The experimental results demonstrate that the combination of control actions enables the optimal integration of a massive penetration of renewable energy.

Keywords: active distribution network; laboratory testbed; renewable energy sources; DC link; centralised control

---

1. Introduction

Massive penetration of renewable energy sources (RES) is unstoppable nowadays because of the need to reduce the dependency of fossil fuels. This new technology of generation assets is being deployed in small units within medium voltage (MV) and low voltage (LV) distribution systems, the so-called distributed generation, in contrast to the conventional connections of large-scale power plants to high voltage (HV) systems. The drivers behind this change in the generation paradigm are threefold: technical because of the maturity of the technology [1], economical due to the related cost reduction [2], and social because of the citizen involvement in decarbonising the electrical consumption [3].

The traditional operation of radial distribution systems cannot be maintained in cases where there is very high RES penetration, because the design of these systems has been done to cope with power flows from primary and secondary substations to the final users [4]. The problems that RES may create have been profusely described in specialised literature [5], for example, higher simultaneity coefficients, reverse power flows, out of control nodal voltages, power quality deterioration, increase of short-circuit power, etc. These technical problems can be released using conventional network reinforcement strategies ranging from increasing the cross-section of existing lines to installing new lines and/or power transformers. However, it has to be questioned as to whether this is the best solution considering the increases in cost and connection time [6] as well as the spare capacity of the new assets over a large number of hours per year [7]. Therefore, new alternatives must be explored to overcome the shortcomings related to this Fit & Forget approach.

Several active network operation approaches have been proposed recently. In general, these can be classified according to the following characteristics: the control assets used to optimise the network operation, the applied control algorithms, and the testing procedure used to validate their performance.

Control assets: Regarding the first issue, HV/MV transformers equipped with on-load tap changers (OLTCs) and step voltage regulators were proposed in [8]. In addition, RES may also contribute to voltage regulation and congestion management by resorting to curtailment [9] or even by using adequate reactive power injections [10–12]. Most of the major RES inverter manufacturers include the possibility of controlling the reactive power in order to fulfil the grid codes imposed by the system operators (either distribution system operators (DSOs) or transmission system operators (TSOs)). These grid codes are becoming more and more restrictive and they include minimal technical requirements for RES connections to occur [13,14], including, among others, voltage regulation issues by means of reactive power injection.It is important to mention that most of the active operation approaches consider several control assets that are managed in a coordinated manner, for example, HV/MV, OLTC, and RES [15–17]; HV/MV, OLTC, and energy storage systems [18,19]; and HV/MV, OLTC, RES reactive power injection, and direct current (DC) links [20].

Control methodology: The active management solutions can be broadly classified into centralised, distributed, and local methodologies. The centralised approaches rely on a control centre in charge of computing the optimal setpoints for all the control assets that considers the available network measurements [11]. The main drawback of this approach is the need for an extensive communication system. Therefore, this solution is suitable for MV distribution systems because of two reasons. On the one hand, the cost of the communication infrastructure is marginal with respect to the cost of the large RES units (in the range of several MVA). On the other hand, nowadays, utilities are equipped with centralised Advanced Demand Management Systems (ADMS) which incorporate monitoring and automation functionalities. However, it should be considered that a failure of part of the communication infrastructure may deteriorate the performance of the controller. For this reason, advanced control strategies providing enhanced system resilience can be found in the specialised literature [21]. Local approaches are just the opposite because the actions taken by the control assets are calculated based on local measurements [10,12,16,22], and therefore, they are suitable for LV distribution systems. Distributed methods can be considered a compromise between the previous alternatives as they have several advantages related to robustness and scalability [23,24].

Testing methodology: The methodologies are usually validated by applying steady-state simulations that consider the daily load and generation profiles. However, other proposals use real-time digital simulators [23] and power hardware-in-the-loop platforms [19].

The results obtained by some of the previous control approaches can be summarised as follows. In [17], a 32% reduction of power losses was reached by using an adequate RES reactive power injection. After including the OLTC as an additional control asset, [16] reported an extra 7% reduction in total daily energy losses. Finally, a similar approach that considered the actual capability curves of the RES units achieved a 14% reduction in power losses [12].

This paper tests the use of centralised control of active assets to manage MV distribution networks with a massive RES penetration. An Optimal Power Flow (OPF) is used in the centralised control to compute the optimal setpoints for three kinds of control assets: (1) HV/MV-transformer OLTCs, (2) RES reactive power injections, and (3) active and reactive power through DC link meshing radial feeders. A high-RES but realistic load/generation scenario is analysed that considers some test cases involving different sets of control assets with the aim of evaluating their performance. These test cases are implemented in a laboratory scaled-down active distribution network including hardware devices, controllers, communication infrastructure, and a real-time monitoring system, as presented in [25]. This testbed can be used to evaluate practical implementation issues of any centralised control algorithm related to the applied control strategy, the required data field, the communication systems, etc., as a step prior to field deployment. Therefore, the main contribution of this paper is the experimental validation of the centralised controller proposed in [20] within an updated version of the testbed described in [25] in which an OLTC transformer, a DC link, and a new control scheme and communication system are incorporated.

The paper is organised as follows. In Section 2, a description of the centralised control to manage high-RES active distribution networks is presented. In Section 3, the benchmark distribution network is described in detail, including its main components and how they are represented in the laboratory scaled-down testing platform. Section 4 depicts and analyses the system's performance in different test cases, comparing them in a quantitative manner by means of key performance indices (KPIs). Finally, Section 5 closes with the main conclusions.

2. Proposed Centralised Control

Smart grids are characterized by extensive measurement, automation, and communication infrastructures which allows a safe and optimized network operation that takes advantage of centralised ADMSs. The main role of any ADMS in this environment is to concentrate the field data to extract the required information about the network status and, in cases where control assets are in operation, compute and send the required control actions to optimize the network operation according to a given criterion.

Figure 1 depicts this centralised control approach. First, the smart meters are in charge of measuring the load demanded by industrial (P_{il} and Q_{il}) and residential (P_{hl} and Q_{hl}) clients. In addition, the RES active power injections, such as the wind turbine (WT) and photovoltaic (PV) plants, P_{wt} and P_{pv}, respectively, are measured.

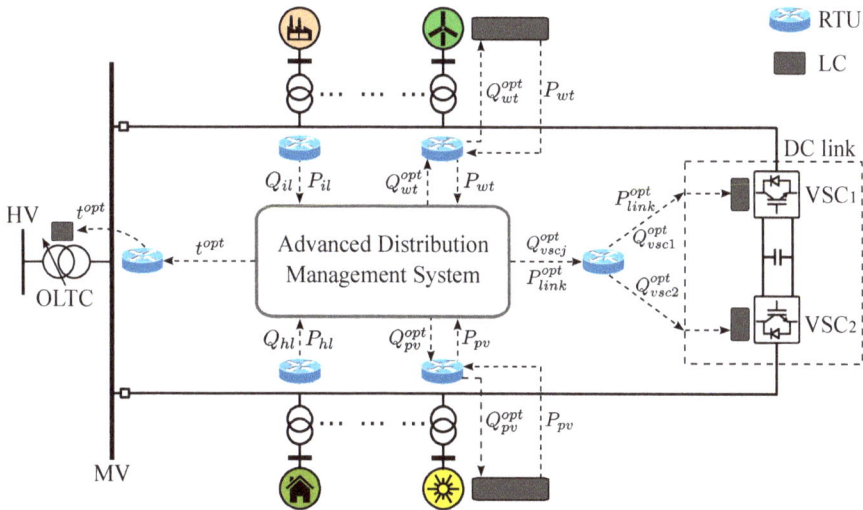

Figure 1. Architecture of the centralised control of an active distribution system.

All field data are sent to the ADMS by means of Remote Terminal Units (RTUs) at regular time intervals (typically 5 to 15 min). Considering all this information, it is possible to compute setpoints for the installed control assets using an OPF to optimize any technical or economic objective. This paper considers the following control assets:

- RES, which can regulate the reactive power injections Q_{wt}^{opt} and Q_{pv}^{opt}.
- Transformer OLTCs, which can adjust the tap position t^{opt}.
- A DC link, which is composed of two Voltage Source Converters (VSCs) in a back-to-back topology connecting two radial feeders. This device can regulate the active power flow between the feeders, P_{link}^{opt}, and two independent reactive power injections, Q_{vscj}^{opt}. It is important to point out that the

DC link is an interesting control asset with proven capability to reduce the network active power losses, maximize the penetration of RES, improve the network voltage profiles, and avoid branch saturations [20,26].

On the other hand, the selected OPF objective is to minimize the active power losses of the system to take advantage of the already available control assets to optimise the operation of the distribution grid, which leads to the following formulation:

$$\min_{\mathbf{x}} P_{\text{loss}}(\mathbf{x}, \mathbf{y}), \tag{1}$$

where $\mathbf{x}$ is the set of control variables (P_{link}^{opt}, Q_{vscj}^{opt}, $Q_{wt,pv}^{opt}$, t^{opt}) and $\mathbf{y}$ is the set of load and generation power injections for a given time interval (P_{il}, Q_{il}, P_{hl}, Q_{hl}, P_{wt}, P_{pv}).

The optimization problem is completed by including the relevant constraints. First, the network operational limits have to be considered. The voltages and currents of the sets of buses, $\mathcal{N}$, and branches, $\mathcal{B}$, have to be within the regulatory boundaries, $[V_i^{min}, V_i^{max}]$, and below the cable ampacities, I_b^{max}, respectively, as stated in (2) and (3):

$$V_i^{min} \leq V_i \leq V_i^{max} \quad \forall i \in \mathcal{N}, \tag{2}$$

$$0 \leq I_b \leq I_b^{max} \quad \forall b \in \mathcal{B}. \tag{3}$$

Second, the OLTC tap has to be within the limits and the apparent power levels of the RES and DC-link VSCs have to be below their rated capability according to (4)–(6):

$$t^{min} \leq t^{opt} \leq t^{max}, \tag{4}$$

$$S_{pv,wt} \leq S_{pv,wt}^{rat}, \tag{5}$$

$$S_{DClink} \leq S_{DClink}^{rat}. \tag{6}$$

Finally, other constraints which are included in the OPF are the active and reactive bus power balances and the power constraints that model the DC link behaviour, which can be found in [26].

3. Laboratory Testing Platform

The objective of building the laboratory testing platform was to faithfully represent the real behaviour of an active distribution system including all of its components to asses the performance of the centralised control strategy outlined in Section 2. In this way, the testing platform was built based on the MV benchmark distribution network proposed by the International Council on Large Electric Systems (CIGRE in french) Task Force C06.04.02 devoted to study the RES integration in MV networks [27]. The main reasons that motivated the selection of this system are detailed below:

- First, this network is based on an actual MV German distribution system, fulfilling the proposed objective of the laboratory testing platform described above.
- Second, an important RES penetration is integrated into the network.
- Third, all the network data, including topology, parameters of lines and cables, loads, RES, and their corresponding daily load/generation curves are available and are well documented.
- Fourth, the benchmark network includes a DC link, a key component of the future active distribution system with high RES penetration.

The next subsections present the MV benchmark distribution system and its scaled-down version built in the laboratory for testing purposes, including the implemented control scheme and the communication infrastructure designed to operate the system as a flexible platform to evaluate the benefits of active distribution networks.

3.1. MV Benchmark Distribution Network

A one-line diagram of the benchmark distribution system is shown in Figure 2 which is composed of two radial subsystems departing from a primary substation where a 40 MVA 110/20 kV transformer equipped with an OLTC is installed. The total network comprises 14 buses grouped in two radial feeders: 11 buses for subsystem 1 and 3 buses for subsystem 2. The total line length of subsystem 1 is about 15 km, while subsystem 2 is just 8 km. In addition, different types of load, involving industrial and domestic customers as well as a large amount of RES, are connected into the different buses. Although [27] considered different types of RES, this work exclusively included PV and WT plants because its current maturity foresees that they will be massive deployed in upcoming years. In addition, the benchmark network includes a DC link to connect both radial subsystems between nodes N8 and N14.

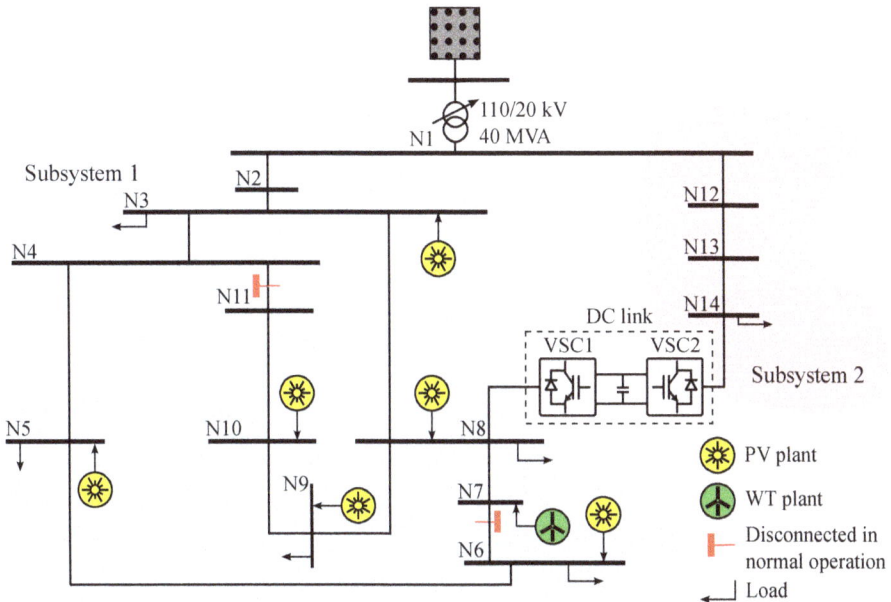

Figure 2. The medium voltage (MV) benchmark distribution network proposed by the CIGRE Task Force C06.04.02.

The 24-h profiles of the total loads and RES of subsystems 1 and 2 are depicted in Figure 3. It is interesting to point out that subsystem 1 was more loaded than subsystem 2. Moreover, most RES were located within subsystem 1 which partially compensate for its higher load with this local generation. It is also worth noting that, in order to analyse a case with a massive RES penetration, the generation was multiplied by 4 and 400 in the case of the WT and PV plants, respectively, with respect to the scenario described in [27]. In this way, the peak generation of the RES units and the peak demands of the loads during the day were established at 0.446 pu and 0.381 pu, respectively (the base power of the MV system was 100 MVA). The ratio between the peak generation and the peak demand was equal to 1.1724—a scenario of high-RES penetration.

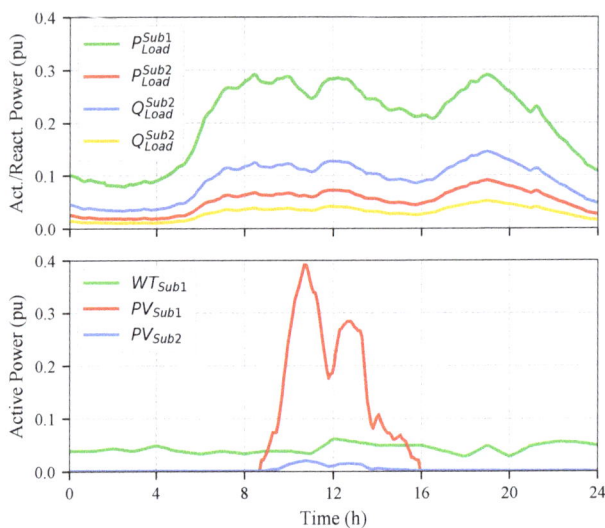

Figure 3. Top: Daily profile of the total loads in subsystems 1 and 2; **Bottom**: Daily profile of the total WT and PV generation in subsystems 1 and 2.

3.2. Laboratory Scaled-Down Distribution Network

This subsection provides a brief outline of the components and functionalities of the scaled-down testbed used to validate the benefits of the centralised controller. Basically, this hardware test rig, depicted in Figure 4, is a three-phase scaled-down 400 V (base/rated voltage) and 100 kVA (base/rated power) representation of the MV benchmark network described in Section 3.1 which is composed of the following components:

- Distribution network branches: The electrical lines of both scaled-down subsystems are represented by a lumped parameter model comprising the series resistor and reactor. The per unit values of these impedances are identical to those of the actual MV system. Therefore, the original line R/X ratios and equivalent lengths are maintained, leading to similar per unit voltage drops and power losses. Table 1 collects the exact values of the resistors and reactors used in the scaled-down network.

- Omnimode Load Emulators (OLEs): These are the building blocks that are responsible for representing any load, generator, or a combination of the two connected to any network node. Basically, each OLE is a VSC with a local controller (LC) whose AC and DC sides are connected to a scaled-down network node and a common DC bus, respectively, as shown in Figure 4. The VSC is a three-phase, three-wire, two-level insulated gate bipolar transistor (IGBT) VSC, rated at 400 V, 20 kVA with a switching frequency of 10 kHz. LCL coupling filters are used to connect the AC-side of the VSC to the scaled-down network. The inductors and the capacitor have the following ratings: L1 = L2 = 2.5 mH and C = 1 μF. Note that all of the OLEs share a common DC bus which is regulated by an extra balancing VSC rated to 100 kVA. This is directly connected to the LV laboratory network by its AC side, providing the net active power required by OLEs: $\sum P_i$. In this way, each OLE may absorb/inject (load/generator) any active power into the AC scaled-down distribution system within the technical constraints imposed by the VSCs. The OLEs are connected to the following nodes: N3, N5, N6, N7, N8, N9, and N10 (subsytem 1), and N14 (subsystem 2). The active and reactive power references to the OLEs are set by a Signal Management System (SMS) which is detailed in the next subsection.

344

A comprehensive description of this scaled-down system can be found in [25]. In addition, two new elements were incorporated with respect to the system described in [25] with the aim of integrating additional active control resources:

- Transformer with OLTC: The underlying idea of this feature is to represent the HV/MV transformers within the primary substations which are equipped with OLTCs to regulate the MV voltage. The transformer used for this purpose is a 400 V ± 5%/400 V, 100 kVA equipped with a thyristor-based tap changer, as shown in Figure 4.
- DC link: This DC link, originally included in the benchmark distribution system [28], is incorporated between N8 and N14 as a suitable device to maximise the RES penetration, as stated previously. Although several topologies can be used to create a flexible loop between radially operated feeders [29], the DC link is based on conventional back-to-back VSCs rated at 400 V and 10 kVA. Note that the DC bus of the DC link is totally independent of the one shared by the OLEs and the balancing VSC.

The optimal setpoints for these two control assets are also managed by the SMS in a similar manner to that of the OLE power references.

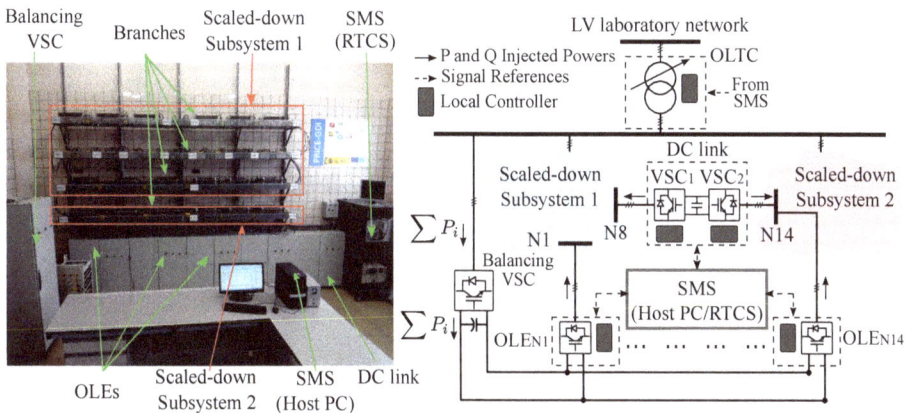

Figure 4. Left: Layout of the laboratory testbed. **Right**: One-line diagram of the updated testbed including the DC link and the transformer with the on-load tap changer (OLTC).

Table 1. Values of the resistors and reactors of each branch of the scaled-down network.

Initial Node	End Node	Resistance (mΩ)	Reactance (mΩ)
N1	N2	60.00	39.25
N2	N3	25.00	15.75
N3	N4	5.00	3.25
N4	N5	10.00	3.25
N5	N6	25.00	7.75
N6	N7	5.00	1.50
N7	N8	25.00	7.75
N8	N9	5.00	1.50
N9	N10	10.00	3.25
N10	N11	5.00	1.50
N11	N12	10.00	3.25
N3	N8	10.00	6.25
N12	N13	60.00	62.50
N13	N14	25.00	15.75

3.3. Control Scheme and Communication System

The control system is a two-level hierarchical structure, as shown in Figure 5. The first control level comprises the SMS, which is in charge of sending the references to the hardware components, whereas the second control level is composed of several LCs attached to the hardware devices (OLEs, DC link and OLTC) that are responsible for tracking these references.

The SMS performs two tasks in a sequential manner which can be summarised as follows:

- Offline tasks: They are carried out by a host PC and mainly consist of the configuration of the setpoint profiles. The OLE active and reactive daily power curves $(P_i^\star, Q_i^\star)$ are defined through two tools developed in the host PC [25]. Once these profiles have been determined, the daily setpoints of the DC link, P_{link}^{opt} and Q_{vscj}^{opt}, the reactive power injected by the RES, $Q_{wt,pv}^{opt}$, and the optimal OLTC tap position, t^{opt}, are automatically computed by the OPF described in Section 2. These setpoints and their computations are new features that are incorporated into the host PC with respect to [25]. Finally, all these data are compiled and uploaded to the Real-Time Control System (RTCS) for real-time operation.

- Online tasks: These are executed by the RTCS which is responsible for two undertakings. On the one hand, the RTCS is in charge of sending the setpoints to the second control level composed of the LCs attached to each hardware controllable component during the online operation according to the profiles previously determined in the offline tasks. On the other hand, the RTCS receives measurements from each each LC attached to the OLEs (V_i, P_i and Q_i), DC-link VSCs (V_{vscj}, P_{link} and Q_{vscj}) and the tap position of the transformer OLTC (t^{opt}). After processing this information, it provides real-time monitoring of the system which is displayed in the host PC.

The second level of the control system is composed of the LCs of each OLE, the DC-link VSCs, and the transformer OLTC which are implemented in Digital Signal Processors. These are in charge of tracking the setpoints sent by the RTCS during the online operation.

The communication infrastructure required to connect the centralised RTCS with the LCs is based on a 100 MBs Ethernet local-area network as a physical layer that implements a communication protocol based on UDP/IP. Finally, an asynchronous communication protocol, TCP/IP, is implemented between the host PC and the RTCS.

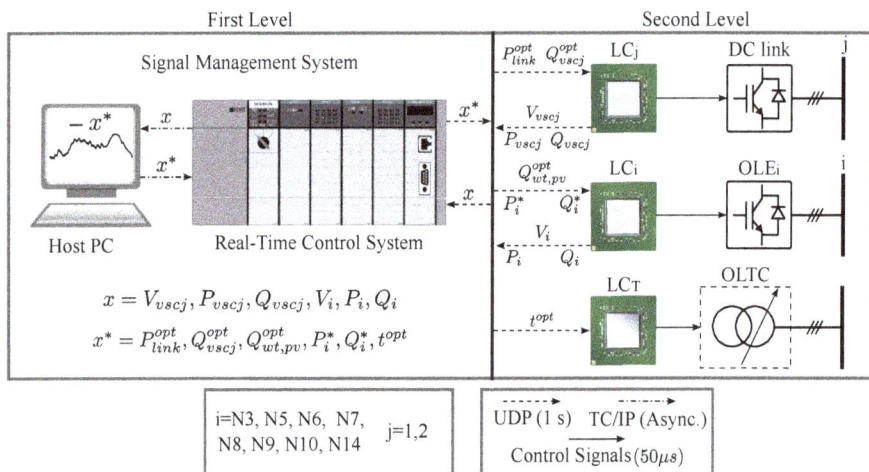

Figure 5. General control scheme of the testing environment.

4. Experimental Assessment of the Proposed Centralised Control

This section describes the analysis of the performance of the centralised control on the scaled-down system under different test cases. These are evaluated through KPIs to quantify the influence of the considered control assets in high-RES active distribution networks.

4.1. Definitions of Test Cases

Table 2 shows the definitions of the designed test cases. The first case, C1, is the base case where no control assets are included in the distribution system and the OLTC is set in the central tap position. The subsequent test cases add the control assets in the centralised control in an incremental manner. In this way, it should be possible to quantify the impact that each control asset has on the system's performance.

Table 2. Definitions of test cases.

Control Assets	C1	C2	C3	C4
OLTC		•	•	•
RES reactive power			•	•
DC link				•

4.2. Definitions of KPIs

The following KPIs were selected to analyse the performance of the centralised control and its related control assets:

- Daily energy loss ($E_{loss}/\Delta E_{loss}$): This KPI measures the daily active energy loss in kWh/day, E_{loss}, and the percentage of loss reduction with respect to the base case, C1, ΔE_{loss}.
- Voltage violation (T_{vv}): This KPI evaluates the percentage of time during the day that which the nodal voltages are outside the technical limits [0.95–1.05 pu].
- Variation of nodal voltages (ΔV): This index provides a global measurement of the daily voltage variations at the nodes of the network. It is computed as the average value of the difference between the maximum and minimum nodal voltages, measured in pu,

$$\Delta V = \frac{\sum_N (V_i^{max} - V_i^{min})}{Ni},$$ (7)

where Ni is the total number of network nodes.

- OLTC operation (N_{OLTC}): This KPI shows the number of OLTC operations that occur during the 24-h testing period.
- RES reactive power injection (Q_{RES}): This index provides a global measurement of the RES collaboration to the network reactive power support. It is computed by dividing the average value of the reactive power injected by the RES during the 24-h period by the total number of RES,

$$Q_{RES} = \frac{\sum_{i,t} Q_{RES_{i,t}}}{N_t \times N_{RES}},$$ (8)

where $Q_{RES_{i,t}}$ is the reactive power injected by RES_i in period t, N_{RES} is the number of RES in the network, and N_t is the number of time periods considered during the 24-h period.

- DC link load (SL_{link}): This evaluates the daily average load of the DC link during the day, and it is computed as

$$SL_{link} = \frac{\sum_{j,t} S_{vscj,t}}{N_t \times S_{DClink}},$$ (9)

where S_{vscj} is the apparent power of each VSC and S_{DClink} is the rated power of the DC link.

- Transformer load (T_L): This represents the daily average load of the transformer as a percentage of its rated power, which can be computed as

$$T_L = \frac{\sum_t S_t^T}{N_t \times S_N},$$ (10)

where S^T is the apparent power through the transformer and S_N is the rated power of the transformer.

4.3. Experimental Results

The objective function proposed for the operation of high-RES active distribution networks is based on an operation with minimal technical losses. This section describes the evaluation of the previously described test cases, which involved the analysis of the following electrical magnitudes: power losses, nodal voltages, and current circulating at the primary substation transformer. In addition, the previously defined KPIs allowed the key magnitudes to be quantified in a comprehensive manner to assess the performance of the proposed control.

Table 3 shows the E_{loss} for the studied test cases and the loss reduction with respect to the base case, C1, ΔE_{loss}, when the load and generation daily profiles presented in Section 3 were implemented into the testing platform. In the laboratory testbed, the 24-h profiles were scaled to the last 48 min and the duration of the tests was reduced.

Table 3. Key performance indices (KPIs) used for the evaluation of the test cases.

	C1	C2	C3	C4
$E_{loss}/\Delta E_{loss}$ (kWh/%)	58.37/−	55.69/4.58	50.17/16.33	46.47/25.59
Q_{RES} (pu)	−	−	0.117	0.095
T_{vv} (%)	38.69	0	0	0
N_{OLTC}	0	2	4	2
ΔV (pu)	0.087	0.061	0.058	0.042
T_L (%)	24.95	24.43	20.62	20.20

C1 presented the greatest daily power losses as no control assets were operating to act on the voltages and power flows to reduce the system losses. The introduction of the OLTC operation in C2 reduced energy losses by almost 5%. The OLTC setpoint was computed in the OPF whose objective function was to reduce the total power losses in the network. Therefore, the tap was established in the −5% position to increase the nodal voltages and to achieve the intended objective.

In test case C3, the RES reactive power capability was also included in the control. This caused the daily energy losses to be reduced by more than 15% with respect to C1. This occurred because the RES were able to provide reactive power to the system. Figure 6 shows the RES reactive power injected at nodes N3 and N8 with respect to their rated power levels for test cases C3 and C4. This is represented using violin plots which allow the distribution of any magnitude as well as its range of variation and frequency of occurrence to be visualized. Note that most of the time, which corresponds to the wider part of the violin plot, the RES were injecting reactive power corresponding to 20% of their rated power levels. This high RES reactive power injection was used to provide part of the reactive power demanded by the loads, thus avoiding the need to supply it from the primary substation, as shown in Figure 7. Note that the reactive power supplied from the primary substation in C3 was lower than 0.05 pu during the 24-h period, helping to reduce the energy losses.

The DC link integration in C4 further reduced the energy losses by up to 25% with respect to C1, as shown in Table 3. This device injected reactive power at the interconnected nodes N8 and N14 by means of VSC1 and VSC2 respectively during the 24-h period, as depicted in Figure 8. This power, added to the RES reactive power, led to almost zero reactive power being supplied from the primary substation, as shown in Figure 7. In this way, the energy losses reduced with respect to C3. An additional effect on the RES reactive power injections was observed. In C4, the RES did not to have to inject as much reactive power as in C3, as can be observed in Figure 6, even becoming zero in some nodes, like N8. This effect was quantified in a global manner with Q_{RES} collected in Table 3, where lower values for this KPI in C4 with respect to those in C3 can be appreciated. Table 4 summarises the rated power and the reactive power injections of the RES units in C3 and C4. The second and third columns indicate the rated power of the RES used in the scaled-down system and the MV system respectively. The two last columns depict the maximum reactive power injected by the VSCs interfacing the RES units during the day in cases C3 and C4. These values refer to the rated power of each device. The RES connected to N5 injected the maximum amount of reactive power, reaching 31.45% of its rated power. With the current technology, these reactive power values are easily reachable due to the combined effect of two actions: (i) the VSC coupling reactance is becoming smaller by using LCL filters, and (ii) the VSC DC voltage is continuously increasing. This extends the VSC reactive power range.

Table 4. Rated power and maximum reactive power injection of the renewable energy source (RES) units in C3 and C4.

RES Connected to Bus	S_{rated} (kVA) Scaled down System	S_{rated} (MVA) MV System	Q_{max}^{C3} (pu)	Q_{max}^{C4} (pu)
N3	12	12	0.3116	0.2584
N5	12	12	0.3145	0.3144
N6	12	12	0.2372	0.2372
N7	7	7	0.0309	0.0309
N8	12	12	0.2552	0.0100
N9	12	12	0.1147	0.1147
N10	16	16	0.1588	0.1588

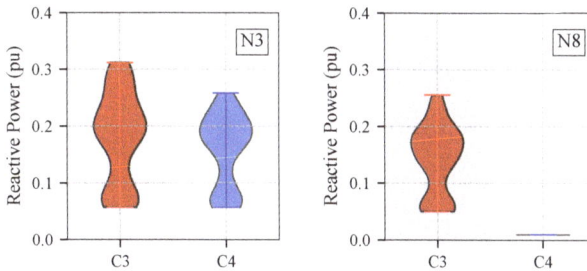

Figure 6. Violin plots of RES reactive power injections for test cases C3 and C4 at nodes N3 and N8.

Notice that the DC link also controlled the active power transferred from subsystem 1 to subsystem 2, as shown in Figure 8. Outside the period of high injection of RES active power (0–10 h and 13–0 h), the DC link absorbed active power from N14 and injected it into N8. This meant that part of the load from subsystem 1 was powered by subsystem 2 which is less loaded and has shorter branches, helping to reduce the total power losses of the system. Conversely, within the hours of high RES active power injection, the active power flow was inverted in the DC link: VSC1 absorbed active power from subsystem 1 and it was injected by the VSC2 into subsystem 2. In this way, part of the power generated by RES in subsystem 1 was transferred to feed the loads in subsystem 2. Therefore, this active power was not supplied by the primary substation, thus reducing the current in this system and the energy losses.

Finally, note that the DC-link loading, SL_{link}, during the day was 49.4%. This means that the DC link was used at half load and there is therefore still a wide margin to take advantage of its flexibility of operation. For example, the RES penetration in subsystem 1 could increase and still be managed by the current DC link.

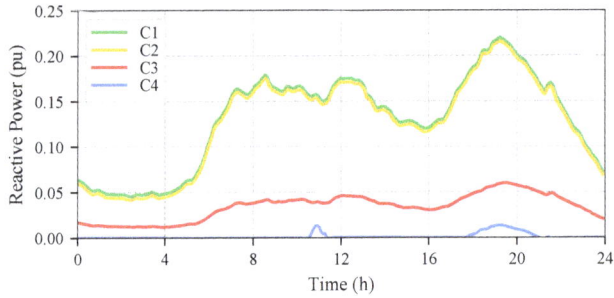

Figure 7. Reactive power flow through the primary substation for test cases C1–C4.

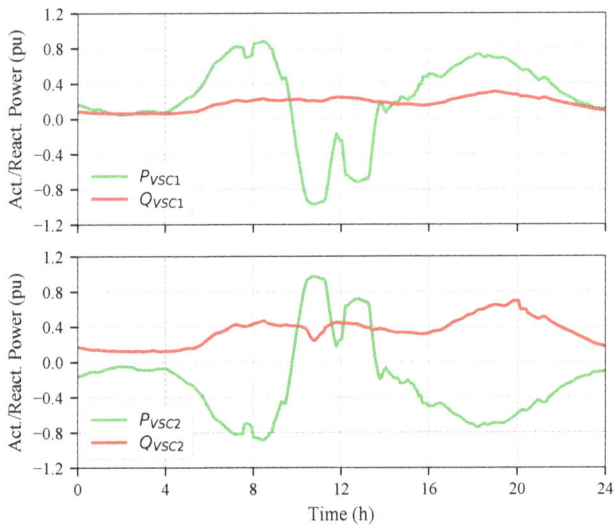

Figure 8. DC link active and reactive power daily profiles.

Figure 9 shows the 24-h nodal voltages at nodes N3, N6, N8, and N14 for the different test cases. These buses were selected to represent the behaviour of nodes nearby (N3) and far from (N6) the primary substation. In addition, nodes N8 and N14 were also included because they are the connection points of the DC link. The analysis of Figure 9 reveals that undervoltage situations—voltages below 0.95 pu—exclusively occurred in the base case, C1, due to the lack of control assets operating in the network. This situation led to a very high T_{vv} value in C1, as shown in Table 3. These voltage violations were more severe at nodes N6 and N8 corresponding to subsystem 1 because of two reasons. First, subsystem 1 was more loaded than subsystem 2, as depicted in Figure 3, especially during the hours without RES generation. This caused greater current flows and, consequently, greater voltage drops along the lines. This effect was especially significant around 8 and 20 h when the RES generation was almost zero and the demand was peaking.

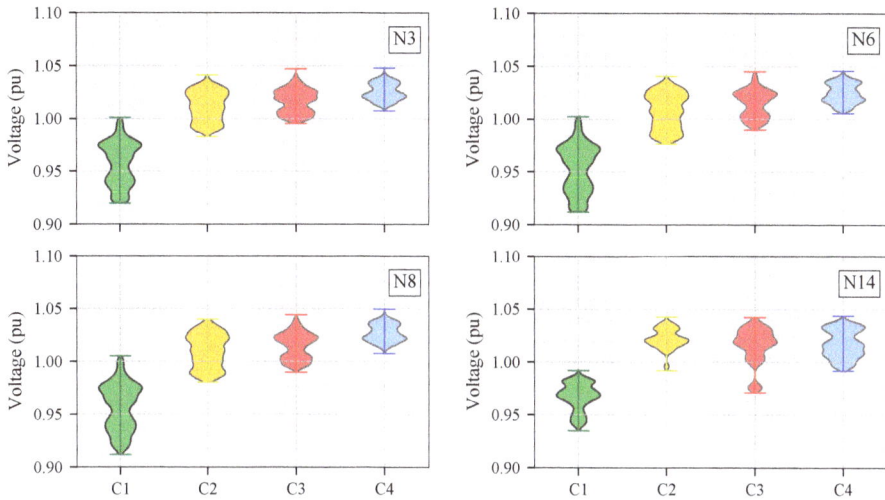

Figure 9. Violin plots of nodal voltages for test cases C1 to C4 at nodes N3, N6, N8, and N14.

The introduction of the OLTC in C2 pushed the voltages within the ±0.05 pu regulatory band around the rated voltage and, consequently, voltage violations were eliminated, as illustrated by its T_{vv}. In C2, the tap was established in the −5% position for most of the day. However, according to the information provided in Table 3, two OLTC operations N_{OLTC} (from −5% to 0% position) over the 24-h period were required to maintain the voltages within the limits. These changes occurred at around 11 h and 13 h when RES generation was maximum, as shown in Figure 3, and the network voltages were excessively high. The range of variation of nodal voltages ΔV was significantly reduced with respect to C1, as shown in Table 3. This effect can also be observed in Figure 9 where the violin plots are shortened, concentrating the nodal voltages within a narrower band. This trend was maintained in C3 due to the contribution of RES to the regulation of voltage with reactive power injections. In addition, it can be seen that the average voltage of nodes N3, N6, and N8 from subsystem 1 increased due to the local effect of the reactive power injections. As a consequence, additional OLTC changes N_{OLTC} (from −5% to 0% position) were required to maintain the voltages within the technical limits. This longer time of the tap within the 0% position caused lower voltages within subsystem 2, as can be observed for the node N14 in Figure 9.

C4 incorporated the operation of the DC link between nodes N8 and N14 allowing the injection of additional reactive power into these nodes and active power transfer between both subsystems. This led to a minimum range of variation in the nodal voltages ΔV and maximum values of these in all the test cases. In fact, in C4, the voltages oscillated in a range between 1 and 1.05 pu over the 24-h period.

Figure 10 shows the daily evolution of the current circulating through the primary substation transformer for the studied test cases. This current reduced as the number of control assets increased. The analysis of C4 revealed that during some periods, the current was almost zero. This means that the generation of RES with adequate management by the control assets is enough to operate the system without the need of supplementary power from the primary substation. Finally, it is worth noting that the state of load of the transformer T_L also progressively reduced in the subsequent test cases, as shown in Table 3. As a consequence, the benefits for the distribution system are clear in this respect: reduction of transformer losses, increment of useful life, and increase of the system loadability, which allows new investment in power assets to be deferred.

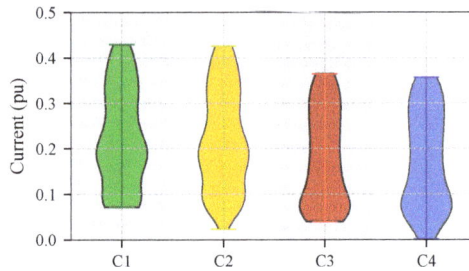

Figure 10. Violin plots of MV current at the primary substation transformer for test cases C1 to C4.

5. Conclusions

This paper assessed the benefits of a centralised controller for active distribution networks with high-RES penetration in an experimental manner. The paper proposed the optimization of the operation of the system through the minimization of active power losses through an OPF with the following control assets: (i) transformers equipped with OLTC, (ii) RES reactive power injections, and (iii) DC links. The assessment of the proposed centralised controlled was carried out on a laboratory scaled-down version of the MV network that was proposed by the CIGRE Task Force C06.04.02. This testing platform was described, including its main components and functionalities as well as the new control assets (transformer OLTC and DC link) which were incorporated into a previous version to improve its testing capabilities. The paper defined a comprehensive design of the testing procedure including some test cases involving different control assets and a set of KPIs to allow a quantintative comparison of performance. The obtained results revealed that a centralised control of high-RES active distribution networks may improve their operation. As a matter of fact, the obtained results, ranging from 15% to 25% of active power loss reduction, are consistent with those of similar works commented on in Section 1. Moreover, this improvement is significant for control assets which are commonly present in distribution networks, i.e., transformers with OLTC and RES reactive power injections. This enhancement could be even larger if uncommon but matured technologies, like DC links, were progressively introduced into the distribution business. This would increase the RES network hosting capacity, contributing to the decarbonization of our society.

Author Contributions: F.P.d.G.-L reviewed the state-of-the-art, designed the test cases and the control algorithms, was responsible for laboratory testing and discussion of the obtained results, and wrote part of the paper; M.B.-V. supported the laboratory testing, contributed to the analysis of the results and wrote part of the paper; A.M.-M. was responsible for the OPF definition, contributed to the analysis of results, and wrote part of the paper; J.M.M.-O. defined the KPIs, contributed to the discussion of results, and wrote part of the paper; J.L.M.-R. supported the OPF definition and practical implementation issues, and discussed the results of the paper.

Funding: The authors would like to acknowledge the financial support of the Spanish Ministry of Economy and Competitiveness under Grants ENE2015-69597-R, PCIN-2015-043 and ENE2017-84813-R.

Conflicts of Interest: The authors declare no conflict of interest.

Abbreviations

The following abbreviations are used in this manuscript:

ADMS	Advanced Distribution Management System
CIGRE	International Council on Large Electric Systems
DC	Direct Current
DSO	Distribution System Operator

HV	High Voltage
IGBT	Insulated Gate Bipolar Transistor
KPI	Keys Performance Index
LC	Local Controller
LV	Low Voltage
MV	Medium Voltage
OLE	Omnimode Load Emulator
OPF	Optimal Power Flow
OLTC	On-Load Tap Changer
PV	Photovoltaic
RES	Renewable Energy Sources
RTCS:	Real-Time Control System
RTU	Remote Terminal Unit
SMS	Signal Management System
TSO	Transmission System Operator
VSC	Voltage Source Converter
WT	Wind Turbine

References

1. Guerrero, J.M.; Blaabjerg, F.; Zhelev, T.; Hemmes, K.; Monmasson, E.; Jemei, S.; Comech, M.P.; Granadino, R.; Frau, J.I. Distributed Generation: Toward a New Energy Paradigm. *IEEE Ind. Electron. Mag.* **2010**, *4*, 52–64. [CrossRef]
2. *Distributed Generation System Characteristics and Costs in the Buildings Sector*; Technical Report; U.S. Energy Information Administration: Washington, DC, USA, 2013.
3. Pérez-Arriaga, I.; Knittel, C. *Utility of the Future. An MIT Energy Initiative Response*; Technical Report; Massachusetts Institute of Technology: Cambridge, MA, USA, 2016.
4. *Power Distribution Planning Reference Book*, 2nd ed.; CRC Press Book: Boca Raton, FL, USA, 2004.
5. Barker, P.P.; Mello, R.W.D. Determining the impact of distributed generation on power systems. I. Radial distribution systems. In Proceedings of the 2000 Power Engineering Society Summer Meeting (Cat. No.00CH37134), Seattle, WA, USA, 16–20 July 2000; Volume 3, pp. 1645–1656.
6. Walling, R.A.; Saint, R.; Dugan, R.C.; Burke, J.; Kojovic, L.A. Summary of Distributed Resources Impact on Power Delivery Systems. *IEEE Trans. Power Deliv.* **2008**, *23*, 1636–1644. [CrossRef]
7. Assessing the Impact of Low Carbon Technologies on Great Britain's Power Distribution Networks. 3 August 2012. Available online: https://www.ofgem.gov.uk/publications-and-updates/assessing-impact-low-carbon-technologies-great-britains-power-distribution-networks (accessed on 5 April 2018).
8. Elkhatib, M.E.; El-Shatshat, R.; Salama, M.M.A. Novel Coordinated Voltage Control for Smart Distribution Networks With DG. *IEEE Trans. Smart Grid* **2011**, *2*, 598–605. [CrossRef]
9. Ueda, Y.; Kurokawa, K.; Tanabe, T.; Kitamura, K.; Sugihara, H. Analysis Results of Output Power Loss Due to the Grid Voltage Rise in Grid-Connected Photovoltaic Power Generation Systems. *IEEE Trans. Ind. Electron.* **2008**, *55*, 2744–2751. [CrossRef]
10. Molina-García, A.; Mastromauro, R.A.; García-Sánchez, T.; Pugliese, S.; Liserre, M.; Stasi, S. Reactive Power Flow Control for PV Inverters Voltage Support in LV Distribution Networks. *IEEE Trans. Smart Grid* **2017**, *8*, 447–456. [CrossRef]
11. Calderaro, V.; Galdi, V.; Lamberti, F.; Piccolo, A. A Smart Strategy for Voltage Control Ancillary Service in Distribution Networks. *IEEE Trans. Power Syst.* **2015**, *30*, 494–502. [CrossRef]
12. Karagiannopoulos, S.; Aristidou, P.; Hug, G. Hybrid approach for planning and operating active distribution grids. *Transm. Distrib. IET Gener.* **2017**, *11*, 685–695. [CrossRef]
13. Puerto Rico Electric Power Authority (PREPA). *Minimum Technical Requirements (MTR) for Photovoltaic Generation (PV) Projects*; Technical Report; Puerto Rico Electric Power Authority (PREPA): San Juan, Puerto Rico, 2012.
14. National Energy Regulator of South Africa (NERSA). *Grid Connection Code for Renewable Power Plants (RPPs) Connected to the Electricity Transmission System or the Distribution System in South Africa*; Technical Report; National Energy Regulator of South Africa (NERSA): Pretoria, South Africa, 2012.

15. Alnaser, S.W.; Ochoa, L.F. Advanced Network Management Systems: A Risk-Based AC OPF Approach. *IEEE Trans. Power Syst.* **2015**, *30*, 409–418. [CrossRef]
16. Kryonidis, G.C.; Demoulias, C.S.; Papagiannis, G.K. A Nearly Decentralized Voltage Regulation Algorithm for Loss Minimization in Radial MV Networks With High DG Penetration. *IEEE Trans. Sustain. Energy* **2016**, *7*, 1430–1439. [CrossRef]
17. Kolenc, M.; Papič, I.; Blažič, B. Minimization of losses in smart grids using coordinated voltage control. *Energies* **2012**, *5*, 3768–3787. [CrossRef]
18. Alnaser, S.W.; Ochoa, L.F. Hybrid controller of energy storage and renewable DG for congestion management. In Proceedings of the 2012 IEEE Power and Energy Society General Meeting, San Diego, CA, USA, 22–26 July 2012; pp. 1–8.
19. Liu, X.; Aichhorn, A.; Liu, L.; Li, H. Coordinated Control of Distributed Energy Storage System with Tap Changer Transformers for Voltage Rise Mitigation Under High Photovoltaic Penetration. *IEEE Trans. Smart Grid* **2012**, *3*, 897–906. [CrossRef]
20. Barragan-Villarejo, M.; Marano, A.; García-López, F.P.; Mauricio, J.M.; Maza-Ortega, J.M. Coordinated control of distributed energy resources and flexible links in active distribution networks. In Proceedings of the International Conference on Renewable Power Generation (RPG 2015), Beijing, China, 17–18 October 2015; pp. 1–6.
21. Marano-Marcolini, A.; Villarejo, M.B.; Fragkioudaki, A.; Maza-Ortega, J.M.; Ramos, E.R.; de la Villa Jaén, A.; Delgado, C.C. DC Link Operation in Smart Distribution Systems With Communication Interruptions. *IEEE Trans. Smart Grid* **2016**, *7*, 2962–2970. [CrossRef]
22. Fazio, A.R.D.; Fusco, G.; Russo, M. Decentralized Control of Distributed Generation for Voltage Profile Optimization in Smart Feeders. *IEEE Trans. Smart Grid* **2013**, *4*, 1586–1596. [CrossRef]
23. Kulmala, A.; Alonso, M.; Repo, S.; Amaris, H.; Moreno, A.; Mehmedalic, J.; Al-Jassim, Z. Hierarchical and distributed control concept for distribution network congestion management. *IET Gener. Transm. Distrib.* **2017**, *11*, 665–675. [CrossRef]
24. Almasalma, H.; Claeys, S.; Mikhaylov, K.; Haapola, J.; Pouttu, A.; Deconinck, G. Experimental Validation of Peer-to-Peer Distributed Voltage Control System. *Energies* **2018**, *11*, 1304. [CrossRef]
25. Maza-Ortega, J.M.; Barragán-Villarejo, M.; García-López, F.d.P.; Jiménez, J.; Mauricio, J.M.; Alvarado-Barrios, L.; Gómez-Expósito, A. A Multi-Platform Lab for Teaching and Research in Active Distribution Networks. *IEEE Trans. Power Syst.* **2017**, *32*, 4861–4870. [CrossRef]
26. Romero-Ramos, E.; Gómez-Expósito, A.; Marano-Marcolini, A.; Maza-Ortega, J.M.; Martínez-Ramos, J.L. Assessing the loadability of active distribution networks in the presence of DC controllable links. *IET Gener. Transm. Distrib.* **2011**, *5*, 1105. [CrossRef]
27. Rudion, K.; Orths, A.; Styczynski, Z.A.; Strunz, K. Design of benchmark of medium voltage distribution network for investigation of DG integration. In Proceedings of the 2006 IEEE Power Engineering Society General Meeting, Montreal, QC, Canada, 18–22 June 2006; p. 6.
28. Benchmark Systems for Network Integration of Renewable and Distributed Energy Resources. Available online: http://www.e-cigre.org/publication/575-benchmark-systems-for-network-integration-of-renewable-and-distributed-energy-resources (accessed on 2 November 2016).
29. Maza-Ortega, J.M.; Gomez-Exposito, A.; Barragan-Villarejo, M.; Romero-Ramos, E.; Marano-Marcolini, A. Voltage source converter-based topologies to further integrate renewable energy sources in distribution systems. *IET Renew. Power Gener.* **2012**, *6*, 435–445. [CrossRef]

energies

MDPI

Article

Robust Allocation of Reserve Policies for a Multiple-Cell Based Power System

Junjie Hu [1], Tian Lan [2,*], Kai Heussen [3], Mattia Marinelli [3], Alexander Prostejovsky [3] and Xianzhang Lei [2]

[1] State Key Laboratory of Alternate Electrical Power Systems with Renewable Energy Sources, North China Electric Power University, Beijing 102206, China; junjiehu@ncepu.edu.cn
[2] Global Energy Interconnection Research Institute Europe GmbH, 10117 Berlin, Germany; xianzhang-lei@sgcc.com.cn
[3] Center for Electrical Power and Energy, DK2800 Lyngby, Denmark; kh@elektro.dtu.dk (K.H.); matm@elektro.dtu.dk (M.M.); alepros@elektro.dtu.dk (A.P.)
* Correspondence: lantian@geiri.sgcc.com.cn; Tel.: +49-176-9570-4134

Received: 6 December 2017; Accepted: 30 January 2018; Published: 7 February 2018

Abstract: This paper applies a robust optimization technique for coordinating reserve allocations in multiple-cell based power systems. The linear decision rules (LDR)-based policies were implemented to achieve the reserve robustness, and consist of a nominal power schedule with a series of linear modifications. The LDR method can effectively adapt the participation factors of reserve providers to respond to system imbalance signals. The policies considered the covariance of historic system imbalance signals to reduce the overall reserve cost. When applying this method to the cell-based power system for a certain horizon, the influence of different time resolutions on policy-making is also investigated, which presents guidance for its practical application. The main results illustrate that: (a) the LDR-based method shows better performance, by producing smaller reserve costs compared to the costs given by a reference method; and (b) the cost index decreases with increased time intervals, however, longer intervals might result in insufficient reserves, due to low time resolution. On the other hand, shorter time intervals require heavy computational time. Thus, it is important to choose a proper time interval in real time operation to make a trade off.

Keywords: linear decision rules; optimal reserve allocation; robust optimization; web of cells

1. Introduction

With increasing concerns regarding global warming and environment pollution, there has been a worldwide movement in the promotion of renewable technologies for electricity generation and for reducing the greenhouse-gas emissions. Many distributed generation units, including wind turbines [1], photovoltaic generators [2,3], fuel cells and fuel cell/gas/steam powered combined heat and power systems[4], are being used and connected to the power systems. However, a large penetration of intermittent and variable generation introduces operational challenges to the power system. To accommodate the variability of the renewables, it is important to harness the flexibility of the newly introduced units, such as batteries and electric vehicles, especially in the distribution network level [5–7]. Enabling this flexibility comes at the cost of an increasingly complex control system, characterized by many state and decision variables [8].

The problems encountered in the power systems have received much attention, and various efforts have been made to address the problems. These range from developing proper control schemes for individual component operation such as hierarchical aggregation method [9,10] to radical rethinking of system operations [11–14]. Traditionally, the system is kept secure by distinguishing the role between transmission system operators (TSO) and distribution system operators (DSO). For example, the TSO centrally controls a few big power plants through a supervisory control and data acquisition system. The DSO centrally manages the status of key devices, such as breakers, reference setting points of on/off load tap changers, capacity banks, etc. However, it is impossible for TSO to control the large number of distributed energy resources found today, as the grid control systems are centralized by design, and do not yet actively integrate distributed energy resources into the operation on a meaningful scale.

To address the aforementioned problem, the European FP7 project ELECTRA IRP proposes and develops a Web-of-Cells (WoC) architecture for operating the future power system [11–14]. In this approach, the power systems operation is divided into connected cells, each responsible for their own balancing and voltage control, thus establishing a robust, decentralized horizontal decomposition as opposed to the conventional centralized and vertical system operation. The WoC concept reformulates the control architecture of electric power systems to accommodate the challenges of fully distributed generation, reduced inertia, storage integration and flexible demand. Cells, which are defined as non-overlapping topological subsets of a power system, are associated with a scale-independent operational responsibility to contribute to system operation and stability. The operating state, including power exchanges and reserve parameters, can then be continuously optimized by coordination across cells.

In this paper, we study how joint energy and reserve scheduling, which are provided by flexible load units such as storage units and electric vehicles (EVs) that have limited power, energy and specific use patterns, could be operated more efficiently in the WoC architecture based power system. To do so, a robust power system reserve allocation approach [15,16] combining a predictive dispatch with optimal control policies in the form of linear decision rules (LDR) [17] is proposed. LDR concept is used in operations research field, where current states, past data or future predictions are combined linearly to make an operational decision. In [15], LDR-based reserve policies consisting of a nominal power schedule with a series of planned linear modification is proposed to accommodate fluctuating renewable energy resources. These policies are time-coupled, which exploits the temporal correlation of these prediction errors. The study showed that LDR-based reserve policies can reduce reserve operation cost compared to existing standard reserve operation method. In [16], the authors proposed an adjustable robust optimization approach to account for the uncertainty of renewable energy sources in optimal power flow. The optimized solution has two part: (1) the base-point generation is calculated to serve the forecast load which is not balanced by RES; and (2) the generation control using participation factors ensures a feasible solution for all realizations of RES output within a prescribed uncertainty set. However, both papers only applied the LDR method for one power system. Compared to stochastic programming, robust optimization method only requires knowledge of the range of variation of the uncertain parameters as opposed to an accurate specification of the uncertain parameter in stochastic programming. Therefore, robust optimization has been gaining popularity for decision making under uncertainty. In [18], the authors applied LDR-based reserve policies to the Web-of-Cells [19]. Firstly, the study shows that the method works fine for a single cell operation, i.e., the power and energy curve of batteries are within the capacity for any realized RES output. Then, three ad hoc cooperation strategies of web-of-cells are studied and compared using the LDR method. The three cooperation strategies include: (a) no cooperation between cells; (b) full cooperation between all cells; and (c) in between these two extreme cases. The study shows that Strategy (a) has a clear disadvantage over other two cooperation strategies.

Building forth on the previously developed work [18], this paper has two advancements: (1) It develops a model that adapts the application of the LDR-based reserve policies to multiple-cell based power systems rather than single-cell based power systems. Based on the proposed model, cross-cell reserve allocation, indicating the cooperation scheme among cells, can be determined. The results show that the involvement of the cross-cell reserve depends on the availability of reserve resources in the local cell. (2) To facilitate the real time operation in the real system, the effects of different time intervals on the LDR control policies are investigated in this paper. The investigation of the effects is made considering energy curves, power curves, and cost index of each discussed case. The remainder of the paper is organized as follows. In Section 2, the optimization formulation for one and multiple cell based power system is proposed, given the basic power system model. Comprehensive case studies are performed in Section 3. Conclusions are drawn in Section 4.

2. Methodology

The methodology used for solving the robust optimal reserve allocation problem is introduced in this section. The methodology is based on LDR, and can be applied to the operation of power systems, which could be single-cell based or multiple-cell based. A single-cell based power system can be considered as an isolated system, which could be an autonomous microgrid or a regional network. A multiple-cell based power system consists of two or more cells linked via interconnectors between each other. In the following subsections, the LDR based methodology is firstly proposed. Then, the optimization problem regarding single-cell and multiple-cell based power systems are, respectively, formulated using proposed methodology.

2.1. Basic Power System Model

A power system with various participants connected to a power grid is considered. The participants of a power system could be production units, loads, or storage units, which either inject power into or extract power from a node in the network. They are categorized into two types in terms of power injection: participants with inelastic power injection and those with elastic power injection. The inelastic power flows indicate the power flows of the participants cannot be regulated by control signals. These participants could be certain loads and renewable generators, such as wind turbines and PV panels. Regarding a participant i of this kind, the power injection into or extraction from a network y_i can be modeled as:

$$y_i = r_i + G_i \delta \tag{1}$$

where $r_i \in \mathbb{R}^T$ indicates a nominal prediction of the power injection or extraction; T is the divided discrete time steps of a planning time horizon, over which electricity can be traded on intra-day markets [20]; $\delta \in \mathbb{R}^{N_\delta T}$ is the random forecast error vector; G_i is a linear function used for mapping the uncertainty δ to power flows; and N_δ is the number of elements in the uncertainty vector at a given time. If the power flows of the participants can be perfectly predicted, the prediction error δ will be zero.

The elastic power injection indicates that the power flows of the participants can be influenced by control signals. In other words, the flexibility of these participants can be exploited and used to mitigate the disturbance in the network. This can be achieved using the control signals determined by the results of proper optimization. According to [15], the elastic power injections can be modeled as $C_j \mathbf{x}_j$, where $\mathbf{x}_j \in \mathbb{R}^{n_j T}$ is a vector of future states of participant j, n_j is the state dimension, and $C_j \in \mathbb{R}^{T \times n_j T}$ is the stacked output matrix used for selecting the needed element of the state vector $\mathbf{x}_j$. The future state vector $\mathbf{x}_j$ is given as:

$$\mathbf{x}_j = A_j x_0^j + B_j \mathbf{u}_j \tag{2}$$

where x_0^j is the current state of the participant j, $\mathbf{u}_j \in \mathbb{R}^T$ is the control input to the elastic power participant j for balancing the system, and A_j and B_j are the corresponding stacked state transition matrices.

2.1.1. Constraints for Production Units

Different participants need to be limited by corresponding constraints. Regarding a production unit at period t, which could be an inelastic or elastic power unit, the upper and lower bounds of the power are imposed by the following constraints:

$$P_i^{\min} \le y_i \le P_i^{\max} \tag{3}$$

$$P_j^{\min} \le C_j \mathbf{x}_j \le P_j^{\max} \tag{4}$$

2.1.2. Constraints for Storage Units

Regarding a storage unit j, which could be a battery, an electric vehicle, etc., whose power injection is elastic, the storage unit's power and energy constraints need to be imposed:

$$P_j^{\min} \le C_j \mathbf{x}_j \le P_j^{\max} \tag{5}$$

$$E_j^{\min} \le E_{j,t-1} + C_j \mathbf{x}_j \Delta t \le E_j^{\max} \tag{6}$$

As all participants are connected to a network, the sum of all the inelastic power injection y_i and elastic power injection $C_j \mathbf{x}_j$ has to be zero at all times $t = 1, ..., T$. This can be achieved using an equality constraint:

$$\sum_{i=1}^{N_{\text{inelas}}} y_i + \sum_{j=1}^{N_{\text{elas}}} C_j \mathbf{x}_j = 0 \tag{7}$$

where N_{inelas} is the number of the inelastic power participants in a network, and N_{elas} is the number of the elastic power participants.

2.2. Linear Decision Rule Based Robust Optimization of Reserve Allocation

Inserting Equations (1) and (2) into Equation (7), the following equation can be obtained:

$$\sum_{i=1}^{N_{\text{inelas}}} (r_i + G_i \delta) + \sum_{j=1}^{N_{\text{elas}}} C_j (A_j x_0^j + B_j \mathbf{u}_j) = 0 \tag{8}$$

As mentioned, to keep power balance in a network, the power injection or extraction of the inelastic power units, $\sum_{i=1}^{N_{\text{inelas}}} (r_i + G_i \delta)$, must be balanced by the power contribution of the elastic power units, $\sum_{j=1}^{N_{\text{elas}}} C_j (A_j x_0^j + B_j \mathbf{u}_j)$, at any point in time. Regarding elastic power units, $\mathbf{u}_j$ is the control input that can regulate the power flow of the elastic power participant j. According to the LDR method, control input signal $\mathbf{u}_j$ can be expressed to policies of the affine form:

$$\mathbf{u}_j = D_j \delta + e_j \tag{9}$$

where $\mathbf{u}_j$ is described by a nominal schedule $e_j \in \mathbb{R}^T$ plus a linear variation $D_j \in \mathbb{R}^{T \times T}$, the nominal schedule e is mainly used for balancing the nominal prediction of the inelastic power flows r, and D is the dynamic response to the prediction errors δ. The matrix D defines a map from the uncertainty into the realization of the power contribution of the elastic power units. In order for the use of future disturbances to be causal, D_j takes the lower-triangular form.

2.3. One Cell-Based Reserve Allocation Model

In a one-cell based power system, the reserve allocation optimization problem is formulated as a cost function as follows:

$$\min \ \mathbb{E}\left\{ \sum_{j \in \phi} (\alpha_j P_j(\delta)^T P_j(\delta) + \beta_j P_j(\delta) + \gamma_j \bar{\mathbf{I}}) \right\} \tag{10}$$

where $j \in \phi$ indicates the elastic power participant; P_i is the power contribution of the participant; α_j, β_j, and γ_j are stacked vectors of quadratic, linear, and constant coefficients of the cost function of the power contribution, respectively; and $\bar{\mathbf{I}}$ is a vector of all of them. This optimization problem is subject to system constraints, power constraints, and energy constraints, as mentioned in the previous section. The power contribution of the elastic power participant can be given as:

$$P_j(\delta) = C_j \mathbf{x}_j(\delta) \tag{11}$$

with the help of Equation (2), $P_j(\delta)$ can be further written as:

$$P_j(\delta) = C_j(A_j x_0^j + B_j \mathbf{u}_j) \tag{12}$$

According to policies of the affine form, as given in Equation (9), the entire optimization problem in Equation (10) becomes tractable due to the restricted variety of candidate $\mathbf{u}_j$.

Substituting Equation (12) into the objective function, i.e. Equation (10), the following form can be obtained:

$$
\begin{aligned}
\min \ \mathbb{E}\{ \sum_{j \in \phi} (\alpha_j (C_j(A_j x_0^j + B_j \mathbf{u}_j))^T (C_j(A_j x_0^j + B_j \mathbf{u}_j)) \\
+ \beta_j C_j (A_j x_0^j + B_j(D_j \delta + e_j)) + \gamma_j \bar{\mathbf{I}}) \}
\end{aligned}
\tag{13}
$$

where $\mathbf{u}_j$ can be further expressed as policies of the affine form, as shown in Equation (9); together with an assumption that $\mathbb{E}[\delta] = 0$, the optimization problem can be written as:

$$\min \ \sum_{j \in \phi} \left(\alpha_j (a_j^T a_j + \langle D_j^T b_j^T b_j D_j, \mathbb{E}[\delta \delta^T] \rangle) + \beta_j a_j + \gamma_j \bar{\mathbf{I}} \right) \tag{14}$$

with

$$a_j = C_j A_j x_0^j + C_j B_j e_j \tag{15}$$
$$b_j = C_j B_j \tag{16}$$

where $\langle X, Y \rangle$ is the trace of product $X'Y$; regarding the reformulation of the equality and inequality constraints, the approach is similar to the one presented in [15,18].

2.4. Multiple Cells-Based Reserve Allocation Model

Compared to the single cell-based power system, the robust optimization of reserve allocation using LDR in multiple cells-based is more complicated. An example of a three-cell power system is shown in Figure 1. Each cell has its respective generations, loads, and storage units. Three cell are connected with tie lines, as depicted in the Figure 1. Regarding the resources with flexibilities, such as electric vehicles and batteries, it is assumed that a portfolio of all flexible energy-constrained resources in one cell can be represented by an aggregator as a single unit.

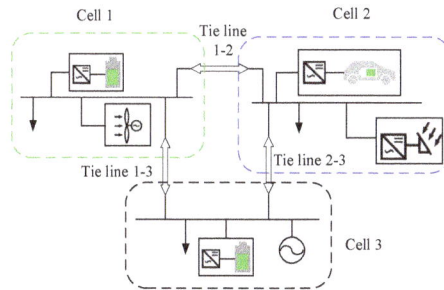

Figure 1. An example of a three-cell small scale power system.

In the WoC concept, local imbalances within a cell are supposed to be solved locally. Cells are not necessarily supposed to be self-sufficient, but are required to have the balancing capability provided by elastic power units for mitigating deviations from a given schedule. The basic concept of reserve allocation in multi-cell-based power systems is that the local reserve provided by local elastic power units is prioritized for handling of local imbalances, which is reflected in the cost function of each available power unit. As long as local reserves can handle the local imbalances, no reserve is needed from other cells. Cross-cell reserve allocation happens when the local resource cannot handle the local imbalance.

Cells are managed by so-called Cell System Operators (CSOs), whose roles incorporate the tasks of traditional Distribution and Transmission System Operators (DSOs and TSOs, respectively) [21]. A dedicated Cell Controller (CC) performs monitoring and automated balancing tasks. Each cell is assigned one CC, which is managed by one CSO using bi-directional communication, as shown in Figure 2. A CSO, on the other hand, may be responsible for more than one CC to allow for flexible adaption to present-day grid partitioning and management schemes.

Among other tasks, CSOs procure reserves and send the schedules to the CCs, which automatically carry out balancing tasks around the given setpoints and trajectories. As indicated in Figure 2, information and measurements from each cell controller, including the prediction of the generation of the renewables, availability of the elastic power units as well as their power and energy constraints, information of loads, etc., are transferred to the cell operator, based on which the cell operator can utilize the proposed optimization to distribute the control signals, i.e., the D and e presented in Equation (9), for each cell controller to allocate the reserve in each cell.

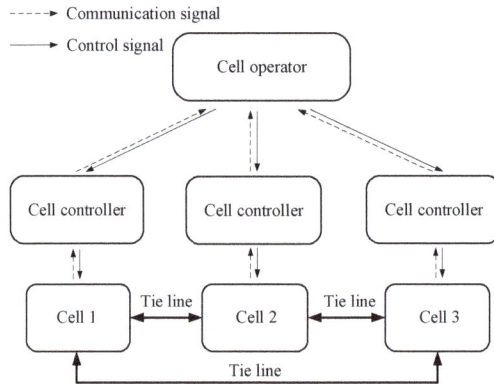

Figure 2. Control system architecture.

The reserve optimization problem for multiple-cell based power system can be generally formulated as:

$$\min \mathbb{E}\Big\{ \sum_{l\in\phi_{\text{cell}}} \sum_{j\in\phi_l} \sum_{m\in\phi_{\text{cell}},l\neq m} \Big(\alpha_{l,l}^j P_{l,l}^j(\delta)^T P_{l,l}^j(\delta) + \beta_{l,l}^j P_{l,l}^j(\delta) $$
$$ + \gamma_{l,l}^j \mathbf{I} + v_{l,m}^j \big(\alpha_{l,m}^j P_{l,m}^j(\delta)^T P_{l,m}^j(\delta) + \beta_{l,m}^j P_{l,m}^j(\delta) + \gamma_{l,m}^j \mathbf{I} \big) \Big) \Big\} \tag{17}$$

where l and m indicate the cell indices; j is the participant involved in the reserve allocation; ϕ_{cell} and ϕ_l are the sets of cell and participants in cell l, respectively; $P_{l,l}$ is the allocated resource in cell l that is reserved for the imbalance in cell l; and $P_{l,m}$ is resource that is reserved from cell l to cell m, also called the cross-cell reserve service. Finally, $v_{l,m}^j$ is a binary decision variable used for determining the involvement of the cross-cell reserve allocation:

$$v_{l,m}^j = \begin{cases} 0 & P_{l,l}^j(\delta_{\max}) >= P_{\max}^j \\ 1 & P_{l,l}^j(\delta_{\max}) < P_{\max}^j \end{cases} \tag{18}$$

Similar to Equation (12), $P_{l,l}$ and $P_{l,m}$ can be expressed as:

$$P_{l,l}^j(\delta) = C_{l,l}^j(A_{l,l}^j x_{0,l,l}^j + B_{l,l}^j \mathbf{u}_{l,l}^j) \tag{19}$$

$$P_{l,m}^j(\delta) = C_{l,m}^j(A_{l,m}^j x_{0,l,m}^j + B_{l,m}^j \mathbf{u}_{l,m}^j) \tag{20}$$

According to the LDR method, $\mathbf{u}_{l,l}^j$ and $\mathbf{u}_{l,m}^j$ can be expressed as:

$$\mathbf{u}_{l,l}^j = D_{l,l}^j \delta + e_{l,l}^j \tag{21}$$

$$\mathbf{u}_{l,m}^j = D_{l,m}^j \delta + e_{l,m}^j \tag{22}$$

where participant j's power schedule in local cell $\mathbf{u}_{l,l}^j$ is determined by a nominal schedule $e_{l,l}^j$, a linear variation $D_{l,l}^j$ and predication errors δ. Similarly, expression of $\mathbf{u}_{l,m}^j$ can be obtained. Together with the assumption that $\mathbb{E}[\delta] = 0$, the optimization problem can be written as:

$$\begin{aligned} \min \sum_{l \in \phi_{\text{cell}}} \sum_{j \in \phi_l} \sum_{m \in \phi_{\text{cell}}, l \neq m} & \left(\left(\alpha_{l,l}^j ((a_{l,l}^j)^T a_{l,l}^j \right. \right. \\ & + \langle (D_{l,l}^j)^T (b_{l,l}^j)^T b_{l,l}^j D_{l,l}^j, \mathbb{E}[\delta\delta^T] \rangle) + \beta_{l,l}^j a_{l,l}^j + \gamma_{l,l}^j \mathbf{I} \right) \\ & + v_{l,m}^j \left(\alpha_{l,m}^j ((a_{l,m}^j)^T a_{l,m}^j + \langle (D_{l,m}^j)^T (b_{l,m}^j)^T b_{l,m}^j D_{l,m}^j, \mathbb{E}[\delta\delta^T] \rangle) \right. \\ & \left. \left. + \beta_{l,m}^j a_{l,m}^j + \gamma_{l,m}^j \mathbf{I} \right) \right) \end{aligned} \tag{23}$$

with

$$a_{l,l}^j = C_{l,l}^j A_{l,l}^j x_{0,l,l}^j + C_{l,l}^j B_{l,l}^j e_{l,l}^j \tag{24}$$

$$b_{l,l}^j = C_{l,l}^j B_{l,l}^j \tag{25}$$

$$a_{l,m}^j = C_{l,m}^j A_{l,m}^j x_{0,l,m}^j + C_{l,m}^j B_{l,m}^j e_{l,m}^j \tag{26}$$

$$b_{l,m}^j = C_{l,m}^j B_{l,m}^j \tag{27}$$

Equation (23) establishes an overall optimization for a multiple-cell based power system. To solve the one-cell and three-cell optimization problems, YALMIP, a toolbox in MATLAB, is used [22,23].

3. Case Studies

In this section, several case studies are carried out. The setup is based on a configuration of SYSLAB, which is a laboratory testing facility for Smart Grid concepts located at the Risø campus of the Technical University of Denmark.

3.1. SYSLAB System

Figure 3 shows the SYSLAB facility [24,25], which is a 400 V three-phase grid designed for studying advanced grid control and communication concepts. The facility has 16 busbars and 116 automated coupling points. A wide range of distributed energy resources, such as wind turbines, solar panels, electric vehicles, etc., can be remotely operated via a distributed monitoring and control platform. Because of the very flexible connection interface of the busbars, various topologies of the system can be configured and operated. Division of the cells in the system can also be flexible and different from the one shown in Figure 3.

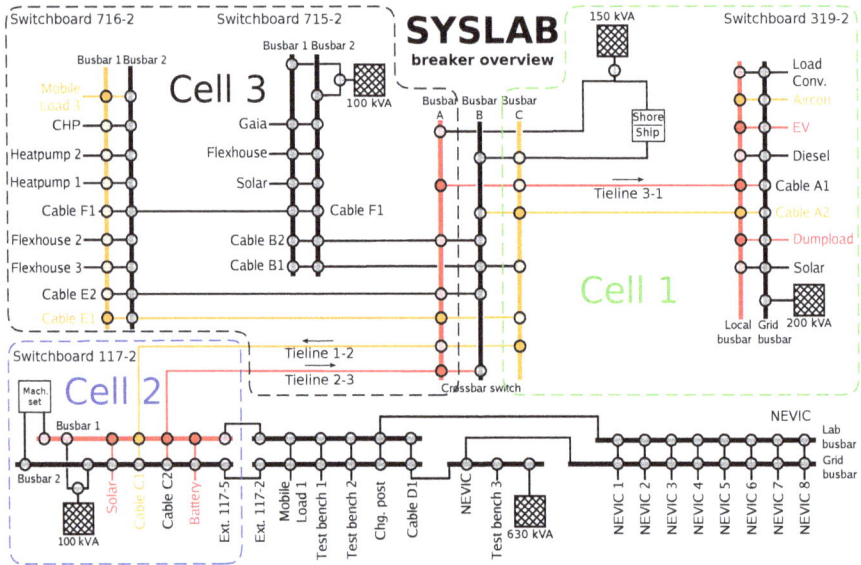

Figure 3. SYSLAB layouts for the two test cases. Devices and lines used in the one-cell case are marked in red. Additional components for the three-cell case are indicated in orange.

3.2. One Cell-Based Simulation

Figure 4 shows one-cell based isolated system, which comprises a battery, a solar PV, an EV, and a mobile load. The mobile load is used for imitating a typical residential electric load profile. The imbalance in the system is mainly caused by the difference between the fluctuated PV generation and the electric load demand. In this system, the inelastic power unit is the PV panel, whose power generation cannot be regulated by control signals. The EV and the battery act as elastic power units, which can provide reserve service due to their controllability of the power flow. The simulation is carried out for 30 min from 11:30 to 12:00. The time interval is chosen to be 3 min, and the horizon length therefore is 10.

One-cell based system

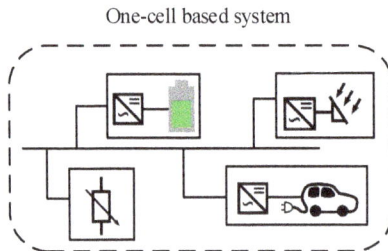

Figure 4. One-cell based system.

From the simulation, the policies for the reserve allocation, i.e., the matrix D introduced in Section 2, of the battery and the EV are presented in Figure 5. As described in Section 2.2, D is a lower-triangular matrix that defines a map from the uncertainty to the realization of power contribution of a reserve participant. The matrix is visualized in Figure 5 for both the battery and the EV. According to Equation (9), the dynamic response to the prediction errors at time instant t is determined by $[D_j]_{t,0}, [D_j]_{t,1}, ..., [D_j]_{t,t}$, which are the elements in row t in the figure. The battery is allocated with more reserve than the reserve allocation contributed by the EV. This is because the battery has a wider power range and larger capacity. Furthermore, the reserve cost of the policy based scheme over the simulation period is compared to that of a flexible-rate scheme. Flexible rate scheme is commonly used for reserve optimization [15,26]. In this paper, the results given by flexible rate scheme act as a reference case. The differences between the two schemes are summarized as follows:

1. Flexible-rate reserves [15]: D_j is a diagonal matrix. This indicates the best possible response to uncertainty without time coupling. The previous uncertainty therefore has no impact on the present operation because the causality of the uncertainty is omitted. The optimization is over the elements of e_j and the diagonal parts of D_j.
2. Policy-based reserves: Compared to the above scheme, this scheme considers the time coupling by taking D_j as the lower-triangular form. It allows full exploitation of the information that will be available at each time step when the reserve is deployed.

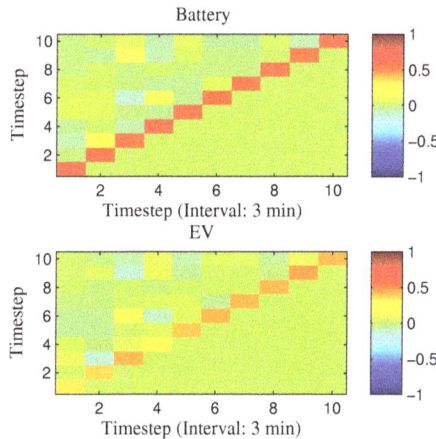

Figure 5. Policy of reserve allocation for one-cell system with time interval of 3 min.

The comparison results regarding the two schemes are given in Table 1. It is shown that the cost index given by the policy-based reserve is smaller than that given by the flexible-rate reserve. This is due to the full exploitation of the covariance matrix using the lower-triangular form of D. From the viewpoint of optimization, the feasibility region of D is larger for policy-based reserves, which gives the optimization more opportunity to find better results.

3.3. Three Cells-Based Simulation

A three-cell based isolated system is presented in Figure 6. Cell 1 has a vanadium battery described in Table 2 and an Aircon wind turbine that has a maximum power generation of 9.8 kW. Cell 2 comprises an EV and a solar panel. Cell 3 only has a mobile load in it. The basic idea of the co-operation is that the local imbalance has priority to be handled by using local reserve, while the cross-cell reserve is allocated only when there is a need. To demonstrate the co-operation of multiple cells, two scenarios are developed by setting the maximum power availability of EV. A three-hour simulation is carried out. The horizon length is set to 12, and the time interval is 15 min.

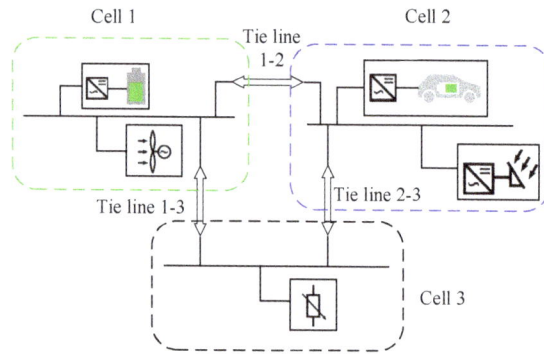

Figure 6. Three-cell based system.

Table 1. Comparison of cost index regarding different time intervals.

Case	Simulation Duration	Time Interval	Horizon Length	Cost Index of Policy-Based Reserve	Cost Index of Flexible-Rate Reserve
1	30 min	2 min	15	30.51	30.62
2	30 min	3 min	10	30.20	30.29
3	30 min	5 min	6	29.97	30.04

Table 2. Properties of devices used in the one-cell system.

Device	Test Case	P_{nom} (kW)	P_{min} (kW)	P_{max} (kW)	Description
Solar	1, 3 Cell	10.1	0.0	10.1	Orientation az. 180°, el. 40°
Battery	1, 3 Cell	0.0	−15.0	15	Vanadium redox flow type 190 kWh, initial state of charge is 50%
EV	1, 3 Cell	0.0	−2.0	2.0	Bidirectional charger 20 kWh, initial state of charge is 50%
Mob. Load	1, 3 Cell	−33.0	−33.0	0.0	Thyristor-contr.
Aircon	3-Cell	9.8	0.0	9.8	Wind turbine

3.3.1. Scenario 1

In this scenario, P_{min} and P_{max} of the EV are set to −15 kW and 15 kW, respectively. With this power availability, the imbalance in Cell 2 caused by the fluctuated power generation of the PV panel can be handled locally. Similarly, Cell 1 can also handle its local imbalance due to the large power capacity of the vanadium battery. Therefore, it is expected that there will be no cross-cell reserve allocation between Cell 1 and Cell 2. Cell 3, however, has only an inelastic power unit, which is the mobile load that represents the residential electric load demand. As it is assumed that the load profile can be precisely predicted, there is no reserve allocation between Cell 3 and other cells, but only nominal schedule e_{13} and e_{23} exist for supporting the residential load. Figure 7 shows the simulation results of Scenario 1. Because of the sufficient flexibility in each cell, the cross-cell reserve allocation does not exist, which leads to $D_{12} = \mathbf{0}$ and $D_{21} = \mathbf{0}$, as depicted in the figure. The imbalances inside Cell 1 and Cell 2 are all handled in their own cell; it can be seen that $D_{11} = \mathbf{I}$ and $D_{22} = \mathbf{I}$, indicating that the predication errors in the two cells are fully compensated using the vanadium battery and the EV in their respective cells.

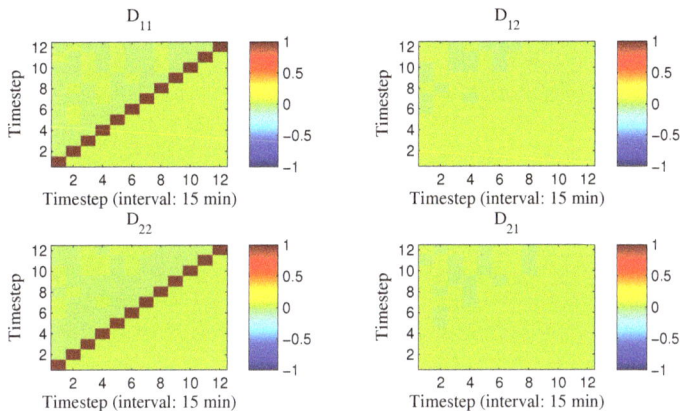

Figure 7. Policy of reserve allocation for three-cell power system in Scenario 1.

3.3.2. Scenario 2

To create the need for cross-cell reserve allocation, in Scenario 2, the P_{min} and P_{max} of the EV in Cell 2 are set to -5 kW and 5 kW, respectively. In this case, the local reserve in Cell 2 cannot handle the fluctuated power generation of the PV. Therefore, reserve allocation from Cell 1 to Cell 2 is necessary. In other words, the elastic resource in Cell 1 needs to be utilized to support the compensation of the predication errors in Cell 2. For that purpose, the vanadium battery in Cell 1 is divided virtually into two parts. One part, corresponding to D_{11}, is used for allocating the reserve for the imbalance in Cell 1. The other part, corresponding to D_{12}, is to reserve the imbalance in Cell 2. This reserve might be the power flow across Cells 1 and 2 via the tie line 1–2. The simulation results are presented in Figure 8. In Cell 1, the local reserve is sufficient for handling the local imbalance, the D matrix for Cell 1 is $D_{11} = \mathbf{I}$. The imbalance in Cell 2, as expected, is covered by local reserve in Cell 2 and some reserve from Cell 1, as can be seen in Figure 8 that $D_{12} + D_{22} = \mathbf{I}$.

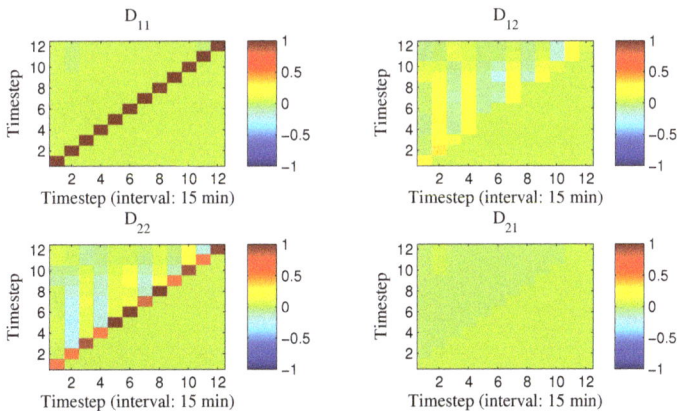

Figure 8. Policy of reserve allocation for three-cell power system in Scenario 2.

3.4. Impact of Time Interval

According to Equations (14) and (23), the optimization formulations depend on the covariances of the historic imbalance signals $\mathbb{E}[\delta\delta^T]$. The covariances of the historic imbalance further depend on the time interval that is used in the simulation and even real time operation. Three cases are carried out based on the one-cell system shown in Figure 4. The configurations of the three cases are given in Table 1. The simulation duration is set to 30 min for all three cases. A simulation with time interval of 3 min is introduced in Section 3.2, while the policy results of a simulation with time intervals of 2 min and 5 min are given in Table 1. Then, a comparison is made based on the time intervals that are less and more than 3 min. Combining the results presented in Section 3.2, the cost indices of the three cases are presented in Table 1. It is noted that the cost index decreases with the increased time interval. This is because the calculation of the covariance of historic imbalance is determined by the time interval. Furthermore, the effects of the time interval on power and energy curves are depicted in Figure 9 based on the renewable inputs and load profile in a specific day. Only the battery curves are presented as an example. It is shown that the power and energy curves contain more information when the time interval is smaller. A longer time interval will smooth the prediction errors of the historic data. This will finally reduce the needed reserve and cost. Longer time intervals can reduce the computation time. However, due to the average operation of the prediction errors, time intervals that are too long might result in inaccurate optimization results, i.e., insufficient reserves. On the other hand, shorter time intervals can make the results more reliable, but the computation time will increase accordingly. In real time operation, it is important to choose a proper time interval to make a trade off.

Table 1 presents the comparison of the cost index given by two schemes. It is shown that the cost index of flexible-rate reserve is higher than that of policy-based reserve for all three cases. The cost saving is important because the presented results are obtained from the 30 min simulation. The cost saving of the policy-based reserve will further increase in accordance with the increasing scale and the operation duration of the system.

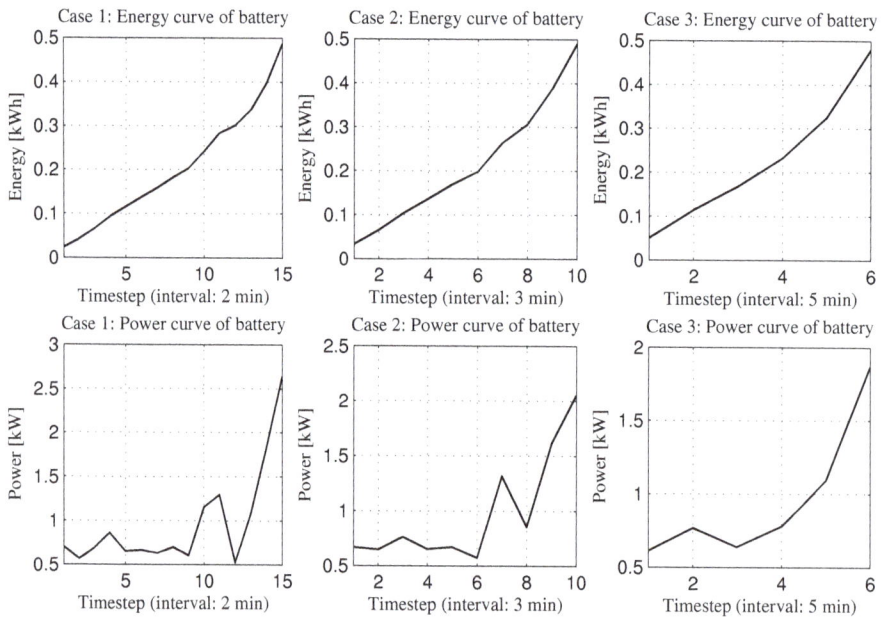

Figure 9. Energy and power curves of battery.

4. Discussion and Conclusions

Linear decision rule-based control policy for coordinating reserve allocation to the ELECTRA Web-of-Cells system architecture has been developed, implemented, demonstrated, and applied to a

Energies **2018**, *10*, 381

three-cell based power system. It has been demonstrated that control policies based on linear decision rules are well suited for integration of distributed energy resources which have flexibility in balancing control, and the robust allocation method is applicable to realistic imbalance signals. Furthermore, it is concluded that a proper time interval selection is important in the linear decision rules-based application. We note that this method relies on several inputs such as: (a) the covariance information of historical data; and (b) the predicted base power production and the prediction error bound of the inelastic power injection, which needs to be improved in the near future. In addition, when adapting the method into real cases, the cost function of elastic power units that represents the balancing service provision should be carefully designed.

Acknowledgments: The research leading to these results has received funding from the European Union Seventh Framework Programme [FP7/2007–2013] under grant agreement No.PADSGCI609687, and includes support from the ELECTRA REX researcher exchange programme. Besides, to accomplish the research, Tian Lan has also received funding from Global Energy Interconnection Research Institute Europe GmbH under the project of "Development of key equipments and engineering technology for large-scale offshore windfarms VSC-HVDC grid connection".

Author Contributions: Junjie Hu, Tian Lan and Kai Heussen conceived the research idea and designed the simulation setup; Tian Lan and Junjie Hu mainly wrote the paper; amd Kai Heussen, Mattia Marinelli, Alexander Prostejovsky proofread the paper.

Conflicts of Interest: The authors declare no conflict of interest.

Nomenclature

Indices

i	Index of inelastic participant.
j	Index of elastic participant.
l, m	Index of cell.

Variable and Parameters

δ	Random forecast error vector.
e_j	Participant j's nominal elastic power.
r_i	Nominal prediction of power injection or extraction of participant i.
$\mathbf{u}_j$	Stacked vector of participant j's future control inputs.
$\mathbf{x}_j$	Stacked vector of participant j's future states.
x_0^j	Vector of participant j's current states.
y_i	Power injection or extraction of inelastic participant i.
A_j	Stacked state transition matrix for participant j.
B_j	Stacked state transition matrix for participant j.
C_j	Stacked output matrix for participant j.
D_j	Matrix adjusting power in response to δ.
G_i	Map from uncertainty to inelastic power injection.
T	Length of time horizon in steps.

References

1. Pineda, I.; Wilkes, J. *Wind in Power: 2014, European Statistics*; Technical Report; WindEurope: Brussels, Belgium, 2015; pp. 1–12.
2. Agency, I.E. *Renewable Energy Medium-Term Market Report 2015*; Market Analysis and Forecasts to 2020; Technical Report; International Energy Agency: Paris, France, 2015.
3. Dabra, V.; Paliwal, K.K.; Sharma, P.; Kumar, N. Optimization of photovoltaic power system: A comparative study. *Prot. Control Mod. Power Syst.* **2017**, *2*, 3.
4. Lan, T.; Strunz, K. Multiphysics Transients Modeling of Solid Oxide Fuel Cells: Methodology of Circuit Equivalents and Use in EMTP-Type Power System Simulation. *IEEE Trans. Energy Convers.* **2017**, *32*, 1309–1321.

5. Hu, J.; Yang, G.; Bindner, H.W.; Xue, Y. Application of Network-Constrained Transactive Control to Electric Vehicle Charging for Secure Grid Operation. *IEEE Trans. Sustain. Energy* **2017**, *8*, 505–515.

6. Hiskens, I.; Callaway, D. Achieving controllability of plug-in electric vehicles. In Proceedings of the 2009 IEEE Vehicle Power and Propulsion Conference (VPPC), Dearborn, MI, USA, 7–10 September 2009; IEEE: Piscataway, NJ, USA, 2009; pp. 1215–1220.

7. Knezović, K.; Martinenas, S.; Andersen, P.B.; Zecchino, A.; Marinelli, M. Enhancing the Role of Electric Vehicles in the Power Grid: Field Validation of Multiple Ancillary Services. *IEEE Trans. Transp. Electr.* **2017**, *3*, 201–209.

8. De Martini, P.; Mani Chandy, K.; Fromer, N. *Grid 2020 Towards a Policy of Renewable and Distributed Energy Resources*; Technical Report; The Resnick Sustainability Institute at Caltech: Pasadena, CA, USA, 2012.

9. Palizban, O.; Kauhaniemi, K.; Guerrero, J.M. Microgrids in active network management—Part I: Hierarchical control, energy storage, virtual power plants, and market participation. *Renew. Sustain. Energy Rev.* **2014**, *36*, 428–439.

10. Bessa, R.J.; Matos, M.A. Economic and technical management of an aggregation agent for electric vehicles: A literature survey. *Eur. Trans. Electr. Power* **2012**, *22*, 334–350.

11. ELECTRA IRP. European Liaison on Electricity Committed towards Research Activity Integrated Research Programm. Available online: http://www.electrairp.eu/ (accessed on 6 February 2018).

12. Marinelli, M.; Pertl, M.; Rezkalla, M.M.; Kosmecki, M.; Canevese, S.; Obushevs, A.; Morch, A.Z. The Pan-European Reference Grid Developed in the ELECTRA Project for Deriving Innovative Observability Concepts in the Web-of-Cells Framework. In Proceedings of the 51st International Universities Power Engineering Conference (UPEC), Coimbra, Portugal, 6–9 September 2016; IEEE: Piscataway, NJ, USA, 2016.

13. D'hulst, R.; Fernández, J.M.; Rikos, E.; Kolodziej, D.; Heussen, K.; Geibelk, D.; Temiz, A.; Caerts, C. Voltage and frequency control for future power systems: The ELECTRA IRP proposal. In Proceedings of the 2015 International Symposium on Smart Electric Distribution Systems and Technologies (EDST), Vienna, Austria, 8–11 September 2015; pp. 245–250.

14. Morch, A.Z.; Jakobsen, S.H.; Visscher, K.; Marinelli, M. Future control architecture and emerging observability needs. In Proceedings of the 2015 IEEE 5th International Conference on Power Engineering, Energy and Electrical Drives (POWERENG), Riga, Latvia, 11–13 May 2015; pp. 234–238.

15. Warrington, J.; Goulart, P.; Mariéthoz, S.; Morari, M. Policy-based reserves for power systems. *IEEE Trans. Power Syst.* **2013**, *28*, 4427–4437.

16. Jabr, R.A. Adjustable Robust OPF With Renewable Energy Sources. *IEEE Trans. Power Syst.* **2013**, *28*, 4742–4751.

17. Bertsimas, D.; Brown, D.B.; Caramanis, C. Theory and applications of robust optimization. *SIAM Rev.* **2011**, *53*, 464–501.

18. Hu, J.; Heussen, K.; Claessens, B.; Sun, L.; D'Hulst, R. Toward coordinated robust allocation of reserve policies for a cell-based power system. In Proceedings of the 2016 IEEE PES Innovative Smart Grid Technologies Conference Europe (ISGT-Europe), Ljubljana, Slovenia, 9–12 October 2016; pp. 1–6.

19. Caerts, C.; Rikos, E.; Syed, M.; Guillo Sansano, E.; Merino-Fernández, J.; Rodriguez Seco, E.; Evenblij, B.; Rezkalla, M.M.N.; Kosmecki, M.; Temiz, A.; et al. *Description of the Detailed Functional Architecture of the Frequency and Voltage Control Solution (Functional and Information Layer)*; Technical Report; Technical University of Denmark: Lyngby, Denmark, 2017.

20. Stoft, S. *Power System Economics: Designing Markets for Electricity*, 1st ed.; Wiley-IEEE Press: New York, NY, USA, 2002.

21. Brunner, H.; Tornelli, C.; Cabiati, M. *The Web of Cells Control Architecture for Operating Future Power Systems*; Technical Report; ELECTRA IRP—Public Deliverables: WP5 Increased Observability; European Union: Brussels, Belgium, 2018, under review.

22. Lofberg, J.Y. YALMIP: A toolbox for modeling and optimization in MATLAB. In Proceedings of the International Symposium on Computer Aided Control Systems Design (CACSD), New Orleans, LA, USA; 2–4 September 2004; IEEE: Piscataway, NJ, USA, 2004; pp. 284–289.

23. Optimization, Gurobi. Inc. *Gurobi Optimizer Reference Manual*; Technical Report; Gurobi Optimization, Inc.: Houston, TX, USA, 2017.

24. Oliver, G.; Bindner, H. Building a test platform for agents in power system control: Experience from SYSLAB. In Proceedings of the International Conference on Intelligent Systems Applications to Power Systems (ISAP), Toki Messe, Niigata, Japan, 5–8 November 2007; IEEE: Piscataway, NJ, USA, 2007; pp. 1–5.

25. Prostejovsky, A.M.; Marinelli, M.; Rezkalla, M.; Syed, M.H.; Guillo-Sansano, E. Tuningless Load Frequency Control Through Active Engagement of Distributed Resources. *IEEE Trans. Power Syst.* **2017**, doi:10.1109/TPWRS.2017.2752962.

26. Energinet. *Ancillary Services to Be Delivered in Denmark Tender Conditions*; Technical Report; Energinet: Erritsø, Denmark, 2017.

![energies logo] *energies*

MDPI

Article

A Network Flow Model for Price-Responsive Control of Deferrable Load Profiles

Juliano Camargo [1,2,*]**, Fred Spiessens** [1,2] **and Chris Hermans** [1,2]

[1] Energy Department, Vlaamse Instelling voor Technologisch Onderzoek (VITO), Boeretang 200, B-2400 Mol, Belgium; fred.spiessens@vito.be (F.S.); chris.hermans@vito.be (C.H.)
[2] Energy Department, EnergyVille, Thor Park, Poort Genk 8130, 3600 Genk, Belgium
*   Correspondence: juliano.camargo@vito.be; Tel.: +32-14-33-5910

Received: 21 December 2017; Accepted: 6 March 2018; Published: 9 March 2018

Abstract: This paper describes a minimum cost network flow model for the aggregated control of deferrable load profiles. The load aggregator responds to indicative energy price information and uses this model to formulate and submit a flexibility bid to a high-resolution real-time balancing market, as proposed by the SmartNet project. This bid represents the possibility of the cluster of deferrable loads to deviate from the scheduled consumption, in case the bid is accepted. When formulating this bid, the model is able to take into account the discretized power profiles of the individual loads. The solution of this type of aggregation problems is necessary for the participation of small loads in demand response programs, but scalability can be an issue. The minimum cost network flow problem belongs to a restricted class of discrete optimization problems for which efficient and scalable algorithms exist. Thanks to its scalability, this technique can be useful in the control of a large number of smart appliances in future real-time balancing markets. The technique is efficient enough to be employed by an aggregation module with limited computational resources. Alternatively, when indicative price information is not made available by the system operator, the technique can be combined with an external forecast in order to minimize possible imbalance costs.

Keywords: demand response; real-time balancing market; elastic demand bids; shiftable loads

---

1. Introduction

Stimulated by financial and regulatory incentives in an effort to reduce carbon emissions, some countries have experienced rapid growth in renewable energy sources such as wind and solar, which are clean, but intermittent. In order to balance the supply and demand of electrical energy, the power grid relies largely on dispatchable generation provided by conventional hydro- or carbon-based technologies. This balance must be kept at all times.

In the initial stages of the energy transition, not much needed to change, since the larger part of the energy mix still consisted of dispatchable generators. This becomes more difficult in a scenario with a very high penetration of intermittent renewable sources. One possible solution is to provide more flexible and controllable consumption in order to compensate for the loss of controllability on the generation side, in what is called demand response (DR) [1,2]. For the purpose of balancing the power grid, a reduction in power consumption is equivalent to an increase in generation. Conversely, an increase in power consumption is equivalent to a decrease in generation. Demand response can be offered in many forms and traded bilaterally or offered to the best buyer in organized energy markets [3].

More efficient energy markets play an important role in this energy transition. Usually, energy is traded separately for each hourly time slot. The higher volatility arising from short-term uncertainties associated with renewable energy sources increases the importance of energy trading being conducted at smaller time slots and closer to delivery, approximating a continuously-traded real-time market (RTM). Without solving the intrinsic volatility problem, these more refined markets at least do not ignore the intra-hour variations in energy prices, exposing all participants to a price that is as close as it is technically possible to an ideal real-time price. In an ideal scenario, the matching between elastic supply and elastic demand of energy would be achieved at every time slot, and this would mitigate price shocks, with dispatchable generators and flexible consumption adjusting to the fluctuations of renewables, following a price indication that is continuously updated.

Unfortunately, flexible demand participation is not yet the current situation in energy markets, as the large majority of consumers are exposed to a fixed price energy contract, which does not reflect the immediate price signals. Many consumers also lack the smart metering capabilities that would enable them to engage in more dynamic energy contracts with their suppliers. There is a need to improve wholesale energy markets so that more energy retailers and their customers can respond to price signals.

In the meantime, flexibility aggregators arose as a new market player that has more freedom to engage in customized agreements with suitable flexibility providers. The aggregator core business is to manage a flexibility pool large enough and controllable enough to participate in existing balancing reserve programs with the system operators. Alternatively, the aggregator can decide to sell its flexibility to balancing responsible parties (BRPs) under bilateral agreements. There are many possible mechanisms for an aggregator to market its flexibility, but there are also technical problems resulting from the uncoordinated activation of flexibility by different actors. An integrated market framework operating close to real-time could serve as a common platform for all these different actors, which could transact flexibility in a transparent manner and on an equal level.

In the context of the the SmartNet project, such a market is proposed [4] and employed in order to simulate the procurement of flexibility from transmission system operators (TSOs) and distribution system operators (DSOs). Different coordination schemes were proposed, described in detail in the SmartNet deliverable D1.3 [5]. Different aggregator techniques were implemented in the project [6], usually associated with a specific device technology, such as battery storage or combined heat and power (CHPs) units. These aggregation models build on the physical characteristics of a specific device or their aggregated characteristics.

The alternative approach is not to delimit a particular device technology, but to define an abstract model used to represent a number of different devices. In this paper, we focus on a model that represents loads with a fixed power profile, which are non-interruptible, but deferrable. The only way to obtain flexibility from these devices is to defer their activation to a different time.

This model can be useful for incorporating wet appliances as flexible loads controlled by an aggregator. Wet appliances consist of domestic appliances such as washing machines, dishwashers and tumble driers. The power profiles can reflect the dynamics of the device, requesting more or less power at different stages of operation, for instance during the spinning cycle of a washing machine.

Similar concepts are described in the literature with the name of shiftable loads, or deferrable loads with deadlines [7]. However, these models allow a device to reallocate energy consumption between different time slots, provided that the total energy required to complete an associated task is attained before a deadline. Some models also incorporate ramping constraints, charging rates or other device-specific constraints.

In the definition of deferrable load profile adopted in this paper, on the other hand, it is not possible to modify the energy consumption profile of a single device following its activation, which is completely determined by its fixed power profile.

The use of a discrete power profile is motivated by the fine-grained resolution of the SmartNet RTM and the increasing availability of disaggregated home appliance data [8]. We see in Figure 1 a representative power profile for a tumble-dryer. Wet appliances can provide flexibility [9], and the cost is low, provided that the delay in finishing their tasks is tolerable for the users.

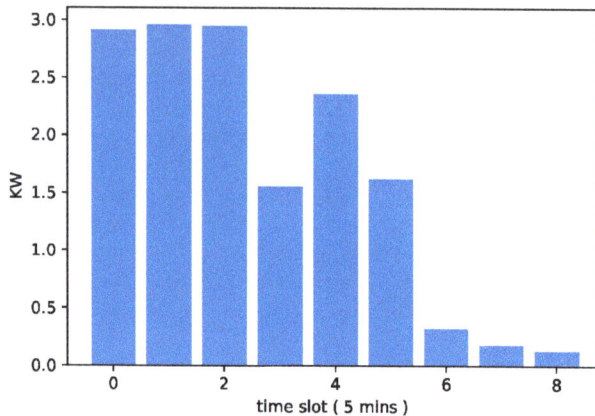

Figure 1. GE-WSM2420D3WW power profile.

In order to handle effectively a large number of loads with similar characteristics, the proposed method groups them by load profile and time of activation and places them in a waiting buffer. This buffer is filled up or emptied according to the objectives of the aggregator and respecting a maximum time delay tolerance. In analogy with a storage device, the aggregator produces flexibility by 'charging' or 'discharging' a buffer of deferrable loads. Different from energy storage, the decision to charge or discharge the waiting buffer will affect the energy consumption not only in the immediate time slot, but also in subsequent time slots, since each activated load will follow its fixed power profile until its termination.

From a market point of view, large storage capacity would allow energy to be treated as a more persistent commodity. The cumulative effect of aggregators and storage owners buying and storing inexpensive energy in order to be sold at a profit at a more expensive time slot would eventually reduce price swings. This would be beneficial to the stability of the power grid, especially in a scenario with a high penetration of renewable generation. A cluster of shiftable loads is capable of behaving in a similar fashion, while avoiding investment in storage technologies.

In order to keep maximum delays tolerable for the users, the model proposed in this paper will enforce deadlines in each shiftable load. The aggregator will cycle loads in the buffer in order to always have enough stored loads, while preventing loads already waiting in the buffer from exceeding their maximum delay tolerance. At every market iteration, the aggregator will formulate a scheduling plan for the expected incoming loads and the loads already in the buffer. It will be shown that this combinatorial problem is a network flow model that can be solved very efficiently. This is a very useful property, given the large number of participating loads necessary in order to produce flexibility in a volume high enough to participate in energy markets.

Paper Organization

The rest of the paper is organized as follows. In Section 2, we will establish a simplified market design that will be assumed by the load control technique. The imbalance settlement conditions on this simplified market will be reviewed. The deferrable load model used to produce flexibility will be described in more detail. Next, the buffer model used by the aggregator to manage a cluster of deferrable load profiles will be described, with the inputs assumed by the aggregator. Finally, some properties of the network flow model will be reviewed in order to give performance indications of the proposed method.

In Section 3, we will illustrate the buffer model employed by the aggregator using indicative price data from an existing real-time balancing market.

We conclude in Section 4, indicating some future lines of investigation.

2. Proposed Framework

2.1. Assumed Market Design

Electrical energy is traded in successive markets well in advance of physical delivery. Starting from long-term bilateral contracts between energy suppliers and retailers, there is a sequence of time-scales in which energy is traded. This complex structure of cascading markets is important to provide continuous adjustments and updates between the involved market participants, especially in a scenario with a high penetration of renewables.

One of the most important energy markets is the day-ahead market (DAM), which takes place one day before delivery and trades energy for a particular country or region at every given hour slot in the next day. The most prevalent time resolution for the DAM in Europe is 1 h, but in some countries, it is possible to trade energy in the DAM in blocks of 30 min, followed by other more granular trading opportunities in continuous intra-day markets (CIM). These two markets are still evolving, and they are the target of many harmonization initiatives, both EU and industry driven.

As we get closer to real-time, grid congestion might cause a price discrepancy between different nodes of the transmission network. In some regions of the U.S., Canada and in Australia, the market operator offers a last trading possibility in a faster nodal real-time balancing market, featuring a rolling window with 5-min dispatch intervals [10]. These markets are more complicated, but also reflect more accurately the true price of energy at a given node of the network.

Without going into the details of energy markets (see more in [11]), we now present a simplified market design similar to what is proposed in SmartNet. This design strikes a balance between existing mechanisms, adopted for day-ahead energy procurement and imbalance settlement, and the proposed SmartNet RTM [4]. Market design decisions are important since they affect the bidding strategy used by the aggregator.

We assume that our aggregator is a demand response provider that offers flexibility from the activation of aggregated loads. There are different arrangements currently in place for demand-side participation in the energy markets. A review of existing DR programs can be found in [12]. The approach adopted in SmartNet is consistent with [13], which recommends that the best mechanism to put DR into the market is by offering it in real-time markets, on the same level with generation resources, respecting the symmetric treatment between supply and demand.

We assume that the aggregator is responsible for managing a nomination position with the system operator, which means that he/she is associated with a balancing responsible party (BRP), a legal entity expected to submit updated energy nominations and be penalized or compensated later for the discrepancies between reported nominations and the realized consumption. We assume that the aggregator has acquired energy in previous markets, which is in line with the forecast consumption of his/her fixed load consumers and his/her flexible consumers. The demand of the fixed load component can be modeled using conventional load forecasting techniques and will not be considered in this paper. We will only consider consumption forecast of the cluster of shiftable profiles, which is assumed to be controllable.

In any case, close to real time, there will be a mismatch between the total energy consumption and the nominated amounts in each time slot, which represents the amount of energy the aggregator is entitled to consume. Close to real time, we assume that the aggregator is provided with new price information from the system operator, and it will have the opportunity to modify its nomination in a sequence of iterations of a fast real-time market (RTM).

Prices and nominations are managed independently for each time slot. We assume the RTM has a rolling time-window, and at frequent intervals, it offers the possibility to update nominations; as a result of each market clearing, it obtains new prices. We assume that the final price for each time slot will be assumed as the final imbalance price, and it will be used to penalize or compensate the aggregator ex post for any deviations between its final nomination and the final consumption. This final price, however, is only known precisely after the corresponding time slot elapses. The RTM mechanism gives only an indication of it.

In Figure 2, we see a diagram illustrating the successive bidding process. Each row indicates a different iteration of the RTM. The first square indicates the time in which the RTM is running. Each RTM iteration only allows trading to be conducted for the time slots indicated by the lighter squares, which define the current trading window. The current bid is formulated only for the current trading window. At every new market clearing, if the bid is accepted, the nominations of those particular slots are updated.

Figure 2. The rolling window mechanism.

In the figure, we indicate prices for one particular time slot. Notice that the energy price in the slot gets updated by the iterations of the RTM represented by the first and second rows. In the third row, the RTM last price is assumed to be final, and it is not possible to trade that particular slot anymore. The final imbalance price will be defined ex post.

For simplicity, we ignore time resolution differences, and we assume that previous nominations from the DAM, nomination updates from the RTM iterations and the final imbalance calculations are all provided in the same resolution. Previous nominations from the DAM will be corrected in the RTM iterations.

We assume the aggregator is a price-taker, unable to influence prices. This assumption must be carefully considered in case this market represents a smaller size of market restricted to a particular node of the transmission grid.

We assume that at each market clearing the system operator shares indicative price information, and this can be used as a look-ahead mechanism by the participants. However, when bidding, the participant is always uncertain of the actual clearing prices. When submitting a bid, the participant must assign a price to it. In the case of a bid requesting to buy (sell) energy, the bid will be accepted only if the clearing price of the current iteration is above (below) the clearing price. This process repeats in every market iteration.

In existing RTMs, the market clearing proceeds independently for each time slot. If the aggregator is interested in bidding for a range of slots, there is a risk that some slots get accepted, and some of them do not. To avoid this situation, in the proposed SmartNet market, it is possible to update the nominations in blocks, in what is known as a block bid. In existing markets, block bids are used in the DAM, but not in RTM, given the higher complexity of the market clearing procedure.

We assume that the market clearing uses a "pay-as-clear" mechanism, remunerating all accepted bids according to the same prices. This mechanism provides an incentive for the aggregators to submit their true costs in their bids and simplifies the bid formation process.

For instance, if in Figure 2, in the first iteration of the market, the aggregator bids to buy energy for Slot 7 with a price of 40.0, the bid will be accepted, and the aggregator will pay 32.1, given that this is the clearing price. The new clearing prices are not known at the beginning of that market iteration. If it attempted to buy energy for the same slot in the next price iteration, the aggregator would have paid 34.3. If the bid price were 33, the bid would be accepted if placed in the first market iteration, but it would be rejected if it were placed only in the second market iteration, since clearing prices increased a bit and are now above the bid price.

In another departure from existing RTMs, the gate closure on the SmartNet RTM is sufficiently close to the operation. This allows the aggregator to submit bids requesting nomination changes only a couple of time steps before execution.

2.2. Conditions on Imbalance Settlement

After the last market clearing, indicated in the second row of Figure 2, there can be further adjustments that will include the costs of conventional reserves, depending on the actual net consumption and generation of all participants. This is indicated for one particular time slot in the fourth row of Figure 2. This will be only known ex post and will form the imbalance prices. Any deviations on the part of the aggregator in relation to its last reported nomination, either voluntary or involuntary, will be settled at these prices. The aggregator will get paid for any excess nomination not consumed and will be requested to pay for any energy used in excess of its nomination for a particular slot.

This opens the possibility that the aggregator decides to mobilize its demand response resources in order to collect payments or minimize any charges on him/her in the imbalance settlement, using the imbalance mechanism as a 'spot market', even after the last gate closure. This possibility implies that when formulating its bids, the aggregator must take into account the opportunity costs or profits that would result if this flexibility were carried to imbalance.

We will assume that these opportunity costs are used to form the bid price. However, these potential gains expected in the imbalance markets must be properly discounted, since the high price volatility and the lack of information about the final state of the system can expose the participant to losses. We assume that this is an incentive for the aggregator to bid in the RTM and update its nomination, in case the bid is accepted, as opposed to maintaining a deliberate imbalance in relation to a previous nomination.

There are some similarities between the SmartNet proposed RTM and real-time balancing markets currently in operation in the U.S. and other countries. Each one of these existing markets, however, has its own design choices and limitations.

Ercot is the ISO responsible for managing most of Texas, and it allows load participation in its real-time security-constrained economic dispatch (SCED) market. Ercot publishes indicative price information for the SCED, which we will use in the next sections to illustrate the proposed technique used by the aggregator on the control of deferrable load profiles.

2.3. Deferrable Load Profile Representation

There are different definitions of responsive loads in the literature. As described briefly in Section 1, we adopted a more restricted definition of load flexibility where the load profiles are fixed, the loads are non-interruptible and only the choice of the time of activation of the device can provide flexibility. In [14], this kind of flexible load is defined as a shiftable profile. In [15], this shiftable load model is classified as non-interruptible and deferrable. Most studies do not consider the shape of these flexible loads, either considering the load profile as any profile that delivers a certain amount of energy before a deadline or as a constant power profile. The choice of a model with a discrete power profile representation suits the resolution of the RTM implemented in SmartNet, which is assumed to use 5-min time slots.

The strategy employed by the aggregator to manage a cluster of loads expected to arrive at different times of the day is to build a buffer of loads, and this buffer is "charged" or "discharged" in order to reduce or increase the power demand for one or more time slots. At the beginning of the bid construction, we assume the buffer has some loads present on it, in order to have some discharge power. Opposite the battery system, the discharge of this buffer will increase the energy consumption. The process is complicated by the inter-temporal power profile of the device.

Other works, such as [15], employ shiftable loads in the context of a residential consumer reacting to real-time energy prices. The small scale of the planning problem allows each deferrable device to be modeled individually, for every possible starting position. The difference in this work is that the aggregator possibly controls hundreds of similar devices arriving at the same time slot and is still able to solve a combinatorial problem efficiently by using the network model described in in Section 2.5. This model will cluster together devices that arrive in each time slot and will provide an optimal load activation schedule in relation to the latest price information.

As happens in general for shiftable loads, the flexibility possible with this kind of device is not easily translated to the interfaces offered by current market design. The value of a task spanning several time steps is associated with the completion of the task, and not its individual energy consumption slots. Current market design usually requires that the participants submit independent price/quantity pairs for each slot.

Shiftable profiles, as well as other sources of flexibility exhibit strong inter-temporal constraints that cannot be easily represented by independent price and quantity pairs. In order to solve this, the proposed SmartNet RTM supports a number of complex bids [4], which we will adopt instead of dealing with the stochastic problems usually employed in optimal bidding [7,16,17].

In this paper, we will focus on a block bid, a bid in which nomination updates at different time slots are submitted together in the same bid, and accepted or rejected in block by the market operator. We will assume the bid is discrete and non-curtailable, that is it cannot be accepted in a fraction. Current market design usually requires bids to be curtailable, being partially accepted within an informed range. We also assume that the price of this update can be established for the whole block, a condition that is similar to minimum income constraints employed in some DAMs. The market clearing procedure is free to accept or reject the whole block. This bid is, however, quite restrictive, and it adds more complexity to the market clearing algorithms utilized by the market operator.

2.4. Initial Setup of Deferrable Load Profiles

The initial setup is as follows. The aggregator starts planning for the current trade window taking into account loads that arrived in a previous iteration of the RTM, but remained inactive. These units were directed to a buffer, to a slot corresponding to their remaining allowed delay, and will now be scheduled to a compatible time slot. The aggregator also estimates the number of deferrable loads that will be joining in the current trade window. These loads, likewise, will be rescheduled to a compatible time slot or carried to the next iteration. Loads can be activated as soon as they join the system, up to a maximum delay Δs.

The current trade window is assumed to have length Δt. For simplicity, we assume that $\Delta s \leq \Delta t$. We assume in this paper homogeneous loads with a fixed load profile L_p, with $L_p > 0$ for $p \in [0, \Delta p)$. We use Δp to represent the length of the power profile, and we assume that $\Delta p \leq \Delta t$.

We will assume that at the start of a new iteration of the RTM, the buffer has enough loads in order to allow a flexibility bid to be formulated in both directions. An empty buffer, for instance, would only allow downward flexibility to be offered in the first time slots. A half-full buffer would allow both directions to be offered.

Our initial data are summarized in Table 1. For simplicity, the index ranges are relative to the time step in which the market is running.

Table 1. Initial data.

Input Data	Description
λ_t	Price information $t \in [0, \Delta t + \Delta p - 1)$
$\tilde{f}_t$	Number of loads expected to join $u \in [0, \Delta t)$
$\tilde{b}_s$	Number of loads in buffer slot $s \in [0, \Delta s)$

The most important input is represented by λ_t, which should contain the indicative prices from a previous real-time market iteration. We assume that this price information extends beyond the end of the current window in order to estimate costs from imbalances carried beyond the currently traded time window. For instance, loads starting in the last time slot at $\Delta t - 1$ will extend a power profile of size $\Delta p - 1$ outside of the currently traded time window.

We assume that the planning is deterministic and that $\tilde{f}_t$ contains the expected number of loads to join in every time slot. In reality, the number of loads joining the system will be random, and this will cause an imbalance position, which can be mitigated by the use of the cluster flexibility. For simplicity, we will not deal with this important aspect in this paper.

The buffer slots $\tilde{b}_s$ keep track of the number of loads that arrived before the current trading window and are waiting to be activated. Deferred loads in the buffer are sorted according to arrival time, with the index s indicating the remaining delay tolerated by loads in that buffer slot. All loads in $\tilde{b}_0$, for instance, must be immediately activated, since they have already waited the maximum allowed delay, Δs.

The nomination at the beginning of the current bidding comes from the energy bought in the DAM. This nomination vector gets modified by successive iterations of the RTM, as established in Section 2.1.

Most of the energy nomination reserved for the day is already committed to loads activated before the current market iteration. In Table 2 a distinction is made between the total nomination and the controllable nomination, still available to be changed. In Figure 3, the controllable nomination is indicated by the lighter bars, represented by $\tilde{Q}_t$. The figure illustrates a market iteration at Time Step 222. The trading window is indicated by the black vertical lines. Past the trading window, there is still controllable nomination corresponding to consumption that can still change depending on the scheduling performed in the current market iteration. Finally, we have a section of the nomination vector indicated by darker bars that represents the consumption of loads expected to arrive only after the end of the trading window.

Table 2. Initial nomination data.

Input Data	Description
$\tilde{N}_t$	Total nomination with $t \in [0, \Delta t + \Delta p - 1)$
$\tilde{Q}_t$	Controllable nomination $t \in [0, \Delta t + \Delta p - 1)$

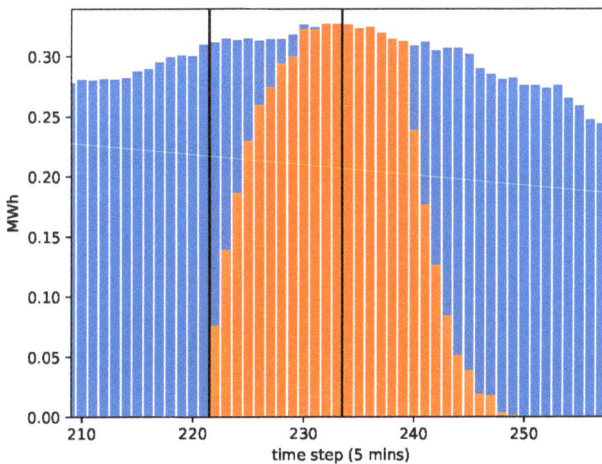

Figure 3. Controllable and non-controllable nomination.

The buffer starts the current market iteration with a certain distribution of units in different slots. In Table 3, we define variables to model the flow of scheduled loads between the initial state of the buffer and the planned schedule. This model is going to be used in the bid formulation to modify $\widetilde{Q_t}$. The variables define how the units waiting in the buffer or expected to arrive in the current time window will be distributed between possible times of activation. In the next section, we will show that this formulation is a type of network flow problem.

Table 3. Main variables.

Variables	Description
b_{st}	Number of loads in the buffer slot s scheduled to be activated in time slot t, $s \in [0, \Delta s), t \in [0, s]$
f_{ut}	Number of loads expected in time slot u rescheduled to a time slot t, with $u \in [0, \Delta t), t \in [u, \max(\Delta t - 1, u + \Delta s)]$
f_{ub}	Number of loads expected in the time slot u that will be used to replenish the buffer, with $u \in [\Delta t - \Delta s, \Delta t)$

In a network problem formulation, we have arcs that connect compatible nodes. In the definition of b_{st}, it is only necessary to consider the case with $t \leq s$. Loads from the buffer slot b_0, for instance, must be activated immediately, while loads from the buffer slot b_s can be scheduled to $t \in [0, \Delta s]$.

In the definition of f_{ut}, loads expected at time slot u can only be rescheduled at a future time slot that does not exceed the maximum delay Δs.

Finally, as explained for $\widetilde{b_s}$, part of the loads arriving at slots u will be able to replenish a buffer slot corresponding to the remaining delay that the load can tolerate. Loads arriving in slot u can be directed to the buffer slot $s = \Delta s - (\Delta t - u)$.

It is convenient to define auxiliary expressions for the final buffer state and the number of loads scheduled to start at a given time slot t. It is also useful to define an expression for the energy flexibility in relation to all possible load rescheduling. These are defined in Table 4.

Table 4. Auxiliary expressions.

Auxiliaries	Description
b_s	Number of loads assigned to final buffer slot s with $s \in [0, \Delta s)$
f_t	Auxiliary variable containing the number of loads activated in slot t of the rolling window, with $t \in [0, \Delta t)$
E_t	Energy consumption flexibility at each time slot t, with $t \in [0, \Delta t + \Delta p - 1)$

2.5. Flow-Based Formulation for the Flexibility

With the variables defined in the previous section, we proceed to define a flow-based model, which will allow us to modulate the flexibility profile according to the indicative price signal.

To avoid the complexity of working with indices, we define $\mathcal{A}_s$ and $\mathcal{A}_t$ indicating compatible arcs by the pairs (s, t) and (u, t), respectively.

$$\mathcal{A}_s = \{(s, t) \mid s, t \in [0, \Delta s), t \leq s\} \tag{1}$$

$$\mathcal{A}_t = \{(u, t) \mid u, t \in [0, \Delta t), 0 \leq t - u \leq \Delta s\} \tag{2}$$

The edge set $\mathcal{A}_s$ represents the connections between compatible buffer indices s and the time slots t in which they can be scheduled to activate. The edge set $\mathcal{A}_t$ plays the same role for incoming loads that are being rescheduled to a future time step in the current time window. We can link the auxiliary expressions with the main variables through the following expressions:

$$b_s = f_{\bar{u}b}, \qquad\qquad \forall s \in [0, \Delta s) \tag{3}$$
$$\text{with } \bar{u} = \Delta t - (\Delta s - s)$$

$$f_t = \sum_{\{s|(s,t)\in\mathcal{A}_s\}} b_{st} + \sum_{\{u|(u,t)\in\mathcal{A}_t\}} f_{ut}, \qquad \forall t \in [0, \Delta t) \tag{4}$$

$$E_t = \widetilde{Q}_t - \sum_{u=0}^{\Delta t-1} \frac{f_u L_{t-u}}{N}, \qquad \forall t \in [0, \Delta t) \tag{5}$$

In (3), for each final buffer slot s, there is a unique arrival time $\bar{u}$ responsible for the loads that contribute to it. Unscheduled loads will replenish the buffer and recover the system for a next market iteration.

The loads activated at t, represented in (4), come from two sources. They are either coming from the buffer or from loads arriving in the current window, and in both cases, they are constrained by the tuple sets $\mathcal{A}_s$ and $\mathcal{A}_t$.

In (5), we define our most important auxiliary expression, characterizing the energy flexibility in terms of the main variables. The number N represents the number of time slots in one hour, necessary to make the conversion from the power profiles to an energy consumption profile.

In order to simplify the summation, we assume that L_p is extended with zeros. We see that the power profile of a cluster of similar devices is the convolution of the power profile of one device with the activation schedule represented by f_t.

The expression for E_t will be expanded as a linear cost in the objective function of the bid construction problem that follows. The objective function will have a fixed component and terms that depend only on the main variables b_{st} and f_{ut}.

$$\max_{b_{st},f_{ut},f_{ub}} \sum_{t=0}^{\Delta t+\Delta p-1} \lambda_t E_t$$

$$\text{s. t.}$$

$$\sum_{\{t|(s,t)\in\mathcal{A}_s\}} b_{st} = \widetilde{b}_s, \forall s \in [0, \Delta s) \tag{6}$$

$$f_{ub} + \sum_{\{t|(u,t)\in\mathcal{A}_t\}} f_{ut} = \widetilde{f}_u, \forall u \in [0, \Delta t) \tag{7}$$

$$b_s = \widetilde{b}_s, \forall s \in [0, \Delta s) \tag{8}$$
$$\text{with } \bar{u} = \Delta t - (\Delta s - s)$$

$$b_{st}, f_{ut}, f_{ub} \in \mathbb{Z}_{\geq 0} \tag{9}$$

In (6) and (7), we require that the loads in the waiting buffer and the loads expected to arrive are spread into arcs that lead to compatible time slots.

The term f_{ub} in (7) can be seen as a remainder. Incoming loads not scheduled to be activated in some compatible time slot will replenish the final state buffer, as already indicated in (3).

In (8), we restrict the search of flexibility profiles within those reschedules that put the final buffer in the same condition as the starting buffer. These cyclic boundary conditions will prevent the bidding process from exploiting the buffer in order to maximize the value of the current bid, while compromising its availability in future market iterations.

Notice that in the objective function, some of the imbalance is expected in time slots beyond the end of the current time window. These extra costs (or profits) represented by $\sum_{t=\Delta t}^{\Delta t+\Delta p-1} \lambda_t E_t$

will not be included in the current bid, but will be carried as a future loss (or future gains) by the aggregator. The other part, represented by $\sum_{t=0}^{\Delta t-1} \lambda_t E_t$, can be seen as opportunity costs arising from the provision of flexibility and will be used to form the cost of the bid. By doing so, the aggregator is proposing to update the nomination vector in that range, provided that the payment is larger than what it expects to obtain in imbalance settlement.

2.6. Network Flow Properties

The previous formulation is a special case of a minimum cost network flow problem [18].

$$\min \sum_{(s,t)\in\mathcal{A}} c_{ij} x_{ij}$$

s. t.

$$\sum_{\{j|(i,j)\in\mathcal{A}\}} x_{ij} - \sum_{\{j|(j,i)\in\mathcal{A}\}} x_{ji} = s_i, \forall i \in \mathcal{N} \tag{10}$$

$$x_{ij} \in \mathbb{Z}_{\geq 0} \tag{11}$$

The problem is uncapacitated, that is there are no constraints on the amount of "flow" through the arcs, apart from requiring that the flow is positive, with $x_{ij} \geq 0$. In (10), s_i is positive at source nodes and negative at sink nodes. The "sources" in our formulation represent available loads expected in the current window or drawn from the buffer. The flow represents the rerouting of these units, and the "sinks" represent the desired final buffer added with one artificial sink.

In Figure 4, we display the corresponding network for the bid planning problem described in Section 2.5 for a small example where $\Delta s = 3$ and $\Delta t = 5$.

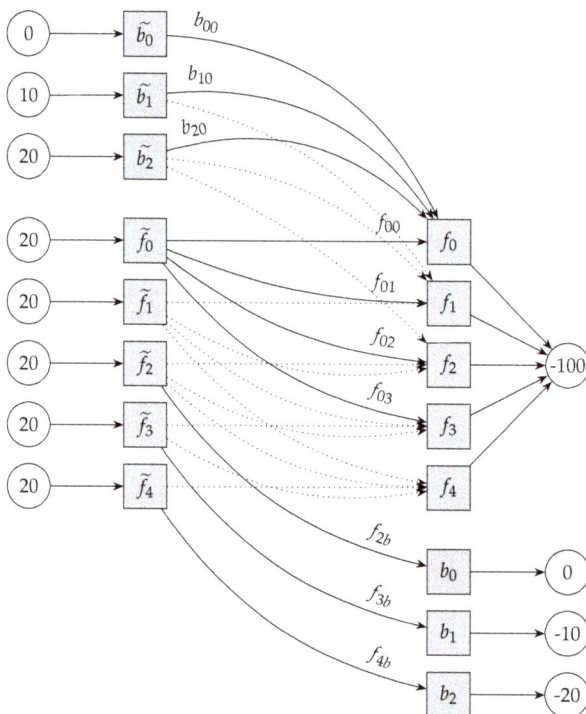

Figure 4. Network flow model for the flexibility bid.

The sources and sinks are represented by circles, and intermediary nodes are represented with square nodes and maintain zero balance. The costs on the arcs can be obtained from the expansion of $\sum_{t=0}^{\Delta t + \Delta p - 1} \lambda_t P_t$ in terms of b_{st}, f_{ut}, f_{ub}. No costs are associated with arcs connecting the square nodes and the circular ones.

Assuming that the problem described is feasible, the integrality theorem on network flows, summarized in Proposition 5.6 in [18], states that since values on sources and sinks are integers, so we will have a primal integer solution for this problem even without imposing integrality constraints explicitly, as in Equations (9) or (11).

This means that although our problem was originally formulated as an integer problem, because of its characteristics, it can be solved directly on its relaxed version, with continuous variables, using any available linear programming solver, or even a specialized network solver. In other words, there is no error introduced by solving the relaxed problem.

In complexity terms, linear programming can be solved in polynomial time, while general integer programming is NP-complete. Network flow algorithms sit in the intersection of these two classes of problems [19].

Mixed integer programming (MIP) solvers generally start by solving the relaxation of the problem, using linear programming, and then proceed to branching and other techniques to solve the problem under integrality constraints. Since as a network flow, we obtain integral solutions automatically, the MIP solver will solve the problem in its relaxed form, avoiding branching entirely.

Even if an MIP solver is used, the problem will be recognized as LP. This gives a good guarantee that the problem will scale well in relation to the number of devices and profiles. On a more practical note, this also gives the option to replace the MIP solver with a more available LP solver.

State-of-the-art MIP solvers are proprietary and expensive software components and because of their computational requirements and high licensing costs may also not be suitable to be deployed in all applications, for instance in embedded systems.

3. Example of the Application of the Proposed Framework

In this section, we use price data from an existing real-time market to illustrate the framework described in the previous section. As described in Section 2.1, the system operator clears the market at periodic intervals. This refreshes an indicative price curve that is used by the aggregator in order to formulate a new flexibility bid. This comes from a new schedule of the cluster of shiftable profiles, found using the network flow method described in Section 2.5. The aggregator will maximize the value of its flexibility by transferring energy consumption from expensive slots to less expensive slots, while respecting the constraints of the model.

There will be a price evolution, and at each new market clearing, the aggregator reformulates a new bid, taking into account any previously accepted bids, as well as the new price information.

In this example, we assume $\Delta t = 12$ as the length of the cleared time window and $\Delta s = 6$ as the maximum allowed delay. We assume that 200 devices arrive at each time slot, all with the same power profile as displayed in Figure 1. The total expected nomination available to be traded is indicated by the lighter bars in Figure 5. This power consumption is limited to load activation decisions that are currently taking place. The darker bars represent consumption corresponding to load activations that already happened or that are beyond the currently traded window. In Figure 6, we show the initial state of the buffer, with the number of waiting units, sorted by the remaining waiting tolerance and the number of time steps they are still allowed to wait before being activated. For instance, Figure 6 indicates that there are 200 loads that are allowed to wait a further five time steps before being activated.

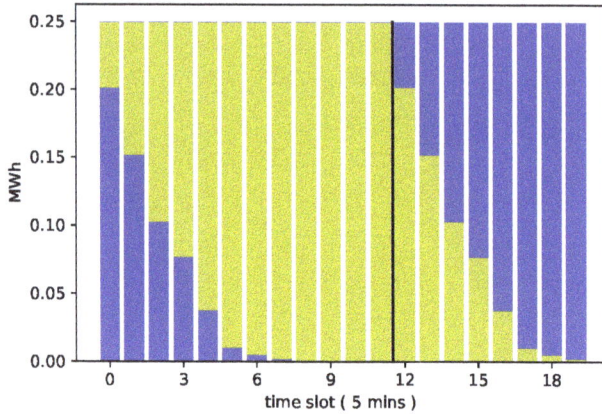

Figure 5. Controllable and non-controllable nomination.

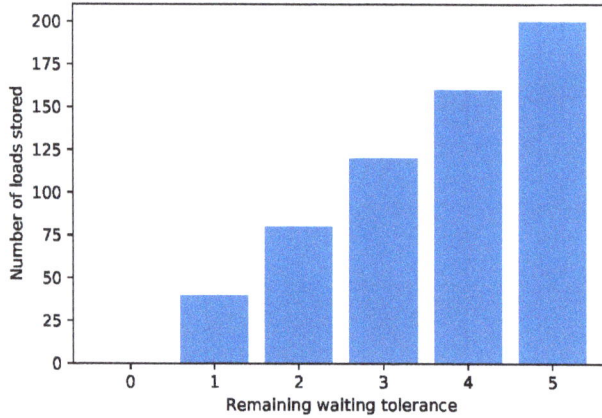

Figure 6. Initial buffer state.

We use indicative price data from Ercot to demonstrate the construction of the flexibility profile, later converted to a block bid. The data are relative to the South Hub, and it was logged on 17 October 2017. In Figure 7, we show indicative price curves collected after sequential market iterations at 18:30, 18:45, 19:00 and 19:15. Each clearing provides prices for each of the 5-min slots, one hour ahead of the collection time. In the absence of price information for the subsequent steps, we assume a simple constant extrapolation. This may be necessary in order to estimate the costs of the imbalances past the current range.

As time evolves, a new market iteration will produce new prices. The new market iteration includes the bid of this aggregator and other bids, representing different aggregators, as well as conventional balancing resources. Price information of future slots is revised up or down, depending on the clearing results. This will also keep extending the range of available data.

The aggregation will produce one flexibility bid at each iteration, and the price volatility is expected to be managed by successive corrections. The data from 18:30, exhibited in Figure 7, indicate that the system operator was expecting a price spike to happen at 19:00 and then again at 19:15. Based on this first information, the aggregator formulates a load activation schedule that shifts consumption away from the most expensive slots. The flexibility profile is represented in Figure 8. Indices are relative, with the first price peak corresponding to Slot 6.

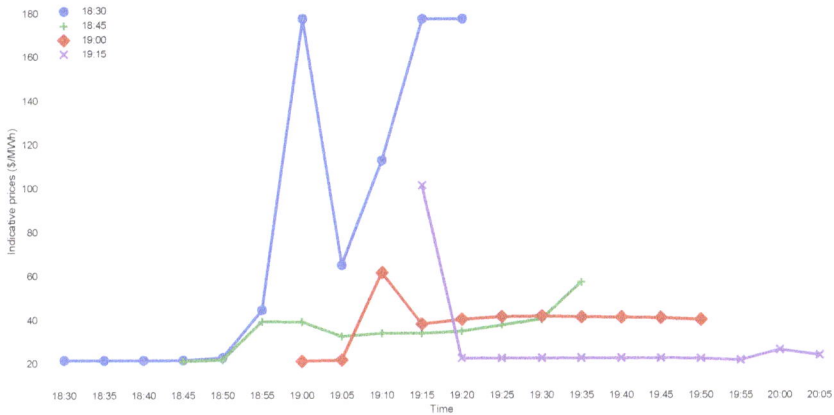

Figure 7. Ercot indicative prices at the South Hub.

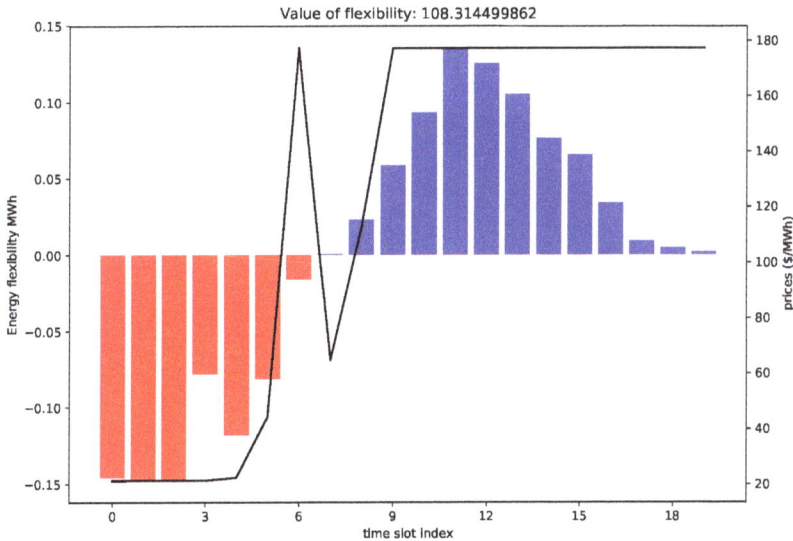

Figure 8. Imbalances allowed beyond the current time window.

The sign convention follows the logic that positive flexibility represents more generation. In the case of demand response, this is equivalent to less consumption, indicated in Figure 8 by the blue bars at the end. Negative flexibility corresponds to less generation, which in the case of demand response corresponds to more consumption.

Figure 8 shows that the aggregator uses the price information from 18:30 to propose an increase in immediate energy consumption, before the price spike arrives. In the final slots in blue, the aggregator is proposing a reduction in energy consumption. Energy consumption was anticipated for the first slots.

Similar to Figure 4, from the solution of the problem in (6)–(9), we obtain the corresponding network flow variables that give the aggregator the number of units to reschedule from each time slot to other compatible time slots.

The selected Ercot data in Figure 7 display a period where price volatility is particularly high. The price peak predicted at 19:00 was actually mitigated. At 18:45, the system operator revealed a new price curve, which shows a price of 40 €/MWh. The final price information collected after the market iteration at 19:00 is even lower.

If this flexibility profile were to be carried as imbalance by the aggregator instead of offered in the RTM, the aggregator would be exposed to the price volatility of the final imbalance prices, and its managed imbalance position could quickly turn into losses. By offering this profile in the RTM instead, the aggregator will have the opportunity to collect a payment, in case the bid is accepted. In this case, its nominations will be updated, and no imbalance will be due if it follows this new baseline strictly.

There is one limitation, however. Since the traded window is limited to one hour, part of the flexibility will not be included in the current iteration of the market.

Since this flexibility would potentially present potential gains (or losses) to the aggregator in the imbalance settlement, the aggregator can use this value as a reference when forming its own bid price. The value must be discounted, and the discounting factor should be suitably modified to reflect the uncertainty of this outcome.

The flexibility profile is ideally submitted as a block bid and with a single price associated with it. If prices in the current market iteration provide advantageous for the system operator to accept this bid, the aggregator will receive a value at least as high as requested, following the pay-as-clear mechanism.

In that case, the aggregator will update its nominations with the system operator. If the aggregator further departs from the updated nominations, it will be penalized according to the final imbalance prices, but taking into account this new updated nominations. If, on the other hand, the bid is not accepted, the aggregator imbalances will still be calculated in relation to the previous nominations.

In the next market iterations, the aggregator is free to form new flexibility offers, to risk accumulating a favorable imbalance position or to reschedule his/her loads in order to reduce his/her own residual imbalance position, if any.

4. Conclusions

The work developed in this paper demonstrates a simple network flow model that can be used to control a cluster of deferrable load profiles in relation to an indicative price signal. The indicative price curve induces periods of lower consumption when indicative prices are higher or vice versa.

We assumed a market design that allows for frequent market clearing that keeps the participants in a close loop with the system operator. This is one possible future market design simulated in SmartNet.

The aggregator uses the network flow model to form a block bid. If the bid is accepted, the aggregator receives a payment and reschedules its load activations in order to follow a new updated nomination. The energy consumption profile will change according to prices indicated by the RTM.

This bid takes into account the discrete power profiles of each individual load and their interruptibility. The network flow model also prevents loads from waiting longer than a maximum time delay.

If the bid is rejected, the aggregator uses the indicated price curve as indication of the imbalance prices and reschedules its loads in order to maintain a favorable imbalance position.

Existing real-time balancing markets are much simpler and asymmetric than the market proposed in SmartNet. The computational burden and the operational constraints of the system operator require that bids are submitted ahead of the current time window being traded, from 45 min to more than one hour. On the other hand, every 5 min, a different quantity is cleared and expected from the aggregators, which need to react in real-time. Complex bids are not available, and usually, there are no block bids and other more computationally-expensive bids. This requires the aggregator to submit a dispatchable bid, signaling to the market its range of capabilities. The construction of a more dispatchable bid requires a different set of techniques than the model exposed here.

Acknowledgments: This research was conducted by Vito-EnergyVille for the SmartNet project. The SmartNet project was funded by the European Union's Horizon 2020 research and innovation program under Grant Agreement No. 691405. This paper reflects only the author's view, and the Innovation and Networks Executive Agency (INEA) is not responsible for any use that may be made of the information it contains.

Author Contributions: J.C. developed the network flow model and wrote the paper. In the course of the project associated with this paper, the model received contributions from F.S. and C.H. The implementation was done in Pyomo/Cplex by J.C. with contributions from C.H.

Conflicts of Interest: The authors declare no conflict of interest.

References

1. Eurelectric. *Designing Fair and Equitable Market Rules for Demand Response Aggregation*; Eurelectric: Brussels, Belgium, 2015; pp. 1–24.
2. Albadi, M.H.; El-Saadany, E.F. Demand response in electricity markets: An overview. In Proceedings of the 2007 IEEE Power Engineering Society General Meeting, PES, Tampa, FL, USA, 24–28 June 2007; pp. 1–5.
3. ERCOT. *Load Participation in the ERCOT Nodal Market*; ERCOT: Austin, TX, USA, 2015.
4. Ashouri, A.; Sels, P.; Leclercq, G.; Devolder, O.; Geth, F.; D'hulst, R. SmartNet—Network and Market Models: Preliminary Report (D2.4), 2017. Available online: http://smartnet-project.eu/wp-content/uploads/2016/03/D2.4_Preliminary.png (accessed on 21 December 2017).
5. Gerard, H.; Energyville, V.; Rivero, E.; Energyville, V.; Six, D.; Energyville, V. Basic Schemes for TSO-DSO Coordination and Ancillary Services Provision, 2017. Available online: http://smartnet-project.eu/wp-content/uploads/2016/12/D1.3_20161202_V1.0.png (accessed on 21 December 2017).
6. Dzamarija, M.; Plecas, M.; Jimeno, J.; Marthinsen, H.; Camargo, J.; Sánchez, D.; Spiessens, F.; Leclercq, G.; Vardanyan, Y.; Morch, A.; et al. SmartNet—Smart TSO-DSO Interaction Schemes, Market Architectures and ICT Solutions for the Integration of Ancillary Services from Demand Side Management and Distributed Generation Aggregation Models: Preliminary Report, 2017. Available online: http://smartnet-project.eu/wp-content/uploads/2017/06/D2.3_20170616_V1.0.png (accessed on 21 December 2017).
7. Mohsenian-Rad, H. Optimal demand bidding for time-shiftable loads. *IEEE Trans. Power Syst.* **2015**, *30*, 939–951.
8. Pipattanasomporn, M.; Kuzlu, M.; Rahman, S.; Teklu, Y. Load profiles of selected major household appliances and their demand response opportunities. *IEEE Trans. Smart Grid* **2014**, *5*, 742–750.
9. Cardinaels, W.; Borremans, I. *Demand Responde for Families—LINEAR Final Report*; EnergyVille: Genk, Belgium, 2014.
10. Vlachos, A.G.; Biskas, P.N. Demand Response in a Real-Time Balancing Market Clearing with Pay-As-Bid Pricing. *IEEE Trans. Smart Grid* **2013**, *4*, 1966–1975.
11. Kirschen, D.S.; Strbac, G. *Fundamentals of Power System Economics*; John Wiley & Sons: Chichester, UK, 2004.
12. Faria, P.; Vale, Z. Demand response in electrical energy supply: An optimal real time pricing approach. *Energy* **2011**, *36*, 5374–5384.
13. Bushnell, J.; Hobbs, B.F.; Wolak, F.A. When it comes to Demand Response, is FERC its Own Worst Enemy? *Electr. J.* **2009**, *22*, 9–18.
14. Ottesen, S.O.; Tomasgard, A. A stochastic model for scheduling energy flexibility in buildings. *Energy* **2015**, *88*, 364–376.
15. Chen, Z.; Wu, L.; Fu, Y. Real-time price-based demand response management for residential appliances via stochastic optimization and robust optimization. *IEEE Trans. Smart Grid* **2012**, *3*, 1822–1831.
16. Conejo, A.J.; Nogales, F.J.; Arroyo, J.M. Price-taker bidding strategy under price uncertainty. *IEEE Trans. Power Syst.* **2002**, *17*, 1081–1088.
17. Kohansal, M.; Mohsenian-Rad, H. Price-maker economic bidding in two-settlement pool-based markets: The case of time-shiftable loads. *IEEE Trans. Power Syst.* **2016**, *31*, 695–705.

18. Bertsekas, D. *Network Optimization: Continuous and Discrete Models*; Athena Scientific: Belmont, MA, USA, 1998.
19. Papadimitriou, C.H.; Steiglitz, K. *Combinatorial Optimization: Algorithms and Complexity*; Prentice Hall: Englewood Cliffs, NJ, USA, 1982.

Article

Sensitivity Assessment of Microgrid Investment Options to Guarantee Reliability of Power Supply in Rural Networks as an Alternative to Underground Cabling

Sanna Uski [1], Kim Forssén [2,*] and Jari Shemeikka [2]

[1]   Ampner Ltd., 66999 Vaas, Finland; sanna.uski@ampner.com
[2]   VTT Technical Research Centre of Finland Ltd., 02044 Espoo, Finland; jari.shemeikka@vtt.fi
*   Correspondence: kim.forssen@vtt.fi; Tel.: +358401951047

Received: 30 August 2018; Accepted: 26 September 2018; Published: 19 October 2018

Abstract: Microgrids could be utilized to improve the distribution network resiliency against weather-related network outages and increase the security of power supply of rural electricity consumers. Whereas underground cabling is expensive for the distribution system operator (DSO), an alternative microgrid investment could benefit the DSO and consumer, provided the necessary changes were made in the network regulation. A rural detached house customer microgrid is analysed in comparison to underground cabling, considering the uncertainties in the calculation parameters through a sensitivity analysis. Adequacy of the microgrid power supply during unexpected network outage for a reasonably long duration is assessed, as well as the economics of the feasible microgrid setup consisting of variable generation, controllable generation, and electric storage. The total costs and benefits for the DSO and consumer/prosumer are considered. A microgrid would likely be a more cost-efficient option overall, but not as-is for the consumer. The battery energy storage system (BESS)-related cost-sharing strategies are suggested in this paper in order to assess possible break-even investment solutions for the related parties. The sensitivities of the microgrid and cabling investments were considered in particular. Cost-sharing strategies under network regulatory framework would need to be developed further in order for both the consumer and DSO to benefit from the solution as a whole.

Keywords: microgrid; resilience; investment; underground cabling; network outage; battery energy storage system (BESS); micro combined heat and power (micro-CHP); electricity distribution; solar photovoltaics (PV); islanded operation

1. Introduction

The resiliency of networks against extreme weather conditions has been studied e.g., in [1]. There are many methods and measures that can be taken to improve the resilience of an electric system such as underground cabling [2], emergency power systems, smart grid technologies, etc. As distributed generation and electricity storages are becoming more common and economic, microgrids are turning into an interesting alternative for improving the reliability and resilience of the electricity system against disturbances and blackouts, e.g., [3–6]. In addition, microgrids can also offer economic and societal value [7,8].

In Finland, extreme weather has caused many severe power outages in recent years and the government has set stricter limits for the outage duration. The Electricity Market Act limits the maximum allowable supply interruptions due to storms or snow load to 36 h in rural areas. Most of the distribution system operators (DSO) have chosen underground cabling as the main solution for improving the reliability of electricity distribution. While an efficient solution for decreasing tree-related faults in the distribution network, underground cabling is also very expensive compared to installing overhead lines (OHL) as it requires digging trenches and may often be complicated by rough terrain or other utility service lines. In some cases, cabling could be unnecessary and electricity storage or a microgrid could very well prove to be a more feasible option.

DSOs, e.g., in Finland, are currently not allowed to own electric storages. Studies, however, have proven that electric storages could be an economically feasible option for increasing the reliability of power supply in rural areas over underground cabling [9,10]. European Union (EU) "winter package" suggests changes in network owner's right to own storages in the future.

Overall, local small-scale generation capacity and bulk power system connected microgrids could have several benefits in the whole electricity system context. Benefits in the system level—as numerous small streams creating large quantities—could be e.g.,

- providing the consumer/prosumer a possibility to influence their own electricity acquisition and its cost;
- increasing the reliability of power supply of the individual consumer;
- increasing the power system total generation capacity;
- slowing down the need to increase network transmission capacity (if electric energy consumption increases, and/or the instantaneous power consumption increases);
- increasing the capacity of units capable of power system and network support, e.g., voltage and reactive power support, frequency control support etc.
- decreasing power transmission losses as generation is located closer to the consumption points.

This study applies the method presented in [11] for a microgrid consisting of a single electricity consumer in a detached house with own partially controllable power production. A full year of hourly data is analysed in order to assess the capability for a suddenly occurring islanded operation during network outage. The duration of the islanded operation capability is also determined for the microgrid in terms of power supply adequacy. The analysis method is here extended to evaluate different microgrid investment cost-sharing options in order to find a break-even investment solution. In addition, sensitivities of the system's feasible technical and economic parameter values are assessed.

The layout principle of the relevant network topology is presented in [10], and the BESS connection topology is presented in [9].

2. Assessment Method

The method used in this paper has been presented in [11]. Normal grid-connected operation of the microgrid is calculated for each hour of the year. A network outage could take place at any time, i.e., any hour of the year. The duration of feasible islanded operation with adequate power supply without adjusting the microgrid electricity consumption for each hour of the year is analysed.

As in [11], the costs of an underground investment and microgrid investment are calculated and compared. The method is used here for demonstrating different cost-sharing strategies for the microgrid investment alternative, as well as assessing the influence of uncertainty and variability of the study case parameters through a sensitivity analysis.

Traditionally the power distribution networks have been the only means of electricity supply to consumers. Distribution network companies' responsibility is to provide the consumers access to the network, as well as guarantee an adequate reliability of power supply.

Over the recent years, the consumers' interest in own power production has increased at the same time the small-scale power generation technologies have developed and become more economic options. Solar power generation is popular already today among small-scale consumers, but its drawbacks are the variability of the production. It is not possible to be controlled in order to balance power production with the electricity demand at all times—instantaneously, nor seasonally. Electricity storage would be needed for balancing the power in short-term (within an hour to within a day) in order to enable the usage of own production of variable generation. Battery energy storage system (BESS) technologies have become more economic over the recent years, and the trend is expected to continue.

The calculation method assumes a minimum charge level in the BESS to be maintained in normal grid connected operation state for the preparation of a sudden and unexpected network outage. During a network outage, the BESS capacity would be fully exploitable.

As seen in the results presented in [11], an electric heated detached house with a battery storage and solar PV production only, would not be a reasonable solution alternative for a microgrid as a means to increase the reliability of supply. In order to enable a relatively long capability of power supply for un-reduced power demand during the network outage and microgrid islanded operation, a PV-BESS combination only, would be a very expensive option if all the equipment was paid by the prosumer. Thus, the prosumer most likely will have to have controllable electricity production capacity.

This paper proposes the BESS investment sharing option as an alternative to DSO underground cabling in those cases when it could be mutually agreed upon and an economic solution for both, the DSO and the consumer/prosumer.

Different BESS investment and ownership strategies could be

1. Prosumer 100% ownership
2. Prosumer/DSO 50/50% ownership
3. DSO 100% ownership
4. Prosumer/DSO 50/50% investment

"Ownership" signifies participation on the investment, ownership of the equipment related responsibilities over the equipment lifetime, as well as beneficiary of the equipment related income. In option 4 above the DSO would remunerate 50% of the BESS procurement to the consumer in order to avoid more expensive underground cabling. An additional prerequisite would be an appropriate determination of the microgrid characteristics, and minimum BESS charging (strategy) for the preparation of the microgrid islanded operation for a sufficient length of time due to OHL network outages.

The method does not consider compensations payable by DSO to the consumer in the case of possible outages because of the related uncertainties.

3. Case Study

The case study analyses a detached house customer in sparsely populated area in rural distribution network.

For the case study, hourly data for a full year is used. As in [11], this time resolution is considered adequate for the purpose of the analysis method.

The major aspects or individual components of the study case are covered in respective sections below. All the parameters are determined a "base case" value, assessed to be the typical or best estimate value.

The sensitivities of the influence of most of the parameters or characteristics on the results are analysed. The varied parameter values for sensitivity analysis are presented in Appendix A.

When relevant, 24% value added tax is used (as typical in Finland). Interest rate of 2% is used for the base case in the investment calculations presented in this paper. The interest rate is varied for sensitivity analysis as specified in Appendix A.

3.1. Network Connection and Underground Cabling

The appropriate distribution network customer in the focus of the study is located in sparsely populated countryside at the end of a rather long-distance distribution network connection. The distance could be a few kilometres. The analysis could be easily applied to a group of customers on a distribution network branch.

The customer is currently supplied with an OHL connection, which is prone to weather-dependent interruptions. The present strategy for increasing the reliability of electricity supply is underground cabling of the network. Underground cabling of individual customers' connections is not in the interest of the DSOs, but the DSOs assess the economic profitability of underground cabling under the network regulation framework. In addition, the return of electricity supply to a single customer is not in the priority of the DSO in major interruption events when the DSO network experiences a large number of outages in a wide area. Thus, individual customer interruption could be extended for a rather long duration of time, i.e., for several hours and even up to a few days.

Actual costs related to underground cabling investments are very case dependent. Here the cabling costs are estimated by assuming ordinary cable trench cost of 25,000 €/km in Finland and 0.4 kV underground cable cost of 10,000 €/km with lifetime of 40 years, approximately in accordance to the regulation price list. The cable length is used as a variable in the simulations determining the total cost of the cable investment. Instead of the cable length, the total cabling investment cost actually is significant in the calculations done in this paper. In the base case, the cable length of 1 km is used, corresponding to 35,000 € cabling investment. The investment cost is varied for the sensitivity analysis by varying the cable length as specified in Appendix A.

3.2. A Detached House Consumer

In this case study, a modern detached house of 150 m^2 in Central Finland (Jyväskylä) is considered. Heating of the house is assumed to be implemented either in the conventional onsite manner or using single house micro-CHP (combined heat and power plant). Thus, the detached house electricity production capability from micro-CHP is dependent on the whole micro-CHP plant characteristics, as well as the momentary heat production of the micro-CHP plant.

The detached house heat demand data series was created by using dynamic building energy simulation. IDA ICE (version 4.7.1, EQUA Simulation AB, Sweden) [12] is a whole-year detailed and dynamic multi-zone simulation application for study of thermal indoor climate as well as the sub-hourly energy consumption of the entire building. The total heat demand for the house amounts to 15.5 MWh/a.

The household electricity consumption (i.e., excluding heating) data series was created by using electric customer type load profiles. The load profiles based on [13], have been partially updated by VTT Technical Research Centre of Finland in 2003 using new measurements. The household annual electricity consumption is a varied parameter (see Appendix A) for the sensitivity analysis, and approximately 5500 kWh/a in the base case.

3.3. Detached House Micro-CHP

Micro-CHP plants in the lower end of the capacity range have not proven profitable yet today in Finland. However, there are ongoing research studies aiming at finding solutions to increase the profitability, especially for the detached houses in distant locations, i.e., the type of houses being in the focus of this study.

The investment cost for a micro-CHP plant is a significant factor in the overall profitability calculations and contains uncertainty. For the base case, 1200 €/kW is assumed for the whole micro-CHP plant in the respective capacity class, and the investment cost is varied for the sensitivity analysis as specified in Appendix A. The electricity share of the costs is assumed according to the electric power share of the plant. A lifetime of 15 years is assumed for the micro-CHP plant and it is varied for sensitivity analysis.

The micro-CHP plant electricity production is assumed to be 20% of the power plant total energy production each hour. Thus, the heat demand determines the electricity production. The micro-CHP is assumed to not have a minimum power, and it is operating throughout the year even at low heat consumption.

The fuel costs for the whole micro-CHP plant (i.e., heat and electricity combined) are assumed a fixed value 1200 €/a, independent of total energy production. The total power production with micro-CHP does not vary significantly in different calculation cases, and thus the fixed annual fuel cost assumption should be acceptable. The fuel costs are, however, varied for the sensitivity analysis, as specified in Appendix A.

By assuming some heat storage capability in the heating system of the house, micro-CHP heating time series was created as 24 h sliding average in order to smooth the heating dynamics. Electricity production time series was derived as a constant 20% share of the total micro-CHP power production in the base case. The required micro-CHP plant total capacity is thus estimated to approximately 8 kW. Also, the option of 30% electricity production share is calculated of a micro-CHP plant with 9 kW total capacity.

In the base case, the micro-CHP microgrid operation is assumed to be continuing similarly to normal grid connected state. During a network outage and microgrid islanded operation, the micro-CHP plant generation could possibly be used differently from normal in order to supply the needed electricity. In this case, the possible excess heat production might need to be dissipated. Thus, for the sensitivity analysis, the micro-CHP electricity generation is varied as specified in Appendix A, allowing a different level of constant electric power production. In these simulations, this specific islanded operation electrical power production is the same value for the whole year in order to illustrate the influence of the parameter setting. In reality, the possible power production level may be dependent on the time of the year and the power might be controllable during the islanded operation.

3.4. Solar PV Production

PV production data series were created for the study case in Central Finland using PV GIS Tool [14]. Crystal-Silicon PVs with 14% system losses are assumed as the most plausible technology. PV panels installed on roof-top are assumed to be facing towards South (i.e., azimuth 0 degr.) and adjacent to the roof, assuming thus a fixed 25 degr. tilt slope installation for the PVs.

A cost of 1800 €/kWp is assumed for the PV panels in the base case, and 25 years as the lifetime for the panels. The PV panel cost and lifetime are varied for the sensitivity analysis.

The data series were created for a number of years and using data of different solar radiation databases. The base case data was created with SARAH solar radiation database. 2015 as an average solar production year is used in the base case. Annual PV production difference between the datasets (i.e., the best production year 2006 and the worst 2008) is almost 20%.

The simulations are done with different PV capacities, and 1.5 kWp capacity is used in the base case. The values for varied parameters are specified in Appendix A.

3.5. Battery Energy Storage, BESS

The BESS is assumed for the study according to the technology available today for a single house scale. The maximum charging ramp of 5 kW is assumed irrespective of the BESS total storage capacity which is varied as specified in Appendix A, and with base case capacity of 13.5 kWh. The BESS minimum charge level is assumed to be 6 kWh in the base case, and it is varied for the sensitivity analysis.

The BESS is assumed to consist of a fixed cost for the installation and system-related equipment, and a capacity dependent investment cost for the batteries. The fixed cost is assumed to be 600 €, and the capacity dependent cost 500 €/kWh in the base case. The BESS lifetime is assumed to be 10 years. The BESS capacity dependent cost and the BESS lifetime are varied for the sensitivity analysis.

The losses in the BESS are omitted. In reality, the losses, i.e., self-discharge of the battery, would be in the order of 1–5 percent per month for Li-ion batteries [15]. Thus, for a 6-kWh permanent minimum charge of the BESS, the annual losses would amount approximately only to 1 kWh. In addition, with Li-ion batteries, up to almost a 100% efficiency could be achieved according to [15].

An annual maintenance cost of 100 €/a is assumed for the microgrid covering any random small cost items.

3.6. Electricity Market Aspects

The consumer electricity purchase from the network is calculated for the electricity price of 5 c/kWh and distribution fee (including electricity tax 2.8 c/kWh, etc.) of 6.5 c/kWh. Electricity price and distribution fee are varied for the sensitivity analysis.

The prosumer can sell their excess power generation to the network. For small-scale producers the present compensation for sold electricity to network in Finland is approximately 2.7 c/kWh and the transmission fee for electricity fed to the grid 0.7 c/kWh. The value in small-scale production is mainly in covering own electricity need by own production, whereas selling electricity to the system is not cost-effective. Furthermore, the compensation for sold electricity is quite marginal and would need to change significantly in order to influence the overall results. Thus, the parameters of sold electricity and related transmission fees were not varied for sensitivity analysis.

The BESS in this study maintains a certain minimum charge at all times in preparation for a sudden and unexpected network outage. Such outage occurrences are very rare. The required minimum charge might vary throughout the year, depending on the microgrid capability of supplying the demand by power production in real-time. Transmission system operator (TSO) acquires frequency containment reserve capacity for disturbances (FCR-D), and this market would be profitable as well as suitable for the microgrid BESS. FCR-D reserve is started to be activated at a certain threshold of the system frequency deviation, and at a specific larger frequency deviation the full FCR-D reserve capacity power would be supplied. The market price of 4.7 €/MWh is assumed in this study for FCR-D capacity.

Avoiding a complex analysis, this study assumes the full BESS power capacity (5 kW) to be able to be sold to the FCR-D market via an aggregator a certain number of hours per year. In the base case, the BESS is assumed to be sold 7000 h/a to the FCR-D market, and the number of hours is varied for the sensitivity analysis as specified in Appendix A.

4. Results

4.1. Base Case Results

For the microgrid with base case parameter values and settings, the power production and consumption are presented in Figure 1, the BESS charging in Figure 2 and microgrid excess power production, i.e., grid fed power, and power take from the grid in Figure 3. The islanded operation capability duration throughout the year are presented in Figure 4.

The base case economics assessment and comparison of the underground cabling and the microgrid options with different cost-sharing alternatives are presented in Figure 5.

4.2. Sensitivity Analysis of Parameters

The influence of different PV database data and different PV production year data in the consumer economic balance is within less than 20 €/a between the high and low PV annual production. Thus, the selection of the PV generation data set to be used in the calculations is not significant.

Islanded operation capability duration depends on the minimum charge of the BESS, and the micro-CHP power production capability during islanded operation. If the micro-CHP is able to produce electricity at the maximum of the rated total capacity (i.e., 1.6 kW electric power), the islanded operation duration, in theory, has no limitations. The maximum power decreased down to 0.7 kW, the minimum islanded operation duration in the calculations was 50 h. This, in many cases, should be quite an acceptable repair time of a network outage even in the case of severe and widespread storm damages.

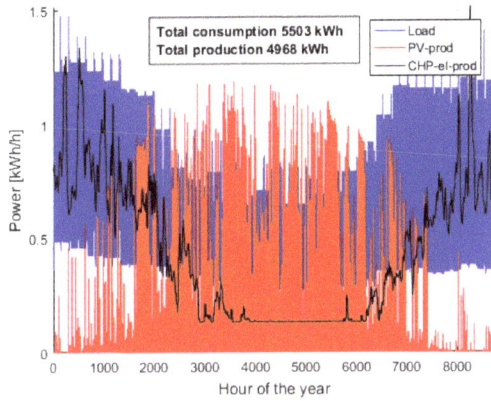

Figure 1. Detached house electricity consumption (Load) and power production (photovoltaics (PV)-prod and combined heat and power (CHP)-el-prod) hourly data series for a year in the base case, as well as the annual totals.

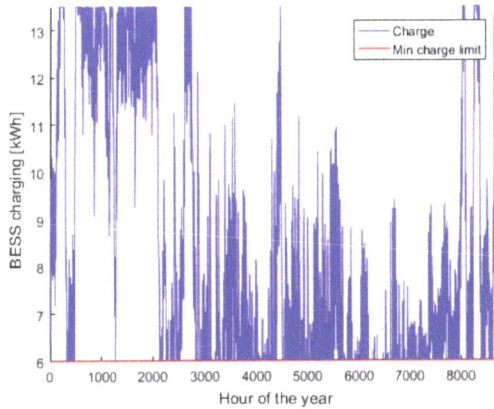

Figure 2. Battery energy storage system (BESS) charging hourly data series in the base case.

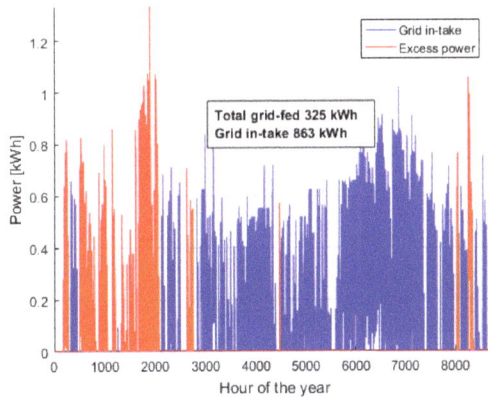

Figure 3. Hourly data series of electricity taken from the network (Grid in-take) and excess electricity fed to the network (Excess power) in the base case, as well as the annual totals.

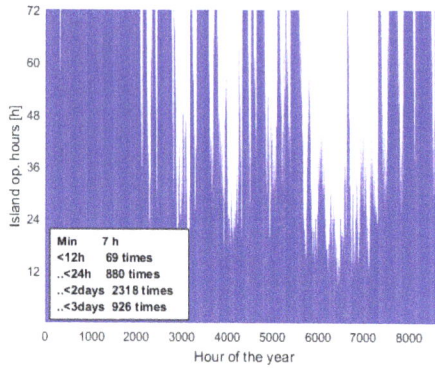

Figure 4. The base case results for microgrid islanded operation capability duration each hour of the year in the case of unexpected network outage taking place at any hour.

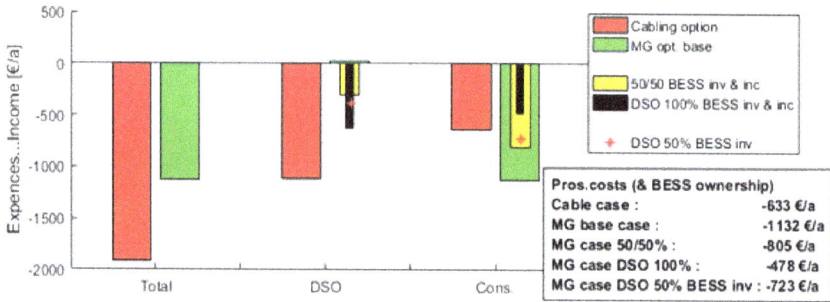

Figure 5. The base case results of economic calculations for the cabling option and microgrid option in different BESS investment and ownership strategies. The comparison of total overall costs, the distribution system operator (DSO) costs, and the consumer/microgrid owner costs in case of different BESS cost-sharing strategies.

Islanded operation capability durations vary from case to case depending on the parameters varied for the sensitivity analysis. The probabilities for minimum islanded operation duration categories are presented in Figure 6. The base case values correspond to those values also presented in the text box in Figure 4.

In the simulation cases, the largest change in the BESS charge between consecutive hours (i.e., average power for the hour) was 1.5 kW, and thus significantly smaller even at largest, than the BESS nominal power 5 kW.

The total costs, i.e., the overall DSO and consumer combined costs, are compared in the cabling option and the microgrid option in all studied cases with varied parameters in Figure 7. The results show that the microgrid option is in all studied cases more cost-efficient than cabling. The interest rate and cable length (i.e., the total cable investment cost) have obvious influence on the costs. The more expensive the cabling option is, the more cost effective the microgrid option seems to be.

The case when microgrid option is closest to the cabling option investment expenses is with largest storage capacity (BESS capacity 20 kWh), i.e., with larger storage investment. The lowest microgrid option costs are in the case with the longest BESS lifetime.

Figures 8 and 9 present the comparison of the cabling option and the microgrid option costs for the DSO and the consumer (i.e., the microgrid owner) respectively. The results show that the microgrid option is the most cost-efficient solution for the DSO in all cases, and for the consumer only in cases when the BESS would either be owned by the DSO, or the DSO compensating a part of the BESS investment.

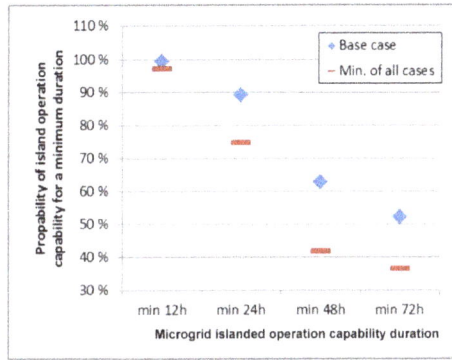

Figure 6. The probabilities of microgrid islanded operation capability duration of the base case and the minimums of all the cases analysed for sensitivity analysis.

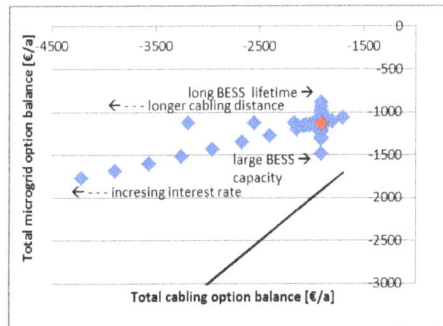

Figure 7. The cabling option and microgrid option total cost comparison with the varied parameter values. The black line represents the equal cost limit, and the base case is marked in red.

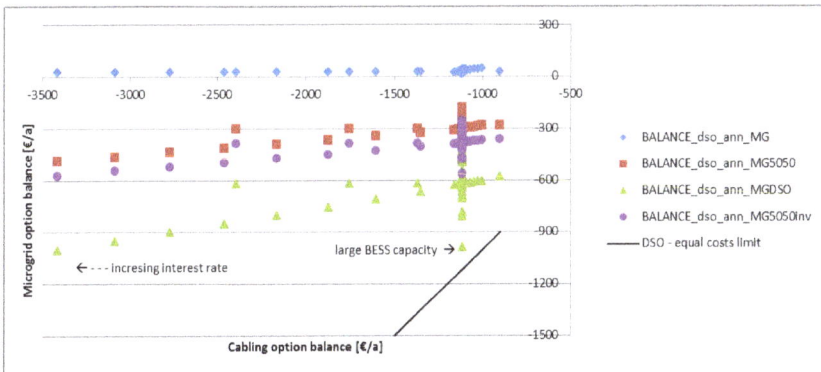

Figure 8. The cabling option and microgrid option DSO cost comparison with the varied parameter values considering different cost-sharing strategies of the microgrid investments. The black line represents the equal cost limit, i.e., when the cabling costs and microgrid costs would be equal. BALANCE_dso_ann_MG, BALANCE_dso_ann_MG5050, BALANCE_dso_ann_MGDSO and BALANCE_dso_ann_MG5050inv refer to prosumer 100% ownership, prosumer/DSO 50/50% ownership, DSO 100% ownership and prosumer/DSO 50/50% investment respectively.

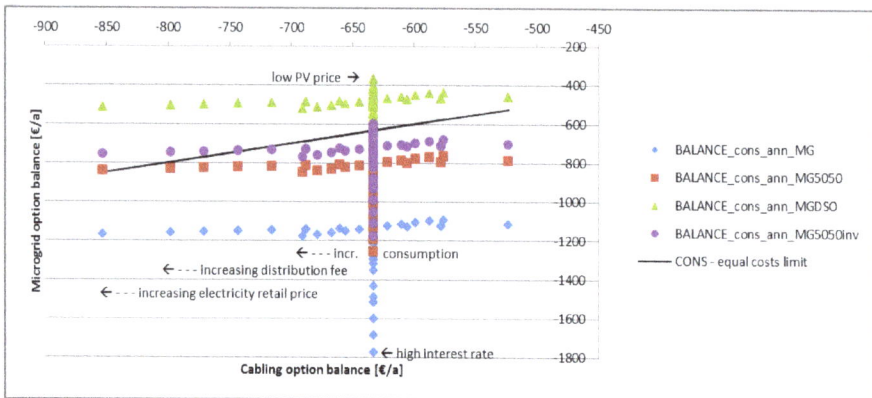

Figure 9. The cabling option and microgrid option consumer/microgrid owner cost comparison with the varied parameter values considering different cost-sharing strategies of the microgrid investments. The black line represents the equal cost limit, i.e., when the cabling costs and microgrid costs would be equal. BALANCE_dso_ann_MG, BALANCE_dso_ann_MG5050, BALANCE_dso_ann_MGDSO and BALANCE_dso_ann_MG5050inv refer to prosumer 100% ownership, prosumer/DSO 50/50% ownership, DSO 100% ownership and prosumer/DSO 50/50% investment respectively.

5. Conclusions

A single detached house microgrid located in rural sparsely populated distribution network was studied as an alternative to underground cabling in order to increase the reliability of power supply. The results demonstrate that with a carefully planned composition of microgrid generation and storage components, a reasonable level of reliability of power supply could be met in an islanded operation during possible unexpected network outages.

The economic calculation results presented in this paper show clearly that an underground cabling option is likely to be overall more expensive an option in a rural sparsely populated network over a microgrid when considering the consumer and DSO costs together.

When cabling is the most expensive option over all, it is clearly the most expensive option also for the DSO alone. The microgrid option, in turn, was more expensive in all cases to the consumer/microgrid owner if the microgrid-related expenses are to be covered solely by the consumer.

Alternative microgrid investment strategies were proposed for cost-sharing between an electric customer, i.e., a microgrid owner, and a DSO. A few shared investment options of the BESS between the customer and the DSO were considered. The overall most cost-efficient microgrid option could be made by cost-sharing an economical option for the consumer/microgrid owner and enable the DSO to avoid a more expensive cabling investment.

A microgrid could be an alternative for rural network underground cabling, provided it was made an optional practice in distribution network planning and operations, and that it was permitted by regulations. In addition, its application would require a willing electricity customer and mutual agreement and cooperation between the customer and DSO.

Author Contributions: The literature review, modeling, simulations, and analysis were done by S.U.; K.F. collected information on micro-CHP-power plants and network resilience; J.S. created the detached house heat demand data series and provided the relevant description for the manuscript; S.U. prepared the manuscript, and K.F. contributed to the editing and revision of the manuscript.

Funding: This work has been supported by the Strategic Research Council at the Academy of Finland, project Transition to a resource efficient and climate neutral electricity system (EL-TRAN) grant number 314319.

Acknowledgments: The authors thank Göran Koreneff for providing advice on the household electric customer load profiles and the used profile data.

Conflicts of Interest: The authors declare no conflict of interest. The funders had no role in the design of the study; in the collection, analyses, or interpretation of data; in the writing of the manuscript; or in the decision to publish the results.

Appendix A

The varied parameter values for the sensitivity analysis are listed here as follows, the bolded value being the base case value.

Cable length [km]:
1.0; 1.2; 1.5; 2.0

Electricity consumption [kWh/a] (in the level of):
5000; 5100; 5200; 5300; 5400; **5500**; 5600; 5700; 5800; 5900; 6000

Micro-CHP total capacity [kW] and share of electric power of total production capacity [%]:
8 kW, 20%; 9 kW, 30%

Micro-CHP production capability during microgrid islanded operation [kW]:
scheduled power; 1.6 (rated/maximum power); 0.8; 0.7; 0.6; 0.5

Micro-CHP power plant investment cost [€/kW]:
800; 1000; **1200**; 1500; 2000; 2500

Micro-CHP energy generation fuel cost [€/a]:
800; 1000; **1200**; 1500; 2000; 2500

Micro-CHP power plant lifetime [a]:
10; **15**; 20

PV capacity [kW]:
0.8; 1.0; 1.2; **1.5**; 1.7; 2.0; 2.2

PV data source solar radiation database:
SARAH; ERA5; COSMO

PV data year:
2005; 2006; 2007; 2008; 2009; 2010; 2011; 2012; 2013; 2014; **2015**; 2016

PV investment cost [€/kW]:
800; 1000; 1200; 1500; **1800**; 2000

PV panels' lifetime [a]:
15; 20; **25**; 30

BESS energy storage capacity [kWh]:
10; **13.5**; 15; 20

BESS minimum charge [kWh]:
5; **6**; 7; 8; 9; 10; 11; 12

BESS lifetime [a]:
8; **10**; 12; 15

Capacity dependent share of BESS investment cost [€/kWh]:
400; 450; **500**; 550

Interest rate [%]:
1; **2**; 3; 4; 5; 6; 7; 8; 9; 10

Electricity retail purchase price [c/kWh]:
3; 4; **5**; 6; 7; 8; 9

Electricity distribution fee [c/kWh]:
5.5; 6.0; **6.5**; 7.0; 7.5; 8.0; 8.5; 9.0

BESS power capacity sold to FCR-D market [h/a]:
0; 4000; 5000; 6000; **7000**; 8000; 8760

References

1. Forssén, K. Resilience of Finnish Electricity Distribution Networks against Extreme Weather Conditions. Master's Thesis, Aalto University, Helsinki, Finland, 2016.
2. Haakana, J. Impact of Reliability of Supply on Long-Term Development Approaches to Electricity Distribution Networks. Ph.D. Thesis, Lappeenranta University of Technology, Lappeenranta, Finland, 2013.
3. Li, Z.; Shahidehpour, M.; Aminifar, F.; Alabdulwahab, A.; Al-Turki, Y. Networked Microgrids for Enhancing the Power System Resilience. *Proc. IEEE* **2017**, *105*, 1289–1310. [CrossRef]
4. Liu, X.; Shahidehpour, M.; Li, Z.; Liu, X.; Cao, Y.; Bie, Z. Microgrids for Enhancing the Power Grid Resilience in Extreme Conditions. *IEEE Trans. Smart Grid* **2017**, *8*, 589–597. [CrossRef]
5. Costa, P.; Matos, M. Assessing the contribution of microgrids to the reliability of distribution networks. *Electr. Power Syst. Res.* **2009**, *79*, 382–389. [CrossRef]
6. Schneider, K.P.; Tuffner, F.K.; Elizondo, M.A.; Liu, C.-C.; Xu, Y.; Ton, D. Evaluating the Feasibility to Use Microgrids as a Resiliency Resource. *IEEE Trans. Smart Grid* **2017**, *8*, 687–696. [CrossRef]
7. Stadler, M.; Cardoso, G.; Mashayekh, S.; Forget, T.; DeForest, N.; Agarwal, A.; Schönbein, A. Value streams in microgrids: A literature review. *Appl. Energy* **2016**, *162*, 980–989. [CrossRef]
8. Schwaegerl, C.; Tao, L. *Quantification of Technical, Economic, Environmental and Social Benefits of Microgird Operation*; Wiley-IEEE Press: Hoboken, NJ, USA, 2013; Chapter 7; p. 344. ISBN 9781118720677. [CrossRef]
9. Vilppo, O.; Markkula, J.; Järventausta, P. *Energy Storage in Low Voltage (LV) Network for Decreasing Customer Interruption Cost (CIC)*; Tampere University of Technology: Tampere, Finland, 2016.
10. Haakana, J.; Lassila, J.; Kaipia, T.; Partanen, J. Utilisation of energy storages to secure electricity supply in electricity distribution networks. In Proceedings of the CIRED—22nd International Conference on Electricity Distribution, Stockholm, Sweden, 10–13 June 2013; pp. 1–4. [CrossRef]
11. Uski, S.; Rinne, E.; Sarsama, J. Microgrid as a Cost-Effective Alternative for Rural Network Underground Cabling for Adequate Reliability. *Energies* **2018**, *11*, 1978. [CrossRef]
12. IDA Indoor Climate and Energy Building Energy Simulation Tool, EQUA Simulations, Sweden. Available online: https://www.equa.se/en/ida-ice (accessed on 1 December 2017).
13. SLY. *Suomen Sähkölaitosyhdistys r.y*; Kuormitustutkimus 1992: Helsinki, Finland, 1992; p. 172.
14. European Commission PV GIS Tool. Available online: http://re.jrc.ec.europa.eu/pvg_tools/en/tools.html (accessed on 1 December 2017).
15. Mousazadeh, H.; Keyhani, A.; Javadi, A.; Mobli, H.; Abrinia, K.; Sharifi, A. Evaluation of alternative battery technologies for a solar assist plug-in hybrid electric tractor. *Transp. Res. Part D* **2010**, *15*, 507–512. [CrossRef]

MDPI

St. Alban-Anlage 66

4052 Basel

Switzerland

Tel. +41 61 683 77 34

Fax +41 61 302 89 18

www.mdpi.com

International Journal of Neonatal Screening Editorial Office

E-mail: neonatalscreening@mdpi.com

www.mdpi.com/journal/neonatalscreening

www.ingramcontent.com/pod-product-compliance
Lightning Source LLC
Chambersburg PA
CBHW051707210326
41597CB00032B/5393